FUZZY THEORY SYSTEMS

Techniques and Applications

VOLUME 1

FUZZY THEORY SYSTEMS

Techniques and Applications

VOLUME 1

Edited by

Cornelius T. Leondes

Professor Emeritus
University of California
Los Angeles, California

ACADEMIC PRESS

San Diego   London   Boston   New York   Sydney   Tokyo   Toronto

Academic Press
a division of Harcourt Brace & Company
525 B Street, Suite 1900, San Diego, California 92101-4495, USA
http://www.apnet.com

Academic Press
24-28 Oval Road, London NW1 7DX, UK
http://www.hbuk.co.uk/ap/

Library of Congress Catalog Card Number: 98-89309

International Standard Book Number: 0-12-443870-9 (set)
International Standard Book Number: 0-12-443871-7 Volume 1
International Standard Book Number: 0-12-443872-5 Volume 2
International Standard Book Number: 0-12-443873-3 Volume 3
International Standard Book Number: 0-12-443874-1 Volume 4

PRINTED IN THE UNITED STATES OF AMERICA
99 00 01 02 03 04 QW 9 8 7 6 5 4 3 2 1

■ CONTENTS

CONTRIBUTORS xxvii
FOREWORD xxxv
PREFACE xxxix

■ I ■ MODELING TECHNIQUES

I Optimization Techniques in the Design of Fuzzy Models

WITOLD PEDRYCZ AND JOSE VALENTE DE OLIVEIRA

   I. Introductory Remarks: Modeling Vis à Vis Fuzzy Modeling   4
  II. Information Granularization   6
 III. Design of Input and Output Interfaces: A Review of Matching Techniques   8
 IV. The Requirements of Semantic Integrity   10
  V. Information Equivalence Criterion   11
 VI. Encoding and Decoding: An Inverse Problem in Relational Calculus   14

VII. Constructing Information Granules 15
VIII. Clustering Techniques in the Design of Linguistic Interfaces 16
IX. Design of the Processing Module 22
X. Optimization Policies in Fuzzy Models 24
XI. Conclusions 24
References 25

2 Modeling Relationships in Data: From Contingency Tables to Fuzzy Multimodels

WITOLD PEDRYCZ

I. Introductory Comments 28
II. Contingency Tables, Multimodels, and Information Granularity 30
III. Generic Architecture of the Multimodel 32
IV. Development of Fuzzy Multimodels 34
V. Numerical Studies 39
VI. Conclusions 44
References 46

3 Fuzzy Dynamical Modeling Techniques for Nonlinear Control Systems and Their Applications to Multiple-Input, Multiple-Output (MIMO) Systems

YA-CHEN HSU AND GUANRONG CHEN

I. Introduction 48
II. Direct Multiple-Input, Multiple-Output Fuzzy Controller Design 49
III. Indirect Multiple-Input, Multiple-Output Fuzzy Controller Design 64
IV. Conclusions 80
References 84

4 Techniques and Applications of Fuzzy Set Theory to Difference and Functional Equations and Their Utilization in Modeling Diverse Systems

ELIAS DEEBA, ANDRÉ DE KORVIN, AND SHISHEN XIE

I. Introduction 87
II. Motivation 88
III. Background 90
IV. Description of the Method 93

V. Fuzzy Analog of Cauchy Functional Equations 96
VI. Fuzzy Analog of a First-Order Difference Equation   102
VII. Conclusion 109
    References 109

5 Techniques in Neural-Network-Based Fuzzy System Identification and Their Application to Control of Complex Systems

YAOCHU JIN AND JINGPING JIANG

I. Introduction 112
II. Takagi and Sugeno's Fuzzy Model 113
III. Neural-Network-Based Identification of Fuzzy Systems 114
IV. Interpretability Considerations 119
V. An Application Example: Decoupled Control of Robot Manipulators 120
VI. Conclusions 126
    References 128

6 Fuzzy Control with Reference Model-Following Response

C. M. LIAW AND Y. S. KUNG

I. Introduction 130
II. Statement of the Problem 131
III. The Indirect Field-Oriented Induction Motor Drive 133
IV. Output Feedback Linear Model-Following Controller 136
V. Conventional Fuzzy Controller 139
VI. Fuzzy Controller with Reference Model-Following Response 145
VII. Design of the Proposed Controller for Induction Motor Drive 149
VIII. Conclusions 155
    References 158

7 Fuzzy-Set-Based Models of Neurons and Knowledge-Based Networks: Techniques and Applications

WITOLD PEDRYCZ

I. Introduction 160
II. Logic-Based Neurons 161
III. Logic Neurons and Fuzzy Neural Networks with Feedback 166

IV. Referential Logic-Based Neurons   169
V. Architectures of Fuzzy Neural Networks   171
VI. Learning in Fuzzy Neural Networks   172
VII. Interpretation of Fuzzy Neural Networks   174
VIII. Case Studies   176
IX. Conclusions   186
References   186

8 Identifying Fuzzy Rule-Based Models Using Orthogonal Transformation and Backpropagation

LIANG WANG, REZA LANGARI, AND JOHN YEN

I. Introduction   188
II. Fuzzy Models   189
III. Method   190
IV. Results   193
V. Conclusion   201
Appendix: Factorizable Property of Gaussian Membership Functions   201
References   202

9 Evolutionary Neuro-Fuzzy Modeling

CHUEN-TSAI SUN AND HAO-JAN CHIU

I. Introduction   205
II. ANFIS: Adaptive-Network-Based Fuzzy Inference System   206
III. RBFN: Radial Basis Function Network   208
IV. Niche: Diversity of Species   209
V. The Evolutionary Model   210
VI. Experimental Results   216
VII. Conclusion   221
References   221

10 Techniques and Applications of Fuzzy Theory in Quantifying Risk Levels in Occupational Injuries and Illnesses

PAMELA R. MCCAULEY-BELL, LESIA L. CRUMPTON-YOUNG, AND ADEDEJI BODUNDE BADIRU

I. Introduction   224
II. Predictive Models for Occupational Injuries   228
III. Fuzzy Modeling   230

IV. Review of Fuzzy Applications to Modeling
Occupational Injuries 237
V. Principles for Applying Fuzzy Modeling to Modeling CTDs 239
VI. Conclusion 259
References 260

11 Fuzzy Inference with Control Schemes for Fuzzy Constraint Propagation

KIYOHIKO UEHARA AND KAORU HIROTA

I. Introduction 268
II. Fuzzy Inference Based on α-Level Sets and
Generalized Means 269
III. Efficient Inference Operations 274
IV. Convex and Resolutionally Identical Forms in
Deduced Consequences 276
V. Control of Fuzzy Constraint Propagation 277
VI. Learning Algorithm for Fuzzy Exemplars 280
VII. Fuzzy Constraint Propagation Control Reflecting Forms of
Given Facts 281
VIII. Simulation Studies 282
IX. Conclusion 288
References 289

II SYSTEM CONTROL METHODS AND APPLICATIONS

12 Intelligent Proportional, Integral, Derivative Controllers

KA CHING CHAN

I. Introduction 293
II. Intelligent Approaches to PID Controller Tuning 294
III. Neural Network-Based Tuner 297
IV. Integrated Fuzzy-Neural Network Tuner 309
V. Adaptive Virtual Fuzzy Tuner 322
VI. Summary 338
References 342

13 Techniques and Applications of Fuzzy Theory via H_∞ Control Techniques for Tracking Algorithms for Uncertain Nonlinear Systems

BOR-SEN CHEN AND YU-MIN CHENG

    I. Introduction   345
    II. Problem Formulation   347
    III. A Review of Fuzzy Logic Systems   349
    IV. Adaptive Fuzzy Control Design   351
    V. Illustrative Examples   362
    VI. Conclusion   369
        References   372

14 Techniques and Applications of Fuzzy Smoothing Algorithms for Control Systems

YEAN-REN HWANG

    I. Introduction   375
    II. Fuzzy Sets and Fuzzy Rule-Based Control   377
    III. Signed Distance of Variable Structure Systems   382
    IV. Fuzzy tuning Low-Pass Filtering Control   385
    V. Fuzzy Regions and Fuzzy Boundaries   394
    VI. Conclusion   402
        References   402

15 Techniques and Applications of Fuzzy Theory in the Validity of Complexity Reduction by Means of Decomposition of Multivariable Fuzzy Systems

P. G. LEE, G. J. JEON, AND K. K. LEE

    I. Introduction   406
    II. Fuzzy Relation Matrix   406
    III. Complexity Reduction   408
    IV. Index of Applicability   411
    V. Applications   415
    VI. Summary   428
        References   429

16 Techniques and Applications of Control Systems Based on Knowledge-Based Interpolation

FRANK KLAWONN AND RUDOLF KRUSE

I. Introduction   431
II. Fuzzy Sets, Vague Environments, and Indistinguishability   433
III. Application to Fuzzy Control   444
IV. A Regression Technique for Fuzzy Control   450
V. Conclusions   457
    References   459

17 Techniques and Applications of Fuzzy Theory to an Elevator Group Control System

HYUNG LEE-KWANG AND CHANGBUM KIM

I. Elevator Group Control System   461
II. Area Weight Generation   465
III. Fuzzy Classification of Passenger Traffic   471
IV. Hall Call Assignment   475
V. Conclusions   480
    References   480

CONTENTS OF VOLUME 2

CONTRIBUTORS xxvii

III INDUSTRIAL, MANUFACTURING, AND COMMERCIAL SYSTEMS

18 Fuzzy Neural Network Systems Techniques and Their Applications to Nonlinear Chemical Process Control Systems

ATSUSHI AOYAMA, FRANCIS J. DOYLE III, AND
VENKAT VENKATASUBRAMANIAN

I. Introduction   486
II. Unique Characteristics of Chemical Process Control   486

III. Neural Network Techniques and Applications in Nonlinear Chemical
Process Control Systems   488
IV. Fuzzy Logic Techniques and Applications in Nonlinear Chemical
Process Control Systems   496
V. Fuzzy Neural Networks   501
VI. Case Studies (Fuzzy Neural Network Model in Internal
Model Control)   511
VII. Summary   523
References   524

19 Techniques and Applications of Fuzzy Theory to Material Selection in Mechanical Design Problems

JEAN-LUC KONING

I. Introduction   528
II. The Materials Selection Problem   531
III. Representation of the Charts   534
IV. Description of the Materials Selection System   540
V. Description of the Program   552
VI. Conclusion   556
References   557

20 Applications of Fuzzy System Theory to Telecommunications

GIRIDHAR D. MANDYAM AND M. D. SRINATH

I. Introduction   559
II. Signal Detection in Non-Gaussian Communications Channels   560
III. Channel Equalization   565
IV. Power Control   572
V. Call Acceptance and Routing   573
VI. Other Applications of Fuzzy System Theory
to Telecommunications   574
VII. Conclusions and Directions for Future Work   575
References   575

21 Fuzzy Control System Techniques and Their Application in Hydraulically Actuated Industrial Robots

N. SEPEHRI, T. CORBET, AND P. D. LAWRENCE

I. Introduction   578
II. Experimental Test Station   579

III. Fuzzy Logic Controls   581
IV. Fuzzy Controller for Hydraulic Robots   585
V. Steady-State Error Elimination   596
VI. Concluding Remarks   602
   References   607

22 Techniques in the Design and Stability Analysis of Fuzzy Proportional-Integral-Derivative Control Systems and Their Industrial Applications

HEIDAR A. MALKI AND GUANRONG CHEN

I. Introduction   610
II. Conventional PID Controllers   610
III. Fuzzy PID Controllers   612
IV. Autotuning and Adaptive PID Controllers   621
V. Applications of Fuzzy PID Controllers   622
VI. Conclusions and Discussion   626
   References   628

23 Fuzzy Theory Techniques and Applications in Data-Base Management Systems

PATRICK BOSC, LUDOVIC LIÉTARD, AND OLIVIER PIVERT

I. Introduction   632
II. Gradual Queries against Regular Data   632
III. Precise Data, Functional Dependencies, and Gradual Rules   654
IV. Imperfect Data   664
   References   686

24 Techniques and Applications of Fuzzy Theory in Document Retrieval Systems

SHYI-MING CHEN

I. Introduction   691
II. Fuzzy Set Theory   692
III. Concept Networks   693
IV. Concept Matrices   696
V. Query Processing Techniques   697
VI. Extended Fuzzy Concept Networks   704
VII. Relation Matrices and Relevance Matrices   707
VIII. Fuzzy Query Processing for Document Retrieval Based on Extended Fuzzy Concept Networks   709

IX. Conclusions 714
    References 714

25 Techniques and Applications of Fuzzy Logic Control of Solar Power Plants

MANUEL BERENGUEL, FRANCISCO R. RUBIO, EDUARDO F. CAMACHO, AND FRANCISCO GORDILLO

    I. Introduction 717
   II. Brief Description of the Plant 719
  III. Experience-Based Fuzzy Logic Controller 721
   IV. Automatic Design Using Input–Output Data 727
    V. Application of GAs to Tuning of the FLC 730
   VI. Plant Results 736
  VII. Conclusions 744
       References 745

26 Fuzzy Metrology

H. JOHN CAULFIELD, JACQUES LUDMAN, AND JOSEPH SHAMIR

    I. Introduction 747
   II. Classical Metrology and Fuzzy Logic 748
  III. Bayesian Logic 750
   IV. The Fuzzy Metrology Concept 751
    V. Fuzzy Metrology in Nature 752
   VI. Fuzzy Metrology in Technology 754
  VII. Back to the Problems of Classical Metrology 755
 VIII. Extending Fuzzy Metrology 755
   IX. Designing the Fuzzy Membership Sets 757
       References 757

27 Techniques and Applications of Fuzzy Statistics in Digital Image Analysis

C. V. JAWAHAR AND A. K. RAY

    I. Introduction 759
   II. Background 762
  III. Histogram and Co-occurrence in Fuzzy Setting 765
   IV. Fuzzy Statistics 767
    V. Applications 771
   VI. Summary 777
       References 777

IV ADAPTIVE FUZZY SYSTEMS

28 Adaptive Fuzzy Logic Control Synthesis without a Fuzzy Rule Base
BRANKO M. NOVAKOVIC

I. Introduction   781
II. Synthesis of Fuzzy Logic Control   784
III. Synthesis of an Adaptive Fuzzy Logic Control Using Parameter β-Adaptation   793
IV. Adaptive Fuzzy Logic Control of a Robot of RRTR Structure   797
V. Conclusion   806
References   807

29 Fuzzy Adaptive Control Techniques for Nonlinear Systems and Their Application
CHUN-YI SU AND YURY STEPANENKO

I. Introduction   809
II. Problem Statement   811
III. Fuzzy Systems and Function Approximation   812
IV. Adaptive Control with Fuzzy Logic   818
V. An Illustrative Example   825
VI. Conclusion   828
References   828

30 Online Self-Organizing Fuzzy Logic Controller Using a Fuzzy Auto-Regressive Moving Average Model and Its Application
YOUNG MOON PARK, UN-CHUL MOON, AND KWANG Y. LEE

I. Introduction   832
II. The Farma Fuzzy Logic Controller   833
III. Application to Power System Stabilization   845
IV. Simulation Results and Discussion   852
V. Conclusions   862
References   867

31 Techniques and Applications of Fuzzy Theory in Generalized Defuzzification Methods and Their Utilization in Parameter Learning Techniques

TAO JIANG AND YAO LI

I. Introduction   872
II. General Concept of the Transformation for Defuzzification Strategy   873
III. The Polynomial Transformation-Based Defuzzification Strategy, a Parameter Learning Procedure, and a Single-Mode Numerical Example   876
IV. Multimode-Oriented Polynomial Transformation-Based Defuzzification Strategy, a Parameter Learning Procedure, and a Multimode Numerical Example   883
V. Conclusion   894
References   894

32 Optimal Adjustment of Scaling for Fuzzy Controllers Using Correlation Techniques

RAINER PALM

I. Introduction   897
II. Input–Output Correlation for a Fuzzy Controller   900
III. Internal Representation of Correlation Functions within a Fuzzy Controller   907
IV. Application to a Redundant Manipulator Arm   909
V. Conclusion   913
References   914

33 Translation and Extraction Problems for Neural and Fuzzy Systems: Bridging over Distributed Knowledge Representation in Multilayered Neural Networks and Local Knowledge Representation in Fuzzy Systems

HIROSHI NARAZAKI AND ANCA L. RALESCU

I. Introduction   918
II. Translation Problem   921
III. Extraction Problem   927
IV. Conclusion   935
References   936

34 Fuzzy Set System Application to Medical Diagnosis: A Diagnostic System for Valvular Heart Diseases

JIRO ANBE AND TOSHIKAZU TOBI

    I. Introduction 938
   II. Method of Diagnosis by Discrimination Analysis 938
  III. Method of Diagnosis by Connectivity Analysis 941
  IV. Valvular Heart Diseases 946
   V. Patient Records 948
  VI. Uncertainty and Noise of Data 948
 VII. Selection of Training and Testing Data 949
VIII. Measurement of Prototypicalness 949
  IX. Results of the Experiments 952
   X. Conclusions 954
     References 955

CONTENTS OF VOLUME 3

CONTRIBUTORS xxvii

V   IMAGE, VISUAL, AND PATTERN PROCESSING SYSTEMS

35 Fuzzy Theory Techniques and Their Application in Digital Image Transformation

YUEFENG ZHANG

    I. Introduction 959
   II. Fuzzy Vertex Correspondence 960
  III. Fuzzy Affine Transformation 963
  IV. Algorithm 966
   V. Conclusion 970
     References 971

36 Techniques and Comparative Analysis of Neural Network Systems and Fuzzy Systems in Medical Image Segmentation

KUO-SHENG CHENG, JZAU-SHENG LIN, AND CHI-WU MAO

    I. Introduction   974
   II. Clustering Based Methods   976
  III. Competitive Learning Networks   978
  IV. Competitive Hopfield Neural Network   981
   V. Fuzzy Hopfield Neural Network   986
  VI. Experimental Results and Discussion   989
 VII. Concluding Remarks   997
      References   1006

37 The Applications of Fuzzy Techniques to Image Processing Based on the Human Visual System Model

KWEI-ANN WEN, FA-SHEN LEOU, AND AN-YI CHEN

    I. Introduction   1009
   II. Fuzzy Sets and the Human Visual Model   1010
  III. Fuzzy Sets and Image Processing   1015
  IV. A Human Visual Model-Based Image-Processing System   1021
   V. Conclusions   1027
      References   1027

38 Fuzzy Logic-Based Visual Feedback Control

IL HONG SUH AND TAE WON KIM

    I. Introduction   1029
   II. What Can Fuzzy Logic Do in Visual Feedback Control?   1031
  III. A Fuzzy-Neural Network-Based Visual Servoing Algorithm   1036
  IV. Concluding Remarks   1049
      References   1049

39 Techniques in Fuzzy Rules Determination and Their Application to Pattern Classification

SHIGEO ABE

    I. Introduction   1051
   II. A Fuzzy Classifier with Hyperbox Regions   1052

III. A Fuzzy Classifier with Polyhedron Regions   1060
IV. A Fuzzy Classifier with Ellipsoidal Regions   1067
 V. Performance Evaluation   1075
VI. Conclusions   1078
      References   1079

40 Techniques and Applications of Genetic Algorithm-Based Methods for Designing Compact Fuzzy Classification Systems

HISAO ISHIBUCHI, TOMOHARU NAKASHIMA, AND TADAHIKO MURATA

  I. Introduction   1082
 II. Fuzzy Rule-Based Classification   1085
III. GA-Based Rule Selection   1091
IV. Michigan Approach   1095
 V. Pittsburgh Approach   1098
VI. Performance Evaluation   1101
VII. Conclusion   1106
      References   1106

41 Combination of Handwritten-Numeral Classifiers with Fuzzy Integral

TUAN D. PHAM AND HONG YAN

  I. Introduction   1111
 II. Fuzzy Measures and Fuzzy Integrals   1113
III. Calculation of Fuzzy Densities   1115
IV. Fuzzy-Integral Model for Fusion of Handwritten-
      Character Classifiers   1117
 V. Experimental Results   1118
VI. Computing Weights with Genetic Algorithms   1125
VII. Conclusions   1126
      References   1127

42 Neural Network and Fuzzy Logic Systems: Techniques and Applications in Telerobot Systems

DONG-HYUK CHA AND HYUNG SUCK CHO

  I. Introduction   1129
 II. Telerobot Systems   1132

III. Applications of Neural Network and Fuzzy Logic in
Telerobot Systems   1144
IV. Conclusions   1176
References   1177

VI FUZZY NEURAL SYSTEMS

43 Implementation Techniques and Applications of Fuzzy Neural Network Systems

YAU-HWANG KUO AND JANG-PONG HSU

I. Introduction   1184
II. Connectionist Fuzzy Classifiers   1185
III. Reinforcement Parallel Fuzzy Inference Model   1211
IV. Fuzzy CMAC Model   1226
V. Conclusions   1237
References   1239

44 Techniques in Fuzzy Inference Neural Networks for Fuzzy Model Improvement and Their Application

KEON-MYUNG LEE AND HYUNG LEE-KWANG

I. Introduction   1242
II. Fuzzy Models   1243
III. Fuzzy Modeling   1246
IV. Fuzzy Inference Neural Networks   1250
V. An Example of Fuzzy Inference Neural Networks   1252
VI. Conclusions   1261
References   1263

45 Techniques and Applications of Integrated Neural Network-Based Fuzzy Logic Control Systems

CHIA-FENG JUANG, HSI-WEN NEIN, AND CHIN-TENG LIN

I. Introduction   1265
II. Structure of the SONFIN   1268
III. Learning Algorithms for the SONFIN   1272
IV. Temperature Control of the RTP System   1284

V. Conclusion   1294
References   1296

46 Neural Fuzzy Systems Techniques and Applications for Production Quality Control

SHING I. CHANG

I. Introduction and Background   1299
II. Review of Neural Network and Fuzzy Set-Related Applications in Quality Control   1302
III. A Neural Fuzzy System for Decision Making   1305
IV. Fuzzy Neural Systems Applications for Quality Control   1311
V. Concluding Remarks   1320
References   1320

47 Fuzzy Neural Network Techniques in Robotic Object Manipulation

KAZUO KIGUCHI AND TOSHIO FUKUDA

I. Introduction   1323
II. Object Manipulation   1326
III. Controller Structure   1327
IV. Object Trajectory Control   1328
V. Robot Manipulator Force Control   1332
VI. Conclusions   1338
References   1338

48 Modeling Cognition with Fuzzy Neural Nets

AMIT KONAR AND SANJUKTA PAL

I. Introduction   1342
II. Model of Cognition   1343
III. Reasoning and Unsupervised Learning in Cognitive Maps   1347
IV. Refinement of Knowledge in a Cognitive Map   1357
V. Supervised Learning by a Cognitive Map   1364
VI. Generating Control Commands in a Cognitive System   1374
VII. Task Planning and Coordination   1380
VIII. Putting It All Together   1381
IX. Applications   1383
X. Conclusions and Future Directions   1384
XI. Appendix   1386
References   1388

49 Techniques in Neural Fuzzy Systems and Their Applications in Processing both Numerical and Linguistic Information

CHIN-TENG LIN, I. FANG CHUNG, AND YA-CHING LU

    I. Introduction 1393
   II. Representation of Linguistic Learning 1396
  III. Basic Structure of the Neural Fuzzy System 1403
  IV. Supervised Learning of the Basic Neural Fuzzy System 1407
   V. Structure of Fuzzy Language Acquisition Networks 1415
  VI. Supervised Learning of the FLAN 1420
 VII. Illustrative Examples 1424
VIII. Conclusion 1434
     References 1435

CONTENTS OF VOLUME 4

CONTRIBUTORS xxvii

VII IMPLEMENTATION TECHNIQUES

50 Implementation Techniques for Fuzzy Theory Systems and Their Applications

A. DE GLORIA, P. FERRARI, D. GROSSO, M. OLIVIERI, AND L. PUGLISI

    I. Introduction 1440
   II. Computational Complexity of Fuzzy Inferences 1440
  III. Implementation Alternatives 1442
  IV. Development of a Dedicated Architecture for Fuzzy Operation 1452
   V. A Fuzzy-Oriented Microcontroller 1455
  VI. An Industrial Application Example 1464
 VII. Analytical Model of the Problem 1465
VIII. Development of the Fuzzy Control Algorithm 1468
  IX. Simulation Results 1472
   X. Hardware Implementation 1472
     References 1488

51 Techniques and Applications of Neural Networks for Fuzzy Rule Approximation
HISAO ISHIBUCHI AND MANABU NII

    I. Introduction   1492
   II. Handling of Fuzzy Rules by Membership Values   1496
  III. Handling of Fuzzy Rules by Level Sets   1498
  IV. Handling of Fuzzy Rules by Fuzzy Arithmetic   1499
   V. Fuzzy Rule Base Design from Sparse Fuzzy Rules   1503
  VI. Learning from Numerical Data and Linguistic Knowledge   1504
 VII. Applications to Classification Problems   1508
VIII. Conclusion   1516
     References   1517

52 Techniques and Applications of Fuzzy System Interface Optimizers in Various System Problems
J. VALENTE DE OLIVEIRA AND J. M. LEMOS

    I. Introduction   1522
   II. Concerns with Ergonomics   1523
  III. Constraints for Semantic Integrity   1527
  IV. Methodology and Algorithms   1535
   V. Selected Applications   1542
  VI. Conclusions   1556
     Appendix I   1556
     Appendix II   1557
     References   1558

53 Techniques and Applications of Fuzzy Theory to Critical Path Methods
SOFJAN H. NASUTION

    I. Introduction   1562
   II. Basic Definitions and Formulas   1566
  III. The Assumptions and the Basic Formulas   1569
  IV. Practical Calculation of the Latest Allowable Start Time   1574
   V. Practical Calculation of the Slack   1583
  VI. Calculation of the Activity Slacks   1588
 VII. A Short Note on Further Improvement of the P or Q Table   1595
VIII. Conclusion   1596
     References   1596

54 Fuzzy Sequential Circuits and Automata

JERNEJ VIRANT, NIKOLAJ ZIMIC, AND MIHA MRAZ

    I. Introduction 1600
   II. Fuzzy Switching Functions 1602
  III. Fuzzy Memory Cells 1607
  IV. Fuzzy Automata 1630
    V. Conclusion 1652
      References 1652

55 Techniques of OR / AND Neurons in Fuzzy Systems and Their Applications

KAORU HIROTA AND WITOLD PEDRYCZ

    I. Introduction 1656
   II. OR/AND Neuron: An Architecture and Its Interpretation 1656
  III. Learning in the OR/AND Neuron 1659
  IV. OR/AND Processing and Learning—Selected Areas of Application 1662
    V. Concluding Comments 1671
      References 1672

56 Techniques and Applications of Hybrid Fuzzy Learning Theory to Systems Modeling

MARCO RUSSO

    I. Introduction 1674
   II. Fuzzy Logic 1675
  III. Genetic Algorithms 1679
  IV. Previous Fuzzy Logic Genetic Coding Methods 1686
    V. FuGeNeSys 1688
  VI. GEFREX 1693
 VII. Some Applications Developed 1699
VIII. Conclusions 1707
      References 1707

57 Techniques and Applications of Fuzzy Systems Based on the Petri-Net Formalism

ALBERTO BUGARÍN, PURIFICACIÓN CARIÑENA, AND SENÉN BARRO

    I. Introduction 1712
   II. Theoretical Models: From Petri Nets to Fuzzy Petri Nets 1713

III. Petri-Net-Based Techniques for Fuzzy Systems Modeling
     and Implementation   1718
IV. Applications of Fuzzy Petri Nets   1742
 V. Concluding Remarks   1745
    References   1746

INDEX   1751

CONTRIBUTORS

Numbers in parentheses indicate the pages on which the authors' contributions begin.

Shigeo Abe (1051) Department of Electrical and Electronics Engineering, Kobe University, Kobe, Japan

Jiro Anbe (937) Department of Cardiac Surgery, Fuchu-Ioh Royal-Medical Hospital, Tokyo, Japan

Atsushi Aoyama (485) Research Laboratory of Resources Utilization, Tokyo Institute of Technology, Nagatsuta 4259 Midori-ko, Yokohama 226-8503, Japan

Adedeji Bodunde Badiru (223) School of Industrial Engineering, University of Oklahoma, Norman, Oklahoma 73019

Senén Barro (1711) Department of Electronics and Computer Science, School of Physics, University of Santiago de Compostela, E-15706 Santiago de Compostela, Spain

Manuel Berenguel (717) Departamento de Lenguajes y Computación, Area de Ingenieria de Sistemas y Automática Universidad de Almería, Escuela Poutecnica Superior, Carretera Sacramento s/n, E-04120 La Cañada, Almería, Spain

Patrick Bosc (631) IRISA/ENSSAT Technopole ANTICIPA, BP 447 22305 Lannion Cedex, France

Alberto Bugarín (1711) Department of Electronics and Computer Science, School of Physics, University of Santiago de Compostela, E-15706 Santiago de Compostela, Spain

Eduardo F. Camacho (717) Departamento de Ingeniería de Sistemas y Automática, Universidad de Sevilla, Escuela Superior de Ingenieros, E-41092 Sevilla, Spain

Purificación Cariñena (1711) Department of Electronics and Computer Science, School of Physics, University of Santiago de Compostela, E-15706 Santiago de Compostela, Spain

H. John Caulfield (747) Diversified Research Corporation, Cornersville, Tennessee 37047

Dong-Hyuk Cha (1129) Department of Automation Engineering, Korea Polytechnic University, Shihung, Kyunggi-do 429-450, Korea

Ka Ching Chan (293) School of Mechanical and Manufacturing Engineering, University of New South Wales, Sydney, Australia

Shing I. Chang (1299) Department of Industrial and Manufacturing Systems Engineering, Kansas State University, Manhattan, Kansas 66506

An-Yi Chen (1009) Institute of Electronic Engineering, National Chiao-Tung University, Hsinchu, Taiwan, Republic of China

Bor-Sen Chen (345) Department of Electrical Engineering, National Tsing-Hua University, Hsin-Chu, Taiwan, Republic of China

Guanrong Chen (47, 609) Department of Electrical and Computer Engineering, University of Houston, Houston, Texas 77204

Shyi-Ming Chen (691) Department of Electronic Engineering, National Taiwan University of Science and Technology, Taipei, Taiwan, Republic of China

Kuo-Sheng Cheng (973) Institute of Biomedical Engineering, National Cheng Kung University, Tainan 70101, Taiwan, Republic of China

Yu-Min Cheng (345) Department of Electrical Engineering, National Tsing-Hua University, Hsin-Chu, Taiwan, Republic of China

Hao-Jan Chiu (205) Department of Computer and Information Science, National Chiao Tung University, Hsinchu, Taiwan 30050, Republic of China

Hyung Suck Cho (1129) Department of Mechanical Engineering, Korea Advanced Institute of Science and Technology, Yusong-gu, Taejon 305-701, Korea

I. Fang Chung (1393) Department of Electrical and Control Engineering, National Chiao-Tung University, Hsinchu, Taiwan, Republic of China

T. Corbet (577) Experimental Robotics and Teleoperation Laboratory, Department of Mechanical and Industrial Engineering, University of Manitoba, Winnipeg, Manitoba, Canada R3T 5V6

Lesia L. Crumpton-Young (223) Department of Industrial Engineering, Mississippi State University, Mississippi State, Mississippi 39762

A. De Gloria (1439) Department of Biophysical and Electronic Engineering, University of Genoa, 16145 Genoa, Italy

André De Korvin (87) Computer and Mathematical Sciences, University of Houston–Downtown, Houston, Texas 77002

Elias Deeba (87) Computer and Mathematical Sciences, University of Houston–Downtown, Houston, Texas 77002

Francis J. Doyle, III (485) Department of Chemical Engineering, University of Delaware, Newark, Delaware 19716

P. Ferrari (1439) Department of Electrical Engineering, University of Genoa, 16145 Genoa, Italy

Toshio Fukuda (1323) Center for Cooperative Research in Advanced Science and Technology, Nagoya University, Chikusa-ku, Nagoya 464-0814, Japan

Francisco Gordillo (717) Departamento de Ingeniería de Sistemas y Automática, Universidad de Sevilla, Escuela Superior de Ingenieros, E-41092 Sevilla, Spain

D. Grosso (1439) Department of Biophysical and Electronic Engineering, University of Genoa, 16145 Genoa, Italy

Kaoru Hirota (267, 1655) Department of Computation Intelligence and Systems Science, Tokyo Institute of Technology, 4259 Nagatsuta-cho, Midori-ku, Yokohama 226-8502, Japan

Jang-Pong Hsu (1183) Department of Finance, Tainan Woman's College of Arts and Technology, Yungkang, Tainan, Taiwan, Republic of China

Ya-Chen Hsu (47) Department of Electrical and Computer Engineering, University of Houston, Houston, Texas 77204

Yean-Ren Hwang (375) Department of Mechanical Engineering, National Central University, Chung-Li, Taiwan, Republic of China

Lee-Kwang Hyung (461, 1241) Department of Computer Science, Korea Advanced Institute of Science and Technology, Taejon 305-701, Korea

Hisao Ishibuchi (1081, 1491) Department of Industrial Engineering, College of Engineering, Osaka Prefecture University, Sakai, Osaka 599-8531, Japan

C. V. Jawahar (759) Department of Electronics and Electrical Communication Engineering, Indian Institute of Technology, Kharagpur 721 302, India

G. J. Jeon (405) School of Electronics and Electrical Engineering, Kyungpook National University, Taegu 702-701, Korea

Jingping Jiang (111) Department of Electrical Engineering, Zhejiang University, Hangzhou 310027, People's Republic of China

Tao Jiang (871) Interim Technology, Edison, New Jersey 08833

Yaochu Jin (111) Department of Electrical Engineering, Zhejiang University, Hangzhou 310027, People's Republic of China; and Institute for Neuroinformatics, Ruhr-University Bochum, D-44780 Bochum, Germany

Chia-Feng Juang (1265) Department of Electrical and Control Engineering, National Chiao-Tung University, Hsinchu, Taiwan, Republic of China

Kazuo Kiguchi (1323) Department of Industrial and Systems Engineering, Niigata College of Technology, Niigata-shi, Niigata 950-2076, Japan

Changbum Kim (461) Department of Computer Science, KAIST, Taejon, 305-701, Korea

Tae Won Kim (1029) Precision Instruments R & D Center, Samsung Aerospace Industries, Ltd., Kyungki-Do 462-121, Korea

Frank Klawonn (431) Department of Electrical Engineering and Computer Science, FH Ostfriesland, University of Applied Sciences, D-26723 Emden, Germany

Amit Konar (1341) Department of Electronics and Telecommunications Engineering, Jadavpur University, Calcutta, India

Jean-Luc Koning (527) INPG-Leibniz, BP 54, 26902 Valence Cedex 9, France

Rudolf Kruse (431) Department of Computer Science, Otto-von-Guericke University, D-39106 Magdeburg, Germany

Y. S. Kung (129) Department of Electrical Engineering, Nan-Tai Institute of Technology, Tainan, Taiwan, Republic of China

Yau-Hwang Kuo (1183) Institute of Information Engineering, National Cheng-Kung University, Tainan 70101, Taiwan, Republic of China

Reza Langari (187) Department of Mechanical Engineering, Texas A & M University, College Station, Texas 77843

P. D. Lawrence (577) Robotics and Control Laboratory, Department of Electrical Engineering, University of British Columbia, Vancouver, British Columbia, Canada V6T 1Z5

K. K. Lee (405) Engineering Research Center for Advanced Control and Instrumentation, Seoul National University, Seoul, Korea

Keon-Myung Lee (1241) Department of Computer Science, Chungbuk National University, Cheongju, Chungbuk 361-763, Korea

Kwang Y. Lee (831) Department of Electrical Engineering, The Pennsylvania State University, University Park, Pennsylvania 16802

P. G. Lee (405) Department of Computer Control Engineering, Uiduk University, Kyungju, Kyungpook 780-910

Hyung Lee-Kwang (461) Department of Computer Science, KAIST, Taejon, 305-701 Korea

J. M. Lemos (1521) INESC–Research Group on Control of Dynamic Systems, Apartado 13069, 1000 Lisbon, Portugal

Fa-Shen Leou (1009) Institute of Electronic Engineering, National Chiao-Tung University, Hsinchu, Taiwan, Republic of China

Yao Li (871) NEC Research Institute, Princeton, New Jersey 08540

C. M. Liaw (129) Department of Electrical Engineering, National Tsing Hua University, Hsinchu, Taiwan, Republic of China

Ludovic Liétard (631) IRISA/IUT, rue Edouard Branly, BP 150 22302 Lannion Cedex, France

Chin-Teng Lin (1265, 1393) Department of Electrical and Control Engineering, National Chiao-Tung University, Hsinchu, Taiwan, Republic of China

Jzau-Sheng Lin (973) Department of Electrical Engineering, National Chin-Yi Institute of Technology, Taichung, Taiwan, Republic of China

Ya-Ching Lu (1393) Department of Electrical and Control Engineering, National Chiao-Tung University, Hsinchu, Taiwan, Republic of China

Jacques Ludman (747) Northeast Photosciences, Hollis, New Hampshire 03049

Heidar A. Malki (609) Electrical-Electronics Technology, University of Houston, Houston, Texas 77204

Giridhar D. Mandyam (559) Nokia Research Center, Irving, Texas 75039

Chi-Wu Mao (973) Department of Electrical Engineering, National Cheng Kung University, Tainan 70101, Taiwan, Republic of China

Pamela R. McCauley-Bell (223) Department of Industrial Engineering and Management Systems, University of Central Florida, Orlando, Florida 32816

Un-Chul Moon (831) Manufacturing Development Team, Samsung Data System Co., 7-25 Shinchun-Dong, Songpa-Gu, Seoul 138-731, Korea

Miha Mraz (1599) Faculty of Computer Science and Informatics, University of Ljubljana, 1000 Ljubljana, Slovenija

Tadahiko Murata (1081) Department of Industrial and Information Systems Engineering, Ashikaga Institute of Technology, Ashikaga, Tochigi, 326-8558, Japan

Tomoharu Nakashima (1081) Department of Industrial Engineering, College of Engineering, Osaka Prefecture University, Sakai, Osaka 599-8531, Japan

Hiroshi Narazaki (917) Electronics and Information Technology Laboratory, Kobe Steel, Ltd., 5-5, Takatsukadai 1-chome, Nishi-ku, Kobe, 651-22, Japan

Sofjan H. Nasution (1561) GPT 8th Floor, Nusantara Aircraft Industry (IPTN), Jl. Pajajaran 154, Bandung 40174, Indonesia; and Agency for the Application and Assessment of Technology (BPPT), Jl. M.H. Thamrin 8, Jakarta 10340, Indonesia

Hsi-Wen Nein (1265) Department of Electrical and Control Engineering, National Chiao-Tung University, Hsinchu, Taiwan, Republic of China

Manabu Nii (1491) Department of Industrial Engineering, Osaka Prefecture University, Sakai, Osaka 599-8531, Japan

Branko M. Novakovic (781) FSB-University of Zagreb, Luciceva 5, 10000 Zagreb, Croatia

M. Olivieri (1439) Department of Biophysical and Electronic Engineering, University of Genoa, 16145 Genoa, Italy

Sanjukta Pal (1341) Department of Electronics and Telecommunications Engineering, Jadavpur University, Calcutta, India

Rainer Palm (897) Siemens AG Corporate Technology Information, Communications Department ZT IK 4, D-81730 Munich, Germany

Young Moon Park (831) School of Electrical Engineering, Seoul National University, Seoul 151-742, Korea

Witold Pedrycz (3, 27, 159, 1655) Department of Electrical and Computer Engineering, University of Alberta, Edmonton, Alberta, Canada T6G 2G7

Tuan D. Pham (1111) Faculty of Information Sciences and Engineering, University of Canberra, ACT 2601, Australia

Olivier Pivert (631) IRISA/ENSSAT Technopole ANTICIPA, BP 447 22305 Lannion Cedex, France

L. Puglisi (1439) Department of Electrical Engineering, University of Genoa, 16145 Genoa, Italy

Anca L. Ralescu (917) ECE & CS Department, University of Cincinnati, ML0030, Cincinnati, Ohio 45221

A. K. Ray (759) Department of Electronics and Electrical Communication Engineering, Indian Institute of Technology, Kharagpur 721 302, India

Francisco R. Rubio (717) Departamento de Ingeniería de Sistemas y Automática, Universidad de Sevilla, Escuela Superior de Ingenieros, E-41092 Sevilla, Spain

Marco Russo (1673) Department of Physics, Faculty of Engineering, University of Messina, 98166 Sant'Agata (ME), Italy

N. Sepehri (577) Experimental Robotics and Teleoperation Laboratory, Department of Mechanical and Industrial Engineering, University of Manitoba, Winnipeg, Manitoba, Canada R3T 5V6

Joseph Shamir (747) The Technion, Technion City, Haifa 3200, Israel

M. D. Srinath (559) Department of Electrical Engineering, Southern Methodist University, Dallas, Texas 75275

Yury Stepanenko (809) Department of Mechanical Engineering, University of Victoria, Victoria, British Columbia, Canada, V8W 3P6

Chun-Yi Su (809) Department of Mechanical Engineering, Concordia University, Montreal, Quebec, Canada, H3G 1M8

Il Hong Suh (1029) Intelligent Control and Robotics Laboratory, Department of Electronics Engineering, Hanyang University, Kyungki-Do 425-791, Korea

Chuen-Tsai Sun (205) Department of Computer and Information Science, National Chiao Tung University, Hsinchu, Taiwan 30050, Republic of China

Toshikazu Tobi (937) Technical Research Laboratory, DAI-DAN Co., Ltd., Saitama, Japan

Kiyohiko Uehara (267) Communication and Information Systems Research Laboratories, Research and Development Center, Toshiba Corporation, 1, Komukai Toshiba-cho, Saiwai-ku, Kawasaki 210-8582, Japan

J. Valente de Oliveira (3, 1521) Department of Mathematics and Computer Science, Universidade da Biera Interior, 6200 Covilhã, Portugal

Venkat Venkatasubramanian (485) School of Chemical Engineering, Purdue University, West Lafayette, Indiana 47907

Jernej Virant (1599) Faculty of Computer Science and Informatics, University of Ljubljana, 1000 Ljubljana, Slovenija

Liang Wang (187) Staff Scientist, Casa Incorporated, Los Alamos, New Mexico 87544

Kwei-Ann Wen (1009) Institute of Electronic Engineering, National Chiao-Tung University, Hsinchu, Taiwan, Republic of China

Shishen Xie (87) Computer and Mathematical Sciences, University of Houston–Downtown, Houston, Texas 77002

Hong Yan (1111) Department of Electrical Engineering, University of Sydney, NSW 2006, Australia

John Yen (187) Department of Computer Science, Texas A & M University, College Station, Texas 77843

Yuefeng Zhang (959) Enabling Technologies, Cellular Infrastructure Group, Motorola Inc., Arlington Heights, Illinois 60004

Nikolaj Zimic (1599) Faculty of Computer Science and Informatics, University of Ljubljana, 1000 Ljubljana, Slovenija

■ FOREWORD

Fuzzy Theory Systems: Techniques and Applications, or FTS for short, is a monumental work. Edited by Professor Cornelius T. Leondes, a leading contributor to systems analysis, it is a wide-ranging collection of authoritative and up-to-date expositions of fuzzy set theory and its applications. One cannot help being greatly impressed by the breadth of its coverage. There are chapters on evolutionary neuro-fuzzy modeling, intelligent PID controllers and applications of fuzzy set theory to elevator control, telecommunications, database management, solar plants, and digital image analysis. There are chapters on medical diagnosis, visual feedback control, robotics, production quality control, and critical path methods. And, of course, there are chapters in which the underlying theory is described with authority and insight.

What is the unifying theme of the many facets of fuzzy set theory and its applications as discussed in FTS? I believe that the answer lies in a view of fuzzy logic that I suggested in some of my recent writings. More specifically, in my view fuzzy logic (FL) has four principal facets: the logical facet (FL/L), the set-theoretic facet (FL/S), the relational facet (FL/R), and the epistemic facet (FL/E). It should be noted that these facets overlap to some degree and do not have sharply defined boundaries.

The logical facet of fuzzy logic, FL/L, is a logical system that is, in the main, concerned with modes of reasoning that are approximate rather than exact. The roots of FL/L lie in multivalued logic, but its content and agenda are quite different from those of multivalued logical systems. A common misconception is to equate FL/L and fuzzy logic rather than to see FL/L as one of the facets of FL. To avoid this misconception, FL/L may be viewed as fuzzy logic in a narrow sense, with FL being fuzzy logic in a wide sense. A recently published treatise, "Metamathematics of Fuzzy Logic," by Professor Petr Hajek, deals with FL/L in great depth and with high authority.

The set-theoretic facet, FL/S, is concerned in the main with classes that do not have sharply defined boundaries. My 1965 paper on fuzzy sets dealt with this facet and was unconcerned with logic in its usual sense. The mathematical literature of fuzzy logic is focused for the most part on this facet of FL. This is true of fuzzy set theory, fuzzy topology, fuzzy measure theory, fuzzy arithmetic, fuzzy group theory, and other generalizations of mathematical systems in which generalization is achieved by replacing the concept of a set with that of a fuzzy set.

The relational facet of fuzzy logic, FL/R, is concerned in the main with dependencies that are fuzzy rather than crisp. One of the principal components of FL/R is the calculus of fuzzy if–then rules (CFR). It is this calculus that is used in most of the practical applications of fuzzy logic, and it is this calculus that plays a major role in FTS. What is frequently not recognized is that—at least at this juncture—FL/L plays a much more limited role than FL/R in practical applications of fuzzy logic.

The epistemic facet of fuzzy logic, FL/E, is concerned for the most part with knowledge, meaning representation, and information processing. The key components of FL/E are possibility theory and generalized versions of probability theory. Most of the applications of FL/E lie in the realms of knowledge-based systems, expert systems, and knowledge-discovery techniques. A particularly important application area for FL/E is the conception, design, and utilization of information/intelligent systems. In essence, the epistemic facet of fuzzy logic is focused on meaning, knowledge, and decision.

To see the contents of FTS in proper perspective, it is important to recognize that much of fuzzy logic may be viewed as the product of a group of generalizations—generalizations of crisp concepts, methods, and theories. The two principal modes of generalization are (a) fuzzification, which replaces a crisp set by a fuzzy set; and (b) granulation, which partitions an object into a collection of granules, with a granule being a clump of objects (points) drawn together by indistinguishability, similarity, proximity, or functionality. In most cases, fuzzification and granulation act in concert, leading to the concept of fuzzy granulation or, equivalently, $f*g$-generalization. Fuzzy granulation plays a pivotal role in human cognition, perception, and reasoning, reflecting as it does the finite ability of the human mind to resolve detail and store information. In effect, fuzzy granulation is a human way of achieving data compression. A concept that plays a central role in fuzzy granulation is that of a linguistic variable. The concept of a linguistic variable is employed—directly or indirectly—in most of the applications described in FTS. In fact, the importance of fuzzy logic and, especially, FL/R, derives in large measure from the fact that no methodology other than fuzzy logic provides a machinery for fuzzy granulation.

The machinery for fuzzy granulation serves as a foundation for the methodology of computing with words (CW). In CW, the objects of computation are words or propositions drawn from a natural language. There are two principal rationales for the use of words in place of numbers. First, the use of words is necessary when the values of variables are not known with sufficient precision to justify the use of numbers; second, it is an advantage when there is a tolerance for imprecision that can be exploited to achieve tractability, robustness, and low solution cost. This is the rationale that underlies the extensive use of fuzzy logic in the realm of consumer products.

A recent important development in fuzzy logic that is not yet reflected in FTS is the initiation of what may be called the computational theory of perceptions (CTP). CTP is inspired by the remarkable human capability to perform a wide variety of physical and mental tasks without any measurements and any computations. Everyday examples of such tasks are parking a car, driving in heavy traffic, playing golf, and summarizing a story. Underlying this capability is the brain's crucial ability to manipulate perceptions—perceptions of time, distance, direction, force, shape, likelihood, intent, and truth, among others.

The main difference between measurements and perceptions is that, in general, measurements are crisp whereas perceptions are fuzzy. In CTP, the point of departure is a description of perceptions in a natural language. This opens the door to manipulation of perceptions through the use of the machinery of computing with words. More specifically, in CTP the so-called constraint-centered semantics is employed to represent as a generalized constraint the meaning of a perception that is described as a proposition or propositions in a natural language. Then, generalized constraint propagation is used for reasoning with perceptions and question answering. I believe that eventually CTP will develop into an important theory in its own right.

My comments and observations are intended to place in perspective the contents of FTS. What should be noted is that FTS covers more than fuzzy logic. In the spirit of soft computing, this book offers informative chapters dealing with combinations of fuzzy logic with neurocomputing and evolutionary computing methodologies. Such combinations are likely to grow in importance and visibility in the years ahead.

It is difficult to exaggerate the importance of the vast panorama of methods and applications described in FTS. Professor Leondes, the contributors to FTS, and the Academic Press have produced a volume that should be on the desk of anyone who is interested in developing an understanding of fuzzy logic and its synergistic relation with the allied methodologies of neurocomputing, evolutionary computing, and probabilistic computing. The creators of FTS deserve our thanks and congratulations.

Lotfi A. Zadeh
April 1, 1999

PREFACE TO THE FOUR VOLUMES

Fuzzy theory is an interesting name for a method that has been highly effective in a wide variety of significant, real-world applications. A few examples make this readily apparent. As the result of a faulty design in the method of computer-programmed trading, the biggest stock market crash in history was triggered on October 19, 1987, by a small fraction of a percent change in the interest charged in a Western European country. A fuzzy theory approach would have weighed a number of relevant variables and the ranges of values for each of these variables. Another example, which is rather simple but pervasive, is that of an electronic thermostat that turns on heat or air conditioning at a specified temperature setting. In fact, actual comfort level involves other variables such as humidity and the location of the sun with respect to windows in a home, among others. Because of its great applied significance, fuzzy theory has generated widespread activity internationally. In fact, institutions devoted to research in this area have come into being.

This four-volume set presents a rather comprehensive treatment of this subject and its broad array of practical applications. The structure of these volumes is as follows:

Volume 1
Section I: Modeling Techniques
Section II: System Control Methods and Applications

Volume 2
Section III: Industrial, Manufacturing, and Commercial Systems
Section IV: Adaptive Fuzzy Systems

Volume 3
Section V: Image, Visual, and Pattern Processing Systems
Section VI: Fuzzy Neural Systems

Volume 4
Section VII: Implementation Techniques

The authors are all to be highly commended for producing a significant and unique reference source of lasting value to students, research workers, practitioners, computer scientists, and others on the international scene.

Cornelius T. Leondes

I

MODELING TECHNIQUES

OPTIMIZATION TECHNIQUES IN THE DESIGN OF FUZZY MODELS

WITOLD PEDRYCZ

Department of Electrical and Computer Engineering, University of Alberta, Edmonton, Alberta, Canada T6G 2G7

JOSE VALENTE DE OLIVEIRA

Department of Informatics, Universidade da Beira Interior, 6200 Covilhã, Portugal

I. INTRODUCTORY REMARKS: MODELING VIS À VIS
   FUZZY MODELING   4
II. INFORMATION GRANULARIZATION   6
   A. Spatial Granularization   6
   B. Temporal Granularization   7
III. DESIGN OF INPUT AND OUTPUT INTERFACES: A
   REVIEW OF MATCHING TECHNIQUES   8
   A. Possibility and Necessity Measures   9
   B. Compatibility Measure   10
IV. THE REQUIREMENTS OF SEMANTIC INTEGRITY   10
V. INFORMATION EQUIVALENCE CRITERION   11
VI. ENCODING AND DECODING: AN INVERSE PROBLEM
   IN RELATIONAL CALCULUS   14
VII. CONSTRUCTING INFORMATION GRANULES   15
VIII. CLUSTERING TECHNIQUES IN THE DESIGN OF
   LINGUISTIC INTERFACES   16
   A. Fuzzy C-Means Clustering Algorithm   16
   B. Context-Oriented Fuzzy C-Means   18
   C. Design of the Linguistic Interface: An
      Equalization Approach   21
IX. DESIGN OF THE PROCESSING MODULE   22
X. OPTIMIZATION POLICIES IN FUZZY MODELS   24
XI. CONCLUSIONS   24
   REFERENCES   25

Fuzzy models are regarded as linguistic nonnumeric modeling structures with well-defined functional blocks of input and output interfaces along with a processing module. This chapter elaborates on the essence of fuzzy models, examines the functions of the underlying modules of such models, and specifies the relevant optimization tasks emerging at the level of fuzzy system identification. We address a fundamental question of information granularization along with the issues of nonnumeric–numeric information processing and pertinent encoding–decoding practices. The associated encoding and decoding tasks are formulated and discussed in detail. Considering several distinct levels of conceptual memorization realized within the realm of fuzzy models (subsequently resulting in establishing short-, medium-, and long-term memories), the ensuing learning policies are discussed.

I. INTRODUCTORY REMARKS: MODELING VIS À VIS FUZZY MODELING

Generally speaking, the current literature on fuzzy modeling clearly exhibits a wealth of diverse approaches to fuzzy modeling, supporting various methodological points of view and embracing distinct classes of models (cf. [1]). The ideas of fuzzy models and fuzzy modeling have been formulated and analyzed from both the methodolgical and the experimental standpoints. There is no doubt that the methodology of fuzzy modeling becomes vital to any application of fuzzy sets. This becomes particularly important when it comes to an extensive and thorough *"what-if"* type of analysis carried out within the framework of fuzzy modeling. As has been recognized already, the quality of fuzzy models becomes crucial to any further utilization no matter what type of application we will be dealing with. The primary questions one should then formulate in this setting are: What makes the fuzzy models so distinct from other widely used classes of the models being described by numerical mappings? What are the commonalities between fuzzy and nonfuzzy models? Which part of the existing modeling methodology should be preserved as being universal to all the available modeling techniques?

Next arise questions of a far more pragmatic flavor: How can the fuzzy models be optimized? How can one evaluate them efficiently and comprehensively?

Being extremely concise and not moving at this point into minute technical details, we regard fuzzy models as the models that operate on fuzzy sets (linguistic labels, information granules) by eliciting interrelationships among these quantities. Following acceptance of this definition, the following two features are worth underlining:

1. Because the linguistic labels can be easily modified, one should become aware that by embracing some collections of the elements of the universe of discourse and encapsulating them into manageable and highly coherent chunks of information (fuzzy sets), the processing level at which the fuzzy model operates becomes substantially affected. There is no doubt that this selection should be carried out with discernment. The information granularization helps shape up any fuzzy model and get the most suitable view of the experimental data.

2. As the interrelationships between the linguistic labels are specified at a more abstract set-based level rather than that recognized at the level of a

vast number of the individual numerical values, these dependencies should be quantified eventually at the level of more logic-inclined fuzzy relations instead of function-type dependencies. Subsequently, fuzzy models are not governed by some underlying relationships of physics but rather allude to the cognitive facet of the modeling activities.

The general architecture of the fuzzy model, as portrayed in Fig. 1, is comprised of three basic functional blocks. Two of them constitute information interfaces (the input and the output) linking the fuzzy model (or, more precisely, its processing core) with the modeling environment. All the computations are aimed at building more detailed relationships among the linguistic labels, which are specified for the interfaces, are carried out by the processing module. These constitute the "internal" functional block of the overall structure (refer again to Fig. 1).

Bearing this topology in mind, the applicability of the general modeling methodology is valid to a certain extent but does require some further enhancements and modifications. Due to the heterogeneous form of the available data, identification and optimization can be realized at quite different levels, being narrowed down, if necessary, to the different functional modules. Furthermore, these blocks can be optimized subject to the various information-oriented optimization criteria. The optimization itself could be realized either separately or together. These optimization aspects give rise to a visible diversity of the identification techniques that need to be designed.

The aim of this chapter is to formulate and study the essential optimization problems arising in the realm of fuzzy models and fuzzy modeling and to propose some solutions to the overall design problem. The material is organized as follows: First we elaborate on the question of information granularization (both spatial and temporal) as an omnipresent pursuit occurring in any fuzzy-set-oriented endeavor (Section II). In the sequel, we discuss several matching techniques involving possibility, necessity, and compatibility measures and review their role in the design of the linguistic interfaces. The related issues, such as semantic integrity of the linguistic interfaces and information equivalence, raised there are studied afterward (Sections IV and

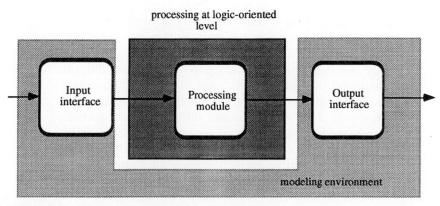

FIGURE I General functional layout of a fuzzy model.

V). A significant portion of the study is devoted to the design of information granules (Section VII). Here we provide a general taxonomy of the methods helpful in the organization of various algorithmic approaches and analysis of their suitability. Some of those related to fuzzy clustering are analyzed in detail (Section VIII).

II. INFORMATION GRANULARIZATION

As underlined by Zadeh [2] (see also [1]), an effect of information granularization plays a primordial role in all endeavors of fuzzy modeling, if not all applications of the technology of fuzzy sets. Roughly speaking, by a linguistic granule we mean a clump of objects (data points) drawn together by indistinguishability, similarity, or alikeness. This issue has been signalled already in some previous publications devoted to modeling of fuzzy systems. Two main ways of dealing with the issue of information granularization can be anticipated, namely:

- Spatial granularization
- Temporal granularization

Although conceptually different, both are aimed at perceiving information from a certain perspective and summarize it according to the specific requirements imposed by the modeling environment. The clear intent of all these activities is to focus on essentials and eliminate all unnecessary computational burden. Before getting into more detailed analysis and design issues, it is instructive to elaborate on the essence of these approaches and to identify the key aspects of the ensuring modeling practices.

A. Spatial Granularization

The concept of *spatial* granularization is eventually far more visible in fuzzy system modeling; in fact, it constitutes the backbone of any fuzzy model. What the granularization of this character delivers is a representation of each variable (or group of variables) in terms of some meaningful chunks of information—linguistic terms. These fuzzy granules make sense of the variables and establish a basic vocabulary through which we focus attention when describing the system under analysis. The changes in the number of linguistic terms, and their distribution across the universe of discourse, make a difference in system modeling leading to another quite different look at the overall simulation environment.

As mentioned, there are a number of formal definitions of the resulting constructs. They appear under different terms, including such well-known names as fuzzy partitions and frames of cognition. Interestingly, this type of granularization occurs also beyond the boundaries of fuzzy set technology. For instance, radial basis functions encountered in neurocomputing are representative examples of information granularization. Here the intent is to focus on the most interesting and promising regions of the search space. This selective search quite often contributes to the enhanced performance of the learning activities. On a more technical note, spatial granularization can be viewed as a useful and powerful process of nonlinear *normalization* of the variables of interest. As such, the role of information granularization has been fully acknowledged in the realm of numeric processing, including optimization, signal processing, and neural networks. Spatial granularization has been pursued at the level of both the individual variables and ensembles of some variables. In the first case we are concerned with fuzz sets; in the latter, fuzzy relations.

B. Temporal Granularization

In contrast to spatial granularization, its *temporal* counterpart is concerned with information granularization (summarization) completed on a temporal basis. This naturally implies an emergence of some events (or fuzzy events) around which fuzzy models tend to revolve. In this sense, this type of fuzzy modeling is far more closely related to the way linguistic models are constructed by a human being. What this entails comes as an example of fuzzy *event* modeling. As mentioned, the granularization effect taking place at a temporal level has not been investigated very intensively. This, in fact, opens up a number of alternatives worth exploring. To become more constructive, let us concentrate on the time series portrayed in Fig. 2.

To come up with a nonnumeric and thus summarized piece of data, we introduce a sliding window (time frame) moving across the time variable. Once moved along the time axis, the successive windows overlap. The data elements embraced by the individual window are then summarized: with each time frame we associate an aggregate of the sample falling therein. In the simplest scenario we can use just an average of these samples and treat is as a reasonable representative of the entire population. A better representative would be the same average equipped with a relevant confidence interval. This means that we produce a construct $[x - \sigma, x + \sigma]$, where σ is the confidence range of the average interval. Let us emphasize that the summarization effect has led us to a nonnumeric interval-like type of information. This form of granularization is well known in statistics, with a vast number of pertinent estimation algorithms.

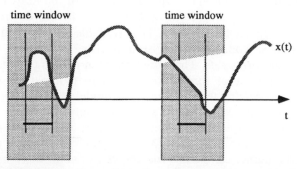

FIGURE 2 Time series under analysis along with a series of set-based windows.

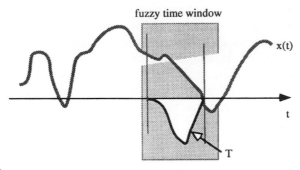

fuzzy time window

FIGURE 3   A time series with a fuzzy window T.

One can envision an interesting fuzzy set counterpart of the previous construct by introducing a fuzzy window T. The fuzzy window becomes a fuzzy set defined in the time line. As before, we let the fuzzy time window slide along the time axis and produce the results of summarization of the time series that is now interpreted in the sense of temporal granularization. To reveal computational details, let us refer to Fig. 3.

The computations of the temporally induced information granules follow the well-known extension principle, meaning that the fuzzy set of signal A implied by T is computed as

$$A = f(T),$$

that is,

$$A(x) = \sup_{t:\, f(t)=x} T(t),$$

where f stands for the time series under discussion.

III. DESIGN OF INPUT AND OUTPUT INTERFACES: A REVIEW OF MATCHING TECHNIQUES

The essential role of the input and output inferences of the fuzzy model is to transfer an information coming from or being sent to the environment in which the fuzzy modeling takes place. The available information could be highly heterogeneous, combining precise numerical quantities (measurements), intervals, and fuzzy sets. The transformation of this external form of information into an internal format acceptable to (compatible with) that being used at the processing level of the fuzzy model is carried out through various matching procedures employing so-called referential fuzzy sets. Quite frequently those procedures hinge on an extensive use of the possibility, necessity, and compatibility measures [3, 1]. In our case we are concerned exclusively with the numerical data so the choice of one of these measures is not that critical. The two of them coincide under these circumstances, yielding the equality $\text{Poss}(X|A) = \text{Nec}(X|A) = A(x_0)$, which holds for any $X = \{x_0\}$.

A. Possibility and Necessity Measures

For the sake of completeness, let us also briefly elaborate on the general case of nonnumerical input data. Let A be one of the referential fuzzy sets of the input interface. Furthermore, let X constitute a nonnumerical input datum. Both X and A are defined in the same universe of discourse $\mathbf{X}$. The possibility measure, $\text{Poss}(X|A)$, as given [1] and [3], is defined as

$$\text{Poss}(X|A) = \sup_{z \in \mathbf{X}} \left[\min(X(z), A(z)) \right]$$

and the necessity measure is given by [1] and [3]

$$\text{Nec}(X|A) = \inf_{z \in \mathbf{X}} \left[\max(\overline{X}(z), A(z)) \right],$$

where $\overline{X}(z) = 1 - X(z)$ for all z in $\mathbf{X}$.

These two measures capture a mutual distribution of X and A. A degree of overlap is expressed by the possibility measure. The necessity measure expresses a degree of containment of X in A (Fig. 4).

Let us now consider the simplest case of an uncertain nonnumerical input, that is, an interval X situated in $\mathbf{X}$. In this case these two measures become different, as illuminated in Fig. 4. It should be stressed that, for the nonpointwise information, the use of only one type of the measures is no longer sufficient due to an evident lack of sufficient discriminatory abilities. Figure 5a makes this claim more visible: the two different intervals X and Y, representing data of quite different levels of uncertainty, are characterized by the same value of the possibility measure. Subsequently, Fig. 5b illustrates the case where two different numerical intervals exhibit the same value of the necessity measure. These considerations remain valid when we study two or more referential fuzzy sets of the interface. In other words, the notion of uncertainty is definitely context-dependent.

To sum up, an important feature of the input interface is to provide a precise representation of uncertain input data. In this sense, for the general case of nonnumerical data, both the possibility and necessity measures are required to discriminate among different levels of granularity of the processed data. Note also that the level of discrimination depends upon a

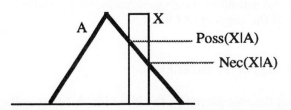

FIGURE 4 Computations of possibility and necessity measures.

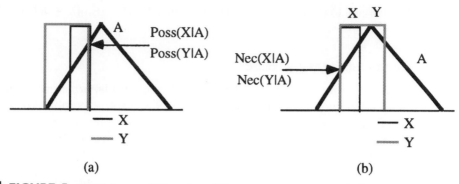

(a)   (b)

FIGURE 5 Discriminatory abilities of possibility and necessity measures in describing nonnumeric datum.

relative granularity of the input data and the linguistic labels of the input interface.

B. Compatibility Measure

The compatibility measure provides a far more comprehensive measure to describe relationships between two fuzzy sets defined in the same universe of discourse. By definition, the compatibility measure of X with respect to A reads as

$$\text{Comp}(X, A)(u) = \sup_{x:\, A(x)=u} X(x), \tag{1}$$

where $u \in [0, 1]$. In comparison with the possibility and necessity measures, the compatibility measure is a function over the unit interval than a single scalar quantity (which, in fact, is true for the possibility and necessity measures).

Two properties of the compatibility measure are worth underlying:

- Monotonicity [If $X' \subset X$ then $\text{Comp}(X', A) \leq \text{Comp}(X, A)$.]
- Boundary conditions: [If $X = \{x_0\}$ then the compatibility assumes 1 at the point where $A(x_0) = u$; for all remaining elements of the unit interval it returns zero. Likewise, if X covers the entire universe of discourse **X** then $\text{Comp}(X, A)(u) = 1$ over the entire unit interval.]

Interestingly, both possibility and necessity values are included in the support of the fuzzy set of compatibility. More specifically, they constitute the upper and lower bounds of the support of $\text{Comp}(X, A)$.

IV. THE REQUIREMENTS OF SEMANTIC INTEGRITY

Taking into account the role played by the input interface, we will enumerate some features that are deemed crucial to the development of the entire fuzzy model. (cf. [1] and [5–7]). The list of these requirements involves the following:

• Focus of attention (or distinguishability)—each linguistic label should have a transparent semantic meaning, whereas the corresponding referential fuzzy sets should clearly define a certain range in the universe of discourse. This implies a clear distinction of the membership functions that represent the resulting referential fuzzy sets.

• A justifiable number of elements—the number of the linguistic terms should be compatible with the number of conceptual entities a human being can efficiently store and utilize in his inference activities. Therefore this number should not exceed the well-known limit of 7 ± 2 distinct terms.

• Natural zero positioning—whenever required one of the linguistic terms should represent the *"around zero"* conceptual entity, that is, its membership function should be convex and centered at zero.

• Coverage—the entire universe of discourse should be "covered" by the linguistic terms: this feature implies that any numerical datum should be matched to a nonzero degree with a least one of the referential fuzzy sets.

• Normalization—because each linguistic label has a clear semantic meaning, at least one numerical datum of the universe of discourse should yield full matching with each label. On the other hand, to utilize the full range of the unit interval of the membership values these referential fuzzy sets should be normal.

Fulfillment of these semantic requirements assures us that the transformation provided by an interface constitutes a one-to-one mapping (cf. [7]).

V. INFORMATION EQUIVALENCE CRITERION

The information equivalence criterion considers an equivalence between the two involved formats of information, that is, numerical (external) information and linguistic (internal) information. Let us remind the reader that fuzzy models inherently deal with nonnumeric data and hence the role of the encoding–decoding machinery deserves special attention. More specifically, the design method is based on satisfaction of the following mapping criterion:

$$\mathscr{F}^{-1}[\mathscr{F}(y)] = y, \quad \text{that is,} \quad \mathscr{F}^{-1}[\mathbf{y}] = y, \quad (2)$$

where $\mathbf{y} \in [0, 1]^n$ stands for a vector of matching values (membership degrees) of the linguistic labels of the interface and y denotes its numerical representative. Especially, $\mathscr{F}(\cdot)$ will be used to denote the numeric-to-linguistic transformation implemented by the input interface (N/L interface, for short). Similarly, $\mathscr{F}^{-1}(\cdot)$ will denote the linguistic-to-numeric conversion provided by the output interface (hence L/N interface); see Fig. 6.

A pair of input and output interfaces satisfying (2) are referred to as optimal interfaces (cf [7]). In fact, when exercising this criterion we can look

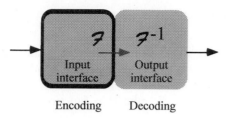

Encoding   Decoding

FIGURE 6   Encoding–decoding mechanisms in fuzzy modeling.

at the underlying structure as a fuzzy model with a substantially reduced processing part (which might be conveniently viewed as a plain identity transformation). The criterion described by (2) will be referred to as a distortion-free property of the linguistic–numerical channel assuring a loss-less reconstruction of information.

The preceding criterion can be specified depending upon the accepted linguistic–numeric transformation. Here we should stress that this transformation should be simultaneously computationally effective. In particular, it should be computationally "transparent" with regard to the remaining modules of the fuzzy model.[1]

For instance, the commonly used averaging transformation (known as the center-of-gravity method) takes on the form

$$\hat{u} = \frac{\sum_{i=1}^{m} A_i(u)\bar{a}_i}{\sum_{i=1}^{m} A_i(u)},$$

where $\hat{u}$ denotes the modal value of the ith membership function A_i and n stands for the numer of labels of this interface. Alternatively, one can design the output interface exploiting the following linear combination of the modal values of the labels including their activation levels as the coefficients of proportionality (weight):

$$\hat{u} = \sum_{i=1}^{m} A_i(u)\bar{a}_i.$$

This mapping applies to the fuzzy sets A_i that constitute a fuzzy partition of the universe of discourse in a strict sense of this term (which means that the equality $\sum_{i=1}^{n} A_i(u) = 1$ becomes satisfies for every element of the universe of discourse). The lossless transformation of information requires that the obtained output $\hat{u}$ be equal to the original u and this holds for all the elements of the universe of discourse. From a practical and application-oriented perspective we can consider that the universe of discourse is represented by a finite collection of numerical points $u_1, u_2, \ldots, u_N$ forming a sampled version of its continuous version. Therefore the information equiva-

---

[1] The computational transparency connotes a situation in which one can access any element of the processing module and optimize it through the use of one of the popular learning schemes, such as, for example, backpropagation. Note, for instance, that the center-of-gravity method is computationally transparent whereas the height method is not.

lence criterion gives rise to the following optimization task driven by the performance index (sum of squared errors):

$$J_1 = \tfrac{1}{2} \sum_{k=1}^{N} \left(u_k - \hat{u}_k \right)^2.$$

More formally, this optimization problem can be stated as follows:

- Minimize J_1 with respect to the parameters of the membership functions and subject to the distinguishability constraint.

The problem can be formulated this way because a proper selection of the membership functions assures the satisfaction of the remaining requirements of semantic integrity. The simulation studies reported later will illustrate this problem in more detail.

Let us concentrate now on the distinguishability constraint and a way of representing it as a part of the optimization problem. Consider the fuzzy representation in the n-dimensional unit hypercube, $\mathbf{x} \in [0,1]^n$ of a given numerical datum specified in the universe of discourse. As noted before, each coordinate of $\mathbf{x}$ represents a degree to which the datum belongs to the corresponding linguistic label (fuzzy set) of the fuzzy partition. If the two consecutive linguistic terms are situated too close each other (become quite indistinguishable), this results in similarly *high* values of the corresponding entries of $\mathbf{x}$ (cf. Fig. 7).

One possible way of "decoupling" these fuzzy sets and making them more distinguishable is to constrain the sigma count (cardinality) of $\mathbf{x}$, $M_p(\mathbf{x})$ [8] to be less than or equal to 1, that is, $M_p(\mathbf{x}) \le 1$ or $\sqrt[p]{x_1^p + x_2^p + \cdots x_n^p} \le 1$, where p is used to specify the strength of the distinguishability requirement imposed on the linguistic terms. The situation where $p = 1$ exhibits a strong constraint satisfaction whereas $p = \infty$ describes a weak constraint. From the optimization point of view, the distinguishability constraint can be represented by the following nonlinear constraint:

$$J_2 = \sum_{j=1}^{N} \left(M_p(x_j) - 1 \right)^2 \mathbf{1}\!\left(\left(M_p(x_j) - 1 \right) \right),$$

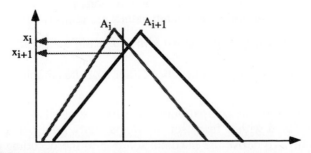

FIGURE 7 The lack of distinguishability in forming linguistic terms.

where $\mathbf{1}(u)$ is a standard unit (Heaviside) function, namely,

$$\mathbf{1}(u) = \begin{cases} 1, & \text{if } u > 0, \\ 0, & \text{otherwise.} \end{cases}$$

VI. ENCODING AND DECODING: AN INVERSE PROBLEM IN RELATIONAL CALCULUS

The possibility, necessity, and compatibility measures are regarded as useful encoding mechanisms that help translate any input datum into an internal, logically inclined entry that becomes utilized afterwards at the level of information processing completed through the processing module. The process dual (inverse) to this encoding concerns decoding of information. By doing this we produce an object compatible with those encountered at the original physical layer of the model. Formally speaking, the decoding is nothing but an inverse process to be completed for the respective encoding mechanism. More formally, we spell it out as follows:

- For the given possibility, necessity, or compatibility value and fuzzy set A, determine X.

There is no unique solution to the problem due to the very nature of the composition operations employing both supremum and infimum—note that these operations are very much selective, leaving out all but a single value. The maximal solution for the possibility-based encoding mechanism reads as

$$\hat{X}(x) = A(x) \, \varphi \, \lambda = \begin{cases} 1, & \text{if } A(x) \leq \lambda, \\ \lambda, & \text{if } A(x) > \lambda. \end{cases}$$

For the necessity-based encoding, we derive the least bound of decoding in the format

$$\tilde{X}(x) = A(x) \, \beta \, \mu = \begin{cases} 1, & \text{if } A(x) \geq \mu, \\ 1 - \mu, & \text{if } A(x) < \mu, \end{cases}$$

These two results are straightforward consequences of the findings originating from the theory of fuzzy relational equations.

It is essential to underline that the computations produce quite limited results as far as the resulting membership values are concerned:

- The possibility calculations return either the original possibility value or 1.
- The necessity computations produce either 0 or the complement of the necessity value.

The inverse problem for the compatibility measure is somewhat more complicated as we are faced with a fuzzy set of compatibility rather than a single numerical quantity. To derive the maximal solution to the decoding problem, let us rewrite the expression for the compatibility relation in an equivalent form by introducing a two-argument Boolean relation R such that

$$R(x,u) = \begin{cases} 1, & \text{if } A(x) = u, \\ 0, & \text{otherwise}. \end{cases}$$

This leads to a well-known form of the sup–min fuzzy relational equation

$$\text{Comp}(X, A) = X \square R$$

(here $\square$ stands for the sup–min composition operation). Also, the maximal solution is immediately derived to be of the form

$$\hat{X}(x) = \inf_{u \in [0,1]} [R(x,u) \varphi \text{Comp}(u)].$$

VII. CONSTRUCTING INFORMATION GRANULES

In the light of the existence of a significant number of algorithmic approaches to the design of the linguistic interface, it is highly desirable to propose a taxonomy of these methods. This will put the entire area into a useful perspective and emphasize the main research agenda along with some open questions that become more articulated in this way.

The first classification splits all methods into two essential groups:

• *Prescriptive* methods of interface development—following this method, we concentrate on identifying the most essential features of any sound linguistic interface that should be satisfied despite the specificity of the problem as well as the format of data available under current application (dictated circumstances). As a matter of fact, these correlate with the requirements of semantic integrity outlined in Section IV.

• *Descriptive* methods of interface development—the key facet taken care of when using these algorithms hinges on the available data and proposes the interface that attempts to encapsulate them to the highest extent. To deal with that, one uses a number of optimization criteria. Each of them conveys a certain motivation justifying the relevant optimization activities.

The descriptive methods include such interesting optimization representatives as

• Clustering techniques
• Mapping properties optimization
• Entropy-based optimization
• Histogram equalization

The descriptive methods also exhibit another general and useful taxonomy. This one takes into account the way in which the interfaces are built. In the first category we place all the methods that look into a single variable (or a block of them), analyze the data therein, and optimize the interface based on a predetermined performance index. The second class of descriptive methods tend to cast the problem of the linguistic interface in a global perspective of fuzzy modeling. In a nutshell, fuzzy modeling leads to some directional architectures, namely, a mapping (transformation) from some input to output variables. In such a sense, any interface developed for the input variable(s) should take into consideration a distribution of the output variable. Practically, if we witness a region in the input space that is associated with the output variables of high variability, it is very likely that the interface granules set up in this input region should be made finer to cope with the details of the eventual mapping.

In what follows, we elaborate on some selected algorithms coming from the main groups indicated above. This will help us underline main functional features of the proposed techniques.

VIII. CLUSTERING TECHNIQUES IN THE DESIGN OF LINGUISTIC INTERFACES

The fuzzy clustering methods are ideal candidates for constructing linguistic interfaces based on experimental data. In fact, what these algorithms tend to accomplish is a formation of a partition of the data into several (usually a fixed number of) categories. The well-known fuzzy C-means (FCM), or Fuzzy Isodata, algorithm is an example falling into this category. As already stated, the methods takes into consideration some variables or include some directional aspects of the fuzzy model. The latter version comes in the form of so-called context-oriented clustering. In what follows, we briefly summarize some interesting points of the method.

A. Fuzzy C-Means Clustering Algorithm

In this form of clustering, the method hinges solely upon a certain performance index Q whose minimization should reveal some meaningful structures in the data set. Quite commonly, one adopts a sum-of-variance criterion that assumes the following form:

$$Q = \sum_{i=1}^{c} \sum_{k=1}^{N} u_{ik}^m \|\mathbf{x}_k - \mathbf{v}_i\|^2 = \sum_{i=1}^{c} \sum_{k=1}^{N} u_{ik}^m d_{ik}^2,$$

where c denotes the number of clusters. The partition matrix $(U = [u_{ik}])$ is used to store all results of clustering (partitioning) the patterns into clusters. Depending on whether we are interested in set-oriented or fuzzy-set-oriented partitioning (clustering), the partition matrix satisfies the following conditions:

- For set-oriented clustering $u_{ik} \in \{0, 1\}$,

$$0 < \sum_{k=1}^{N} u_{ik} < N \quad \text{for } i = 1, 2, \ldots, c,$$

$$\sum_{i=1}^{c} u_{ik} = 1 \quad \text{for } k = 1, 2, \ldots, N.$$

- For fuzzy-set-oriented clustering (fuzzy clustering) we get $u_{ik} \in [0,1]$ with the same two conditions for set-oriented clustering.

These conditions give rise to a family of partition matrices; we will be using the notation $\mathscr{U}$ to describe the entire family,

$$\mathscr{U} = \left\{ U \,\middle|\, 0 < \sum_{k=1}^{N} u_{ik} < N, \, 0 < \sum_{k=1}^{N} u_{ik} < N \right\}.$$

Note that these conditions assume a straightforward and intuitively legitimate interpretation:

- Each cluster is nonempty.
- The element belongs to only one cluster (set-oriented clustering) or the sum of membership values of all clusters is equal to 1 (fuzzy clustering).

The corresponding rows of the partition matrix form fuzzy sets over the entire data set:

$$U = \begin{bmatrix} \mathbf{u}_1 \\ \mathbf{u}_2 \\ \mathbf{u}_c \end{bmatrix},$$

where $\mathbf{u}_i$ stands for the membership function of the ith cluster. These are the constructs we are interested in. The optimization is realized through an iterative procedure that can be described as a series of steps:

Initialization step. Define the number of clusters (c), fix the distance function, and decide upon the value of the power factor (m) in the objective function.

Iterative computation step. Compute the prototypes ($\mathbf{v}_i$) and update the partition matrix (U) based upon the first-order conditions of the minimized objective function. The computations are stopped when some termination criterion is satisfied.

The computations of membership values for any element of the universe of discourse ($\mathbf{x}$) result directly from the FCM model. In particular, we get

$$u_i = \frac{1}{\sum_{j=1}^{c} \left(\|\mathbf{x} - \mathbf{x}_i\| / \|\mathbf{x} - \mathbf{v}_j\| \right)^2},$$

$j = 1, 2, \ldots, c$, with the same distance function as encountered in the original objective function under minimization.

There are two essential design components when it comes to the clustering method:

- The distance function $\| \cdot \|$
- The fuzzification parameter (m)

Here we do not tackle the issue of the number of clusters (c); their choice is very much experiment-driven. The distance addresses an important issue of defining a similarity between two elements (data). In general, a Minkowski distance is often used, with the Hamming, Euclidean, and Tschebyschev distances being viewed as its special cases. The fuzzification factor (m) affects the form of the clusters being produced (or, equivalently, the form of membership function); see Fig. 8. With the increasing values of m there is a profound rippling effect where the membership functions tend to show more local minima. This, in fact, stands in conflict with the requirement of modality of the linguistic terms (see Section IV). Hopefully, for lower values the membership function tend to resemble characteristic functions of sets, meaning that we are getting fewer elements with intermediate membership values. It is interesting to underline that $m = 2$ constitutes a reasonable compromise between setlike membership functions and those with excessive oscillations in the membership values.

B. Context-Oriented Fuzzy C-Means

The standard FCM method concentrates on a collection of itemized variables. Especially, this may concern the input variables specified for the model. Let us emphasize that under such circumstances we are exclusively guided by the distribution of the input data, completely ignoring the fact that they will be used as a part of the larger entity of the fuzzy model and need to be associated in some way with the output variables. The concept of context-based clustering attempts to alleviate this shortcoming by reflecting on the output variable while clustering the remaining input data. This means we first agree upon some granularization of the output variable of the model (which again could be guided by some semantically sound criterion) and afterwards produce some information granules being, in fact, induced by the successive fuzzy sets already formed for the output variable. This phenomenon of focus of attention (invoking only a subset of data in the input space) is illustrated in Fig. 9.

The linguistic granule A defined in the output space is usually referred to as a context. More specifically, f_k describes a level of involvement of $\mathbf{x}_k$ in the assumed context, $f_k = A(k)$. In other words, A acts as a fuzzy modeling filter by focusing attention on some specific subsets of the data set. The way in which f_k can be associated with or allocated among the computed membership values of $\mathbf{x}_k$, at $u_{1k}, u_{2k}, \ldots, u_{ck}$, is not unique. Two possibilities are worth exploring:

- We admit f_k to be distributed additively across the entries of the kth column of the partition matrix, meaning that

$$\sum_{i=1}^{c} u_{ik} = f_k,$$

   $k = 1, 2, \ldots, N.$
- We may request that the maximum of the membership values within the corresponding column equals f_k,

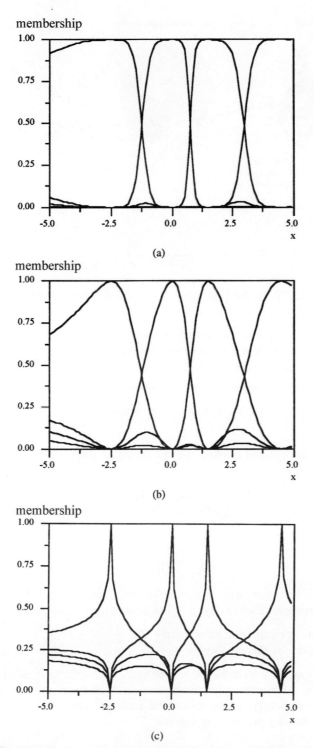

FIGURE 8 Examples of membership functions generated for selected values of *m* for the same prototypes: (a) *m* = 1.5, (b) *m* = 2, (c) *m* = 5.

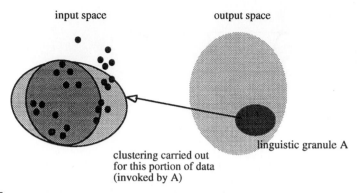

input space

output space

linguistic granule A

clustering carried out
for this portion of data
(invoked by A)

FIGURE 9   Context-based clustering for fuzzy modeling.

$$\max_{i=1}^{c} u_{ik} = f_k,$$

$k = 1, 2, \ldots, N$. In general, one can envision some other alternatives, that is, regard f_k as a function of the resulting membership values.

In the sequel, we confine ourselves to the first way of distributing the conditioning variable. Bearing this in mind, let us modify the requirements to be met by the partition matrices and define the family

$$\mathscr{U}(\mathbf{f}) = \left\{ u_{ik} \in [0,1] \,\middle|\, \sum_{i=1}^{c} u_{ik} = f_k \ \forall_k \text{ and } 0 < \sum_{k=1}^{N} u_{ik} < N \ \forall_i \right\}.$$

The optimization problem guiding the clustering of data is now reformulated accordingly,

$$\min_{U, \mathbf{v}_1, \mathbf{v}_2, \ldots, \mathbf{v}_c} Q$$

subject to

$$U \in U(\mathbf{f}).$$

Let us now proceed with a derivation of a complete solution to this optimization problem. Essentially, it splits into two subproblems:

- optimization of the partition matrix U
- optimization of the prototypes

As these tasks can be handled independently of each other, we start with the partition matrix. Moreover, one can notice that each column of U can be optimized independently, so let us fix the index of the data point (k) and rephrase the resulting problem as follows:

$$\min \sum_{i=1}^{c} u_{ik}^{m} d_{ik}^{2}$$

subject to

$$\sum_{i=1}^{c} u_{ik} = f_k$$

(thus with the fixed data index, we have to solve N independent optimization problems). To make the notation more concise, we have introduced the notation d_{ik} to describe the distance between the pattern and the prototype, namely $d_{ik}^2 = \|\mathbf{x}_k - \mathbf{v}_i\|^2$.

As the preceding is an example of optimization with constraints, we can easily convert this into unconstrained optimization by considering the technique of Lagrange multipliers. Not getting into details, the iterative optimization involves the calculations of the partition matrix and the prototypes completed in the following way:

$$u_{ik} = \frac{f_k}{\sum_{j=1}^{c}\left(\|\mathbf{x}_k - \mathbf{v}_i\|/\|\mathbf{x}_k - \mathbf{v}_j\|\right)^2}$$

and

$$\mathbf{v}_i = \frac{\sum_{k=1}^{N} u_{ik}^m \mathbf{x}_k}{\sum_{k=1}^{N} u_{ik}^m},$$

$i = 1, 2, \ldots, c$, $k = 1, 2, \ldots, N$. The convergence conditions for the method are the same as discussed in the case of the original FCM algorithm [9].

C. Design of the Linguistic Interface: An Equalization Approach

This design method relies on the use of available data. The linguistic terms are fitted into the data so that each of them absorbs the same amount of statistical variability of the data. In other words, the linguistic terms equalize the data. In regions where there is a substantial amount of data, the resulting fuzzy set patched over this area is made narrow and specific. In some other regions where data points are distributed quite sparsely, we accommodate a fuzzy set with a relatively broad membership function. The utilization of the method requires a knowledge of the statistical characteristics of the data, especially their probability density function.

To illustrate this idea, we concentrate on triangular membership functions with $1/2$ overlap between the successive terms. Let $p(x)$ denote a probability density function of the data. We start from a fuzzy set situated at the left side of the universe of discourse; the goal is to optimize its modal value (m_1). Once completed, we proceed in the same way with m_2, m_3, and so on.

The performance index V is introduced in the form

$$V = \left[\int_a^{m_1} A_1(x) p(x)\, dx - \delta\right]^2,$$

where the positive constant δ depends upon the number of the linguistic

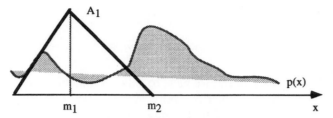

FIGURE 10   Optimization of the linguistic interface — the equalization approach.

terms to be distributed; this value has to be fixed in advance. Refer to Fig. 10 for further details and notation.

The necessary minimum condition

$$\frac{\partial V}{\partial m_1} = 0$$

translates into the form

$$\frac{\partial V}{\partial m_1} = \left[\int_a^{m_1} A_1(x)p(x)\,dx - \delta \right] \frac{\partial}{\partial m_1} \int_a^{m_1} A_1(x)p(x)\,dx = 0,$$

which can be resolved once we confine ourselves to a specific form of the probability density function. This leads to the required location of the modal value of the first linguistic term. Subsequently, one distributed the successive linguistic terms by solving the optimization problems for m_2, m_3, and so on.

IX. DESIGN OF THE PROCESSING MODULE

The processing module realizes computations at the level of the linguistic labels distinguished at the level of the interfaces of the developed model. Its construction calls for the determination of the logical relationships between these labels. We will be concerned with the $s-t$ and $t-s$ composition operators as the basic computational means used in the formation of the model. Let us recall that these constitute a generalization of the two well-known operations on fuzzy sets, namely, the max–min and min–max composition, respectively. More formally, let X and Y constitute the levels of activation of the linguistic labels of the input and output interface, respectively. The $s-t$ composition of X and R, where $R: X \times Y \to [0,1]$, reads as

$$Y = X \cdot R,$$

which coordinatewise translates into

$$y_j = \overset{n}{\underset{i=1}{S}} \left(x_i \, t \, r_{ij} \right)$$

$j = 1, 2, \ldots, m$. Similarly, the t–s composition is given by

$$Y = X \circ R,$$

namely,

$$y_j = \mathop{\mathsf{T}}_{i=1}^{n} (x_i \, s \, r_{ij}).$$

The symbols S (or s) and T (or t) stand for s norm and t norm, respectively.

One can look at these two expressions as examples of logic-based neurons (cf. [10]), where the final aggregation is either completed OR-wise (s–t composition) or AND-wise (t–s composition).

For the fuzzy processing modules with multiple (M) inputs, the generic expression extends to the following form:

$$Y = X_1 \cdot X_2 \cdot \cdots \cdot X_M \cdot R,$$

with X_i, $i = 1, 2, \ldots, M$, forming the successive inputs to the processing module. Using the equivalent pointwise notation, the previous expression reads, accordingly,

$$y_j = \mathop{\mathsf{S}}_{i_1=1}^{n_1} \mathop{\mathsf{S}}_{i_2=1}^{n_2} \cdots \mathop{\mathsf{S}}_{i_M=1}^{n_M} \left(x_{1i_1} \, t x_{2i_2} \, t \cdots t x_{Mi_M} \, t r_{i_1 i_2 \cdots i_M j} \right),$$

where R now becomes a multidimensional fuzzy relation, $R = [r_{i_1 i_2 \cdots i_M j}]$.

The optimization of the processing module is accomplished through a sequence of adjustments of the entries of the fuzzy relation R. These updates are developed through the process of supervised learning. The procedure requires that the fuzzy relation R is modified by changing its individual entries according to the gradient of a given performance index Q,

$$R = R + \Delta R$$

with

$$\Delta R = -\alpha \frac{\partial Q}{\partial R},$$

where $\alpha \in [0, 1]$ constitutes the learning step.

Rewriting the preceding increment expression in a coordinatewise format results in the formula

$$\Delta r_{i_1 i_2 \cdots i_M j} = -\alpha \frac{\partial Q}{\partial y_j} \frac{\partial y_j}{\partial r_{i_1 i_2 \cdots i_M j}},$$

$i_1 = 1, 2, \ldots, n_1$, $i_2 = 1, 2, \ldots, n_2, \ldots$, $i_M = 1, 2, \ldots, n_M$, $j = 1, 2, \ldots, m$. The detailed formulas can be derived once the triangular norms have been specified. For more computational details the reader is referred to [1], [10], and [7].

X. OPTIMIZATION POLICIES IN FUZZY MODELS

In general, similarly to a taxonomy of short- and long-term memories encountered in the standard theory of cognition [11], a clear distinction can be made here by identifying several categories of the memories existing in the fuzzy model:

1. Short-term memory (Formed by the inputs and current mapping errors occurring at the level of the output interface; these are viewed as different, yet temporally short, stimuli.)
2. (Micro)feature coders [The stimuli coded through the interfaces may then potentially effect many elements in a long-term memory. Here it should be stressed that, as some of them are presented more frequently to the model, the feature code (interface) tends to be more and more dedicated or tuned to these particular situations.]
3. Long-term (associative) memory [In the architecture discussed, the memory is composed of the connections (entries) of the processing part of the model (i.e., the entries of the fuzzy relation).]

In this light, the model's architecture conveys a significant amount of psychological evidence. The clearly distinguished levels of memorization suggest also different dedicated learning policies. The following four learning scenarios are worth analyzing:

A. Optimization of the input and output interfaces are carried out in advance and realized separately from the processing module. The parameters of the membership functions of the interfaces are adjusted to achieve a distortion-free reconstruction of information.
B. This option is similar to that described in scenario A: the parameters of the interfaces are learned in advance as in scenario A. In comparison with the previous scenario, they are allowed to be optimized even further during the optimization phase of the processing module.
C. A simultaneous optimization of the interfaces along with the processing module is performed. This option is quite frequently exploited in the existing literature.
D. A global optimization of all the blocks of the model (as in scenario C) is performed, with the parameters of the interfaces now subject to the constraints of semantic integrity.

XI. CONCLUSIONS

We have addressed the problems of optimization of fuzzy models. As shown, and further enhanced by the detailed numerical studies, these optimization procedures should be carried out globally and definitely involve all three functional blocks of the model. Several specific learning policies have been exploited and illustrated with the aid of the relevant numerical experiments. It is worth noting that these scenarios are closely related to the several distinguished levels of the memories realized by the fuzzy model.

The crucial role of the input and output interfaces has been emphasized from a standpoint of system identification. This aspect should also be raised while discussing any further applicational facets of fuzzy models, such as prediction (viz., the task translating itself into a series of iterations using the fuzzy model) or control problems formulated in the framework of fuzzy modeling.

ACKNOWLEDGMENT

Support from the Natural Sciences and Engineering Research Council of Canada NSERC (W. Pedrycz) is gratefully acknowledged.

REFERENCES

1. Pedrycz, W. *Fuzzy Sets Engineering*. CRC Press, Boca Raton, FL, 1995.
2. Zadeh, L. A. Fuzzy sets and information granularity. In *Advances in Fuzzy Set Theory and Applications* (M. M. Gupta, R. K. Ragade, and R. R. Yager, Eds.), pp. 3–18. North-Holland, Amsterdam, 1979.
3. Dubois, D. and Prade, H. *Possibility Theory—An Approach to Computerized Processing of Uncertainty*. Plenum, New York, 1988.
4. Zadeh, L. A. Fuzy sets as a basis for a theory of possibility. *Fuzzy Sets Systems* 1:3–28, 1978.
5. Pedrycz, W. and Valente de Oliveira, J. An alternative architecture for application driven fuzzy systems. In *Proceedings of the 5th IFSA World Congress*, Seoul, 1993, pp. 985–988.
6. Pedrycz, W. and Valente de Oliveira, J. Optimization of fuzzy relational methods. *Proceedings of the 5th IFSA World Congress*, Seoul, 1993, pp. 1187–1190.
7. Valente de Oliveira, J. Neuron-inspired learning rules for fuzzy relational structures. *Fuzzy Sets Systems* 57:41–53, 1993.
8. Kosko, B. Counting with fuzzy sets. *IEEE Trans. Pattern Anal. Machine Intelligence* PAMI-8:556–557, 1986.
9. Bezdek, J. C. *Pattern Recognition and Fuzzy Objective Function Algorithms*. Plenum, New York, 1981.
10. Pedrycz, W. Fuzzy neural networks and neurocomputations. *Fuzzy Sets Systems* 56:1–28, 1993.
11. Anderson, J. A. Cognitive and psychological computation with neural networks. *IEEE Trans. Systems Man Cybernet.* SMC-13:799–815, 1986.
12. Valente de Oliveira, J. A design methodology for fuzzy systems interfaces. *IEEE Trans. Fuzzy Systems* 1996.

2

MODELING RELATIONSHIPS IN DATA: FROM CONTINGENCY TABLES TO FUZZY MULTIMODELS

WITOLD PEDRYCZ

Department of Electrical and Computer Engineering, University of Alberta, Edmonton, Alberta, Canada T6G 2G7

I. INTRODUCTORY COMMENTS 28
II. CONTINGENCY TABLES, MULTIMODELS, AND
INFORMATION GRANULARITY 30
III. GENERIC ARCHITECUTRE OF THE MULTIMODEL 32
IV. DEVELOPMENT OF FUZZY MULTIMODELS 34
   A. Fuzzy Clustering and Its Directionality Aspects 34
   B. Construction of Local Model Induced by Clusters 39
V. NUMERICAL STUDIES 39
VI. CONCLUSIONS 44
   REFERENCES 46

In any endeavor of exploratory data analysis we are concerned with revealing relationships between variables. These relationships can be modeled through rules, linear or nonlinear functions, and neural networks, to name a few alternatives. An interesting and useful scenario arises when the data are governed by relational and one-to-many mappings rather than being confined to purely functional mechanisms. Fuzzy multimodeling introduced in this study is concerned with the design and utilization of families of models rather than single models. The intent is to approximate data that are originated by phenomena whose nature is more relation-based than function-oriented. In general, fuzzy multimodels comprise a collection of local models $M_1, M_2,$ $\ldots, M_c$ along with the relevant mechanisms of their triggering and aggregating aimed at assuring a suitable interaction among these models. The idea of multimodeling is contrasted with some other approaches to fuzzy modeling available in the current literature. In particular, we relate a multimodel approach based on functions and elaborate on an alternative relation–noise perspective assumed during data analysis. The algorithmic details are laid down and illustrated through several detailed simulation studies.

I. INTRODUCTORY COMMENTS

Most, if not all, models encountered in fuzzy modeling [1] are concerned with representing or approximating relationships between input and output variables in the form of some functional dependencies. A notable exception comes in the form of fuzzy relational models [2–5]. As the name stipulates, these models are oriented toward capturing the dependencies between the variables in the form of relations. There are, however, numerous situations in which function-oriented, or even relation-focused, models cannot be sufficient. The intent of this study is to come up with a new modeling framework termed fuzzy multimodeling. The objective of fuzzy multimodels is to deliver an environment assuring a successful interaction among several relational or functional constructs and to allow for their efficient utilization.

In general, the multimodel can be thus regarded as a collection of models $M_1, M_2, \ldots, M_c$ equipped with some mechanisms aimed at a relevant triggering among the models or, if necessary, aggregating the results furnished by the individual models. It is worth contrasting the proposed approach with those existing in the literature, pointing out their main differences. The concept of "local" fuzzy models that is fundamental to the idea outlined by Sugeno and Yasukawa [6] has nothing to do with the aspect of multimodeling addressed in this study. Even though the architecture of the multimodel to be studied here relies on a sort of local model, the identification methodology is evidently very distinct. Moreover the switching mechanism considered by Sugeno and Yasukawa [6] is set up in such a way that only a single result becomes released by the entire model. In contrast, as will be clarified in depth, we allow for several models to be simultaneously invoked and actively pursued. Similarly, the other concept of regression clustering studied by Hathaway and Bezdek [7] looks into certain related aspects but does not tackle explicitly the facet of multimodeling on which the entire discussion in this paper is focused.

The material is organized as follows. First, in Section II, we concisely characterize the fundamental notions supporting the concept of multimodeling and underline crucial links with generic notions such as information granularity and noise versus relations, and we study dependencies between those and introduce the architecture of multimodels and elaborate on its functioning. Section III concentrates on the architectures of multimodels. Here particular attention is focused on the method of directional clustering, forming a key component of the multimodel approach. Subsequently, in Section IV, we cover some experimental studies.

A few examples give useful insight into the very essence of the discussed multimodeling problem and highlight the importance of an operational continuum of modeling that is spread between functions and relations.

EXAMPLE 1. Consider the series of data sets given in Fig. 1a–d. It is evident that in Fig. 1a the relationships between x and y constitute a "pure" function—there seem to be very few noisy data points and they could be easily identified as genuine outliers. In Fig. 1c we are faced with a more difficult and conceptually intriguing question. One can still attempt to build a functional form of dependencies between x and y by treating the region Ω as the one containing points corrupted by a high level of noise. But is this fully legitimate? It could well be that in this region one should rather structure the entire model in terms of a *relation* rather than a *function* between x and y. Similarly, in Fig. 1b we are again faced with relations—in fact M_2 and M_3 treated together form a relation. Roughly speaking, by admitting the language of fuzzy relations one captures some structural noise of a nonprobabilistic nature. The data in Fig. 1d are definitely represented as a relation (what is shown here is known as an s-curve in chemical engineering, portraying a relation between temperature and Damköhler number, d). The data can be split into three separate segments in the sequel, giving rise to three models M_1, M_2, and M_3. Again the data in Fig. 1c exhibit quite a high variety as we can distinguish between a function (models M_1 and M_3) and one apparent fuzzy relation (M_2). Whether the form of the last model (M_2) is functional or relational depends on the level of the structural noise one is willing to accept. If the answer is affirmative then M_2 can still be regarded as function (not a relation).

The next example shows two situations in which the use of fuzzy models becomes even more indispensable.

EXAMPLE 2 (Obstacle avoidance). An autonomous robot is faced with an obstacle and has to avoid it. Not knowing the details of the navigation algorithm exercised by the robot, our intent is to build a respective model of obstacle avoidance by monitoring the behavior of the robot. To accomplish this we record a distance of the robot from the obstacle and the correspond-

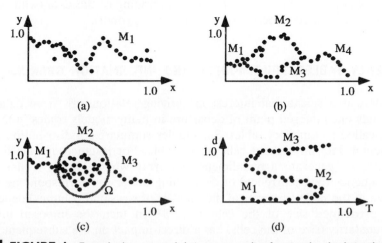

FIGURE I Example data sets and their corresponding functional and relational models.

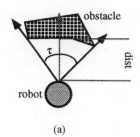

(a)

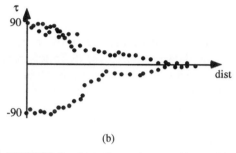

(b)

FIGURE 2 Obstacle avoidance problem and relational character of experimental data.

ing angle of a turn taken at this location (Fig. 2a). The example log of these data is portrayed in Fig. 2b. It is obvious that while being rather close to the obstacle the robot takes a sharp turn right *or* left (see Fig. 2b). Definitely, as the data are highly relational, the multimodel facet of modeling becomes highly required.

These problems constitute a genuine conceptual challenge to standard modeling techniques as being inherently relational. It is not the form of the model to be used that really matters but the way in which they are constructed and put together. Imagine, for instance, the use of a single neural network, no matter how complex, in describing the problem of obstacle avoidance. This will lead to a perfect averaging of the data (with subsequent highly undesired effects for the autonomous robot).

II. CONTINGENCY TABLES, MULTIMODELS, AND INFORMATION GRANULARITY

There is a substantial interest in deriving relationships from data. A commonly encountered point of departure in many models comes in the form of so-called contingency tables. Such tables summarize a distribution of experimental data throughout cells of the table. For instance, in two-dimensional data, we generate a two-dimensional array whose entries (cells) contain a frequency (or occurrence) of data falling under the corresponding cells. The size of the table (in terms of the number of rows and columns) depends upon a predefined size of the cells (Fig. 3). In fact, the imposed information granularity (size of the cells) has a direct impact on the subsequent phases of system modeling.

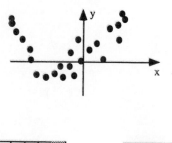

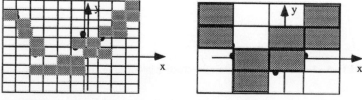

FIGURE 3 Contingency tables developed at various levels of information granularity.

Once the contingency table has been formed, one can proceed with the determination of functionality of data. The algorithm, called the functionality test [8] comprises several phases:

Given a finite data set, pairs $(x(k), y(k))$, $k = 1, 2, \ldots, N$, the following functionality test concerns a function-like relationship between x and y (thus we regard y as a function of x). The data are arranged in a contingency table including counts of the data set. Denote by AV an average number of records per cell.

- for each value x in the data set
  find all groups of cells with adjacent values of y and counts greater than AV; if the number of groups is greater than a then return no_function.
  if the average number of groups is greater than b then return no_function else return function.

The algorithm is controlled by two parameters a and b that measure local and global uniqueness in the dependent variable (y). The default values used by the system developed there and called 49er were

Local uniqueness, 2 (experimental data), 3 (databases)
Global uniqueness, 1.5

As stated, the contingency table can be built at different levels of information granularity. Furthermore we may granulate the contingency table using fuzzy cells formed by fuzzy sets (linguistic terms) defined over both variables. It is intuitively appealing that information granules close to numeric values exhibit the highest levels of information granularity. This level goes down as the fuzzy sets become less specific (Fig. 4).

There is an interesting phenomenon that can be further quantified. If the granularity of the linguistic terms becomes lower, then we derive function-like structure. If these terms are finer (the granularity increases), then we reveal a

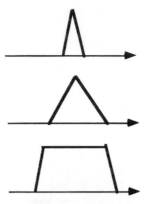

FIGURE 4   Examples of fuzzy sets of decreasing information granularity.

relation-like structure. This can be qualitatively illustrated as visualized in Fig. 5.

In the sequel, we move on to the general architecture of the multimodel and discuss pertinent algorithms.

III. GENERIC ARCHITECTURE OF THE MULTIMODEL

The main idea is to treat the multimodel as a collection of interacting models with the topology outlined in Fig. 6.

For completeness of the discussion we assume $\mathbf{x} \in [0, 1]^n$ and a one-dimensional output $y \in [0, 1]$. The input datum $\mathbf{x}$ is accepted by each model, which in turn determines the corresponding output $y_1, y_2, \ldots, y_c$. Note that we have not specified any particular form of the model; in fact one should stress that the concept is model-independent. If necessary, at a later stage of the multimodel development each of the models M_i, $i = 1, 2, \ldots, c$, can be particularized as a fuzzy relational equation, linear or polynomial function, neural network, and so on. With each model is associated a matching module (match) whose role is to express the extent to which $\mathbf{x}$ should be handled by the ith model. In other words the outcome of this matching reflects a degree

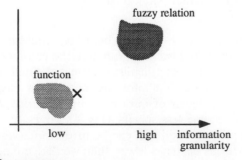

FIGURE 5   Function and relation structures versus information granularity.

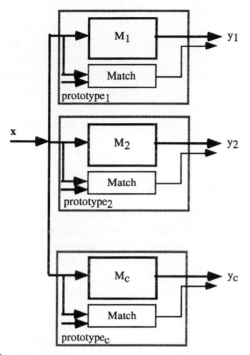

FIGURE 6 General architecture of multimodeling.

of confidence to be associated with the result produced by the corresponding model. For instance, if there is only a single dominant level of matching γ linked with the i_0th model, then this model does count and the result one should rely on is the one provided by it. On the other hand, if two or more models return similar levels of matching then the results of the corresponding models need to be considered—in this case the effect of multimodeling becomes evident as the output comes in the form of a collection of several admissible values, say $y_{i_1}, y_{i_2}, y_{i_p}$. The number of admitted alternatives (the models taken into account) can be easily controlled by the predefined level of matching, where the pertinent rule of model participation reads

$$\text{consider model } M_i \text{ if } \gamma_i \geq \gamma^*,$$

where γ^* is a certain threshold (called admission level) and γ_i denotes the degree of confidence achieved for the ith model, $\gamma_i, \gamma^* \in [0, 1]$.

Noting that even though the selection of γ^* is not overly critical, still some caution should be exercised: too low γ^*s can lead to a fairly numerous collection of output results (as too many models become involved). On the other hand, by moving the value of γ^* too high the multimodel could easily become incomplete as none of its contributing models can be regarded as relevant for some regions of the input variable(s). Selection of the threshold level emerges quite commonly in the compromises between specificity and completeness encountered in many other decision-making situations.

IV. DEVELOPMENT OF FUZZY MULTIMODELS

In what follows we describe the construction of the fuzzy multimodel. Before getting into details, it is important to note that multimodels heavily exploit a concept of "locality" of their individual models. From the operational standpoint, it is essential to elaborate on how the partition of the space of the input variables can be carried out efficiently. The main idea is to reveal the structural relationship between the input and output variables via a method of specialized fuzzy clustering.

A. Fuzzy Clustering and Its Directionality Aspects

As the data used in developing the fuzzy multimodel constitute collectively elements (vectors) in an $(n + 1)$-dimensional space, fuzzy clustering can be regarded as a suitable tool aimed at revealing a structure within the given data set. In contrast with the standard formulations of most of the currently available clustering problems [2, 3, 7], here we are interested in a direction of the dependencies in the data that is evidently from x to y (but not the other way around).

To illustrate the essence of the problem, let us analyze a simple two-dimensional data set in Fig. 7. As seen there, the differences in the x and y coordinates are very similar, $\Delta x \approx \Delta y$.

For the time being, let us neglect our intent to build a model and consider these points as distributed in a two-dimensional cube. If the clusters Ω_1 and Ω_2 were used to construct the two local models M_1 and M_2, despite the particular form of the model, the results would be quite unacceptable—the models will end up averaging data points 1 and 2 (3 and 4, respectively, by producing Ω_1 and Ω_2; Fig. 7a). By identifying the directionality in the problem we simply acknowledge that some points should be dealt with separately (that is, 1 and 2; 3 and 4) by being placed into separate clusters. By doing that it is recognized that the dispersion of 1 and 2 is not merely an effect of noise. Instead, we admit that these data should exhibit a relational

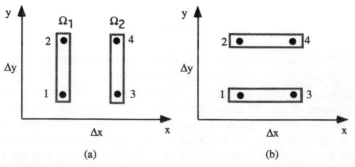

(a)   (b)

FIGURE 7   Four data in $x - y$ space and their clustering: (a) no directionality component included; (b) directionality component included.

character. By accepting this position the multimodel needs to be constructed upon the two clusters indicated in Fig. 7b.

The preceding discussion prompts us to carefully inspect the performance index (objective function) to be used in the construction of the clusters.

I. Performance Index

As usual in any clustering method, we consider a data set consisting of ordered pairs $(\mathbf{x}_k, y_k)$, $k = 1, 2, \ldots, N$, where $\mathbf{x}_k \in [0, 1]^n$ and $y_k \in [0, 1]$. The formation of the objective function (clustering criterion) used to describe the extent to which two pairs of data $(\mathbf{x}_k, y_k)$ and $(\mathbf{x}_1, y_1)$ could be regarded as elements of the same cluster should be guided by the following commonsense observations:

 i. $(\mathbf{x}_k, y_k)$ and $(\mathbf{x}_1, y_1)$ treated as two candidates to be included in the same cluster should be similar coordinatewise, namely, the corresponding coordinates of $\mathbf{x}_k$ and $\mathbf{x}_1$ as well as y_k and y_1 should be *similar*.

 ii. The "directionality" component of the performance index should reflect the functional direction needed to be discovered within the data (meaning that "$\mathbf{x}$ implies y"); this fact should be reflected by the character of the elements assigned to the cluster. In such a sense, with $\mathbf{x}_k$ almost equal to $\mathbf{x}_1$ but quite different y_k and y_1, these two data should not be placed in the same cluster. On the other hand, when $\mathbf{x}_k$ and $\mathbf{x}_1$ do not differ very significantly while having similar y_k and y_1, there is a high likelihood that these patterns could be allocated to the same cluster.

Let us first explicitly express how to quantify the notion of similarity. Because $\mathbf{x}_k$ and y_k are just fuzzy sets defined in the corresponding finite-dimensional spaces, one can adopt the equality index as originally defined in [9]. Recall briefly that the membership values a and b match to the degree $a \equiv b$ equal to

$$a \equiv b = \tfrac{1}{2}\left[\min(a \to b, b \to a) + \min(\bar{a} \to \bar{b}, \bar{b} \to \bar{a})\right], \tag{1}$$

where $a, b \in [0, 1]$, "$\to$" denotes a multivalued implication (pseudocomplement; cf. [9]), $a \to b = \sup\{c \in [0, 1] \mid atc \le b\}$, and the overbar symbol stands for the complement, $\bar{a} = 1 - a$. Interestingly enough, the equality index equals 1 if and only if the two arguments (membership values) are the same, $a = b$. In particular, the implication can be specified in many different ways, including several specific cases:

 • Lukasiewicz implication,

$$a \to b = \begin{cases} 1 & \text{(if } a \le b) \\ 1 - a + b & \text{(otherwise)} \end{cases} = \min(1, 1 - a + b) \tag{2}$$

- Gödel implication,

$$a \rightarrow b = \begin{cases} 1 & \text{(if } a \leq b) \\ b & \text{(otherwise)} \end{cases} \tag{3}$$

- Gaines

$$a \rightarrow b = \begin{cases} 1 & \text{(if } a \leq b) \\ \dfrac{b}{a} & \text{(otherwise)} \end{cases} = \min\left(1, \frac{b}{a}\right), \tag{4}$$

$a, b \in [0, 1]$.

The multidimensional version of (1) could be easily derived by averaging the results of the coordinatewise comparison obtained for the individual coordinates of $\mathbf{x}_k$ and $\mathbf{x}_1$. This yields

$$\mathbf{x}_k \equiv \mathbf{x}_1 = \frac{1}{n} \sum_{i=1}^{n} (x_{ki} \equiv x_{1i}). \tag{5}$$

The necessary directionality aspect of the objective function is captured in the form

$$(\mathbf{x}_k \equiv \mathbf{x}_1) \rightarrow (\mathbf{y}_k \equiv y_1). \tag{6}$$

As the clusters should be formed on a basis of both (i) and (ii), we aggregate (1) and (6) in a multiplicative way that gives rise to the formula

$$Q = \tfrac{1}{2}[(\mathbf{x}_k \equiv \mathbf{x}_1) + (y_k \equiv y_1)][(\mathbf{x}_k \equiv \mathbf{x}_1) \rightarrow (y_k \equiv y_1)] \tag{7}$$

that from now on will be used as the clustering objective function.

To visualize the role of the implication operator in the identification of the directionality of the data, Fig. 8 shows the contour plots of (7) along with its reduced version (without the component of directionality)

$$Q = \tfrac{1}{2}[(\mathbf{x}_k \equiv \mathbf{x}_1) + (y_k \equiv y_1)] \tag{8}$$

(the brighter the color, the higher the value of the objective function). Note that both x and y are just numbers in the unit interval. In both formulas the implication is defined by (4). The differences are profound as the implication component delivers a very visible deformation of the objective function favoring the direction parallel to the x axis.

The effect of directionality can be easily tuned by augmenting (7) by some adjustable parameters. The generalization to be proposed along this line reads as

$$Q = \tfrac{1}{2}[(\mathbf{x}_k \equiv \mathbf{x}_1) + (y_k \equiv y_1)][\alpha(\mathbf{x}_k \equiv \mathbf{x}_1) \rightarrow \beta(y_k \equiv y_1)],$$

where α and β are two parameters defined in $[0, 1]$. If both of them are equal to 1 then this generalization reduces to the original performance index (7).

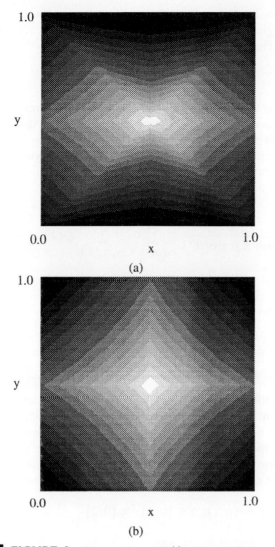

FIGURE 8   Contour plots of Q (a) and its reduced version (b).

The decrease in α reduces the focus of the objective function on the directionality component. The opposite effect can be achieved by keeping the value of α constant and increasing the second parameter (β), starting from values higher than α. In this case the directionality aspect of the performance index gets gradually reduced.

2. The Clustering Algorithm

The clustering procedure guided by the objective function (7) is applied successively to the individual elements of the data set. The process is carried out bottom-up agglomeratively: we start with N clusters each consisting of a single pair of the input–output patterns and merge them successively based on the values of the objective function produced via this combination. Starting from N single-element clusters $\{(\mathbf{x}_1, y_1)\}, \{(\mathbf{x}_2, y_2)\}, \ldots, \{(\mathbf{x}_N, y_N)\}$, the new two-element cluster

$$\left\{ (\mathbf{x}_{i_0}, y_{i_0}), (\mathbf{x}_{j_0}, y_{j_0}) \right\} \tag{9}$$

is formed in such a way that it achieves a maximum of Q determined over all possible mergings of the available data points. Subsequently, the clusters to be expanded (merged) at the successive stages are guided by the maximal value of the performance index Q averaged over the corresponding cluster. This leads to the general merging rule:

- Merge clusters $\mathcal{X}$ and $\mathcal{X}'$ for which the sum

$$\frac{1}{\text{card}(\mathcal{X})\text{card}(\mathcal{X}')} \sum_{(\mathbf{x}_k, \mathbf{y}_k) \in \mathcal{X}} \sum_{(\mathbf{x}_1, \mathbf{y}_1) \in \mathcal{X}'} Q(k, 1) \tag{10}$$

attains a maximal value among all possible merging options. The index $Q(k, 1)$ specifies explicitly the data $(\mathbf{x}_k, \mathbf{y}_k)$ and $(\mathbf{x}_1, \mathbf{y}_1)$ being involved in the determination of (7).

The fundamental question that usually occurs when using any clustering technique is the one about a "plausible" number of clusters to be distinguished in the data set. Because the proposed method is agglomerative, one should be able to control the process of merging by terminating it when the produced clusters cannot sufficiently represent the data. The deficient representation phenomenon may occur due to an excessive variety of objects placed within the same cluster. This in turn calls for a formal definition of the representation capabilities of the clusters. More precisely, we will be interested in expressing how well the prototypes represent the elements of the generated clusters. Let us introduce the following notion. A prototype

$$\mathbf{p} = (\mathbf{p}_x, p_y) \in [0, 1]^n \times [0, 1]$$

of cluster $\mathcal{X}$ is taken as one of its elements

$$(\mathbf{x}_{i_0}, y_{i_0})$$

such that it maximizes the objective function (7) where the computations concern all the elements belonging to this cluster,

$$\mathbf{p}_x = \mathbf{x}_{i_0} \quad \text{and} \quad p_y = y_{i_0} \quad \text{if} \max_{(\mathbf{x}_1, y_1)} \sum_{k=1}^{\text{card}(\mathcal{X})} Q(k, 1) = \sum_{k=1}^{\text{card}(\mathcal{X})} Q(k, i_0) \tag{11}$$

(note that the indices in Q are used to emphasize the data being discussed).

The resulting value of Q, say $Q(\mathcal{X})$, is used as a measure of the representation capabilities of the prototype taken with respect to $\mathcal{X}$. For the single-element clusters, $\text{card}(\mathcal{X}) = 1$, the elements of the clusters are obviously ideal prototypes and the preceding expression always equals 1. In sequel, the global sum of Q taken over all the clusters c,

$$V(c) = \sum_{\text{all clusters}} Q(\mathfrak{X}),\qquad(12)$$

could be admitted as an indicator of representativeness of the data conveyed by their clusters (more precisely, their prototypes). For $c = N$ one has $V(c) = N$. Generally speaking, $V(c)$ is a nondecreasing function of the number of clusters, namely $V(c_1) \leq V(c_2)$ for $c_2 > c_1$. The analysis of the behavior of $V(c)$ plotted versus c could be used to detect the most "plausible" number of clusters: the minimal value of c, say, c^*, that does not lead to a substantial and abrupt decrease in V can be accepted as a viable candidate for the structure in this set of patterns. Similarly, as envisioned in the case of cluster validity indices [7], $V(c)$ should be viewed as a measure indicating a range (rather than a single specific number) of plausible clusters worth considering in the data set.

B. Construction of Local Model Induced by Clusters

The data grouped in the ith cluster are approximated in the obvious way by the corresponding local model

$$y = M_i(\mathbf{x}, \mathbf{a}_i),\qquad(13)$$

$i = 1, 2, \ldots, c$, where $\mathbf{a}_i$ denotes a collection (vector) of the parameters of the model. As the hierarchical clustering is two-valued (the clusters are sets rather than fuzzy sets of data points), the identification of the corresponding local model embraces all these data points belonging to this particular cluster. As mentioned, the form of the model is not restricted in any way so finally the multimodel can include a variety of linear and nonlinear models, fuzzy relational equations, neural networks, fuzzy neural networks, and so on. This means that in general the mapping capabilities of the multimodel can be made as high as required.

When using the multimodel, each local model becomes invoked with a certain confidence μ_i depending on a position of $\mathbf{x}$ with respect to the prototype of this model $\mathbf{p}_{xi}$. The pertinent calculations of the confidence level are realized by the matching module associated with this local model. More specifically,

$$\mu_i = \frac{1}{n} \sum_{k=1}^{n} (x_k \equiv p_{kxi}),\qquad(14)$$

where $\mathbf{p}_{xi} = [p_{1xi}\ p_{2xi}\ \cdots\ p_{nxi}]$ and $\mathbf{x} = [\mathbf{x}_1\ \mathbf{x}_2\ \cdots\ x_n]$.

V. NUMERICAL STUDIES

The following examples including several collections of synthetic data serve as an illustration of the proposed method. In all these experiments, the objective function exploits the implication and equality index based on the implication given by (4). This choice is somewhat arbitrary; any other realization of the implication operator could lead to slightly different numerical results. Similarly we do not modify any optimizing parameters (α and β), keeping both of them equal to 1. The reason is that they could be used efficiently if we are given some data whose class assignment is available. Otherwise there are not enough specialized guidelines on how to select α and β except the qualitative ones contained in Section IV.A. For illustrative reasons all the examples are two-dimensional.

EXAMPLE 3.　Here we discuss six data points,

$$\begin{bmatrix} 0.30 & 0.45 \\ 0.50 & 0.43 \\ 0.70 & 0.42 \\ 0.40 & 0.55 \\ 0.60 & 0.56 \\ 0.80 & 0.58 \end{bmatrix};$$

see also Fig. 9.

The values of the objective function V for several c as shown in Fig. 10 point to $c = 2$ as the most plausible number of the models in the multimodel (this happens for clustering equipped with the component of directionality as well as without it).

In the case of no directionality component involved [the objective function specified by (8)] the two clusters are shown in Fig. 11. With the directionality component in place the results are diametrically different (Fig. 12). In both cases the local models are linear with the parameters obtained through a standard regression.

EXAMPLE 4.　The data analyzed here (Fig. 13), are distributed very regularly, forming a simple geometrical construction. By inspecting the plot of

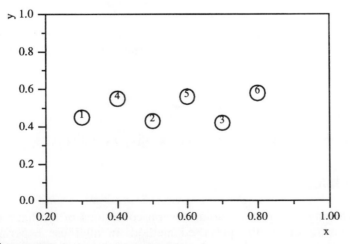

FIGURE 9　Two-dimensional data set.

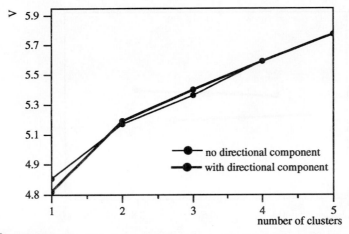

FIGURE 10 $V(c)$ for clustering with an objective function with directional component and without it.

the objective function obtained (Fig. 14), we can admit $c = 4$ equals the minimal number of clusters identified in the data set considered.

The prototypes of the clusters assume the values

$$\begin{bmatrix} 0.35 & 0.35 \\ 0.65 & 0.35 \\ 0.50 & 0.20 \\ 0.60 & 0.50 \end{bmatrix},$$

and the obtained partition of the data is illustrated in Fig. 15.

The models (M_1, M_2, M_3, M_4) are constructed locally within each cluster (all the models are linear except the one for M_4 described as $y = 1.204x^3 -$

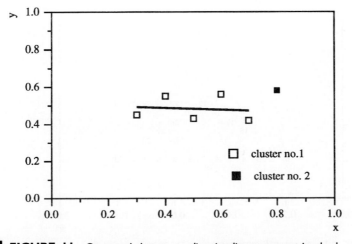

FIGURE 11 Generated clusters, no directionality component involved.

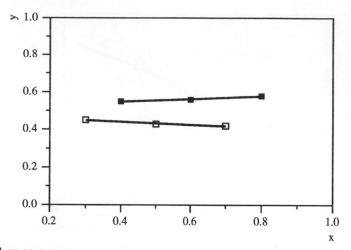

FIGURE 12 Generated clusters, directionality component included.

$2.668x^2 + 1.99x$). The models obtained are outlined in Fig. 16. As emphasized before, they apply only to a certain region of the universe of discourse.

Assuming the activation rule of multimodeling with $\gamma^* = 0.8$ (the choice
of threshold is primarily experiment-driven; some auxiliary guidelines are
provided in the concluding section) the results are visualized in Fig. 17,
providing on the whole an excellent match with the original data—note that
the relational nature of the data is fully retained.

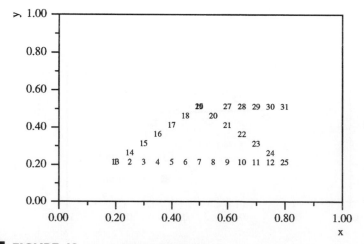

FIGURE 13 A two-dimensional data set.

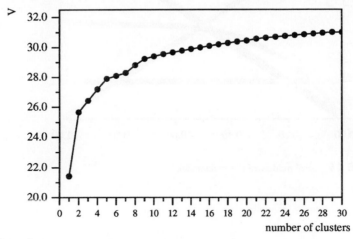

FIGURE 14 *V* versus number of clusters.

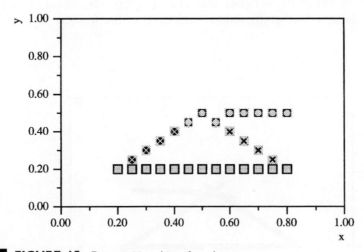

FIGURE 15 Data partitioned into four clusters.

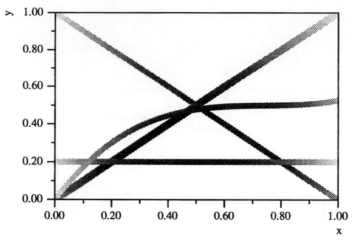

FIGURE 16   Local models of the multimodel.

VI. CONCLUSIONS

The concept of fuzzy multimodeling opens up a new avenue of system modeling with fuzzy sets or, even more generally, system modeling. Within this framework we treat the multiplicity of possible output values of the multimodel not as a deficiency of the modeling technique but rather as its genuine advantage. The multimodel places the idea of randomness in a novel context of structural varieties that are operationally captured in terms of fuzzy relations. The multimodels considered in the framework of rule-based

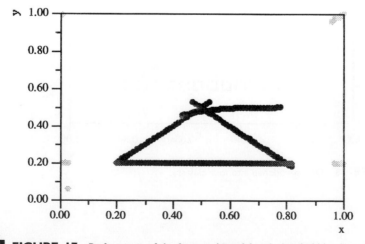

FIGURE 17   Performance of the fuzzy multimodel with threshold level equal 0.8.

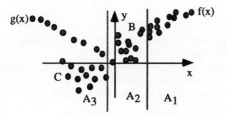

FIGURE 18 Experimental data and a multiplicity of associated local models.

systems give rise to rules with many conclusions,

if condition then conclusion$_1$ or conclusion$_2$ *or* $\cdots$ *or* conclusion$_c$

or certainty-quantified rules of the type

if condition then conclusion$_1$ with certainty$_1$ *or* conclusion$_2$ with certainty$_2$

or $\cdots$ *or* conclusion$_c$ with certainty$_c$.

Interestingly enough, one can easily come up with a multiplicity of models. For instance, the data in Fig. 18 give rise to the rules

- if x is A_1 then y is y is $f(x)$
- if x is A_2 then y is B
- if x is A_3 then y is C or y is $g(x)$

where A_1, A_2, A_3, B, and C are fuzzy sets (relations), and f and g are functions defined over some linguistic terms defined over the input space.

Although this study has focused primarily on the fundamentals of fuzzy multimodeling, some essential design issues are still left open:

1. The choice of the admission threshold γ^* has to be investigated in depth along with some parametric adjustments (parameters α and β) of the objective function. This could eventually involve some scenarios of limited supervision involving samples of data with some indications about the admitted level of structural variability. Similarly, more detailed guidelines should be established with respect to the number of clusters (models) contributing to the multimodel.

2. One can exploit various types of local models and experimentally quantify their computational features as well as pay particular attention to the emerging generalization capabilities achieved by the entire multimodel.

ACKNOWLEDGMENT

Support from the Natural Sciences and Engineering Research Council of Canada (NSERC) is gratefully acknowledged.

REFERENCES

1. Zadeh, L. A. Outline of a new approach to the analysis of complex systems and decision processes. *IEEE Trans. Systems Man Cybernet.* 3:28–44, 1973.
2. Pedrycz, W. Construction of fuzzy relational models. In *Cybernetics and Systems Research*, (R. Trappl, Ed.), Vol. 2, pp. 545–549. North-Holland, Amsterdam, 1984.
3. Pedrycz, W. *Fuzzy Control and Fuzzy Systems*, 2nd extended ed. Research Studies Press/Wiley, Taunton/New York, 1993.
4. Pedrycz, W. *Fuzzy Sets Engineering*. CRC Press, Boca Raton, FL, 1995.
5. Pedrycz, W. *Computational Intelligence: An Introduction*. CRC Press, Boca Raton, FL, 1997.
6. Sugeno, M. and Yasukawa, T. A fuzzy logic based approach to qualitative modeling. *IEEE Trans. Fuzzy Systems* 1:7–31, 1993.
7. Hathaway, R. J. and Bezdek, J. C. Switching regression models and fuzzy clustering. *IEEE Trans. Fuzzy Systems* 1:195–204, 1993.
8. Zytkow, J. and Zembowicz, R. Database exploration in search of regularities. *J. Intelligent Inform. Systems* 2:39–81, 1993.
9. Pedrycz, W. Direct and inverse problem in comparison of fuzzy data. *Fuzzy Sets Systems* 34:223–236, 1990.

3

FUZZY DYNAMICAL MODELING TECHNIQUES FOR NONLINEAR CONTROL SYSTEMS AND THEIR APPLICATIONS TO MULTIPLE-INPUT, MULTIPLE-OUTPUT (MIMO) SYSTEMS

YA-CHEN HSU
GUANRONG CHEN*

Department of Electrical and Computer Engineering, University of Houston, Houston, Texas 77204

I. INTRODUCTION 48
II. DIRECT MULTIPLE-INPUT, MULTIPLE-OUTPUT FUZZY CONTROLLER DESIGN 49
  A. Derivation of Fuzzy Single-Input, Single-Output Controller 52
  B. Multiple-Input, Multiple-Output Controller Structure and Adaptation Laws 54
  C. Simulations 58
III. INDIRECT MULTIPLE-INPUT, MULTIPLE-OUTPUT FUZZY CONTROLLER DESIGN 64
  A. System Representation 64
  B. Fuzzy Model Identification 65
  C. Selective Rule Activation Technique 70
  D. Control Law Design 74
  E. Simulation Examples 76
IV. CONCLUSIONS 80
  REFERENCES 84

* Corresponding author. This research was supported by U.S. Army Research Office Grant DAAG55-98-1-0198 and by the Energy Lab/ISSO at the University of Houston.

This chapter proposes a new approach to fuzzy adaptive controller design using only system input–output data. The design procedure consists of three steps: First, a fuzzy ARMAX model is identified using the available data; then, a fuzzy controller is derived based on a combination of sliding mode control (SMC) theory and fuzzy control methodology; finally, the controller is equipped with an adaptive mechanism to update all control parameters online. The resulting control systems generally possess strong model-adaptation capability, good tracking control performance, and robustness against perturbations and uncertainties. This approach is applicable to general multiple-input, multiple-output (MIMO) nonlinear and uncertain systems. Computer simulations of tracking control of uncertain systems including multilink robot arms are shown as verification and validation of the new design, demonstrating the superiority of the controller in many real-world applications with no system models but with measurement data available.

I. INTRODUCTION

System modeling plays an important role in automatic control, pattern recognition, and many different areas of engineering. The basic objective of system modeling is to establish an input–output representative mapping that can satisfactorily describe the system behavior over the entire operating space. Conventional modeling technique suggests constructing a model using observations of phenomena based upon empirical or physical knowledge; this usually leads to the derivation of a set of state differential (difference) equations for a system. This kind of approach is effective only when simple and well-defined systems are considered. Conventional modeling approaches often fail when dealing with complex and/or uncertain systems because they try to find a precise function or structure of a system; however, the resulting differential (difference) equations usually cannot globally approximate system behavior, especially in the nonlinear case.

Different from the conventional approaches, fuzzy systems [1] use if–then rules to describe physical systems, which give a more meaningful expression of the qualitative aspect of human perspective. A number of rules will be fired with various strengths corresponding to the match between the inputs and the antecedents of the fuzzy rules. The invoked fuzzy rule actions are combined by a defuzzification mechanism to generate a final output. For this reason, use of fuzzy systems can also be recognized as a local modeling technique because it is based upon the spirit of "divide and conquer." The antecedents of the fuzzy rules decompose the range of the input into a number of fuzzy regions; then the consequents approximate the system in each region via a simpler (local) model. This makes fuzzy models capable of aggregating the local actions to globally describe the system behavior.

Usually, a fuzzy system can be constructed by capturing human-expert knowledge and transforming such expertise into rules and membership functions. The fuzzy system's performance is optimized by adjusting the parameters through some leaning methods. Incorporating learning ability into fuzzy systems has led to the development of neuro-fuzzy modeling. However, depending entirely upon human knowledge is impractical and dangerous. First, even if the ideal expert knowledge exists, the knowledge base is usually incomplete or partially incorrect. There is no objective way to extract all correct human knowledge; often different experts give different rules, which may even have conflicts. Moreover, knowledge acquisition is difficult, especially in the MIMO case. The representation of multivariate systems is substantially more complex than that of single-input, single-output (SISO) systems. It is rarely possible that human knowledge can satisfactorily describe the complex couplings between the various inputs and outputs. The problem of knowledge acquisition becomes more formidable when the controlled systems are nonlinear and uncertain.

For these reasons, it has been proposed to integrate several successful systematic approaches such as the sliding mode control strategy to construct a rule base and to reduce the number of rules. For this kind of combination to succeed, some theoretical assumptions have to be made about the systems; that implies certain properties of the systems have to be known in advance. Notice that, in many real-world applications, all we can obtain is a set of input–output data. This leads to the attempt of automating the modeling procedure based on the input–output measurements. Designing an MIMO fuzzy system can be viewed as fitting a complex surface in a higher dimensional space. Given a set of input–output data, the task of modeling is equivalent to constructing a system that can satisfactorily fit the given data set according to some criteria. This can be done through training–learning. After the learning phase, the resulting system should be able to interpolate new sets of data pairs. A self-contained approach of this type is introduced in this chapter, whereby a direct fuzzy controller design method is developed.

II. DIRECT MULTIPLE-INPUT, MULTIPLE-OUTPUT FUZZY CONTROLLER DESIGN

The primary difficulties encountered in the fuzzy controllers design for multivariate systems can be attributed to (1) a huge number of variables to be taken into consideration and (2) lack of complete knowledge of the complex input–output couplings. To resolve these problems, the sliding mode control method deserves considerable attention. Sliding mode control (SMC) was proposed in the early 1950s [2]. The most distinguished feature of SMC is its robustness to system uncertainty and external disturbance. In an SMC system, control laws are designed to drive the system error-dynamics states toward a specific sliding hyperplane (stable manifold). As the system error-dynamics states are maintained in the plane, the system response is governed by the plane, leading to the desired tracking performance such as asymptotic stability and invariance to uncertainty as well as disturbance.

For a large class of systems, SMC design provides a systematic approach to maintaining not only system stability but consistent performance in the presence of model and measurement imprecision. However, it is well known that there is a major drawback in the sliding mode control approach: the undesired phenomenon of chattering due to high-frequency switching, which often excites undesired dynamics. Yet it is interesting to note that fuzzy control strategy may provide an effective way to resolve this problem. Recently, combinations of fuzzy control and SMC approaches, the fuzzy sliding mode control (FSMC) methods, have achieved superior performance [3–8]. For example, Hwang and Lin [4] developed a nonadaptive fuzzy controller, and Wu and Liu [8] used the switching manifold as a reference, whereas sliding modes are used to determine the optimal values of parameters in fuzzy control rules. Ohtani and Yoshimura [5] also suggested a fuzzy control law using the concept of sliding mode, where fuzzy rules are tuned by learning. A self-organizing and adaptive fuzzy sliding mode controller was proposed by Hsu and Fu [3]. Chen and Hsu [9] have also successfully developed an FSMC for MIMO systems. Based upon the FSMC methodology, not only are the resulting fuzzy sliding mode controllers designed with guaranteed stability, but also the number of fuzzy rules are significantly reduced [10].

To introduce the FSMC strategy, consider a class of MIMO nonlinear systems of the form

$$\mathbf{x}^{(n)} = f(\mathbf{x}, \dot{\mathbf{x}}, \ldots, \mathbf{x}^{(n-1)}, t) + G(\mathbf{x}, \dot{\mathbf{x}}, \ldots, \mathbf{x}^{(n-1)}, t)u \qquad (1)$$

where $\mathbf{x}$ is the system state (output) vector, $\mathbf{X} = \{\mathbf{x}, \dot{\mathbf{x}}, \ldots, \mathbf{x}^{(n-1)}\}^T$, and u is the control input, $f\colon R^{l \times l} \times R_+ \to R^l$, $G \in R^{l \times l}$.

The following assumptions are needed:

Assumption 1. The function $f(\mathbf{X}, t)$ need not be exactly known, but is bounded by a continuous function $f^U(\mathbf{X}, t)$; that is, $f^U \geq |f|$, for all $\mathbf{X}$ and all $t \geq 0$.

Assumption 2. $G(\mathbf{X}, t)$ is positive-definite and confined in a certain range:

$$0 < G_L \leq G(\mathbf{X}, t) \leq G^U, \qquad \text{for all } \mathbf{X} \text{ and all } t \geq 0,$$

where G_L and G^U are known positive-definite constant matrices.

Let $\mathbf{x}_d$ be a desired target and let $\tilde{\mathbf{x}} = \mathbf{x} - \mathbf{x}_d$ be the tracking error, with error vector

$$\tilde{\mathbf{X}} = \mathbf{X} - \mathbf{X}_d = \left[\tilde{\mathbf{x}}, \dot{\tilde{\mathbf{x}}}, \ldots, \tilde{\mathbf{x}}^{(n-1)}\right]^T. \qquad (2)$$

To apply the SMC strategy, we first define a time-varying sliding surface $S(t)$ in the state space R^n by the vector equation $s(\tilde{\mathbf{X}}) = 0$, with

$$s(\tilde{\mathbf{X}}) = \left(\frac{d}{dt}\mathbf{I} + \lambda\right)^{n-1} \tilde{\mathbf{x}} = \tilde{\mathbf{x}}^{(n-1)} + A_1\tilde{\mathbf{x}}^{(n-2)} + \cdots + A_{n-1}\tilde{\mathbf{x}}, \qquad (3)$$

where λ is a positive-definite diagonal matrix, and $A_1, \ldots, A_{n-1}$ are constant matrices in terms of λ. Since (3) satisfies the Hurwitz stability criterion, maintaining system states on surface $S(t)$ for all $t > 0$ is equivalent to the tracking problem $\mathbf{x} \to \mathbf{x}_d$. Indeed, it will force $\tilde{\mathbf{x}}$ to approach zero given any bounded initial condition $\tilde{\mathbf{x}}(0)$. Thus, we have in effect transformed an nth-order tracking problem to a first-order stabilization problem [11]. To keep vector $s(\tilde{\mathbf{X}})$ at zero, the control law u is designed to satisfy the following sliding condition:

$$\frac{d}{dt} s^T \Gamma s < -\varepsilon, \qquad \text{if } s \neq 0, \text{ where } \varepsilon > 0, \tag{4}$$

in which Γ is a positive-definite matrix, probably a function of $\mathbf{X}$. This approach eventually leads to a sliding mode switching scheme [12]. Essentially, (4) states that the weighted distance to the sliding surface decreases along all system trajectories eventually. Thus, the states are driven from any initial state to the sliding surface on which sliding mode control takes place. Differentiating $s(\tilde{\mathbf{X}})$ with respect to time, we obtain

$$\dot{s} = \bar{f} + Gu,$$

$$\bar{f} = F - \mathbf{x}_d^{(n)} + A_1 \tilde{\mathbf{x}}^{(n-1)} + \cdots + A_{n-1} \dot{\tilde{\mathbf{x}}}. \tag{5}$$

In the simplest case, $\Gamma = \mathbf{I}$, and if the functions f and G are known, the control law u can be chosen as

$$u = -G^{-1}\bar{f} - Ks. \tag{6}$$

By plugging u into (5), we have $\dot{s} = -Ks$. The sliding condition (4) can then be easily verified. However, f and G are unknown; only their bounds can be used to construct u. In this case, the control law is chosen to be

$$u = -F \operatorname{sgn}(s) - Ks, \tag{7}$$

where

$$F = \left\{ G_L^{-1} \left[f^U + |\mathbf{x}_d^{(n)} - A_1 \mathbf{x}^{(n-1)} - \cdots - A_{n-1} \dot{\tilde{\mathbf{x}}}| \right] \right\},$$

$$F \operatorname{sgn}(s) = \begin{bmatrix} F_1 \operatorname{sgn}(s_1) \\ \vdots \\ F_1 \operatorname{sgn}(s_l) \end{bmatrix}.$$

Substituting (7) into (5), we have

$$\dot{s} = f - \left\{ GG_L^{-1} \left[f^U + |\mathbf{x}_d^{(n)} - A_1 \tilde{\mathbf{x}}^{(n-1)} - \cdots - A_{n-1} \dot{\tilde{\mathbf{x}}}| \right] \right\} \operatorname{sgn}(s) - GKs, \tag{8}$$

$$s^T \dot{s} = s^T f - s^T \left\{ GG_L^{-1} \left[f^U + |\mathbf{x}_d^{(n)} - A_1 \tilde{\mathbf{x}}^{(n-1)} - \cdots - A_{n-1} \dot{\tilde{\mathbf{x}}}| \right] \mathrm{sgn}(s) \right\}$$

$$- s^T GKs + s^T \left(-\mathbf{x}_d^{(n)} + A_1 \tilde{\mathbf{x}}^{(n-1)} + \cdots + A_{n-1} \dot{\tilde{\mathbf{x}}} \right)$$

$$\leq -s^T GKs + |s^T| |\bar{f}| - |s^T|$$

$$\times \left\{ GG_L^{-1} \left[f^U + |\mathbf{x}_d^{(n)} - A_1 \tilde{\mathbf{x}}^{(n-1)} - \cdots - A_{n-1} \dot{\tilde{\mathbf{x}}}| \right] \right\}$$

$$\leq -s^T GKs \leq 0. \tag{9}$$

Although this sliding mode control law can achieve zero tracking error and asymptotic stability, it has several drawbacks. In most cases, to account for the system uncertainties, f^U and G_L^{-1} are conservatively estimated. Consequently, applying $F \, \mathrm{sgn}(s)$ constantly increases the implementation cost; sometimes, implementation becomes practically impossible. Moreover, F may not even be available in some applications. Also it can be seen from (7) that the control input will be discontinuously crossing the sliding surface due to the switching of the sign function, and this may cause undesired chattering. For these reasons, a new fuzzy controller is proposed next to improve the drawbacks, particularly eliminating the chattering, to achieve practical applicability of the SMC design.

A. Derivation of Fuzzy Single-Input, Single-Output Controller

In this section, a new fuzzy adaptive controller of the sliding mode type is first developed to overcome the aforementioned disadvantages. It is clear from the description given in the preceding section that the sliding mode control law explains the intuitive feedback control strategy "if the error is negative, push the system output in the positive direction." On the other hand, the control term $-F_i \, \mathrm{sgn}(s_i)$ can be interpreted as follows:

$$\text{IF} \quad s_i > 0 \quad \text{THEN} \quad \text{output} = -F_i.$$

$$\text{IF} \quad s_i < 0 \quad \text{THEN} \quad \text{output} = F_i, \qquad i = 1, \ldots, l.$$

Hence, it will not be surprising that simple fuzzy if–then rules associated with proper membership functions can be used to construct a control law that emulates the action of $-F \, \mathrm{sgn}(s)$ and utilizes control energy more efficiently. To design a fuzzy SISO controller, we extend the fuzzy controllers developed in [13] and [14]. In a standard procedure, a fuzzy controller design consists of three main components: (i) fuzzification, (ii) control rule base establishment, and (iii) defuzzification. We will follow this procedure to design the new fuzzy controller.

1. Fuzzification

As shown in Fig. 1, the fuzzy controller employs two inputs: the sliding signal s and the direction of change of the sliding signal, Δs, with

$$\Delta s = \frac{s(t) - s(t - d)}{d}, \tag{10}$$

FIGURE 1 Fuzzy controller.

where d is a small constant. This fuzzy controller has only one control output u. The membership functions for the inputs of the fuzzy controller are chosen to be the same, as shown in Fig. 2, where two positive scaling factors K_s and $K_{\Delta s}$ are used for s and Δs, respectively. In other words, $K_s s$ and $K_{\Delta s} \Delta s$ will be used to replace s and Δs, respectively, in the control formulation. Finally, parameter L in the membership functions is also used as a tunable parameter.

2. Fuzzy Rule Base

Based on (9) and the chosen membership functions, then fuzzy control rules are designed as follows:

(R1) IF $K_s s = $ P AND $K_{\Delta s} \Delta s = $ P THEN output = NB,

(R2) IF $K_s s = $ P AND $K_{\Delta s} \Delta s = $ N THEN output = NS,

(R3) IF $K_s s = $ N AND $K_{\Delta s} \Delta s = $ P THEN output = PS,

(R4) IF $K_s s = $ N AND $K_{\Delta s} \Delta s = $ N THEN output = PB,

where P and N stand for positive and negative, B and S for big and small, respectively. In these rules, "AND" is defined by algebraic multiplication:

$$\mu_A \text{ AND } \mu_B = \mu_A \times \mu_B$$

for any two membership values μ_A and μ_B over fuzzy subsets A and B, respectively.

The preceding four rules can be understood as follows. For rule (R1), condition $K_s s = $ P implies that the system states are above the sliding surface, and $K_{\Delta s} \Delta s = $ P means that the system states are moving away from the sliding surface. Hence, the control action should be set to be negative and big enough to turn the system states downward. For rule (R3), the system states are below the sliding surface and moving toward the surface. In this

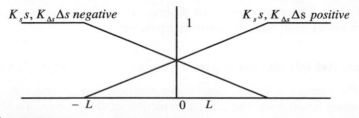

FIGURE 2 Input membership functions.

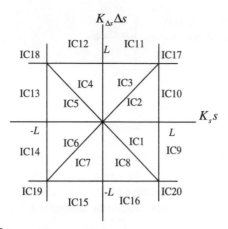

FIGURE 3   The IC regions.

case, only small positive control action is needed to accelerate the convergence. Rules (R2) and (R4) are similarly understood.

3. Defuzzification

In the defuzzification step, the commonly used center of mass (COM) formula [13, 14] is applied to defuzzify the fuzzy control laws. By employing the COM formula, we obtain

$$u = \frac{\sum_{j=1}^{4} \mu^j \xi^j}{\sum_{j=1}^{4} \mu^j} = \phi^T \xi, \qquad (11)$$

where $\xi = [\xi^1, \xi^2, \xi^3, \xi^4]^T$ is the consequent vector with ξ^j the output of the jth rule, μ^j is the membership value of the jth rule, and $\phi = [\phi^1, \phi^2, \phi^3, \phi^4]^T$ with

$$\phi^j = \frac{\mu^j}{\sum_{k=1}^{4} \mu^k}, \qquad j = 1, \ldots, 4. \qquad (12)$$

Actually, this analytical expression can also be obtained if we divide the $s - \Delta s$ plane into 20 input combination (IC) regions as shown in Fig. 3 (see [13] and [14] for more details).

To this end, the basic structure of the fuzzy SISO controller has been established. The next step is to design an MIMO structure and suitable adaptive laws to update the controller parameters.

B. Multiple-Input, Multiple-Output Controller Structure and Adaptation Laws

In MIMO cases, a system can be decomposed into several interconnected subsystems based upon the physical structure of the process. To obtain good performance, coupling effects cannot be neglected, especially in a nonlinear system. A parallel-connected fuzzy control structure [15] is adapted to the control systems under consideration. Since f and G are unknown, we replace the term $-F\,\text{sgn}(s)$ in (7) by a small fuzzy system $\hat{F}(s, \Delta s)$ in the form of (12). The overall control law is

$$u = \hat{F} - Ks + u_s = \hat{F}(s, \Delta s | \Theta) - Ks + u_s,$$

$$\hat{F}_i = \sum_{j=1}^{n} C_{ij} = \sum_{j=1}^{n} w_{ij}^T \theta_{ij}, \qquad i = 1, \ldots, n, \tag{13}$$

where each C_{ij}, with s_j and Δs_j fed into it, stands for an SISO fuzzy system described in the previous section, and Θ is the adjustable controller parameter vector (specifically, Θ is the collection of K_{sij}, $K_{\Delta sij}$, L_{ij}, and θ_{ij}). The fuzzy control scheme is shown in Fig. 4. There are several reasons for using such a control scheme. First, because the system model is coupled, each fuzzy control force $\hat{F}_i$ is constructed to contain information from all states. Second, it was proven in [16] and [17] that the fuzzy systems are universal approximators; that is, they are capable of approximating any continuous function uniformly over a compact set. Therefore, the fuzzy subsystems are qualified as basic building blocks of controllers for nonlinear MIMO systems. Finally, the fuzzy controller is constructed from the intuitive if–then rules; thus, human experience can be directly incorporated into the controller. The additional control term u_s is called a supervisory controller [16]; it will be determined later to satisfy the sliding condition. Our next task is to develop an adapting law for adjusting the controller parameters, to force the tracking error to converge to zero. First, let us define

$$\Theta^* = \arg\min_{\Theta \in \Omega} \left[\sup_{X \in R^n} \left| F - \hat{F}(s, \Delta s | \Theta) \right| \right], \tag{14}$$

where Ω is a closed constraint set for Θ, in which each component of Θ is confined in a certain range specified by the designer. Furthermore, we define the minimum approximation error

$$\omega = -F\,\text{sgn}(s) - \hat{F}(s, \Delta s | \Theta^*). \tag{15}$$

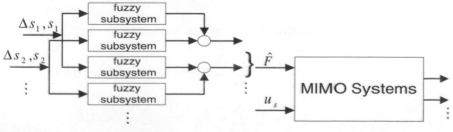

FIGURE 4 The structure of FSMC.

The system equation can be rewritten as

$$\dot{s} = Gu_s + \left(\hat{F}(s|\Theta^*) - \hat{F}(s|\Theta) \right)$$
$$+ (G - \mathbf{I})\left[\hat{F}(s|\Theta^*) - \hat{F}(s|\Theta) \right] + G\omega + \bar{f}. \tag{16}$$

Now, we want to approximate $\hat{F}$ by using Taylor series expansion. Taking the Taylor series expansion of $\hat{F}(s|\Theta^*)$ around Θ, we obtain

$$\hat{F}(s|\Theta^*) - \hat{F}(s|\Theta) = \psi^T\left(\frac{\partial \hat{F}(s|\Theta)}{\partial \Theta} \right) + O(|\psi|^2), \tag{17}$$

where $\psi = \Theta^* - \Theta$, and $O(|\psi|^2)$ represents the higher order terms. Substituting (17) into (16), the system dynamics becomes

$$\dot{s} = Gu_s + \psi^T\left(\frac{\partial \hat{F}(s, \Delta s|\Theta)}{\partial \Theta} \right) + \bar{f} + v, \tag{18}$$

where

$$v = G\omega + (G - \mathbf{I})\left[\hat{F}(s|\Theta^*) - \hat{F}(s|\Theta) \right] + O(|\psi|^2).$$

Next, consider the Lyapunov function candidate

$$V = \tfrac{1}{2}s^T s + \tfrac{1}{2}\left(\sum_{i=1}^{n} \sum_{j=1}^{n} \rho_{ij}\tilde{K}_{sij}^2 + \sum_{i=1}^{n} \sum_{j=1}^{n} \sigma_{ij}\tilde{K}_{\Delta sij}^2 \right.$$
$$\left. + \sum_{i=1}^{n} \sum_{j=1}^{n} \alpha_{ij}\tilde{L}_{ij}^2 + \sum_{i=1}^{n} \sum_{j=1}^{n} \eta_{ij}\tilde{\theta}_{ij}^T\tilde{\theta}_{ij} \right), \tag{19}$$

where

$$\tilde{K}_{sij} = K_{sij}^* - K_{sij},$$

$$\tilde{K}_{\Delta sij} = K_{\Delta sij}^* - K_{\Delta sij},$$

$$\tilde{L}_{ij} = L_{ij}^* - L_{ij},$$

$$\tilde{\theta}_{ij} = \theta_{ij}^* - \theta_{ij}.$$

The time derivative of V along the trajectory is

$$\dot{V} = s^T\dot{s} + \sum_{i=1}^{n} \sum_{j=1}^{n} \rho_{ij}\tilde{K}_{sij}\dot{\tilde{K}}_{sij} + \sum_{i=1}^{n} \sum_{j=1}^{n} \sigma_{ij}\tilde{K}_{\Delta sij}\dot{\tilde{K}}_{\Delta sij}$$
$$+ \sum_{i=1}^{n} \sum_{j=1}^{n} \alpha_{ij}\tilde{L}_{ij}\dot{\tilde{L}}_{ij} + \sum_{i=1}^{n} \sum_{j=1}^{n} \eta_{ij}\tilde{\theta}_{ij}^T\dot{\tilde{\theta}}_{ij}, \tag{20}$$

where $\dot{\tilde{K}}_{sij} = -\dot{K}_{sij}$, $\dot{\tilde{K}}_{\Delta sij} = -\dot{K}_{\Delta sij}$, $\dot{\tilde{L}}_{ij} = -\dot{L}_{ij}$, $\dot{\tilde{\theta}}_{ij} = -\dot{\theta}_{ij}$. If we choose the adaptive law to be

$$\dot{K}_{sij} = \rho_{ij}\frac{\partial C_{ij}}{\partial K_{sij}}s_i, \tag{21-1}$$

$$\dot{K}_{\Delta sij} = \sigma_{ij}\frac{\partial C_{ij}}{\partial K_{\Delta sij}}s_i, \tag{21-2}$$

$$\dot{L}_{ij} = \alpha_{ij}\frac{\partial C_{ij}}{\partial L_{ij}}s_i, \tag{21-3}$$

$$\dot{\theta}_{ij} = \eta_{ij}\frac{\partial C_{ij}}{\partial \theta_{ij}}s_j = \eta_{ij}w_{ij}s_i, \tag{21-4}$$

where C_{ij} is defined in (13), then from (20) we have

$$\dot{V} = -s^T GKs + s^T Gu_s + s^T(v + \bar{f}). \tag{22}$$

The next step is to design the supervisory control term u_s. Motivated by the boundary layer approach [11], we define u_s as follows:

$$u_{si} = -K_i\left(\frac{|s_i|}{\delta_i}\right)^\beta \operatorname{sgn}(s_i), \qquad 0 < \beta < 1, i = 1, \dots, n, \tag{23}$$

where δ_i is a small positive constant, and $K_i > 0$ is a function of s and $\mathbf{X}$. In most cases, K_i can be chosen to be a constant because the fuzzy system can approximate $-F\operatorname{sgn}(s)$ closely. Consequently, (22) becomes

$$\dot{V} = -\sum_{i=1}^{n} K_i\left(\frac{|s_i|}{\delta_i}\right)^\beta |s_i| + s^T(v + \bar{f}). \tag{24}$$

By choosing parameters properly, according to the argument given in [18] and [17], the asymptotic tracking error will be small, if not equal to zero. Some simulation results will be shown later to justify this.

If we update the controller parameters using (21), it cannot guarantee $\Theta \in \Omega$. If we can keep $\Theta \in \Omega$ then $\hat{F}$ and u_s will be bounded and the error will also be bounded. To constrain Θ within the set Ω, the parameter projection algorithm [19] can be adopted. If the parameter vector Θ is within or on the boundary of the constraint set, then use adaptation law (23); otherwise, we use the projection algorithm to modify adaptation law (23) such that the parameter vector will remain inside the constraint set, namely,

$$\dot{K}_{sij} = \begin{cases} \rho_{ij}\dfrac{\partial C_{ij}}{\partial K_{sij}}s_i, & \text{if } |K_{sij}s_j| \le L_{ij}, \\ 0, & \text{otherwise}, \end{cases} \tag{25-1}$$

$$\dot{K}_{\Delta sij} = \begin{cases} \sigma_{ij} \dfrac{\partial C_{ij}}{\partial K_{\Delta sij}} s_i, & \text{if } |K_{\Delta sij} \Delta s_j| \le L_{ij}, \\ 0, & \text{otherwise,} \end{cases} \tag{25-2}$$

$$\dot{L}_{ij} = \begin{cases} \alpha_{ij} \dfrac{\partial C_{ij}}{\partial L_{ij}} s_i, & \text{if } \underline{M}_L \le L_{ij} \le \overline{M}_L, \\ 0, & \text{otherwise,} \end{cases} \tag{25-3}$$

$$\dot{\theta}_{ij} = \begin{cases} \eta_{ij} \dfrac{\partial C_{ij}}{\partial \theta_{ij}} s_i = \eta_{ij} w_{ij} s_i, & \text{if } |\xi_{ij}| \le M_\xi, \\ \eta_{ij} w_{ij} s_i - \eta_{ij} s_i \dfrac{\theta_{ij} \theta_{ij}^T}{\theta_{ij}^T \theta_{ij}} \phi_{ij}, & \text{otherwise.} \end{cases} \tag{25-4}$$

There are three remarks about this controller design: (1) With adaptive law (21), each C_{ij} is a function of (s_i, s_j). In this way, the coupling effect can be dealt with. (2) As can be seen from (23), the supervisory control term will increase the approaching speed when the states are far away from the sliding manifold, but it reduces the rate when the states are near the sliding manifold. This will lead to a fast approaching with less chattering. Actually, this supervisory control term can utilize energy more efficiently and result in a finite approach time [12]. (3) Further modification of the control law design depends on the systems under consideration.

C. Simulations

EXAMPLE 1.   We first consider a regulation problem, with the following system:

$$\dot{x}_1 = 2\sin(0.5t)x_1 + [1 + 3\sin(0.5t)]x_2 + \sin(0.5t)x_3 + u_1,$$

$$\dot{x}_2 = 2\sin(1.2t)x_1 + 3\sin(1.2t)x_2 + [1 + \sin(1.2t)]x_3 + u_2,$$

$$\dot{x}_3 = [\sin(x_3) + 2\sin(t)]x_1 + [1 + 3\sin(t)]x_2$$

$$+ [\cos(t) + \sin(t)]x_3 + \psi(t) + u_3,$$

where $\psi(t) = 2\sin(t) + 0.5\cos(t)$. This is a time-varying system. Let the K_{sij}s and $K_{\Delta sij}$s start with 0.1, the L_{ij}s start with 100, and the H_{ij}s start with $[10, 0, 0, -10]^T$. Also, let $K_1 = 3$, $K_2 = 5$, $K_3 = 3$, $\delta_i = 0.01$, and $\beta = 0.4$. The initial state is chosen to be $[2, -1.5, 1.5]^T$. Figure 5 shows the simulation results.

EXAMPLE 2.   Consider a general case of an n-link robot arm that takes into account the friction forces, unmodeled dynamics, and disturbances, with equation of motion given by [20]

$$\mathbf{M}(q)\ddot{q} + \mathbf{C}(q, \dot{q})\dot{q} + \mathbf{G}(q) + \mathbf{F}_d \dot{q} + F_S(q) + T_d = \tau,$$

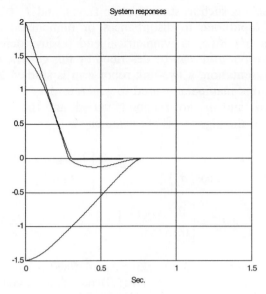

(a) system responses

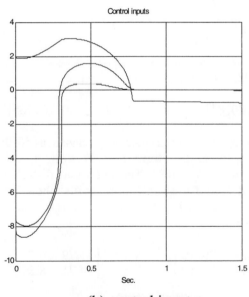

(b) control inputs

FIGURE 5 Simulation results for Example 1: (a) system responses; (b) control inputs.

where $q \in R^n$ denotes the joint angular position vector of the arms, $\tau \in R^n$ is the applied joint torques (or forces), $\mathbf{M}(q) \in R^{n \times n}$ is the inertia matrix, $\mathbf{C}(q, \dot{q})\dot{q} \in R^n$ represents the effect of Coriolis and centrifugal forces, $\mathbf{G}(q) \in R^n$ is the gravitational torques (or forces), $\mathbf{F_d} \in R^{n \times n}$ is a diagonal matrix of viscous and/or dynamic friction coefficients, $F_S \in R^n$ is the vector of unstructured friction effects such as static friction terms, and $T_d \in R^n$ is the vector of generalized input due to disturbances or unmodeled dynamics. Because the inertia matrix $\mathbf{M}(q)$ is symmetrical and positive-definite, it is obvious that all n-link robot arms can be described by this generic system.

To simplify the presentation, a two-link robot arm is studied here. The parameters of the two-link manipulator used to generate data in the simulation were as follows: m_1 and m_2 are randomly varied, and the friction and disturbance (unknown to the algorithm) are given as

$$\mathbf{F_d} = \begin{bmatrix} 5\cos(\dot{q}_1) & 0 \\ 0 & 3\cos(\dot{q}_2) \end{bmatrix}, \qquad F_s = \begin{bmatrix} 1.8\,\mathrm{sgn}(\dot{q}_1) \\ 1.2\,\mathrm{sgn}(\dot{q}_2) \end{bmatrix},$$

$$T_d = \begin{bmatrix} 0.3\cos(t) \\ 0.15\cos(t) \end{bmatrix}.$$

The two desired joint trajectories for tracking are defined as $q_{1d} = \pi/4 + (1 - \cos(3t))$ rad, and $q_{2d} = \pi/6 + (1 - \cos(5t))$ rad, respectively. We assigned the initial values to the controller parameters according to the human experience. Let

$$K_s = \begin{bmatrix} 265 & 90 \\ 10 & 160 \end{bmatrix}, \qquad K_{\Delta s} = \begin{bmatrix} 12 & 14 \\ 24 & 10 \end{bmatrix}, \qquad L = \begin{bmatrix} 150 & 288 \\ 300 & 260 \end{bmatrix},$$

$$\theta_{11} = [-408, -308, 280, 380]^T, \qquad \theta_{21} = [-112, -10, 16, 106]^T,$$

$$\theta_{21} = [-87, -27, 20, 84]^T, \qquad \theta_{22} = [-167, -67, 65, 165]^T,$$

$K = \mathrm{diag}\{10\}$, $\Gamma = \mathrm{diag}\{10\}$. The initial state was chosen as $[0, 0]^T$ deg. The supervisory control term is in the form of (29) with $\tau_i = 10$, $\beta = 0.2$, $\rho_{ij} = \sigma_{ij} = \alpha_{ij} = 15$, $\eta_{ij} = 20$, $\delta_i = 0.01$. The mass and disturbance profiles are shown in Figs. 6 and 7, and the simulation results when human experience is incorporated are shown in Fig. 8, respectively. The final values of the controller are the following:

$$K_s = \begin{bmatrix} 294.29 & 116.05 \\ 37.45 & 178.03 \end{bmatrix}, \qquad K_{\Delta s} = \begin{bmatrix} 11.79 & 6.74 \\ 14.71 & 8.39 \end{bmatrix},$$

$$L = \begin{bmatrix} 91.67 & 280 \\ 298.33 & 248.87 \end{bmatrix},$$

$$\theta_{11} = [-441.03, -341, 379.93, 439.67]^T,$$

$$\theta_{12} = [-109.39, -7.68, 71.34, 139.29]^T,$$

$$\theta_{21} = [-88.74, -32.40, 31.91, 91.97]^T,$$

$$\theta_{22} = [-181.55, -81.61, 91.66, 180.26]^T.$$

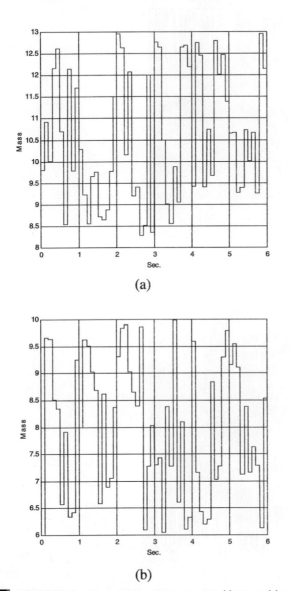

(a)

(b)

FIGURE 6 Mass profiles of links 1 and 2: (a) link 1; (b) link 2.

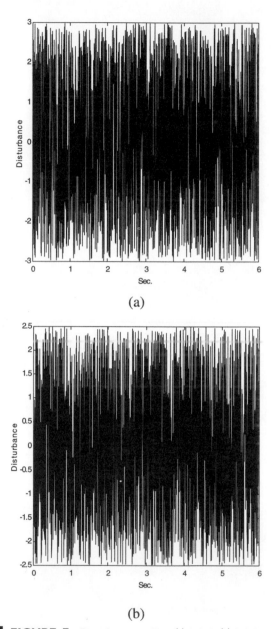

(a)

(b)

FIGURE 7   Disturbance profiles: (a) link 1; (b) link 2.

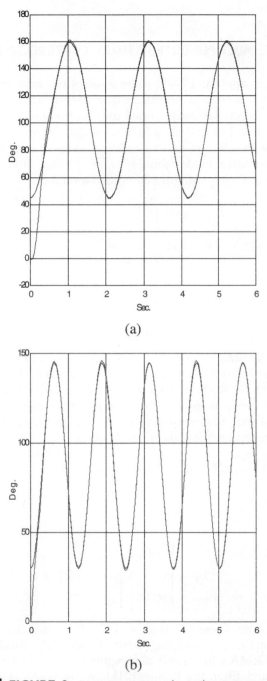

(a)

(b)

FIGURE 8 Desired trajectories (dashed) and output responses (solid): (a) link 1; (b) link 2.

As can be seen from the simulation results, the proposed FSMCs have shown satisfactory performance and robustness; they are also very effective for eliminating chattering.

III. INDIRECT MULTIPLE-INPUT, MULTIPLE-OUTPUT FUZZY CONTROLLER DESIGN

As mentioned before, to apply an FSMC, some assumptions have to be made. However, some needed condition or information is not always available. Therefore, an indirect controller design based only on the input–output data is more practical, which will be described in this section. Because attempts to control such systems with arbitrary initial conditions always result in instability, a suitable approximated model should be established first; then the controller is designed indirectly using the identified model.

A. System Representation

When using input–output data to approximate a system, a discrete-time model is a natural choice for system representation. Consider a nonlinear MIMO discrete-time system

$$\Sigma: \quad x(k+1) = f[x(k), u(k)],$$
$$y(k) = h[x(k)], \tag{26}$$

where $x(k) \in R^n$ is the state vector at the current step k, and $u(k) \in R^m$ and $y(k) \in R^l$ are the input and output vectors, respectively, with $f: R^n \times R^m \to R^n$ and $g: R^n \to R^l$. Without any loss of generality, it is assumed that $f(0,0) = 0$, so that the origin is an equilibrium point of the given system.

One popular input–output description that represents the current output in terms of the past inputs and outputs is described by

$$y(k+1) = F\big[y(k), \ldots, y(k - d_y + 1), u(k), \ldots, u(k - d_u + 1),$$
$$e(k), \ldots, e(k - d_e + 1)\big], \tag{27}$$

where

$$y(k) = \begin{bmatrix} y_1(k) \\ \vdots \\ y_1(k) \end{bmatrix}, \quad u(k) = \begin{bmatrix} u_1(k) \\ \vdots \\ u_m(k) \end{bmatrix}, \quad e(k) = \begin{bmatrix} e_1(k) \\ \vdots \\ e_p(k) \end{bmatrix},$$

$F: R^1 \times R^m \to R^l$ is a nonlinear function, and d_y, d_u, d_e are the maximum lags in the output, input, and modeling error, respectively. Model (27) is referred to as the NARMAX (**n**onlinear **a**uto**r**egressive **m**oving **a**verage with e**x**ogenous inputs) model. Chen and Billings [21] have rigorously shown that a nonlinear discrete-time time-invariant system can always be represented by a NARMAX model in a region around an equilibrium point. To increase the flexibility in the model structure, the lags for each output and input may be different, so that, for the jth row,

$$y_1(k + 1) = F_i\big[y_1(k),\ldots,y_1(k - d_y^1 + 1),\ldots,y_l(k),\ldots,$$

$$y_l(k - d_y^l + 1), u_1(k),\ldots,u_1(k - d_u^1 + 1),\ldots,$$

$$u_m(k),\ldots,u_m(k - d_u^m + 1), e_1(k),\ldots,e_p(k - d_e^p + 1)\big],$$

$$i = 1,\ldots,l. \quad (28)$$

If the form of function F_i is known, then the task of identifying input–output relations is to determine the unknown parameters. However, in many systems, the exact form of F_i is difficult to obtain. Therefore, a suitable approximation is desired.

The simplest structure of (28) is the linear difference equation given by

$$\hat{y}_i(k) = a_0 + \sum_{j=1}^{l} \sum_{t=1}^{d_y^l} a_{jt}\hat{y}_j(k - t) + \sum_{j=1}^{m} \sum_{t=1}^{d_u^j} b_{jt}u(k - t) + e_i(k). \quad (29)$$

This linear model may be applied, provided that the system is operated close enough around the desired operating point. As stated in [21], the linear ARMAX structure may change when the inputs vary over different regions. This gives the idea of combining the fuzzy model with the ARMAX structure. If F_i^j takes the form of the ARMAX model, the resulting combination can be referred to as FARMAX (fuzzy ARMAX). The NARMAX model can also be incorporated into neural networks. Narendra and Mukhopadhyay [22, 23] have rigorously developed a neural-networks-based adaptive controller for nonlinear systems. However, this kind of controller cannot take linguistic information from human experts. This gives fuzzy systems an advantage over neural networks.

B. Fuzzy Model Identification

Now consider the problem of identification using fuzzy models. Identification is divided into two parts: structure identification and parameter identification. For structure identification using a fuzzy model, two important issues should be considered: (1) variable selection; (2) feature-space partition.

I. Variable Selection

The behavior of a dynamical system is modeled by analyzing time series data of certain system variables. The success of the model identification heavily relies on identifying the underlying structure of the time series; it is advantageous to know in advance the embedding dimension (output lags), the most relevant inputs, and their lags. One has to include enough variables to explain the process, but on the other hand too many variables will increase the complexity of the model and overfit the data. In addition, the size of the fuzzy rule base depends heavily on the number of selected variables. Therefore, variable selection is very important especially when real-time implementation is required.

Savit and Green [24] developed a conditional probability approach in which the degree of variable dependence may be quantified. Inspired by their work, a method called the δ test was proposed by Pi and Peterson [25] from a geometric point of view, which exploits the potential of function continuity. Based upon the same function continuity principle, Molina and Niranjan [26] presented a geometrical technique to determine the embedding dimension of a time series. Another method based upon the statistical study of the derivatives of the outputs of the model with regards to its inputs was studied by Czernichow [27]. The basic idea is that the smaller these derivatives are, the more irrelevant the corresponding input is. Because fuzzy systems can be reformed into adaptive networks [28, 29], the selection methods developed for neural networks can also be applied to fuzzy systems. Various approaches based on weight pruning have been proposed, which have led to some new methods called optimal brain damage (OBD) [30] and optimal cell damage (OCD) [31], among others. Such variable selection is achieved during the training phase of the identification, to ensure that the selected set of variables is adequate for the system complexity.

2. Feature-Space Partition

The construction of a rule base can be viewed as partitioning the multidimensional input space. Since it has been shown that an MIMO form of a fuzzy rule can be decomposed into a number of multiple-input, single-output (MISO) rules [32], several commonly used MISO partitioning methods are described here by highlighting their basic ideas and characteristics.

a. Grid Partition

Figure 9a illustrates a typical grid partition in a two-dimensional input space. Wang and Mendel [33] used fuzzy grids to generate fuzzy rules from input–output training data. They proposed a one-pass build-up procedure that avoids the time-consuming learning process. The performance depends heavily on the definition of grids. In general, the finer the grid, the better the performance obtained. However, it is likely that the fuzzy regions used in this approach will not cover all training data and some regions remain undefined. Adaptive fuzzy grids can be used to optimize the system mapping. First, a uniformly partitioned grid is used as the initial state. As the process goes on,

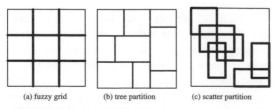

(a) fuzzy grid   (b) tree partition   (c) scatter partition

FIGURE 9 Some typical feature-space partitions: (a) fuzzy grid; (b) tree partition; (c) scatter partition.

the parameters in the antecedent membership function will be adjusted. Consequently, the fuzzy grid evolves. The gradient descent method optimizes the size and location of the fuzzy regions and the degree of overlap among them. No mater what grid partition method is used, there exists one major drawback in this scheme: the performance suffers from an exponential explosion as the number of inputs or the membership functions (MFs) on each input increases. For example, a fuzzy model has five inputs, and five MFs on each input would result in $5^5 = 3125$ if–then rules. It is usually referred to as the "curse of dimensionality."

b. Tree Partition

Figure 9b shows a basic tree partition scheme. The tree partition results from a series of guillotine cuts. Each region is generated by a guillotine cut, a cut that is made entirely across the subspace to be partitioned. At the $(k - 1)$st iteration step, the input space is partitioned into k regions. Then a guillotine cut is applied to one of the regions to further partition the entire space into $k + 1$ regions. There are several strategies for deciding which dimension to cut, where to cut at each step, and when to stop [34, 29]. This flexible tree-partition algorithm relieves the problem of an exponential increase of the number of rules. However, more MFs will be needed for each input variable, and these MFs usually do not have clear linguistic meanings. The resulting fuzzy model consequently is less descriptive.

c. Scatter Partition

Abe and Lan [35] presented a method for extracting fuzzy rules directly from numerical data. Suppose that we have a one-dimensional output y and an m-dimensional input vector x. First, we divide the universe of discourse of y into n intervals as follows:

$$[y_0, y_1]: \qquad y_0 \leq y \leq y_1,$$

$$(y_1, y_2]: \qquad y_1 < y \leq y_2,$$

$$(y_{n-1}, y_n]: \qquad y_{n-1} < y \leq y_n,$$

where the ith interval is called the output interval i. At first, activation hyperboxes are determined, which define the input region corresponding to the output interval i, by calculating the minimum and maximum values of input data for each output interval. If the activation hyperbox for the output interval i overlaps with activation hyperbox for the output interval j, the overlapped region is defined as an inhibition hyperbox. If the input data for output intervals i and/or j exist in the inhibition hyperbox, within this inhibition hyperbox one or two additional activation hyperboxes will be defined. Moreover, if two activation hyperboxes are defined and they overlap, an additional inhibition hyperbox is further defined. This procedure is repeated until the overlapping is resolved.

d. Clustering Algorithm

The clustering method, the fuzzy clustering algorithm in particular, provides a flexible partition technique [36, 37]. Basically, this is because we have a lot of data but no more information is available. Clustering can be used to detect the possible groups that exist, groups that have similarity in their behavior, and groups that can be used to establish some hypothesis about the structure presented in the data. However, this approach also has its own problems. First, the number of clusters has to be predetermined. While the resulting performance is not satisfactory and we want to increase the number of rules, the whole clustering algorithm has to be applied from the beginning. In several studies [37], the number of clusters is obtained as follows: Calculate some validity measures $S(c)$ over a range of number of rules, in which the number c^* minimizing $S(c)$ is the number of rules. This is supposed to be a local optimum as usual. Although alternatives such as progressive cluster identification algorithm [36] can be used, this method allows one to select the "*best*" clusters in a progressive way. The computational cost tends to be high because the efficiency of convergence is not guaranteed. The lack of a linguistic interpretation for the rules is another drawback. There are some suggestions for projecting the clusters onto the coordinate axes of input variables, so that linguistic labels can be attached to each input variable. However, several problems will occur when the input and output clusters are not all convex [37]. Basically, clustering algorithms result in the most flexible partitions and least number of rules. On the contrary, a fuzzy grid might generate a large number of rules, yet its resulting fuzzy rule base has the greatest descriptive ability. To keep the explanatory power, a fuzzy grid will be chosen to partition the input space in our approach. Inspired by the idea of node activation proposed in [38], a selective rule activation technique will be developed later to limit the size of the rule base.

3. Initial Parameter Determination

To introduce the rule activation technique, let us first describe the parameter determination needed by this method. Actually, with slight modification, this parameter determination procedure can be applied to a rule base constructed by any partition method. Suppose we have a set of input–output data pairs with N points generated by system (27). To construct the initial set of fuzzy rules, we begin by assigning the centers of membership functions on each input dimension, where inputs include the past outputs and the exogenous inputs as well as their past values. By the center of an MF, we mean the position of the MF that reaches its maximum:

$$\text{center } C_j^i = \left\{ x \,\middle|\, \max_{x \in x_i} \left[\mu_j^i(x) \right] \right\}, \tag{30}$$

where superscript i indicates the ith and subscript j the jth MF on input x_i, which can be a number or a linguistic label such as *Large, Small*. Thus, we have C_{Large}^i and C_{Small}^i. Similarly, we can define the center of rule b, $\mathbf{v}^b$, which is a vector consisting of C_j^i. Initially, the C_j^i are evenly distributed on each input; through each center in every dimension, grid lines can be drawn.

Each intersection of the grid lines is a mesh point that represents a center of a fuzzy rule. The ith mesh point is denoted by $\mathbf{v}^i$. This formulation of the multivariate fuzzy sets results in a lattice partition of the input space. An examples of a two-dimensional model with five centers defined on each axis is shown in Fig. 10.

Note that the membership function used can be triangular, trapezoidal, gaussian, and so on, and they can be different on different axes (e.g., triangular on one axis and gaussian on another). The total number of rules is $H = \prod_{i=1}^{n} N_i$, where N_i is the number of centers on input x_i. The fuzzy rule centered at $\mathbf{v}^b$ can b described as follows:

$$(\mathrm{R}_b) \quad \text{IF } \mathbf{x} \text{ is } \mathbf{v}^b \quad \text{THEN} \quad y^b = F^b(\hat{\mathbf{x}}), \tag{31}$$

where

$$\mathbf{x} = \left[y_1(k), \ldots, y_1(k - d_y^1 + 1), \ldots, y_l(k), \ldots, y_l(k - d_y^l + 1), \right.$$

$$u_1(k-1), \ldots, u_1(k - d_u^1 + 1), \ldots,$$

$$\left. u_m(k-1), \ldots, u_m(k - d_u^m + 1) \right]^T,$$

$$\bar{\mathbf{x}} = \left[1, \mathbf{x}^T \right]^T, \qquad \hat{\mathbf{x}} = \left[\bar{\mathbf{x}}^T, u_1(k), \ldots, u_m(k) \right]^T,$$

$$F_j^b(\hat{\mathbf{x}}) = \theta_{b,j}^T \hat{\mathbf{x}}, \qquad j = 1, \ldots, l,$$

in which $\theta_{b,j}$ are the parameters to be determined.

For each rule center $\mathbf{v}^b$, we wish to minimize the squared error between $F_j^b(\hat{\mathbf{x}})$ and the output portion of the training data pairs. We apply an optimal

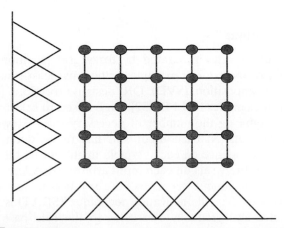

FIGURE 10 An example of lattice partition of the input space.

output defuzzification [32] to determine the parameters. The cost function to be minimized is given by

$$J_j^b = \sum_{k=1}^{N} \left[\mu_b(\mathbf{x}^k) \right]^2 \left[y_j^k - \theta_{b,j}^T \hat{\mathbf{x}}^k \right]^2, \qquad b = 1, \ldots, H, j = 1, \ldots, l, \quad (32)$$

where $\mu_b(\mathbf{x}^k)$ is the membership value of the kth input data for rule b, and y^k is the kth output data. The weighted least-squares problem can be solved by calculating

$$\theta_{b,j} = \left(\hat{\mathbf{X}}^T \mathbf{D}_b^2 \hat{\mathbf{X}} \right)^{-1} \hat{\mathbf{X}}^T \mathbf{D}_b^2 \mathbf{Y}_j, \tag{33}$$

where

$$\hat{\mathbf{X}} = \left[\hat{\mathbf{x}}^1, \ldots, \hat{\mathbf{x}}^N \right]^T,$$

$$\mathbf{Y}_j = \left[y_j^1, \ldots, y_j^N \right]^T,$$

$$\mathbf{D}_b = \mathrm{diag}\{ \left[\mu_b(\mathbf{x}^1), \ldots, \mu_b(\mathbf{x}^k) \right] \}.$$

Consequently, the approximate output is calculated by

$$\hat{y}_j = \frac{\sum_{b=1}^{H} \mu_b(\mathbf{x}) \left(\theta_{b,j}^T \hat{\mathbf{x}} \right)}{\sum_{b=1}^{H} \mu_b(\mathbf{x})} \tag{34}$$

and the total approximation error is

$$E = \sum_{k=1}^{N} \sum_{j=1}^{l} \left(y_j^k - \hat{y}_j^k \right)^2. \tag{35}$$

As mentioned earlier, the number of fuzzy rules will increase exponentially if the grid partition is used, and some of them may be redundant. Therefore, a rule activation technique is developed next to overcome this troublesome problem.

C. Selective Rule Activation Technique

To reduce the number of rules generated by fuzzy grid partition, several approaches have been proposed. For example, orthogonal least-squares [39], the singular value decomposition (SVD) QR design procedure [40], and modified learning from examples (MLFE) [32] can be used to select the most significant rules, thus reducing the number of necessary rules. However, the initial number of fuzzy sets on each input still has to be determined in advance, because the dependence on each variable might not be the same. It is inappropriate to assign fuzzy sets on each input arbitrarily. Adoption of the following rule activation technique seems to be a solution to this problem. The basic idea of the selective rule activation technique (SRAT) is that one can start with no rules and gradually add rules to the rule base if some conditions are being satisfied. In this way, the rule base grows sequentially to ensure that the function being approximated can be learned up to a certain required accuracy while only a minimal number of rules will be generated. In this approach, each intersection thus created represents a "potential" rule in the input space, from where a fuzzy rule might be constructed as needed. Fuzzy rules are created on the *activated* intersections as in (31); the rule base grows as more of the intersections are activated. The entire scheme is briefly described as follows: First, define a closeness function as

$$\psi(\mathbf{x}) = \frac{1}{1 + \left(\|\mathbf{x} - \mathbf{v}^i\|/a\right)^c}, \tag{36}$$

where $\mathbf{x}$ is the input data vector, $\mathbf{v}^i$ is the rule center defined as before, and a and c are constants. This closeness function computes the degree of closeness between the input data $\mathbf{x}$ and the rule center $\mathbf{v}^i$. Conceptually, this technique is based on the simple fact that the closer the input $\mathbf{x}$ is to $\mathbf{v}^i$, the larger the firing strength of rule i will be. Consider the case in which a rule i is activated at the current time. If it was not previously activated and its closeness function value exceeds a certain threshold, namely,

$$\psi(\mathbf{x}) = \frac{1}{1 + \left(\|\mathbf{x} - \mathbf{v}^i\|/a\right)^c} \geq \delta_{\min}, \tag{37}$$

where $\delta_{\min}$ stands for an activation threshold, which is a constant and satisfies $0 < \delta_{\min} < 1$, then the output of a fuzzy rule will be considered negligible at the current time. It follows that, for a fuzzy rule centered at $\mathbf{v}_i$ to be activated, the following condition needs to be satisfied:

$$\|\mathbf{x} - \mathbf{v}_i\| \leq a\sqrt[c]{\delta_{\min}^{-1} - 1}. \tag{38}$$

Geometrically, (38) represents a region inside an m-dimensional ball of radius $a\sqrt[c]{\delta_{\min}^{-1} - 1}$. Thus, the selective activation technique can be carried out in the following way (as shown in Fig. 11): at the current state, all rule centers within the ball must be activated, unless they have been activated at the previous states. Because each activated rule center will account for an

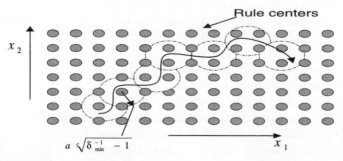

FIGURE 11   Illustration of the selective activation technique.

activated fuzzy rule, as the process goes on, the rule base will grow until no more rule centers are activated. The larger the activation threshold δ_{min} is, the smaller the radius will be; and so fewer fuzzy rules will be activated at every time instant. By adjusting δ_{min}, the size of the fuzzy rule base can be restricted.

Using this selective rule activation technique results in a dynamically structured fuzzy rule base. The rule base is dynamic in the sense that it can grow in size in correspondence to which rules are fired. Furthermore, the number of rules in such a dynamic rule base will be significantly less than that created by the grid partition method described previously because the inactivated mesh points (rules) are ignored here. This technique gives resolutions to the problems in the preceding section, by reducing the number of rules in the rule base.

After the selective rule activation technique is applied, the parameters and the approximated output are calculated by Eqs. (33) and (34).

In summary, the rule-base construction is completed in the following steps:

Step 1.   Set $t = 1$, assign N_i, $i = 1, \ldots, n$, determine C_j^i, $i = 1, \ldots, n$, $j = 1, \ldots, N_i$, calculate $\mathbf{V}_i = \{\mathbf{v}_i^i\}$, and apply SRAT to activate the rules.

Step 2.   Based on the activated rules, compute the error measurement E_t using Eqs. (31)–(35). If $E_t \leq \varepsilon$ then stop; otherwise proceed, where ε is chosen by the designer.

Step 3.   For each d from 1 to n, perform the following tasks:

- If $i = d$, $N_i^{t+1} = N_i^t + \lambda$; otherwise $N_i^{t+1} = N_i^t$, where λ is to be decided (default = 1), $i = 1, \ldots, n$.
- With the new set $\mathbf{V}_{t+1}^d$, and $H_{t+1}^d = \prod_{i=1}^n N_i^{t+1}$, use SRAT and compute the approximation error E_{t+1}^d.

Step 4.   Let $s = \{d|\min_{d=1,\ldots,n}(E_{t+1}^d)\}$; and let $E_{t+1} = E_{t+1}^s$. If $E_{t+1} \leq \varepsilon$ then stop; otherwise proceed.

Step 5.   Set $t = t + 1$, and repeat Steps 3 and 4 until $|E_{t+1} - E_t| \leq \varepsilon_t$, where $\varepsilon_t > 0$ is chosen by the designer.

Note that this construction procedure is progressive, where the number of rules does not have to be determined in advance. Also, the degree of dependence on each input can be detected by repeating Step 3. For instance, it is possible that the number of centers on one input keep increasing while the others remain unchanged. By applying the SRAT, some redundant rules can be removed from the rule base.

After the first approximation rule base has been created, several learning algorithms can be adopted to fine-tune the system to achieve better performance. The learning phase should include both offline and online learning. For instance, the widely used backpropagation (BP) algorithm can be used to adjust all the design parameters in the fuzzy model, such as the location of centers, the membership functions, and the parameters in the consequent parts. This will be very helpful, especially in the case where there exist complex interactions between the model variables. Note that the BP method is more suitable for offline learning. On the other hand, least-squares (LS) approaches [32] are suitable to adapt the consequents for both learning phases. Thus, both BP and recursive least-squares (RLS) [41] methods will be applied here. First, according to the given input–output data, the BP algorithm is used to adjust the centers and shapes of the membership functions and then the LS method is used to calculate the parameters in the consequents. While operating online with the true system, only the RLS algorithm will be applied to update the consequent parameters. One popular recursive least-squares formula used in our cases is given by

$$\hat{\Theta}_j(k) = \hat{\Theta}_j(k-1)$$

$$+ \frac{P_j(k-1)\phi(k)\left[y(k+1) - \phi^T(k)\hat{\Theta}_j(k-1)\right]}{\lambda_j + \phi^T(k)P_j(k-1)\phi(k)}, \quad (39)$$

$$P_j(k) = \frac{1}{\lambda_j}\left[P_j(k-1) - \frac{P_j(k-1)\phi(k)\phi^T(k)P_j(k-1)}{\lambda_j + \phi^T(k)P_j(k-1)\phi(k)}\right],$$

where

$$\hat{\Theta}_j = \left[\theta_{1,j}^T, \theta_{2,j}^T, \ldots, \theta_{H,j}^T\right]^T,$$

$$\phi(k) = \left[w_1\hat{\mathbf{x}}^T, w_2\hat{\mathbf{x}}^T, \ldots, w_H\hat{\mathbf{x}}^T\right]^T,$$

$$w_i = \frac{\mu_i(\mathbf{x})}{\sum_{b=1}^H \mu_b(\mathbf{x})}, \qquad i = 1, \ldots, H,$$

in which λ_j, the so-called forgetting factor, can be a function of the approximate error, or a constant. Basically, the smaller λ_j is, the faster the convergence will be.

While BP is used to update the location of rule centers and the membership functions, the fuzzy grid evolves. Consequently, some rule centers may become too close to each other. In this case, the closeness function can be used again to simplify the rule base further. For example, we can use a tolerance criterion $\sigma_{\max} > 0$ to define

$$\psi_j(\mathbf{v}^i) = \frac{1}{1 + \left(\|\mathbf{v}^j - \mathbf{v}^i\|/a\right)^c} \geq \sigma_{\max}. \quad (40)$$

If the closeness function value between rule centers i and j exceeds the threshold $\sigma_{\max}$, the one that produces a larger prediction error will be removed.

D. Control Law Design

After the introduction of model identification, we are now in a position to discuss our primary goal of designing a control law to achieve satisfactory performance. Note that, from Eqs. (33) and (34), y^b can be expressed as

$$y_j^b = \left[\theta_{b,j}^{\bar{x}}\right]^T \bar{x} + \left[\theta_{b,j}^u\right]^T u(k), \qquad \theta_{b,j} = \begin{bmatrix} \theta_{b,j}^{\bar{x}} \\ \theta_{b,j}^u \end{bmatrix}. \tag{41}$$

Consequently, the overall output of the system can be written as

$$\hat{y}(k+1) = \frac{\sum_{b=1}^{H} \mu_b(\mathbf{x}) y^b}{\sum_{b=1}^{H} \mu_b(\mathbf{x})} = \mathbf{A}\bar{x} + \mathbf{B}\mu(k), \tag{42}$$

where

$$\mathbf{A} = \frac{\sum_{b=1}^{H} \mu_b(\mathbf{x})\mathbf{A}_b}{\sum_{b=1}^{H} \mu_b(\mathbf{x})}, \qquad \mathbf{A}_b = \left[\theta_{b,1}^{\bar{x}}, \ldots, \theta_{b,l}^{\bar{x}}\right]^T,$$

$$\mathbf{B} = \frac{\sum_{b=1}^{H} \mu_b(\mathbf{x})\mathbf{B}_b}{\sum_{b=1}^{H} \mu_b(\mathbf{x})}, \qquad \mathbf{B}_b = \left[\theta_{b,1}^{u}, \ldots, \theta_{b,l}^{u}\right]^T.$$

Thus, the desired control law can be calculated as

$$u(k) = \mathbf{B}^+ (y_d(k+1) - \mathbf{A}\bar{x}), \tag{43}$$

where $\mathbf{B}^+$ is the pseudoinverse of $\mathbf{B}$.

As can be seen, the control design is very straightforward, which is another advantage of constructing an identification model in the FARMAX form. To compensate the discrepancy between the desired output and the system output, the conventional sliding mode control strategy described earlier is applied. Although the development of the aforementioned controller is devoted to continuous-time systems, a similar methodology can be applied to develop discrete-time sliding mode controllers (DSMC).

In the MIMO case, a DSMC has to be designed based upon the identified model because the true system is unavailable. Let us first define the tracking error e and the switching function s as follows:

$$\bar{e}(k) = \hat{y}(k) - y_d(k), \tag{44}$$

$$s(k) = \left(\mathbf{I} + \Gamma_1 z^{-1} + \Gamma_2 z^{-2} + \cdots + \Gamma_d z^{-d}\right)\bar{e}(k), \tag{45}$$

where $d = \max_{d_y^i}\{d_y^i\}$, and $\Gamma_1, \ldots, \Gamma_d$ are positive-definite diagonal matrices chosen so that $(\mathbf{I} + \Gamma_1 z^{-1} + \Gamma_2 z^{-2} + \cdots + \Gamma_d z^{-d})$ satisfies the Hurwitz stability criterion.

For discrete-time cases, several approaching conditions can be used to construct the control law. Dote and Hoft [42] first used an equivalent form of the continuous approach condition

$$[s(k + 1) - s(k)]s(k) < 0. \tag{46}$$

Sarpturk *et al.* [43] proposed the following approach conditions:

$$[s(k + 1) - s(k)]\text{sgn}(s(k)) < 0,$$
$$[s(k + 1) + s(k)]\text{sgn}(s(k)) > 0. \tag{47}$$

In [44], Gao *et al.* generalized the preceding approach conditions and proposed

$$s(k + 1) = (\mathbf{I} - \mathbf{K}T)s(k) - \varepsilon \, \text{sgn}[s(k)], \qquad \mathbf{I} - \mathbf{K}T > 0. \tag{48}$$

For example, adopting the approach condition proposed in [36] and using the identified model $\hat{y}(k + 1) = \mathbf{A}\bar{x} + \mathbf{B}u(k)$, the DSMC is constructed as follows:

$$\begin{aligned} \text{DSMC} = \mathbf{B}^+\big\{-\big(\Gamma_1 z^{-1} + \cdots + \Gamma_d z^{-d}\big)e(k) \\ + (\mathbf{I} - \mathbf{K}T)s(k) - \varepsilon T \, \text{sgn}[s(k)]\big\}. \end{aligned}$$

As can be seen from the design steps, the performance of the proposed controller depends heavily upon the accuracy of the identified model. Thus, to enhance the dynamical capability of the fuzzy model and the accuracy, the following sequential–parallel model is used

$$\begin{aligned} \hat{y}(k + 1) = a_1[y(k) - \hat{y}(k)] + a_2[y(k - 1) - \hat{y}(k - 1)] + \cdots \\ + a_d[y(k - d + 1) - \hat{y}(k - d + 1)] + \mathbf{A}\bar{x} + \mathbf{B}u(k). \end{aligned} \tag{49}$$

Figure 12 shows the structure of the fuzzy identifier, where

$$W(z) = \big(1 + a_1 z^{-1} + a_2 z^{-2} + \cdots + a_d z^{-d}\big)^{-1}. \tag{50}$$

Finally, the desired control law becomes

$$\begin{aligned} u(k) = \mathbf{B}^+\big[y_d(k + 1) - \mathbf{A}\bar{x} - a_1\hat{e}(k) - a_2\hat{e}(k - 1) \\ - \cdots - a_d\hat{e}(k - d + 1)\big] + \text{DSMC}, \end{aligned} \tag{51}$$

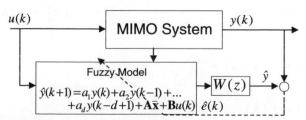

FIGURE 12 The fuzzy identifier.

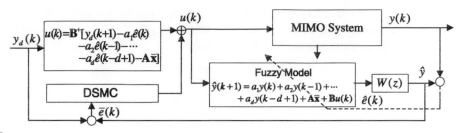

FIGURE 13   The closed-loop modeling and control scheme

where $\hat{e}(k) = \hat{y}(k) - y(k)$, and the DSMC is designed to satisfy a certain sliding condition. The entire control system is shown in Fig. 13.

E. Simulation Examples

In this section, we describe the application of the proposed design procedure for MIMO systems by simulations. According to the preceding discussion, the following assumptions are made about the systems used for simulation:

    i. The order of the system is known.
    ii. The maximum lag of each output is known.
    iii. The variables input to the fuzzy system are known (because the preceding two pieces of information are available).

Moreover, each membership function used in the simulation is described by

$$\mu_A(x) = \begin{cases} \dfrac{1}{1 + \left(\|x - c\|/a_1\right)^b}, & x \le c, \text{ if } A \text{ is leftmost, } \mu_A(x) = 1, \\[3ex] \dfrac{1}{1 + \left(\|x - c\|/a_2\right)^b}, & x > c, \text{ if } A \text{ is rightmost, } \mu_A(x) = 1, \end{cases}$$

where the membership function center c as well as the parameters a_1, a_2, and b will be adjusted using the BP algorithm. The details concerning the simulations are not presented here, only the simulation results are shown.

EXAMPLE 3.   Let

$$x_1(k + 1) = 0.1x_1(k) + \frac{2}{1 + x_1^2(k) + x_2^2(k)}\left[u_1(k) + u_2(k)\right],$$

$$x_2(k + 1) = 0.2x_2(k) + \left(2 + \frac{x_1}{1 + x_1^2}\right)u_2(k),$$

$$y_1(k) = x_1(k), \qquad y_2(k) = x_2(k).$$

For this example, the final rule base consists of 15 rules and the predicted output is calculated by (49), with

$$\mathbf{x} = \left[y_1(k), y_2(k), y_1(k-1), y_2(k-1) \right]^T,$$

$$\hat{\mathbf{x}} = \left[1, \mathbf{x}^T, u_1(k), u_2(k) \right]^T.$$

Figure 14 shows the prediction results of the identified fuzzy model, where

$$u_1(k) = \cos(k\pi/25) \times \sin(e^{k/300}\pi/3),$$

$$u_2(k) = \sin(k\pi/30) \times \cos(e^{k/300}\pi/3).$$

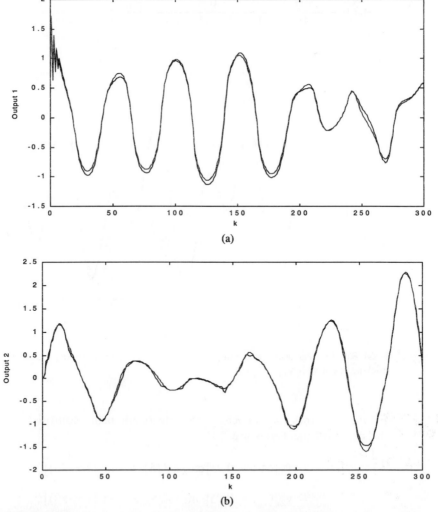

(a)

(b)

FIGURE 14 The prediction results of the fuzzy identifier: (a) output 1 of the identified fuzzy model; (b) output 2 of the identified fuzzy model.

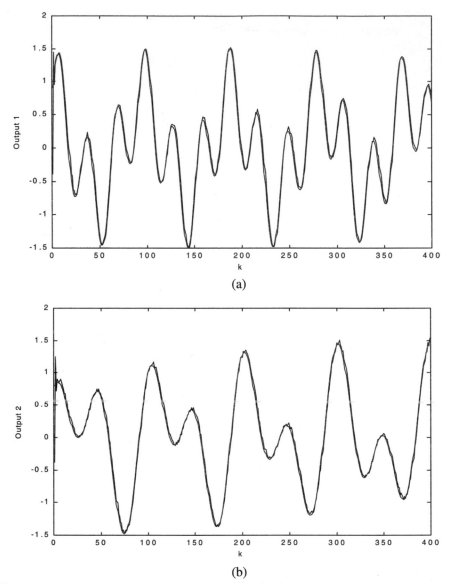

FIGURE 15 Performance of the proposed controller: (a) the tracking result with y_{1d}; (b) the tracking result with y_{2d}.

Figure 15 shows the trajectory tracking result using control law (51), where the desired trajectories are

$$y_{1d}(k) = 0.75[\cos(k/15) + \sin(2k\pi/30)],$$

$$y_{2d}(k) = 0.75[\sin(k/15) + \cos(2k\pi/50)].$$

Figures 16 and 17 show the control signals and tracking errors, respectively.

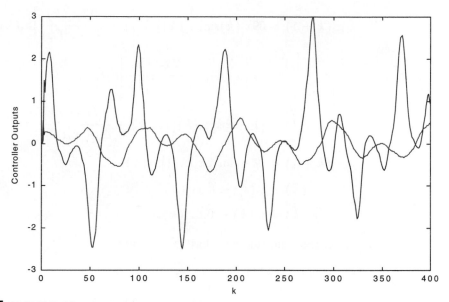

FIGURE 16   The simulated control signals.

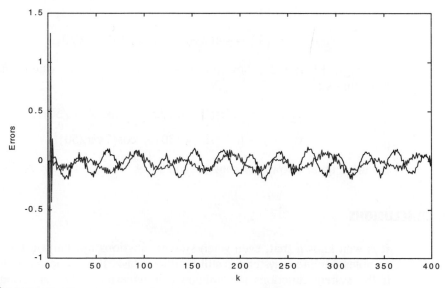

FIGURE 17   Tracking errors using the proposed controller.

EXAMPLE 4. Consider a third-order system with two inputs and two outputs:

$$x_1(k+1) = 09x_1(k)\sin[x_2(k)] + \left[1 + \frac{0.5x_1(k)}{1 + x_1^2(k)}\right]u_1(k)$$

$$+ \left[x_1(k) + \frac{2x_2(k)}{1 + x_2^2(k)}\right]u_2(k),$$

$$x_2(k+1) = \{1 + \cos[x_1(k)]\}x_3(k) + \frac{x_1(k)x_3(k)}{1 + x_3^2(k)},$$

$$x_3(k+1) = \{2 + \sin[2x_1(k)]\}u_2(k),$$

$$y_1(k) = x_1(k) + 0.5x_3(k),$$

$$y_2(k) = x_2(k) - 0.5x_3(k).$$

In this case, the final rule base has 35 rules and

$$\mathbf{x} = [y_1(k), y_2(k), y_1(k-1), y_2(k-1), y_1(k-2), y_2(k-2),$$

$$u_1(k-1), u_2(k-1), u_1(k-2), u_2(k-2)]^T,$$

$$\hat{\mathbf{x}} = [1, \mathbf{x}^T, u_1(k), u_2(k)]^T.$$

Figure 18 shows the prediction result using the identified fuzzy model, where

$$u_1(k) = \cos(k\pi/25) \times \sin(e^{k/300}\pi/3),$$

$$u_2(k) = \sin(k\pi/30) \times \cos(e^{k/300}\pi/3),$$

and Figs. 19 and 20 show the tracking performance, where the desired trajectories are

$$y_{1d}(k) = 0.75[\sin(2k\pi/50) + \cos(2k\pi/30)],$$

$$y_{2d}(k) = 0.75[\sin(k/30) + \cos(2k\pi/50)].$$

Finally, Fig. 21 shows the control signals

IV. CONCLUSIONS

It is well known that, even when system functions are known, the control of complex nonlinear systems is still a difficult task. It becomes more formidable if the system functions are unknown or uncertain. Such situations exist in many real-world applications such as manufacturing, aerospace engineering, robotic systems, and aircraft control. To control those ill-defined systems, using fuzzy systems to generate satisfactory approximation models of complex systems is a very natural approach, because fuzzy systems not only can

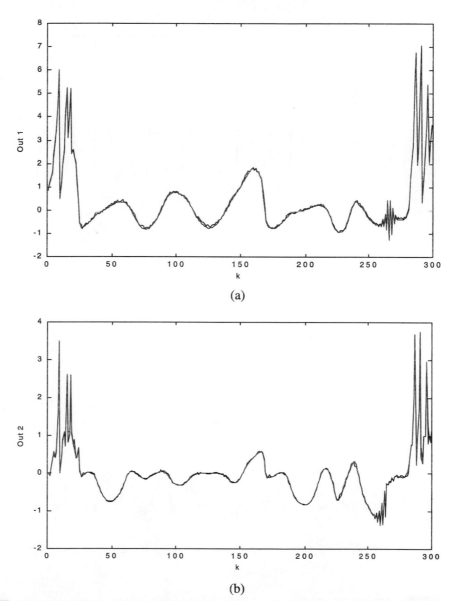

FIGURE 18 The prediction results of the fuzzy identifier: (a) output 1 of the identified fuzzy model; (b) output 2 of the identified fuzzy model.

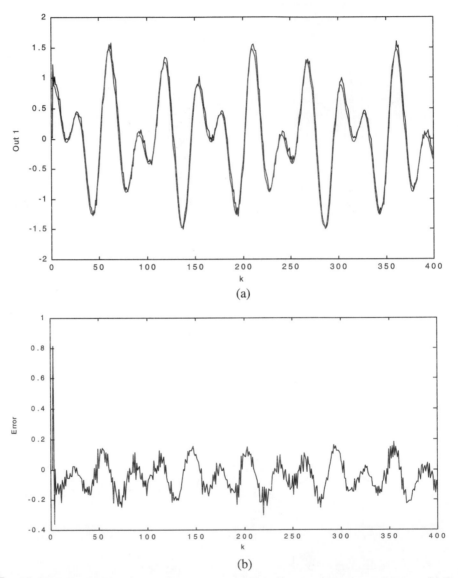

FIGURE 19 Performance of the proposed controller: (a) the system response with y_{ld}; (b) the tracking error with y_{ld}.

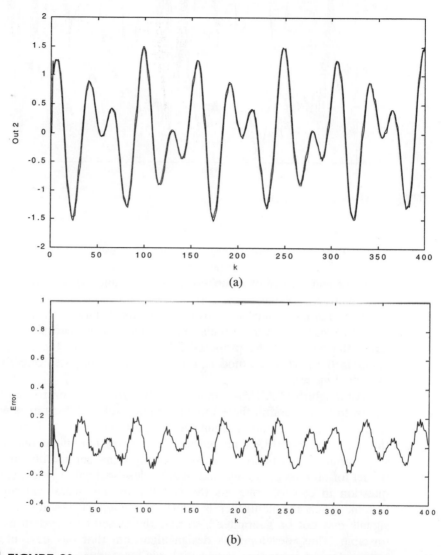

FIGURE 20 Performance of the proposed controller: (a) the system response with y_{2d}; (b) the tracking error with y_{2d}.

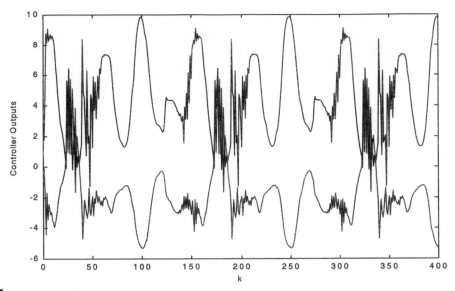

FIGURE 21   The simulated control signals.

incorporate human experience, but also can use numerical data to adapt themselves to achieve better performance. In this paper, the basic problem of modeling and control of an unknown multivariable dynamic system is studied with analysis, design, and simulations. The new controller only used input–output data, but has a strong prediction ability and satisfactory tracking performance. Several simulation examples have been shown to demonstrate the merits of the proposed FARMAX. Because the control input is linear in the FARMAX models, it can be easily computed directly from the identified model.

Although the FARMAX model and the proposed control law have been shown to be successful, there remain several problems for future work: (1) It is impossible to include all input–output patterns in the training data. Thus, the identified FARMAX model and the control law might fail to work for some inputs. Determining a suitable training signal set and designing a global model for all possible inputs and outputs deserve further study. (2) A critical question in control concerns the stability of the overall adaptive control system. When the controller is adjusted online, the boundedness of some signals may not be guaranteed, so that the closed-loop system may become unstable. Thus, developing a design algorithm that has guaranteed online robust stability is another important topic for study.

REFERENCES

1. Zadeh, L. A. Outline of a new approach to the analysis of complex systems and decision processes. *IEEE Trans. Systems Man Cybernet.* 3:28–33, 1973.
2. Utkin, V. I. *Sliding Modes and Their Applications*. Mir, Moscow, 1978.

3. Hsu, F. Y. and Fu, L. C. Adaptive robust fuzzy control for robot manipulators, In *Proceedings of the International Conference on Robotics and Automation*, 1994, Vol. 1, pp. 649–654.

4. Hwang, G. C. and Lin, S. C. A stability approach to fuzzy control design for nonlinear systems. *Fuzzy Sets Systems* 48:279–287, 1992.

5. Ohtani, Y. and Yoshimura, T. Fuzzy control of a manipulator using the concept of sliding mode. *Internat. J. Systems Sci.* 27:179–186, 1996.

6. Sun, F. C., Sun, Z. Q., and Feng, G. Design of adaptive fuzzy sliding mode controller for robot manipulators. *Proceedings of the 5th IEEE International Conference on Fuzzy Systems*, 1996, Vol. 1, pp. 62–67.

7. Ting, C. S., Li, T. H. S., and Kung, F. C. An approach to systematic design of fuzzy control system. *Fuzzy Sets Systems* 77:151–166, 1996.

8. Wu, J. C. and Liu, T. S. A sliding mode approach to fuzzy control design. *IEEE Trans. Control Systems Technol.* 4:141–148, 1996.

9. Hsu, Y. C. and Chen, G. Design of a fuzzy PD-sliding mode adaptive controller. In *Proceedings of the Joint Conference on Comput. & Intel. Sys.*, Durham, NC, March 2–5, 1997, pp. 255–258.

10. Glower, J. S. and Munighan, J. Design fuzzy controllers from a variable structure standpoint. *IEEE Trans. Fuzzy Systems* 5:138–144, 1997.

11. Slotine, J. J. E. and Li, W. *Applied Nonlinear Control.* Prentice-Hall, Englewood Cliffs, NJ, 1991.

12. Hung, J. Y., Gao, W., and Hung, J. C. Variable structure control: A survey. *IEEE Trans. Industrial Electronics* 2–22, 1993.

13. Malki, H. A., Li, H., and Chen, G. New design and stability analysis of fuzzy proportion-derivative control systems. *IEEE Trans. Fuzzy Systems* 2:245–254, 1994.

14. Misir, D., Malki, H. A., and Chen, G. Design and analysis of a fuzzy proportional-integral-derivative controller. *Fuzzy Sets Systems* 79:297–314, 1995.

15. Gupta, M. M., Kiszka, J. B., and Trojan, G. M. Multivariable structure of fuzzy control systems. *IEEE Trans Systems Man Cybernet.* 638–656, 1986.

16. Wang, L. X. *Adaptive Fuzzy Systems and Control: Design and Stability Analysis.* PTR Prentice-Hall, Englewood Cliffs, NJ, 1994.

17. Wang, L. X. Stable adaptive fuzzy controllers with application to inverted pendulum tracking. *IEEE Trans. Systems Man and Cybernet.* 677–691, 1996.

18. Neo, S. S. and Er, M. J. Adaptive fuzzy controllers of a robot manipulator. *Internat. J. System Sci.* 519–532, 1996.

19. Goodwin, G. C. and Sin, K. S. *Adaptive Filtering, Prediction, and Control.* Prentice-Hall, Englewood Cliffs, NJ, 1984.

20. Qu, Z. and Dawson, D. M. *Robust Tracking Control of Robot Manipulators.* IEEE Press, NJ, 1996.

21. Chen, S. and Billings, S. A. Representation of nonlinear systems: The NARMAX model. *Internat. J. Control* 49:1013–1032, 1989.

22. Narendra, K. S. and Mukhopadhyay, S. Adaptive control of nonlinear multivariable systems using neural networks. *Neural Networks* 7:737–752, 1994.

23. Narendra, K. S. and Mukhopadhyay, S. Adaptive control using neural networks and approximate models. *IEEE Trans. Neural Networks* 8:475–485, 1997.

24. Savit, R. and Green, M. Time series and dependent variables. *Physica D* 50:95–116, 1991.

25. Pi, H. and Peterson, C. Finding the embedding dimension and variable dependencies in time series. *Neural Comput.* 6:509–520, 1994.

26. Molina, C. and Niranjan, M. Finding the embedding dimension of time series by geometrical techniques. Technical Report CUED/FINFENG/TR.221, Cambridge University, 1995.

27. Czernichow, T. Architecture selection through statistical sensitivity analysis. In *Proceedings of ICANN'96*, Germany, 1996.

28. Jang, J. S. and Sun, C. T. Neuro-fuzzy modeling and control. *Proc. IEEE* 83:378–406, 1995.

29. Sun, C. T. Rule base structure identification in an adaptive-network-based fuzzy inference system. *IEEE Trans. Fuzzy Systems* 2:64–73, 1994.

30. Cun, Y. L., Denker, J. S., and Solla, S. A. Optimal brain damage. In *Proceedings of NIPS'89*, 1990, Vol. 2, pp. 598–605.

31. Cibas, T., Soulie, F. F., Gallinari, P., and Raudys, S. Variable selection with optimal cell damage. In *Proceedings of ICANN'94*, pp. 727–730, 1994.
32. Passino, K. and Yurkovich, S. *Fuzzy Control*. Addison-Wesley, Menlo Park, CA, 1997.
33. Wang, L. X. and Mendel, J. M. Generating fuzzy rules by learning from examples. *IEEE Trans. Systems Man Cybernet.* 22:1414–1427, 1992.
34. Lin, Y., Cunninghan, G. A., and Coggeshall, S. V. Using fuzzy partitions to create fuzzy systems from input–output data and set the initial weights in a fuzzy neural network. *IEEE Trans. Fuzzy Systems* 5:614–621, 1997.
35. Abe, A. and Lan, M. S. Fuzzy rules extraction directly from numerical data for function approximation. *IEEE Trans. Systems Man Cybernet.* 25:119–129, 1995.
36. Delgado, M., Gomez-Skarmeta, A. F., and Martin, F. A fuzzy clustering-based rapid prototyping for fuzzy rule-base modeling. *IEEE Trans. Fuzzy Systems* 5:223–233, 1997.
37. Sugeno, M. and Yasukawa, T. A fuzzy-logic-based approach to qualitative modeling. *IEEE Trans. Fuzzy Systems* 1:7–31, 1993.
38. Fabri, S. and Kadirkamanathan, V. Dynamic structure neural networks for stable adaptive control of nonlinear systems. *IEEE Trans. Neural Networks* 7:1151–1167, 1996.
39. Chen, S., Cowan, C. F. N., and Grant, P. M. Orthogonal least squares learning algorithm for radial basis function networks. *IEEE Trans. Neural Networks* 2:302–309, 1991.
40. Mendel, J. M. and Mouzouris, G. C. Designing fuzzy logic systems. *IEEE Trans. Circuits Systems* 44:885–895, 1997.
41. Wang, L. and Langari, R. Identification and control of nonlinear dynamic systems using fuzzy models. *Internat. J. Intelligent Control and Systems* 1:247–260, 1996.
42. Dote, Y. and Hoft, R. G. Microprocessor based sliding mode controller for dc motor drives. Presented at the Ind. Applicat. Society Annual Meeting, Cincinnati, OH, 1980.
43. Sarpturk, S. Z., Isefanopulos, Y., and Kanak, O. On the stability of discrete-time sliding mode control systems. *IEEE Trans. Automat. Control* 32:930–932, 1987.
44. Gao, W., Wang, Y., and Homaifa, A. Discrete-time variable structure control system. *IEEE Trans. Industrial Electronics* 42:117–122, 1995.
45. Narendra, K. S. and Parthasarathy, K. Identification and control of dynamical systems using neural networks. *IEEE Trans. Neural Networks* 1:4–27, 1990.

4

TECHNIQUES AND APPLICATIONS OF FUZZY SET THEORY TO DIFFERENCE AND FUNCTIONAL EQUATIONS AND THEIR UTILIZATION IN MODELING DIVERSE SYSTEMS

ELIAS DEEBA*
ANDRÉ DE KORVIN[†]
SHISHEN XIE[‡]

Computer and Mathematical Sciences, University of Houston–Downtown, Houston, Texas 77002

I. INTRODUCTION 87
II. MOTIVATION 88
III. BACKGROUND 90
IV. DESCRIPTION OF THE METHOD 93
V. FUZZY ANALOG OF CAUCHY FUNCTIONAL EQUATIONS 96
VI. FUZZY ANALOG OF A FIRST-ORDER DIFFERENCE EQUATION 102
VII. CONCLUSION 109
REFERENCES 109

I. INTRODUCTION

This paper deals with the study of difference and functional equations in the setting of fuzzy theory. This study has been initiated in [1], [2], [3], and [4]. In the analysis of many systems governed by such equations, one is faced with uncertainties due to imprecise measurements or lack of complete informa-

* Partially supported by UHD ORC Grant 1996.
† Partially supported by Grants ARO DAAH-0495-10250 and NSF CDA-95-22157.
‡ Partially supported by UHD ORC Grant 1997.

tion. Indeed, the closer one looks at a real-world problem, the more apparent inherent uncertainties become. Parameters such as probability distributions, metabolic rates, genetic fitness, and other measurements that arise in the modeling of problems in economics, genetics, and population dynamics are rarely precisely known. Fuzzy set theory then becomes a natural setting for the definition of such quantities.

In Section II we expand our motivation of this study. We present diverse systems and show why their study in the fuzzy setting renders a better understanding of their behavior. In Section III we introduce the elements of fuzzy theory that are needed for the development of this chapter. Section IV deals with a formulation of the method for casting a functional or a difference equation in the fuzzy setting. Indeed, we will introduce the steps needed to fuzzify a functional or a difference equation and the method for solving the resulting fuzzy equations. Section V deals with the Cauchy functional equations; Section VI deals with a linear system of difference equations and its fuzzy analog.

II. MOTIVATION

In this section we provide the motivation for studying various models governed by functional equations in the setting of fuzzy theory. Difference equations and, in general, functional equations arise in the study and the modeling of many applied problems. Inventory analysis, learning models, system theory, information theory, genetics, ecological systems, and blood-flow modeling form a small sample of the numerous situations conveniently modeled by such equations (see [5] and [6]).

Why will one consider the fuzzy analog in modeling such phenomena? In many cases, more is assumed than is really known. For example, consider the amount of information generated by two independent events A and B with probability p and q, respectively. The corresponding amount of information I that is to be determined is a function of p and q and may be modeled using the Cauchy functional equation (see [5] and [7]),

$$I(pq) = I(p) + I(q). \tag{1}$$

In the real world it is often the case that p and q are not precisely known. Indeed, the probabilities p and q may reflect the observer's subjective opinion. Thus, this imprecise information may be better handled if Eq. (1) is cast in the realm of fuzzy set theory.

Another related example could be stated as follows: If we make two measurements yielding mass x and mass y, what is the combined mass? A rough answer is of course $x + y$. However, if we do not necessarily assume that measurements yield a precise value, the answer becomes somewhat more complex. Again, this may be modeled by the Cauchy equation (see [5] and [7]),

$$f(x + y) = f(x) + f(y), \tag{2}$$

where f is an unknown function to be determined and represents the measure of the underlying mass. The imprecise measurement associated with the mass again motivates us to consider Eq. (2) in the fuzzy setting.

An example along different lines may be phrased as follows: Consider the difference equation (see [8])

$$x_{n+1} = \frac{w_{11}x_n^2 + w_{12}x_n}{w_{12}x_n + w_{22}}. \tag{3}$$

This difference equation arises in the study of selection problem of genotypes where w_{11}, w_{12}, and w_{22} describe "the relative fitness" of the genotypes AA, Aa, and aa, respectively, and x_n describes the gene frequency for generation n. The w-values measure the relative contribution of the genotypes to the next generation. In the real world, the fitness factors of the genotypes are not completely known. This incomplete information may be better handled if one considers (3) in a fuzzy setting.

Along somewhat similar lines of thought one may consider the following differential equation (see [8]):

$$y' = \beta y - \gamma y^2. \tag{4}$$

This equation is used to model a singular population ecological system with growth term βy modified by an inhibiting term γy^2. A possible discretization of the equation leads to the difference equation

$$u_{k+1} - u_k = h(\beta - \gamma u_k)u_{k+1}, \tag{5}$$

where h is the step size in the discretization of (4). Equation (5) describes the change in population of a single species from the kth generation u_k to the $(k + 1)$th generation u_{k+1} under growth with inhibition. The growth factor and the inhibition factors may depend on environmental conditions and this may render their determination imprecise. Thus, the parameters h, β, γ and the initial data u_0 may not be completely known. Again, this incomplete information is yet another reason to cast and study models such as (5) in the fuzzy setting.

As a last example of a real-world situation where not all of the parameters may be precisely known, we look at a possible model for the carbon dioxide (CO_2) level in the blood. The level of CO_2 may be described by the equations (see [6])

$$C_{n+1} = C_n - L(C_n, V_n) + m,$$

$$V_{n+1} = F(C_n),$$

where C_{n+1} is the concentration of CO_2 level in the $(n + 1)$th cycle, the function V_n describes the rate of ventilation in the nth cycle, and the function F relates V_n and C_n. Often the model is simplified by making the nonlinear operator $L(C_n, V_n)$ directly proportional to V_n (and independent of C_n). The parameter m denotes the amount of CO_2 in the blood due to metabolism. The metabolism measurements as well as the approximation to $L(C_n, V_n)$ are inherently imprecise.

Many additional examples could be given where parameters of different equations describing real-world situations are not precisely known. Some mathematical tools need to be developed to address these situations. The main idea of this chapter is to assume that the parameters in the underlying equations are fuzzy and to develop techniques to solve these equations. In the next section we give the background necessary to understand the fuzzy operations needed to obtain the appropriate solutions.

III. BACKGROUND

Let X be a set. A fuzzy subset A of X is a function from X into $[0, 1]$. The fuzzy subset A is thus characterized by a membership function $\mu_A: X \to [0, 1]$, which associates with each element $x \in X$ a number $\mu_A(x)$ in the interval $[0, 1]$. This function represents the grade of membership of x in the set A. In this chapter, we shall denote the membership of x in A by $A(x)$.

EXAMPLE 1. If X denotes positive integers, the fuzzy subset of X called *few* could be defined as follows:

$$few(2) = 0.4, \qquad few(3) = 0.8, \qquad few(4) = 0.9,$$
$$few(x) = 0, \qquad \text{if } x \notin \{2, 3, 4\}.$$

This definition of *few* says that 2, 3, and 4 are examples of what is meant by *few*. The best example is 4, because its membership in the fuzzy set *few* is 0.9. Any positive integer greater than 4 is too large to have any membership other than 0 in *few*, and 1 is too small to have any positive membership in *few*.

It should be noted that if $B \subset X$, then B can be considered to be a fuzzy subset by identifying B with its characteristic function, that is,

$$B(x) = \begin{cases} 1, & \text{if } x \in B, \\ 0, & \text{if } x \notin B. \end{cases}$$

We now define operations on fuzzy sets. Let A and B be fuzzy subsets of X. The *union*, *intersection*, and *complements* are, respectively, defined by

$$(A \vee B)(x) = \max\{A(x), B(x)\},$$
$$(A \wedge B)(x) = \min\{A(x), B(x)\},$$
$$(\neg A)(x) = 1 - A(x).$$

EXAMPLE 2. Let $X = \{1, 2, 3, 4, 5\}$ and define the two fuzzy subsets A and B via

$$A(1) = 0.3, \qquad A(2) = 0.4, \qquad A(3) = 0.7,$$
$$A(x) = 0, \qquad \text{if } x = 4 \text{ or } x = 5.$$

and

$$B(2) = 0.2, \qquad B(3) = 0.4, \qquad B(4) = 0.8,$$
$$B(x) = 0, \qquad \text{if } x = 1 \text{ or } x = 5.$$

One may compute

$$(A \vee B)(2) = \max\{A(2), B(2)\} = 0.4,$$
$$(A \wedge B)(2) = \min\{A(2), B(2)\} = 0.2,$$
$$(\neg A)(2) = 1 - A(2) = 0.6.$$

If A and B are two regular subsets of X and if the previous operations are applied to their characteristic functions, it is clear that we obtain the usual union, intersection, and complement operations on sets.

One major difference between fuzzy and regular sets is that the intersection of a set and its complement need not be empty. In Example 2, $(A \wedge \neg A)(2) = \min\{0.4, 0.6\} = 0.4$, and the union of a fuzzy set with its complement need not be the whole space. Again, in Example 2, $(A \vee \neg A)(2) = \max\{0.4, 0.6\} = 0.6$. This has major implications when fuzzy expert systems are considered. Whereas in classical settings only one rule from a competing set of rules is considered, in the fuzzy setting the situation is different. Several competing rules are considered in parallel. For example, in dealing with "fuzzy controllers," the fuzzy approach generates better controllers than the traditional approach. In fact, it is better to control a room's temperature by constant adjustment of the thermostat rather than switching it on and off whenever the temperature falls below or above some fixed number.

A fuzzy subset A of X is called *convex* if the following relation holds:

$$A(sx_1 + (1 - s)x_2) \geq \min\{A(x_1), A(x_2)\},$$

for any $x_1, x_2 \in X$ and $s \in (0, 1)$.

EXAMPLE 3. An example of a *convex fuzzy* set is given by the *triangular function* in $[4, 6]$:

$$A(x) = \begin{cases} 0, & x < 4, \\ x - 4, & 4 \leq x < 5, \\ -x + 6, & 5 \leq x < 6, \\ 0, & x \geq 6. \end{cases} \tag{6}$$

By an α-*cut* $(0 < \alpha \leq 1)$ of a fuzzy subset A we mean the regular set

$$[A]_\alpha = \{x | A(x) \geq \alpha\}.$$

Remark 1. There is an easy characterization of convex fuzzy sets. (See [9] and [10].) A fuzzy set A can be shown to be convex if and only if its α cuts $[A]_\alpha$ are intervals for all α $(0 < \alpha \leq 1)$. Of course, if A is continuous, then it follows that A is convex if and only if $[A]_\alpha$ are closed intervals.

Remark 2. It can be shown that the fuzzy set A is totally defined by its α cuts. In fact,

$$A = \sup_{0 < \alpha \leq 1} \min\{\alpha, \chi_{[A]_\alpha}\},$$

where $\chi_{[A]_\alpha}$ denotes the characteristic function of the set $[A]_\alpha$.

If X is the set of real numbers, by a *fuzzy number* N_x we mean a fuzzy subset of X where

    i. $N_x(y) = 1$, if and only if $y = x$.
    ii. $N_x(\cdot)$ is continuous.
    iii. $N_x(\cdot)$ is convex.
    iv. $N_x(\cdot)$ vanishes at infinity.

$N_x(y)$ denotes the degree of belief that the value of x is in fact y.

After the collection of these definitions and properties, we discuss the *extension principle* [9], which is heavily utilized in this chapter. The extension principle roughly states that any function of several variables $x_1, \ldots, x_n$ can be extended to a function of the fuzzy subsets $A_1, A_2, \ldots, A_n$ of $X_1, X_2, \ldots, X_n$, where $x_i \in X_i$ $(1 \leq i \leq n)$. More precisely, let f be a function from $X_1 \times X_2 \times \cdots \times X_n$ into Y. Let A_i be a fuzzy subset of X_i. Then the extension of f, still denoted by f, is defined by

$$f(A_1, A_2, \ldots, A_n)(y) = \sup \min\{A_1(x_1), \ldots, A_n(x_n)\},$$

where the sup is taken over all $(x_1, x_2, \ldots, x_n)$ such that

$$f(x_1, x_2, \ldots, x_n) = y.$$

Note that $f(A_1, A_2, \ldots, A_n)$ is a fuzzy subset of Y. Also, if each A_i is taken to be the characteristic function of x_i, then

$$f\big(\chi_{\{x_1\}}(\cdot), \ldots, \chi_{\{x_n\}}(\cdot)\big)(y) = \chi_{\{y\}}(\cdot),$$

where $y = f(x_1, x_2, \ldots, x_n)$.

Remark 3. In particular, if f is a function of one variable, then the extension is defined by

$$f(A)(y) = \sup\{A(x)\},$$

where the sup is over all x for which $f(x) = y$.

Remark 4. How do we extend binary operations? For example, consider the operation "$+$". Set $f(a, b) = a + b$. The extension for each fixed x yields

$$f(A, B)(x) = \sup \min(A(x_1), B(x_2)),$$

where the sup is over all x_1 and x_2 for which $f(x_1, x_2) = x_1 + x_2 = x$, and A and B are fuzzy subsets. Likewise, for $f(a, b) = ab$, the extension for each fixed x yields

$$f(A, B)(x) = \sup \min(A(x_1), B(x_2)),$$

where the sup is over all x_1 and x_2 for which $f(x_1, x_2) = x_1 x_2 = x$, and A and B are fuzzy subsets.

A main idea that is essential for developing our technique is to be able to address the following question: How do we relate the α cuts $[f(A_1, \ldots, A_n)]_\alpha$

to $f([A_1]_\alpha, \ldots, [A_n]_\alpha)$, when f is a continuous function from $X_1 \times X_2 \times \cdots \times X_n$ into Y and $A_1, A_2, \ldots, A_n$ are fuzzy subsets of $X_1, X_2, \ldots, X_n$, respectively? In particular, when does the relation

$$[f(A_1, \ldots, A_n)]_\alpha = f([A_1]_\alpha, \ldots, [A_n]_\alpha) \tag{7}$$

hold?

It was shown by Nguyen [11] that the equality in (7) holds if and only if, for every $y \in Y$, there exist $x_1^*, x_2^*, \ldots, x_n^*$ such that

$$f(A_1, A_2, \ldots, A_n)(y) = \min\{A_1(x_1^*), \ldots, A_n(x_n^*)\};$$

that is, the sup is attained for every fixed y. This result will play an important role in our method of solving fuzzy equations. Indeed, since a fuzzy set is completely determined by its α cuts, it is natural to use Nguyen's theorem to rewrite fuzzy difference or functional equations as equations involving cuts of appropriate sets. Therefore, this theorem will provide the means to go back and forth between fuzzy and classical models. The next section details this process.

IV. DESCRIPTION OF THE METHOD

In this section we discuss the method of casting a general equation in the fuzzy setting. We present the steps needed to solve the resulting fuzzy equation. In Sections V and VI we further illustrate this method through two examples. In Section V, we consider a class of functional equations of the form

$$f(x \mathbin{\Box} y) = f(x) * f(y), \tag{8}$$

where f is unknown, and $\Box$ and $*$ may be either addition or multiplication of real numbers; in Section VI, we will consider a first-order difference equation of the form

$$x_{n+1} = Ax_n + b, \tag{9}$$

where A is an $m \times m$ known matrix, b is an $m \times 1$ known matrix, and x is an $m \times 1$ unknown vector whose components represent the state of a system under consideration.

Equations (8) and (9) may be put in a general form

$$w = F(x_1, x_2), \tag{10}$$

where $F: \Re \times \Re \to \Re$ is a function relating the unknown quantities x_1 and x_2. In the functional equation (8), w stands for the left-hand side $f(x \mathbin{\Box} y)$ and $F(x_1, x_2)$ stands for $f(x) * f(y)$. In other words, with an appropriate matching, say $x_1 = f(x)$, $x_2 = f(y)$, then $F(x_1, x_2) = x_1 * x_2$. In the difference equation (9), w stands for x_{n+1} and $F(x_1, x_2)$ stands for the right-hand side with the proper matching $x_1 = Ax_n$, $x_2 = b$, and $F(x_1, x_2) = x_1 + x_2$.

We note that many other functional or difference equations can also be put into the form of Eq. (10). We further note that conceptually the extension of the results to higher dimensions presents no difficulty.

To obtain the fuzzy analog of Eq. (10), we ask the following questions:

Question 1.   How do we cast Eq. (10) in the realm of fuzzy theory?

Let $N_x(y)$ denote the degree of belief that x has the value y as noted in the previous section. Using the extension principle, we can put Eq. (10) in the fuzzy form

$$N_w(y) = \sup \min\{N_{x_1}(y_1), N_{x_2}(y_2)\}, \tag{11}$$

where the sup is taken over all y_1 and y_2 for which $F(y_1, y_2) = y$. Equivalently, we may extend Eq. (10) as

$$F(N_{x_1}, N_{x_2})(y) = \sup \min\{N_{x_1}(y_1), N_{x_2}(y_2)\}, \tag{12}$$

where the sup is taken over all y_1 and y_2 for which $F(y_1, y_2) = y$.

Question 2.   What are the technical conditions that F must satisfy?

C1. Equation (10) with real arguments must have a known solution.
C2. F is nondecreasing in its argument.
C3. F is continuous.
C4. The solution of (10) is continuous and nondecreasing.
C5. The inverse image $F^{-1}(t)$ is bounded for every t

These conditions are needed to verify the following lemmas. The first lemma is to show that the sup in (12) is attained.

LEMMA 1.   *If conditions* (C3) *and* (C5) *are satisfied, then the* sup *in Eq.* (12) *is attained; that is, there exist* y_1^* *and* y_2^* *for which*

$$F(N_{x_1}, N_{x_2})(y) = \min\{N_{x_1}(y_1^*), N_{x_2}(y_2^*)\},$$

where $F(y_1^*, y_2^*) = y$.

Proof.   Because F is continuous, it follows that $F^{-1}(\{y\})$ is a closed subset of $\Re$. Condition (C5) will then imply that $F^{-1}(\{y\}) = \{(y_1, y_2)|F(y_1, y_2) = y\}$ is a compact subset. Hence, there exist y_1^* and y_2^* for which the sup in (12) is attained, that is, $F(N_{x_1}, N_{x_2})(y) = \min\{N_{x_1}(y_1^*), N_{x_2}(y_2^*)\}$, where $F(y_1^*, y_2^*) = y$.   ∎

Lemma 1 together with the result due to Nguyen [11] implies the following:

LEMMA 2.   *For* $\alpha \in (0, 1]$, *the* α *levels are equal:*

$$[w]_\alpha = [F(N_{r_1}, N_{r_2})]_\alpha = F([N_{r_1}]_\alpha, [N_{r_2}]_\alpha).$$

Remark 1 implies that a fuzzy number A is convex and continuous if and only if the α level of A, $[A]_\alpha$, is a closed interval. Therefore, we have

LEMMA 3. *The α level of the fuzzy number w is a closed and bounded interval. Hence, $[w]_\alpha = [L_{w,\alpha}, R_{w,\alpha}]$, where $L_{w,\alpha}$ and $R_{w,\alpha}$ are the left endpoint and the right endpoint associated with the α level of the fuzzy number w.*

We will now show that the left endpoint and the right endpoint satisfy the classical equation (10).

LEMMA 4. *If F satisfies condition (C2) and (C3), then the following two relations hold*:

$$L_{w,\alpha} = F\left(L_{N_{x_1},\alpha}, L_{N_{x_2},\alpha}\right), \tag{13}$$

$$R_{w,\alpha} = F\left(R_{N_{x_1},\alpha}, R_{N_{x_2},\alpha}\right), \tag{14}$$

Proof. It is enough to verify

$$F\left(\left[L_{N_{x_1},\alpha}, R_{N_{x_1},\alpha}\right] \times \left[L_{N_{x_2}\,\alpha}, R_{N_{x_2},\alpha}\right]\right)$$

$$= \left[F\left(L_{N_{x_1},\alpha}, L_{N_{x_2},\alpha}\right), F\left(R_{N_{x_1},\alpha}, R_{N_{x_2},\alpha}\right)\right].$$

Let $z \in F([L_{N_{x_1},\alpha}, R_{N_{x_1},\alpha}] \times [L_{N_{x_2},\alpha}, R_{N_{x_2},\alpha}])$. Then there exist $z_1 \in [L_{N_{x_1},\alpha}, R_{N_{x_1},\alpha}]$ and $z_2 \in [L_{N_{x_2},\alpha}, R_{N_{x_2},\alpha}]$, such that $z = F(z_1, z_2)$. Since F is nondecreasing function of its argument [condition (C2)], it follows that

$$F\left(L_{N_{x_1},\alpha}, L_{N_{x_2},\alpha}\right) \leq F(z_1, z_2) \leq F\left(R_{N_{x_1},\alpha}, R_{N_{x_2},\alpha}\right).$$

Hence, $F(z_1, z_2)$ belongs to $[F(L_{N_{x_1},\alpha}, L_{N_{x_2},\alpha}), F(R_{N_{x_1},\alpha}, R_{N_{x_2},\alpha})]$.

Conversely, let $z \in [F(L_{N_{x_1},\alpha}, L_{N_{x_2},\alpha}), F(R_{N_{x_1},\alpha}, R_{N_{x_2},\alpha})]$. Since F is continuous, the image of the set $[L_{N_{x_1},\alpha}, R_{N_{x_1},\alpha}] \times [L_{N_{x_2},\alpha}, R_{N_{x_2},\alpha}]$ is a connected interval. Thus, there exist $z_1 \in [L_{N_{x_1},\alpha}, R_{N_{x_1},\alpha}]$ and $z_2 \in [L_{N_{x_2},\alpha}, R_{N_{x_2},\alpha}]$, such that $z = F(z_1, z_2)$. This implies that $z \in F([L_{N_{x_1},\alpha}, R_{N_{x_1},\alpha}] \times [L_{N_{x_2},\alpha}, R_{N_{x_2},\alpha}])$. This proves the lemma. ∎

Finally, we solve Eq. (12). Equations (13) and (14) imply that the left endpoint and the right endpoint satisfy the classical equation. This means that we know the solutions to the following two equations:

$$L_{w,\alpha} = F\left(L_{N_{x_1},\alpha}, L_{N_{x_2},\alpha}\right),$$

$$R_{w,\alpha} = F\left(R_{N_{x_1},\alpha}, R_{N_{x_2},\alpha}\right).$$

Moreover, the solutions to Eqs. (13) and (14), say functions H_1 and H_2, respectively, are continuous and nondecreasing by conditions (C1) and (C4). Thus

$$y \in [w]_\alpha \quad \Leftrightarrow \quad L_{w,\alpha} \leq y \leq R_{w,\alpha}$$

$$\Leftrightarrow \quad H_1\left(L_{N_{x_1},\alpha}, L_{N_{x_2},\alpha}\right) \leq y \leq H_2\left(R_{N_{x_1},\alpha}, R_{N_{x_2},\alpha}\right)$$

$$\Leftrightarrow \quad y \in \left[H_1\left(L_{N_{x_1},\alpha}, L_{N_{x_2},\alpha}\right), H_2\left(R_{N_{x_1},\alpha}, R_{N_{x_2},\alpha}\right)\right].$$

THEOREM 1. *The solution of Eq.* (12) *is given by*

$$[w]_\alpha = \left[H_1\left(L_{N_{x_1},\alpha}, L_{N_{x_2},\alpha}\right), H_2\left(R_{N_{x_1},\alpha}, R_{N_{x_2},\alpha}\right)\right], \tag{15}$$

where H_1 and H_2 are the solutions of (13) *and* (14), *respectively.*

Remark 5. The solution of (15) may also be represented by

$$w = \sup \min\left\{\alpha, \chi_{[H_1(L_{N_{x_1}},\alpha, L_{N_{x_2}},\alpha), H_2(R_{N_{x_1}},\alpha, R_{N_{x_2}},\alpha)]}\right\},$$

where the sup is taken over all $\alpha \in (0,1)$ and χ is the characteristic function over the interval $[H_1(L_{N_{x_1},\alpha}, L_{N_{x_2},\alpha}), H_2(R_{N_{x_1},\alpha}, R_{N_{x_2},\alpha})]$.

V. FUZZY ANALOG OF CAUCHY FUNCTIONAL EQUATIONS

In this section we consider the four Cauchy equations

$$f(x+y) = f(x) + f(y), \tag{16}$$

$$f(xy) = f(x) + f(y), \tag{17}$$

$$f(x+y) = f(x)f(y), \tag{18}$$

$$f(xy) = f(x)f(y), \tag{19}$$

in the fuzzy setting. Equations of the forms (16)–(19) arise in the formation of problems in information theory. As we noted earlier, f in (17) may represent the amount of information due to two independent events A and B with probabilities x and y, respectively. The exact values of x and y may not be known because not enough data are available or because x and y partially reflect the decision maker's subjective opinion. As such, the decision maker will not exactly know the amount of information, $f(\cdot)$, generated by the independent event A, B, and $A \cap B$. Rather the decision maker has an amount of belief in each possible values of such quantities as $f(x)$. The assignment of beliefs to possible values is at the heart of fuzzy set theory. Likewise, by considering rates of growth and their imprecise measurements one would arrive at an equation of the form (18) and (19). As such we are motivated to study the fuzzy analogs of these equations.

These equations will be considered as special cases of the general fuzzy equation given by

$$A(x * y) = A(x) \; \Box \; A(y), \tag{20}$$

where $A(x)$ is a fuzzy number, and $x, y \in \Re^+$. The operation $*$ is a binary operation on the reals and $\Box$ is a binary operation on numbers properly extended to fuzzy numbers. (See Remark 4.) The assumptions that the operation $\Box$ must satisfy are listed as follows:

C2. It is nondecreasing in its argument.
C3. It is continuous.
C5. The inverse image under $\Box$ is bounded.

The following are basic to understanding how to manipulate and obtain the solution of Eq. (20). We will now show that Lemmas 1–4 are well adapted to Eq. (20).

LEMMA 5.   *If h is a function from $\Re \times \Re$ into $\Re^+$ defined by*

$$h(x, y) = x \; \Box \; y$$

is such that the inverse $h^{-1}(z)$ is bounded, then

$$[A(x * y)]_\alpha = [A(x)]_\alpha \; \Box \; [A(y)]_\alpha.$$

Proof.   Let N and M be two fuzzy subsets of $\Re^+$. By the extension principle,

$$h(N, M)(z) = \sup_{x \, \Box \, y = z} \; (\min\{N(x), M(y)\})$$

$$= \sup_{(x, y) \in h^{-1}(z)} \; (\min\{N(x), M(y)\}). \tag{21}$$

Because $h^{-1}(z)$ is assumed to be bounded and $h^{-1}(\{z\})$ is closed (note that $\Box$ is continuous), it follows that $h^{-1}(z)$ is compact. This implies that there exist x^* and y^* such that the sup in (21) is attained. Hence, by Nguyen's result [11],

$$[h(M, N)]_\alpha = h([N]_\alpha, [M]_\alpha).$$

From this it follows that

$$[N \; \Box \; M]_\alpha = [N]_\alpha \; \Box \; [M]_\alpha.$$

Upon setting $N = A(x)$ and $M = A(y)$, we obtain

$$[A(x) \; \Box \; A(y)]_\alpha = [A(x)]_\alpha \; \Box \; [A(y)]_\alpha,$$

or, equivalently,

$$[A(x * y)]_\alpha = [A(x)]_\alpha \; \Box \; [A(y)]_\alpha.$$

∎

Remark 6. What makes $h^{-1}(z)$ bounded? This is the case if $S = \{(x, y)|x \; \square \; y = z\}$ is bounded. In case $\square = +$, and x and y belong to $\Re^+$, then, for every z, it is clear that x and y belong to $[0, z]$ and therefore S is bounded. However, if $\square = \times$, one has the boundedness property provided that x and y are bounded away from the origin. We shall choose in the latter case $\Re_a^+ = \{x|x \geq a\}$ for $a \in \Re^+$, a fixed, and assume that

$$\lim_{x \to \infty} a \; \square \; x = +\infty.$$

In this section, we shall restrict ourselves to $\Re^+$ or $\Re_1^+$ depending upon the case under consideration.

Because the fuzzy number $A(x)$ satisfies the properties

  i. $A(x)(y) = 1$, if and only if $y = x$
 ii. $A(x)(\cdot)$ is continuous
iii. $A(x)(\cdot)$ is convex
 iv. $A(x)(\cdot)$ vanishes at infinity

it can be easily shown that

LEMMA 6. $[A(x)]_\alpha$ *is a closed set.*

LEMMA 7. *Let $A(x)$ be any fuzzy number. Then the α level of $A(x)$, $[A(x)]_\alpha$, is a bounded and closed interval.*

Proof. From (C1) and Lemma 6, it follows that $[A(x)]_\alpha$ is a closed interval. Because $\lim_{u \to \infty} A(x)(u) = 0$, the right endpoint of this interval is finite. ∎

To simplify the notation, in the remainder of the section we denote by $L(x)$ and $R(x)$ the left and right endpoints respectively, of $[A(x)]_\alpha$, instead of $L_{A(x), \alpha}$ and $R_{A(x), \alpha}$.

LEMMA 8. *The left endpoint $L(x)$ and the right endpoint $R(x)$ of the interval $[A(x)]_\alpha$ satisfy Eq. (20).*

Proof. If $[A(z)]_\alpha = [L(z), R(z)]$, then $[A(x * y)]_\alpha = [A(x)]_\alpha \; \square \; [A(y)]_\alpha$ in Lemma 5 and the assumption that the operation $\square$ is increasing will imply the equations

$$L(x * y) = L(x) \; \square \; L(y), \tag{E-1}$$

$$R(x * y) = R(x) \; \square \; R(y). \tag{E-2}$$

∎

Finally, Lemma 8 can be restated as

LEMMA 9. *The α level of $A(x)$, $[A(x)]_\alpha$, is of the form $[S_1(x), S_2(x)]$, where $S_1(x)$ and $S_2(x)$ are solutions of Eqs. (E-1) and (E-2), respectively.*

We now consider the four Cauchy equations (16)–(19) in the fuzzy setting.

Case 1. Equation (20) is the fuzzy analog of (16) if $*$ and $\square$ are replaced by addition $+$. We assume that x and y belong to $\Re^+$. In this case (16) is given by

$$A(x+y) = A(x) + A(y) \tag{F1}$$

with equations (E-1) and (E-2) reading

$$L(x+y) = L(x) + L(y),$$
$$R(x+y) = R(x) + R(y). \tag{E1}$$

Because the set $S = \{(x, y)|x + y = z\}$ is bounded for every fixed z in $\Re^+$, it follows that Lemmas 7–9 hold. The solution to system (E1) is

$$L(x) = L(1)x \quad \text{and} \quad R(x) = R(1)x.$$

Thus $[A(x)]_\alpha = [L(1)x, R(1)x]$.

Now, $y \in [A(x)]_\alpha$ if and only if

$$A(x)(y) \geq \alpha \quad \Leftrightarrow \quad L(1)x \leq y \leq R(1)x$$

$$\Leftrightarrow \quad L(1) \leq \frac{y}{x} \leq R(1)$$

$$\Leftrightarrow \quad A(1)\left(\frac{y}{x}\right) \geq \alpha$$

$$\Leftrightarrow \quad xA(1)(y) \geq \alpha. \tag{22}$$

This implies, formally, that $A(x) = xA(1)$. To see this, consider the function $h(t) = tx$. By the extension principle, if N is any fuzzy number, then

$$h(N)(z) = \sup_{tx=z} N(t) = N\left(\frac{z}{x}\right). \tag{23}$$

Thus, upon substitution $N = A(1)$ in (23), we obtain

$$h(A(1))(z) = A(1)\left(\frac{z}{x}\right).$$

By (22),

$$(xA(1))(z) = A(x)(z),$$

Hence, we have formally $A(x) = A(1)x$. This is analogous to the solution of the "classical" Cauchy equation (16).

Case 2.   Equation (20) is the fuzzy analog of Eq. (17) if $*$ is replaced by multiplication $\cdot$ and $\square$ is replaced by addition $+$. In this case (17) reads

$$A(xy) = A(x) + A(y) \tag{F2}$$

and equations (E-1) and (E-2) read

$$L(xy) = L(x) + L(y),$$
$$R(xy) = R(x) + R(y). \tag{E2}$$

As in Case 1, for Lemmas 5–9 to hold, we need the set $S = \{(x, y)|xy = z\}$ to be bounded. This is the case if x and y belong to the set, say $\Re_1^+ = \{z|z > 1\}$. The solutions of (E2) are

$$L(x) = c_1 \log x \quad \text{and} \quad R(x) = c_2 \log x.$$

By matching the transformation $x = e^u$ and $y = e^v$ for all u, v in $\Re^+$ (note that $x, y > 1$), Eq. (17) reduces to Eq. (16). Now, $y \in [L(e)\log x, R(e)\log x]$ if and only if

$$L(e) \leq \frac{y}{\log x} \leq R(e).$$

This is so if and only if $y \in [A(e)\log x]_\alpha$. Repeating similar argument, as in Case 1, we obtain a solution of (F2) as $A(x) = A(e)\log x$, which is again analogous to the "classical" solution of Eq. (17).

Case 3.   Equation (20) is the fuzzy analog of Eq. (18) if $*$ is replaced by $+$ and $\square$ is replaced by $\cdot$. For this case, (20) reads

$$A(x + y) = A(x)A(y) \tag{F3}$$

and equations (E-1) and (E-2) read

$$L(x + y) = L(x)L(y),$$
$$R(x + y) = R(x)R(y). \tag{E3}$$

For Lemmas 5–9 to hold, the set $S = \{(x, y)|xy = z\}$ must be bounded. This is the case because we restrict ourselves to the set $\Re_1^+ = \{v|v > 1\}$. We note that we do not know if all solutions of (E3) over $\Re_1^+$ are the restrictions of solutions of (E3) over $\Re^+$ and the question of uniqueness to the best of our knowledge is open.

The solution that will be exhibited for (F3) will be valid for $x > 1$ and in here we will assume that $A(x) = 0$ for $x \leq 1$.

The solutions of (E3) are

$$L(x) = (L(1))^x \quad \text{and} \quad R(x) = (R(1))^x.$$

Thus $[A(x)]_\alpha = [(L(1))^x, (R(1))^x]$. Now $y \in [A(x)]_\alpha$ if and only if

$$A(x)(y) \geq \alpha \quad \Leftrightarrow \quad (L(1))^x \leq y \leq (R(1))^x$$

$$\Leftrightarrow \quad x \log L(1) \leq \log y \leq x \log R(1)$$

$$\Leftrightarrow \quad \log L(1) \leq \frac{1}{x} \log y \leq \log R(1)$$

$$\Leftrightarrow \quad L(1) \leq e^{(1/x)\log y} \leq R(1)$$

$$\Leftrightarrow \quad A(1)(e^{(1/x)\log y}) \geq \alpha$$

$$\Leftrightarrow \quad A(1)(y^{1/x}) \geq \alpha. \tag{24}$$

Now consider the function $h(t) = t^{1/x}$ $(t \geq 1)$. By the extension principle, for any fuzzy subset N of $\{y | y \geq 1\}$,

$$h(N)(z) = \sup_{t^{1/x}=z} N(t) = N(z^x).$$

Applying this with $z = y^{1/x}$, we obtain

$$h(N)(y^{1/x}) = N(y). \tag{25}$$

Upon replacing N by $A(1)$ in (25) and using (24), we get

$$h(A(1))(y^{1/x}) - A(1)(y).$$

That is,

$$(A(1))^{1/x}(y^{1/x}) = A(1)(y),$$

which implies that

$$A(1)(y^{1/x}) = (A(1))^x(y).$$

By (24), $A(x)(y) = (A(1))^x(y)$. Thus, formally $A(x) = (A(1))^x$ is a solution to Eq. (F3). Again, this is analogous to the "classical" solution of Eq. (18).

Case 4. Equation (20) is the fuzzy analog of Eq. (19) if both $*$ and $\square$ are replaced by $\cdot$. For this case Eq. (20) reads

$$A(xy) = A(x)A(y) \tag{F4}$$

and equations (E-1) and (E-2) read

$$L(xy) = L(x)L(y),$$
$$R(xy) = R(x)R(y). \tag{E4}$$

As in Case 3, we restrict ourselves for all x and y in $\mathfrak{R}_1^+$. $\mathfrak{R}_1^+$ forms a semigroup under multiplication and it generates the group $(\mathfrak{R}^+, \cdot)$. Therefore, all solutions of Eq. (F4) over $\mathfrak{R}_1^+$ are the restriction of the solutions over $\Re^+$ (see [10]). The solution to (F4) will be $A(x) = 0$ or $A(x) = x^c$ $[c = \log A(e)]$. To see this, we note that

$$A(x)(y) \geq \alpha \quad \Leftrightarrow \quad y \in \left[x^{\log L(e)}, x^{\log R(e)} \right]$$

$$\Leftrightarrow \quad y \in \left[e^{(\log L(e))\log x}, e^{(\log R(e))\log x} \right]$$

$$\Leftrightarrow \quad (\log L(e))\log x \leq \log y \leq \log y \leq (\log R(e))\log x$$

$$\Leftrightarrow \quad \log L(e) \leq \frac{\log y}{\log x} \leq \log R(e)$$

$$\Leftrightarrow \quad L(e) \leq e^{(\log y)/(\log x)} \leq R(e)$$

$$\Leftrightarrow \quad A(e)\left(e^{(\log y)/(\log x)}\right) \geq \alpha$$

$$\Leftrightarrow \quad A(e)\left(y^{1/\log x}\right) \geq \alpha. \tag{26}$$

Now let $h(t) = t^{\log x}$. By the extension principle, we have

$$h(N)(z) = \sup_{z = t^{\log x}} N(t) = N(z^{1/\log x}).$$

Thus by replacing N by $A(e)$ and z by y, we obtain

$$A(e)\left(y^{1/\log x}\right) = h(A(e))(y).$$

By (26), $A(x)(y) = h(A(e))(y)$.
Hence, $A(x) = h(A(e))$. Formally,

$$A(x) = (A(e))^{\log x} = e^{(\log x)\log(A(e))} = \left(e^{\log x}\right)^{\log A(e)},$$

and this implies that $A(x) = x^{\log A(e)}$, which plays the role of the solution to (F4): $A(x) = x^c$, with $c = \log A(e)$. Again this is analogous to the "classical" solution of Eq. (19).

VI. FUZZY ANALOG OF A FIRST-ORDER DIFFERENCE EQUATION

Many linearized discrete systems may be represented by the first-order difference equation

$$x_n = Ax_{n-1} + B \qquad (x_0, \text{ known initial value}), \tag{27}$$

where x_n is an $m \times 1$ matrix with $x_n^{(i)}$ representing the ith component of the state of the system at time n, A is a $m \times m$ matrix with entries a_{ij} representing the characteristics of the system, and B is an $m \times 1$ matrix with entries b_i representing forcing terms. We will consider two scenarios. The first scenario is to solve system (27) by decoupling it. If the matrix A is similar to a diagonal matrix (i.e., there is a nonsingular matrix P such that $P^{-1}AP = D$, where D is a diagonal matrix with diagonal elements being the eigenvalues of A and the columns of P are the corresponding eigenvectors of

A), then the system in (27) can be decoupled via the transformation $x_n = Pz_n$. Indeed, we obtain, after some algebraic computation, the decoupled system

$$z_n = Dz_{n-1} + P^{-1}B \qquad (z_0, \text{ known initial value}). \qquad (28)$$

From this we can solve for each component $z_n^{(i)}$, $i = 1, 2, \ldots, m$, by setting

$$z_n^{(i)} = \lambda^{(i)}z_{n-1}^{(i)} + \beta^{(i)}, \qquad (29)$$

where $\beta^{(i)}$ is the ith component of the $m \times 1$ matrix $P^{-1}B$, and $\lambda^{(i)}$ is the ith eigenvalue of A. The solution of (29) can be obtained recursively and is equal to

$$z_n^{(i)} = \left(\lambda^{(i)}\right)^n z_0^{(i)} + \left(1 + \lambda^{(i)} + \cdots + \left(\lambda^{(i)}\right)^{n-1}\right)\beta^{(i)}. \qquad (30)$$

Clearly as $n \to \infty$, each $z_n^{(i)}$ approaches the steady state $(1/(1 - \lambda^{(i)}))\beta^{(i)}$, provided that $|\lambda^{(i)}| < 1$ for $i = 1, \ldots, m$. Thus, if decoupling is possible, one can obtain $x_n^{(i)}$, the ith component of the state vector x_n.

The second scenario is to assume that $I - A$ is an invertible matrix. For example, this is the case if we assume the norm of A, $\|A\|$, is less than 1. After n iterations, the solution to (27) may be written as

$$x_n = A^n x_0 + (A^{n-1} + A^{n-2} + \cdots + A + I)B.$$

In the limiting case, as $n \to \infty$, $x_\infty = (I - A)^{-1}B$, where $(I - A)^{-1} = \sum_{n=0}^{\infty} A^n$, provided that $\|A\| < 1$. In this scenario the ith component $x_n^{(i)}$ is an "average of $x_{n-1}^{(j)}$" $(j = 1, 2, \ldots, m)$. So one possible interpretation of the result is that $x_{n-1}^{(1)}, \ldots, x_{n-1}^{(m)}$ are m alternate versions of what state is at time $n - 1$ and the solution x_n of the equation $x_n = Ax_{n-1} + B$ yields m alternate versions at time n.

As noted earlier, Eq. (27) arises in the study of systems. The uncertainty about the systems arises from, say, aging and temperature variations. These variations do not usually follow any known probability distributions and are often quantified in terms of bounds. It is this behavior that motivates us to consider the study of difference equations in the fuzzy setting. The ith component of the state $x_n^{(i)}$ will be treated as a fuzzy number. Based on the two scenarios, we will consider two cases.

Case 1. We assume that the system in (27) can be decoupled. Furthermore, we assume that the states of systems and the forcing term B are fuzzy and that A is a real matrix. We also assume that A and B satisfy the following conditions:

1. A has m distinct eigenvalues.
2. The entries a_{ij} of the matrix A are all positive.
3. The entries b_i of matrix B are positive.

The positive condition on the entries of matrix A does not restrict the class of matrices in applications, assuming that A represents physical characteristics of the system and such quantities are inherently positive.

We note that Eq. (27) can be put into the form

$$x_n = F(A, x_{n-1}, B) \qquad (x_0, \text{ known initial value}), \qquad (31)$$

with $F(A, x_{n-1}, B) = Ax_{n-1} + B$. Clearly F is continuous in its arguments and is increasing provided that $a_{ij} > 0$ for all $i, j = 1, \ldots, m$ and $b_i > 0$ for all $i = 1, \ldots, m$. Furthermore, as we noted earlier, if all eigenvalues $\lambda^{(i)}$ of A are such that $|\lambda^{(i)}| < 1$, then each state of the system approaches the steady behavior $(1/(1 - \lambda^{(i)}))\beta^{(i)}$. The fuzzy analog of (27) can be written as

$$x_n(y) = \sup \min\{Ax_{n-1}(y_1), B(y_2)\}, \qquad (32)$$

where the sup is taken over all y_1 and y_2 for which $F(y_1, y_2) = y$. In (32), $Ax_{n-1}(y_1)$ represents the $m \times 1$ matrix whose ith component $\sum_{j=1}^{m} a_{ij} x_{n-1}^{(j)}(y_1)$ is a fuzzy number. Likewise, $B(y_2)$ represents the $m \times 1$ matrix whose ith component $b^{(i)}(y_2)$ is a fuzzy number. The conditions imposed on A $(a_{ij} > 0,$ for all $i, j)$ and B imply that Lemmas 1 and 2 hold immediately. Thus, for each $\alpha \in (0, 1]$, we have

$$[x_n]_\alpha = [A]_\alpha [x_{n-1}]_\alpha + [B]_\alpha,$$

where $[A]_\alpha = A$. The fact that an α level is a closed bounded interval implies that $[x_n]_\alpha = [L_{x_n, \alpha}, R_{x_n, \alpha}]$, where $L_{x_n, \alpha}$ and $R_{x_n, \alpha}$ are $m \times 1$ matrices representing the left and right endpoints of the interval $[x_n]_\alpha$. Likewise, $L_{B, \alpha}$ and $R_{B, \alpha}$ are $m \times 1$ matrices, and represent the left and the right endpoints of the interval $[B]_\alpha$, respectively. With this, we obtain the coupled system for the left and the right endpoints:

$$L_{x_n, \alpha} = AL_{x_{n-1}, \alpha} + L_{B, \alpha}, \qquad (33)$$

$$R_{x_n, \alpha} = AR_{x_{n-1}, \alpha} + R_{B, \alpha}. \qquad (34)$$

Because A is similar to a diagonal matrix, we follow (28) to write (33) and (34) as

$$L_{z_n, \alpha} = DL_{z_{n-1}, \alpha} + P^{-1}L_{B, \alpha}, \qquad (35)$$

$$R_{z_n, \alpha} = DR_{z_{n-1}, \alpha} + P^{-1}R_{B, \alpha}. \qquad (36)$$

Clearly one can solve the decoupled systems in (35) and (36) for each $L_{z_n^{(i)}, \alpha}$ and $R_{z_n^{(i)}, \alpha}$, $i = 1, 2, \ldots, m$. From this we obtain the solution $L_{x_n^{(i)}, \alpha}$ and $R_{x_n^{(i)}, \alpha}$. In the case of decoupling $x_n^{(i)}$ is essentially a function of $x_{n-1}^{(i)}$. In the limit case, as $n \to \infty$, $L_{x_n, \alpha}$ and $R_{x_n, \alpha}$ will tend to $L_{x_\infty, \alpha}$ and $R_{x_\infty, \alpha}$. The conditions imposed on A and B will guarantee convergence. Thus, in the limit case, the solution of the fuzzy analog of (31) with A being a real matrix is

$$x_\infty(\cdot) = \sup_{0 < \alpha \leq 1} \{\alpha \wedge \chi_{[L_\infty, \alpha, R_\infty, \alpha]}(\cdot)\}. \qquad (37)$$

We will interpret this solution later. First we give an example to illustrate the method. Consider the system

$$\begin{bmatrix} x_n^{(1)} \\ x_n^{(2)} \end{bmatrix} = \begin{bmatrix} \frac{1}{5} & \frac{3}{5} \\ \frac{2}{5} & \frac{2}{5} \end{bmatrix} \begin{bmatrix} x_{n-1}^{(1)} \\ x_{n-1}^{(2)} \end{bmatrix} + \begin{bmatrix} 1 \\ 0 \end{bmatrix},$$

or, in a shorter form, $x_n = A x_{n-1} + B$ with proper identification of A and B. The eigenvalues of the matrix A are $-\frac{1}{5}$ and $\frac{4}{5}$. Compute the matrices

$$P = \begin{bmatrix} -3 & 1 \\ 2 & 1 \end{bmatrix} \quad \text{and} \quad D = \begin{bmatrix} -\frac{1}{5} & 0 \\ 0 & \frac{4}{5} \end{bmatrix}.$$

It follows from Eq. (35) that

$$L_{z_n, \alpha} = \begin{bmatrix} -\frac{1}{5} & 0 \\ 0 & \frac{4}{5} \end{bmatrix} L_{z_{n-1}, \alpha} + \begin{bmatrix} -\frac{1}{5} & \frac{1}{5} \\ \frac{2}{5} & \frac{3}{5} \end{bmatrix} L_{B, \alpha}.$$

This matrix equation can be written in the form of a linear system,

$$L_{z_n^{(1)}, \alpha} = -\tfrac{1}{5} L_{z_{n-1}^{(1)}, \alpha} - \tfrac{1}{5} L_{b^{(1)}, \alpha} + \tfrac{1}{5} L_{b^{(2)}, \alpha},$$

$$L_{z_n^{(2)}, \alpha} = \tfrac{4}{5} L_{z_{n-1}^{(1)}, \alpha} + \tfrac{2}{5} L_{b^{(1)}, \alpha} + \tfrac{3}{5} L_{b^{(2)}, \alpha}.$$

We find the solution of this system from (30),

$$L_{z_n^{(1)}, \alpha} = \left(-\tfrac{1}{5}\right)^n L_{z_0^{(1)}, \alpha} + \left(1 + \left(-\tfrac{1}{5}\right) + \cdots + \left(-\tfrac{1}{5}\right)^{n-1}\right)\left(-\tfrac{1}{5} L_{b^{(1)}, \alpha} + \tfrac{1}{5} L_{b^{(2)}, \alpha}\right),$$

$$L_{z_n^{(2)}, \alpha} = \left(\tfrac{4}{5}\right)^n L_{z_0^{(1)}, \alpha} + \left(1 + \left(\tfrac{4}{5}\right) + \cdots + \left(\tfrac{4}{5}\right)^{n-1}\right)\left(\tfrac{2}{5} L_{b^{(1)}, \alpha} + \tfrac{3}{5} L_{b^{(2)}, \alpha}\right).$$

We can write this result in matrix form,

$$L_{z_n, \alpha} = \begin{bmatrix} \left(-\dfrac{1}{5}\right)^n & 0 \\ 0 & \left(\dfrac{4}{5}\right)^n \end{bmatrix} L_{z_0, \alpha}$$

$$+ \begin{bmatrix} \dfrac{1 - \left(-\frac{1}{5}\right)^n}{\frac{6}{5}} & 0 \\ 0 & \dfrac{1 - \left(\frac{4}{5}\right)^n}{\frac{1}{5}} \end{bmatrix} \begin{bmatrix} -\dfrac{1}{5} & \dfrac{1}{5} \\ \dfrac{2}{5} & \dfrac{3}{5} \end{bmatrix} L_{B, \alpha}.$$

Likewise,

$$
R_{z_n, \alpha} = \begin{bmatrix} \left(-\dfrac{1}{5}\right)^n & 0 \\ 0 & \left(\dfrac{4}{5}\right)^n \end{bmatrix} R_{z_0, \alpha}
$$

$$
+ \begin{bmatrix} \dfrac{1 - \left(-\frac{1}{5}\right)^n}{\frac{6}{5}} & 0 \\ 0 & \dfrac{1 - \left(\frac{4}{5}\right)^n}{\frac{1}{5}} \end{bmatrix} \begin{bmatrix} -\dfrac{1}{5} & \dfrac{1}{5} \\ \dfrac{2}{5} & \dfrac{3}{5} \end{bmatrix} R_{B, \alpha}.
$$

From this we can determine $L_{x_n, \alpha}$ and $R_{x_n, \alpha}$. The solution x_n will then be given by $x_{n, \alpha} = [PL_{z_n, \alpha}, PR_{z_n, \alpha}]$, and in the limiting case $x_{\infty, \alpha} = [PL_{z_\infty, \alpha}, PR_{z_\infty, \alpha}]$.

Interpretation of solution (37).   Equation (37),

$$
x_\infty(\cdot) = \sup_{0 < \alpha \leq 1} \left\{ \alpha \wedge \chi_{[L_{\infty, \alpha}, R_{\infty, \alpha}]}(\cdot) \right\},
$$

expresses to what degree a number u is a solution of the corresponding system. Thus, given u, $x_\infty(u)$ is the degree to which u is the asymptotic solution of the fuzzy analog of (31). Note that if x_n, A, and B are real numbers, then

$$
[A]_\alpha = A, \qquad [B]_\alpha = B
$$

for all $\alpha > 0$, and

$$
\lim_{n \to \infty} L_{n, \alpha} = \lim_{n \to \infty} R_{n, \alpha} = u_\infty
$$

for all $\alpha > 0$. Then

$$
x_\infty(\cdot) = \sup_{0 < \alpha \leq 1} \left\{ \alpha \wedge \chi_{\{u_\infty\}}(\cdot) \right\}.
$$

This implies that

$$
x_\infty(u) = \begin{cases} 1, & \text{if } u = u_\infty, \\ 0, & \text{if } u \neq u_\infty. \end{cases}
$$

Therefore, $x_\infty(\cdot)$ can be identified with the point u_∞. Because u_∞ is the solution in the classical (nonfuzzy) case, it is shown that $x_\infty(\cdot)$ may be viewed as the generalization from the classical to the fuzzy case.

Case 2.   We will now consider the case where A is a matrix with fuzzy entries and cannot necessarily be decoupled. As noted earlier, if $\|A\| < 1$, then the limiting state is given by

$$x_\infty = (I - A)^{-1} B.$$

We propose to find the limiting state of (27) where A and B are matrices with fuzzy entries, thus reflecting some uncertainty as to the values of the parameters defining the system.

We shall denote by $\mathscr{A}$ an $m \times m$ matrix with fuzzy entries. $L_{\mathscr{A}, \alpha}$ and $R_{\mathscr{A}, \alpha}$ denote the $m \times m$ matrices formed by looking at the left and right endpoints of the α cuts of the entries of $\mathscr{A}$. If x_n is the state vector at time n, $L_{x_n, \alpha}$ and $R_{x_n, \alpha}$ denote the $m \times 1$ matrices obtained by taking the left and right endpoints of the α cuts of the coordinates of x_n. $[L_{x_n^{(i)}, \alpha}, R_{x_n^{(i)}, \alpha}]$ denotes the α cut of the ith component of x_n.

We assume the following conditions on matrix $\mathscr{A}$:

1. The entries of $\mathscr{A}$ are *triangular membership functions* reaching 1 at a_{ij}. [See (6) for any example of the triangular membership function.]
2. The norm of the matrix $\mathscr{A}$, $\|\mathscr{A}\|$, is less than 1.
3. The fuzzy entries of $\mathscr{A}$ are bounded away from 0, that is, $\mu_{ij}(x) > 0$ for $x > \varepsilon > 0$, where ε is fixed, (μ_{ij} is of course the characteristic function of the entry a_{ij}).

We shall show that (27) has a limiting state provided that not too much fuzziness is present; that is, the width of the triangular membership is not too large.

Consider two matrices A and B and a function $f: \Re^{2m} \to \Re$ defined by

$$f(t_1, t_2, \ldots, t_m; t_m + 1, \ldots, t_{2m}) = \sum_{k=1}^{m} t_k t_{n+k}.$$

Then $A \cdot B = C$, where $c_{ij} = f[a_{i1}, \ldots, a_{im}; b_{1j}, \ldots, b_{mj}] = \sum_{k=1}^{m} a_{ik} b_{kj}$.

We now use the extension theorem to define the (i, j)th entry of the fuzzy matrix $\mathscr{C} = \mathscr{A} \cdot \mathscr{B}$, where $\mathscr{A}$ and $\mathscr{B}$ are matrices with fuzzy entries,

$$c_{ij}(c) = f[a_{i1}, \ldots, a_{im}; b_{1j}, \ldots, b_{mj}](c)$$

$$= \sup\{a_{i1}(\xi_1) \wedge \cdots \wedge a_{im}(\xi_m) \wedge b_{1j}(\eta_1) \wedge \cdots \wedge b_{mj}(\eta_m)\},$$

where the sup is over all $\xi_1, \ldots, \xi_m, \eta_1, \ldots, \eta_m$ such that $\sum_k \xi_k \eta_k = c$.

The fuzzy matrices $\mathscr{A}$ and $\mathscr{B}$ have positive fuzzy entries bounded away from 0; that is,

$$a_{ik}(\xi_k) > 0 \qquad \text{for } \xi_k > \varepsilon > 0,$$

$$b_{kj}(\eta_k) > 0 \qquad \text{for } \eta_k > \varepsilon > 0.$$

All the conditions necessary for Lemmas 1 and 2 are satisfied. Hence by Nguyen's theorem [11]

$$[c_{ij}(c)]_\alpha = \sum_{k=1}^m [a_{ik}]_\alpha [b_{kj}]_\alpha,$$

or if $[c_{ij}(c)]_\alpha = [\mathscr{L}_{c_{ij}\alpha}, \mathscr{R}_{c_{ij}\alpha}]$,

$$\mathscr{L}_{c_{ij}\alpha} = \sum_{k=1}^m L_{a_{ik}\alpha} L_{b_{kj}\alpha} \quad \text{and} \quad \mathscr{R}_{c_{ij}\alpha} = \sum_{k=1}^m R_{a_{ik}\alpha} R_{b_{kj}\alpha}.$$

In other words, for such "positive and bounded away from 0" matrices, the product is a fuzzy matrix where α-cut endpoints are obtained by taking the regular matrix product of left endpoints and the regular product of right endpoints.

Consider the following fuzzy analog of (27), $x_{n+1} = \mathscr{A}x_n + B$. Then

$$L_{n+1,\alpha} = L_{\mathscr{A},\alpha} L_{n,\alpha} + L_{B,\alpha}, \tag{38}$$

$$R_{n+1,\alpha} = R_{\mathscr{A},\alpha} R_{n,\alpha} + R_{B,\alpha}, \tag{39}$$

where $L_{0,\alpha}$ and $R_{0,\alpha}$ are known. Recursively, we obtain

$$\begin{aligned}
L_{n+1,\alpha} &= L_{\mathscr{A},\alpha} L_{n,\alpha} + L_{B,\alpha} \\
&= L_{\mathscr{A},\alpha} [L_{\mathscr{A},\alpha} L_{n-1,\alpha} + L_{B,\alpha}] + L_{B,\alpha} \\
&= \cdots.
\end{aligned}$$

Therefore,

$$L_{n+1,\alpha} = L_{\mathscr{A},\alpha}^{n+1} L_{0,\alpha} + [I + L_{\mathscr{A},\alpha} + \cdots + L_{\mathscr{A},\alpha}^n] L_{B,\alpha}.$$

A similar expression holds for $R_{n+1,\alpha}$.

In the limiting case if $\|L_{\mathscr{A},\alpha}\| < 1$ and $\|R_{\mathscr{A},\alpha}\| < 1$, then $I - L_{\mathscr{A},\alpha}$ and $I - R_{\mathscr{A},\alpha}$ have inverses and

$$L_{\infty,\alpha} = [I - L_{\mathscr{A},\alpha}]^{-1} L_{B,\alpha},$$

$$R_{\infty,\alpha} = [I - R_{\mathscr{A},\alpha}]^{-1} R_{B,\alpha}.$$

Hence

$$[L_{\infty,\alpha}, R_{\infty,\alpha}] = \begin{bmatrix} [x_\infty^{(1)}]_\alpha \\ \vdots \\ [x_\infty^{(m)}]_\alpha \end{bmatrix}.$$

Thus the limiting state x_∞ is defined by

$$x_\infty^{(i)}(\cdot) = \sup_{0 < \alpha \le 1} \left\{ \alpha \wedge \chi_{[x_\infty^{(i)}]_\alpha}(\cdot) \right\}. \tag{40}$$

Now, when are $I - L_{\mathscr{A}, \alpha}$ and $I - R_{\mathscr{A}, \alpha}$ invertible for all α? If the entries of $\mathscr{A}$ are triangular fuzzy numbers reaching membership 1 at a_{ij} and if a_{ij} are assumed to be such that $A = (a_{ij})$ has $\|A\| < 1$, then if the width of the triangle is small enough, by continuity $L_{\mathscr{A}, \alpha}$ and $R_{\mathscr{A}, \alpha}$ will have norm less than 1, so $I - L_{\mathscr{A}, \alpha}$ and $I - R_{\mathscr{A}, \alpha}$ will be invertible.

Solution (40) may also be interpreted in the same manner as the interpretation given to solution (37).

VII. CONCLUSION

We have presented a technique for dealing with the fuzzy analog of functional and difference equations. The same technique may be applied in the study of many other diverse systems that arise in the formulation of problems in engineering, information theory, decision theory, and computer sciences.

REFERENCES

1. Deeba, E. and de Korvin, A. On a fuzzy difference equation. *IEEE Trans. Fuzzy Systems* 3(3):469–472, 1995.
2. Deeba, E., de Korvin, A., and Koh, E. L. On a fuzzy logistic difference equation. *Differential Equations Dynam. Systems* 4(2):149–156, 1996.
3. Deeba, E. and de Korvin, A. A fuzzy difference equation with an application. *J. Differ. Equations Appl.* 2:365–374, 1996.
4. Deeba, E., de Korvin, A., and Xie, S. Pexider functional equations—their fuzzy analogs. *Internat. J. Math. Math. Sci.* 19(3):529–538, 1996.
5. Aczel, J. *Lectures on Functional Equations and Their Applications*. Academic Press, New York, 1966.
6. Eldestein-Keshet, L. *Mathematical Models in Biology*. McGraw-Hill, New York, 1987.
7. Aczél, J. *Functional Equations in Several Variables*. Cambridge Univ. Press, Cambridge, UK, 1989.
8. Agarwal, R. P. *Difference Equations and Inequalities—Theory, Methods and Applications*. Dekker, New York, 1992.
9. Dubois, D. and Prade H. *Fuzzy Sets and Systems: Theory and Applications*. Academic Press, New York, 1980.
10. Zadeh, L. A. Fuzzy sets. *Inform. Control* 8:338–353, 1965.
11. Nguyen, H. T. A note on the extension principle for fuzzy sets. *J. Math. Anal. Appl.* 64(2):369–380, 1978.
12. Baas, S. *Computer Algorithms*. Addison-Wesley, New York, 1978.
13. Dubois, D. and Prade, H. Operations on fuzzy numbers. *Internat. J. Systems Sci.* 9:613–626, 1978.
14. Fuller, W. A. *Measurement Error Models*. Wiley, New York, 1987.
15. Godunov, S. K. and Ryabenku, V. S. *Difference Schemes*. North-Holland, Amsterdam, 1987.
16. Hersch, H. M. and Caramazza, A. A fuzzy-set approach to modifiers and vagueness in natural languages. *J. Experiment. Psychol. General* 105:254–276, 1976.

17. Kaleva, O. Fuzzy performance of a coherent system. *J. Math. Anal. Appl.* 117:234–246, 1986.
18. Kreinovich, V. With what accuracy can we measure masses if we have an (approximately known) mass standard. Technical Report UTEP-CS-93-40, Computer Science Department, University of Texas–El Paso, 1993.
19. Lakshmikantham, V. and Frigiante, D. *Theory of Difference Equations*. Academic, New York, 1990.
20. Oden, G. C. Integration of fuzzy logical information. *J. Experiment. Psychol. Human Perception Perform* 3(4):565–575, 1977.
21. Zadeh, L. A. The concept of a linguistic variable and its applications to approximate reasoning. Parts I, II, and III. *Inform. Sci.* 8:199–249; 8:301–357; 9:43–80, 1975.
22. Zimmerman, H. J. Results of empirical studies in fuzzy set theory. In *Applied General System Research* (G. J. Klir, Ed.), pp. 303–312. Plenum, New York, 1978.

5

TECHNIQUES IN NEURAL-NETWORK-BASED FUZZY SYSTEM IDENTIFICATION AND THEIR APPLICATION TO CONTROL OF COMPLEX SYSTEMS

YAOCHU JIN

Department of Electrical Engineering, Zhejiang University,
Hangzhou 310027, People's Republic of China and
Institute for Neuroinformatics, Ruhr-University Bochum,
D-44780 Bochum, Germany

JINGPING JIANG

Department of Electrical Engineering, Zhejiang University,
Hangzhou 310027, People's Republic of China

I. INTRODUCTION 112
II. TAKAGI AND SUGENO'S FUZZY MODEL 113
III. NEURAL-NETWORK-BASED IDENTIFICATION OF FUZZY SYSTEMS 114
    A. Hybrid Neural Networks 114
    B. Identification Algorithms Based on Neural Networks 115
IV. INTERPRETABILITY CONSIDERATIONS 119
V. AN APPLICATION EXAMPLE: DECOUPLED CONTROL OF ROBOT MANIPULATORS 120
    A. Decoupling of the Robot Dynamics 121
    B. Simulation Study 123
VI. CONCLUSIONS 126
    REFERENCES 128

Identification of fuzzy systems with artificial neural networks is discussed in this chapter. By using updated version of the pi–sigma neural network, both premise and consequent parameters of the fuzzy system can be efficiently identified online or offline. Learning algorithms for both Gaussian and triangular forms of membership functions are presented. The consequent part of the fuzzy rules is represented by a subnetwork, which enables the algorithm to be applicable to high-order Takagi–Sugeno fuzzy systems. Some measures are taken to preserve the interpretability of the fuzzy system in the course of learning. The proposed method is applied to the nonlinear decoupled control of robot manipulators and satisfactory simulation results are obtained.

I. INTRODUCTION

Two basic forms of fuzzy rules, namely, Mamdani fuzzy rules and Takagi–Sugeno fuzzy rules, have been developed to date. The main difference between these two types of fuzzy rules lies in the fact that the consequent part of the Takagi–Sugeno fuzzy rules is normally a concrete linear function of input variables instead of some fuzzy linguistic variables. Generally speaking, Takagi–Sugeno fuzzy rule systems are more flexible and thus have stronger modeling capability to solve some complex problems. Theoretically, fuzzy rules can be built based either on expert knowledge or on a group of observed data. However, it is difficult, if not impossible, for human beings to establish an acceptable fuzzy rule system when the input dimension of the system becomes high. Therefore, building fuzzy rule systems on the basis of collected data is becoming more and more important.

Here we limit our attention to identification of Takagi–Sugeno fuzzy rule systems. It is noticed that identification methods for such fuzzy systems have been proposed by Takagi and Sugeno [1, 2]. The algorithms are quite complicated and are mainly suitable for fuzzy rules with piecewise linear fuzzy membership functions and linear consequents. Moreover, their algorithms have difficulties in real-time implementation, which has limited its application seriously.

Methodologies in artificial neural networks, fuzzy systems, and evolutionary computation have been successfully combined and new techniques called soft computing or computational intelligence have been developed. These techniques are attracting more and more attention in several research fields because they are able to tolerate a wide range of uncertainty. Use of neural networks to perform the adjustment of membership functions and modification of the consequent of fuzzy rules makes it practical to design adaptive fuzzy models and self-organizing fuzzy control. Different neural networks, such as backpropagation networks [3], radial basis function (RBF) neural networks [4], hybrid pi–sigma networks [5], B-spline networks [6], and neural-like structure [7], have been applied to adaptation of fuzzy membership functions and consequent parameters.

With all these successes, it is also necessary to point out that some important features of fuzzy systems have been lost. One common problem for most neuro-fuzzy algorithms is that the interpretability of fuzzy systems is deteriorated. After adaptation, either the distinguishability among the fuzzy subsets in a fuzzy partitioning is blurred, or the fuzzy partitionings of the input space are incomplete. In fact, most neuro-fuzzy schemes have failed to pay attention to preserving the rule structure of the fuzzy systems in the course of adaptation. This problem has been alleviated in [5] by introducing the hill-climbing method. Further discussions on the interpretability of fuzzy systems, including completeness and consistency considerations, can be found in [8] and [9].

This chapter is an extension of the work in [5]. A systematic approach to identification of Takagi–Sugeno fuzzy systems is described. The algorithm has been applied to a wide range of modeling and control problems, and inspiring results have been achieved.

II. TAKAGI AND SUGENO'S FUZZY MODEL

Takagi and Sugeno [1–2] proposed a new fuzzy model by replacing the linguistic variables in the THEN part of the fuzzy rules with a crisp linear function of the input variables. Because a multi-input multi-output (MIMO) fuzzy systems can always be separated into a group of multi-input single-output (MISO) systems, we discuss here only MISO fuzzy systems, without loss of generality. Given an MISO system with n inputs, the Takagi–Sugeno fuzzy rules have the following form:

$$R_i: \quad \text{If } x_1 \text{ is } A_1^i \text{ and} \dots \text{and } x_n \text{ is } A_n^i, \text{ then } y^i = p_0^i + p_1^i x_1 + \cdots + p_n^i x_n, \quad (1)$$

where R_i $(i = 1, 2, \dots, N)$ denotes the ith rule, x_j $(j = 1, \dots, n)$ are the premise variables, A_j^i are the fuzzy subsets defined by corresponding piecewise linear membership functions such as triangle or trapezoid, and y^i is the consequent of the ith rule. According to Takagi and Sugeno, the final output of the fuzzy system can be written in the following form:

$$y = \frac{\sum_{i=1}^{N} (w^i y^i)}{\sum_{i=1}^{N} w^i}, \quad (2)$$

where w^i is calculated by

$$w^i = \mathop{t}_{j=1}^{n} A_j^i(x_j), \quad (3)$$

where t represents the t-norm operator. Currently, a number of fuzzy t-norms, as well as some extended forms [10] have been proposed. Discussions on some widely used t-norms are given in [11] and [12] among others. In this work, the most widely used minimum operator is considered.

If some pairs of input and output data are available, the parameters in the THEN part of the rules can be estimated using the least square method. However, estimation of the parameters of the membership functions are a little complicated, even if the membership functions are supposed to be piecewise linear [1–2].

The Takagi–Sugeno fuzzy model is believed to be more flexible than the Mamdani fuzzy model and has found wider application. Nevertheless, two general shortcomings still exist. First, identification of the fuzzy system is not trivial and therefore it is hard to apply the fuzzy system to real-time systems. Second, not only are the membership functions limited to piecewise functions, but the consequent part is also assumed to be linear. This problem remains unsolved until neural networks are combined systematically with fuzzy systems and the so-called neuro-fuzzy system theory appears. Such hybrid systems are preferred because they possess the main features of both fuzzy and neural systems. That is to say, they have clear physical meanings and are easy to interpret like conventional fuzzy systems; on the other hand, they have good learning ability and nonlinear mapping capacity. Despite that, special attention should be paid in the course of learning so that both of the merits could be preserved.

III. NEURAL-NETWORK-BASED IDENTIFICATION OF FUZZY SYSTEMS

A. Hybrid Neural Networks

Most researchers use multiplication as the t-norm so that the conventional multiple-layered perceptions or RBF neural networks can be directly implemented. In fact, the Gaussian-based RBF neural networks with some minor conditions are shown to be mathematically equivalent to fuzzy systems. One condition is that the multiplication operator should be used as the fuzzy t-norm. In our work, however, the minimum operator is kept. To this end, the neural network used to identify the fuzzy system should contain not only summing and multiplication neurons, but also fuzzy neurons that are able to perform fundamental fuzzy operations such as the minimum operation. Therefore, a hybrid neural network structure inspired by the pi–sigma neural network [13] is suggested. For clarity, a neural network with two inputs is illustrated in Fig. 1, where a shaded circle denotes a summing neuron with nonlinear activation function, the blank circle represents a neuron without nonlinear activation, Π means the product neurons, and Λ represents the minimum neurons. From Fig. 1, it can also be noticed that the consequent part of the fuzzy rules is implemented by a feedforward subnetwork. Suppose the subnetwork has one hidden layer with H neurons. The overall output of the hybrid neural network can be written

$$y = \frac{\sum_{i=1}^{N}(w^i y^i)}{\sum_{i=1}^{N} w^i}, \tag{4}$$

$$y^i = p_i^{(0)} + \sum_{k=1}^{H}\left(p_{ki}^{(2)} \cdot g\left(\sum_{j=1}^{n} p_{jk}^{(1)} x_j \right)\right), \tag{5}$$

$$w^i = \min_{j=1}^{n} A_j^i(x_j), \tag{6}$$

where $g(\cdot)$ is a sigmoid function and $\mathbf{p}^{(0)}$, $\mathbf{p}^{(1)}$, and $\mathbf{p}^{(2)}$ are the weights in the subnetwork. Notice that the output layer of the subnetwork is linear.

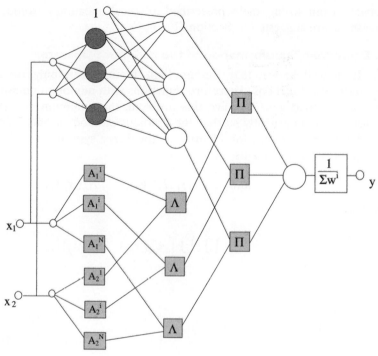

FIGURE 1   The structure of the hybrid neural network.

The hybrid neural network has very clear physical meanings. It is easy to observe that the neural system described by Eqs. (4)–(6) is functionally equivalent to the Takagi–Sugeno fuzzy system expressed in Eqs. (2) and (3), except for the fact that the linear consequent functions in the Takagi–Sugeno fuzzy model has been extended to general nonlinear functions described by a neural network. In addition, the membership functions can be arbitrary continuous functions that satisfy the definitions for fuzzy memberships. In practice, triangular, trapezoidal, and Gaussian functions are most widely used, although spline [14] and polynomial [15] functions have also been suggested as membership functions. However, they sometimes do not satisfy the definitions for fuzzy memberships, and they are in most cases not normal.

B. Identification Algorithms Based on Neural Networks

Because the hybrid neural network system consists of minimal nodes that are not differential, the gradient-method-based backpropagation algorithm cannot be directly applied. There are some alternatives to deal with this problem. In this work, two different approaches are suggested. The first approach carries out an equivalent transformation of the minimum operator, so that it can be treated as differential and thus the gradient method can be applied. The other approach uses a one-step backward searching algorithm to avoid the differential operation on the minimum nodes. To prevent the fuzzy subsets from losing their prescribed physical meanings, some additional measures are suggested in Section III.

1. Equivalent Transformation of the Minimum Operator

It is well known that the gradient method can only be applied to differentiable functions. Therefore, functions with minimum operators do not satisfy this condition. To solve this problem, the minimum operator will be transformed equivalently so that the gradient method can be used to derive the learning algorithm for identifying the fuzzy system. For the minimum operation with n elements,

$$w^i = \min\{A_1^i(x_1), A_2^i(x_2), \ldots, A_n^i(x_n)\}, \tag{7}$$

we have

$$w^i = \sum_{j=1}^{n} \prod_{m \neq j} \bigcup \left[A_m^i(x_m) - A_j^i(x_j) \right], \tag{8}$$

where

$$\bigcup \left[A_m^i(x_m) - A_j^i(x_j) \right] = \begin{cases} 1, & A_m^i(x_m) > A_j^i(x_j), \\ 0, & A_m^i(x_m) \leq A_j^i(x_j). \end{cases} \tag{9}$$

In this way, the learning algorithm can be derived based on the gradient method. Let y^d be the desired output of the system, and define the following quadratic cost function:

$$E = \tfrac{1}{2}(y - y^d)^2. \tag{10}$$

Thus,

$$\frac{\partial E}{\partial p_i^{(0)}} = \frac{\partial E}{\partial y} \frac{\partial y}{\partial y^i} \frac{\partial y^i}{\partial p_i^{(0)}} = \frac{(y - y^d)w^i}{\sum_{i=1}^{N} w^i}, \tag{11}$$

$$\frac{\partial E}{\partial p_{jk}^{(1)}} = \frac{\partial E}{\partial y} \frac{\partial y}{\partial y^i} \frac{\partial y^i}{\partial p_{jk}^{(1)}} = \frac{(y - y^d)w^i \sum_{k=1}^{H} \{p_{ki}^{(2)} g'(\cdot) x_j\}}{\sum_{i=1}^{N} w^i}, \tag{12}$$

$$\frac{\partial E}{\partial p_{ki}^{(2)}} = \frac{\partial E}{\partial y} \frac{\partial y}{\partial y^i} \frac{\partial y^i}{\partial p_{ki}^{(2)}} = \frac{(y - y^d)w^i g(\cdot)}{\sum_{i=1}^{N} w^i}. \tag{13}$$

According to Eqs. (11)–(13), the learning algorithm of the consequent parameters could be expressed as follows:

$$\Delta p_i^{(0)} = -\eta \delta_1^i, \tag{14}$$

$$\Delta p_{jk}^{(1)} = -\eta \sum_{k=1}^{H} \{p_{ki}^{(2)} g'(\cdot) x_j\} \delta_1^i, \tag{15}$$

$$\Delta p_{ki}^{(2)} = -\eta g(\cdot) \delta_1^i, \tag{16}$$

where η is a positive learning rate and the general error δ_1^i is defined by

$$\delta_1^i = \frac{(y - y^d)w^i}{\sum_{i=1}^{N} w^i}. \tag{17}$$

Next, we will derive the learning algorithms for the membership parameters. As indicated, Gaussian and triangular functions are the most widely used fuzzy membership functions. Due to this, we first suppose the membership functions are Gaussians in the following form:

$$A_j^i(x_j) = \exp\left(-\left(x_j - a_j^i\right)^2 / b_j^i\right). \tag{18}$$

Thus we have

$$\frac{\partial E}{\partial a_j^i} = \frac{\partial E}{\partial y} \frac{\partial y}{\partial w^i} \frac{\partial w^i}{\partial A_j^i(x_j)} \frac{\partial A_j^i(x_j)}{\partial a_j^i}, \tag{19}$$

because

$$\frac{\partial A_j^i(x_j)}{\partial a_j^i} = \frac{\partial\left\{\exp\left(-\left(x_j - a_j^i\right)^2 / b_j^i\right)\right\}}{\partial a_j^i} = \frac{2A_j^i(x_j)\left(x_j - a_j^i\right)}{b_j^i}. \tag{20}$$

Recalling the transformation of the minimum operation in Eqs. (8) and (9), it is straightforward that

$$\frac{\partial w^i}{\partial A_j^i(x_j)} = \frac{\partial\left\{\sum_{j=1}^{n} A_j^i(x_j) \Pi_{m \neq j} \cup \left[A_m^i(x_m) - A_j^i(x_j)\right]\right\}}{\partial A_j^i(x_j)}$$

$$= \bigcup_{m \neq j} \left[A_m^i(x_m) - A_j^i(x_j)\right] = \begin{cases} 1, & \text{if } A_j^i(x_j) \text{ minimum,} \\ 0, & \text{else.} \end{cases} \tag{21}$$

Substitute Eqs. (20) and (21) into Eq. (19), and if we define

$$\delta_2^i = \frac{(y - y^d)(y^i - y)}{\sum_{i=1}^{N} w^i} \tag{22}$$

then the learning algorithm for the membership parameters a_j^i can be written

$$\Delta a_j^i = \begin{cases} -2\xi\left(x_j - a_j^i\right)w^i \delta_2^i / b_j^i, & \text{if } A_j^i(x_j) \text{ minimum,} \\ 0, & \text{else.} \end{cases} \tag{23}$$

Similarly, the learning algorithm for b_j^i is

$$\Delta b_j^i = \begin{cases} -\xi w^i\left(x_j - a_j^i\right)^2 \delta_2^i / \left(b_j^i\right)^2, & \text{if } A_j^i(x_j) \text{ minimum,} \\ 0, & \text{else,} \end{cases} \tag{24}$$

where ξ is a positive learning rate.

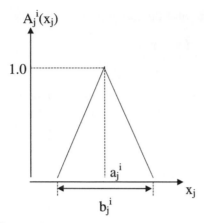

FIGURE 2 Isosceles triangulars as membership functions.

Now we consider the situation where the membership functions are triangular. For the sake of simplicity, we assume the membership functions are all isosceles as shown in Fig. 2. In this case, it can be described by

$$A_j^i(x_j) = 1 - \frac{2|x_j - a_j^i|}{b_j^i}.$$ (25)

Hence, the learning algorithm of a_j^i and b_j^i can be described by the following equations:

$$\Delta a_j^i = \begin{cases} -2\xi \, \mathrm{sgn}(x_j - a_j^i)\delta_2^i/b_j^i, & \text{if } A_j^i(x_j) \text{ minimum,} \\ 0, & \text{else,} \end{cases}$$ (26)

$$\Delta b_j^i = \begin{cases} -\xi[1 - w^i]\delta_2^i/b_j^i, & \text{if } A_j^i(x_j) \text{ minimum,} \\ 0, & \text{else.} \end{cases}$$ (27)

2. One-Step Backward Searching Algorithm

In the previously described method, the minimum operation is equivalently transformed so the gradient method can be used to derive the learning algorithms. However, if we have a close look at the transformation, we find that this transformation leads the gradient search toward the input node of the fuzzy neuron that has the minimum value. This works if the initial distribution of the membership functions agree with the real distributions approximately. However, if the initial distributions have significant differences with the real distributions, this algorithm may fail to find the optimal solutions. To expound this, consider a fuzzy neuron that implements the minimum operation as in Fig. 3, where we have

$$O = \min\{A, B, C\}.$$ (28)

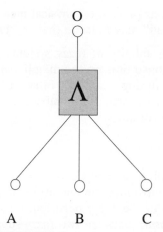

FIGURE 3   Illustration of a fuzzy neuron.

If $A = 0.5$, $B = 0.4$, and $C = 0.3$, then $O = \min(A, B, C) = 0.3$. Suppose the desired value of the node output is 0, then the error will be -0.3. According to the learning algorithm developed here, the input weight of connection C will be adjusted because C has the minimum value. However, it is easy to notice that the error of the fuzzy neuron is not necessarily caused by weight C. It may be the case that the truth value of B should be 0 and the error is fully caused by input weight B. To cope with this situation, we introduce the hill-climbing searching method, which is a counterpart of the gradient method and does not require the differentiability of the cost function. The learning algorithm will be carried out in the following two phases. At first, all the input weights A, B, and C are updated simultaneously if there exists an error of e at the output O. Then comparisons are made to see which modification reduces the error most significantly. As supposed, if the error before learning is -0.3 and after adjusting weights A, B, and C, the new errors are -0.25, -0.05, and -0.2, respectively. It is found that the smallest error is obtained by updating weight B. Consequently, only weight B is actually modified.

IV. INTERPRETABILITY CONSIDERATIONS

Given the learning algorithms introduced previously, the rule parameters can be identified online without any difficulty. However, if no other constraints are imposed, some problems may appear, especially when adjusting the membership parameters. The problems that appear most frequently are:

- Two neighboring fuzzy subsets in a fuzzy partition have no overlap and consequently the partition is incomplete.
- The membership functions of two fuzzy subsets are so similar that the distinguishability of the fuzzy partition is lost. This not only can make the fuzzy system unnecessarily complicated, but also can give rise to difficulties in assigning suitable physical meanings to the membership functions.

- The membership functions lose their prescribed physical meanings. For example, the center value of "small" may precede that of "big."

All these phenomena harm the interpretability of fuzzy systems. To avoid these problems, we suggest here some constraints on the membership parameters that should be checked online during learning. First, to keep the prescribed meanings, the center values of each fuzzy subset in one fuzzy partitioning should satisfy the following condition:

$$a_i^1 < a_i^2 < \cdots < a_i^n, \tag{29}$$

where we suppose the fuzzy partition of x_i is composed of n fuzzy subsets $\{A_i^1(x_i), A_i^2(x_i), \ldots, A_i^n(x_i)\}$. Before we introduce the conditions to guarantee the completeness and distinguishability of the fuzzy partition, we first introduce the concept of *fuzzy similarity measure* between two fuzzy subsets:

$$\text{FSM}(A, B) = \frac{M(A \cap B)}{M(A) + M(B) - M(A \cap B)}, \tag{30}$$

where $M(A)$ is called the size of fuzzy set A and is calculated as

$$M(A) = \int_{x \in U} A(x)\, dx, \tag{31}$$

where U is the universe of discourse of x. Notice that if $\text{FSM}(A, B) = 0$, the two fuzzy sets have no overlap; on the contrary, they are completely equal if $\text{FSM}(A, B) = 1$. Therefore, the fuzzy partition is complete if any of its two neighboring fuzzy subsets satisfy

$$\text{FSM}(A_i^j, A_i^{j+1}) > 0. \tag{32}$$

To ensure good distinguishability of a fuzzy partition, the following condition should hold for two arbitrary fuzzy subsets in the fuzzy partition:

$$\text{FSM}(A_i^j, A_i^k) < \delta, \tag{33}$$

where $0 < \delta < 1$ is a constant to be determined.

V. AN APPLICATION EXAMPLE: DECOUPLED CONTROL OF ROBOT MANIPULATORS

Dynamic control of robot manipulators is a challenging task in the field of system control. Various modern control strategies have been widely investigated to deal with the high nonlinearity and strong coupling of the robot dynamics. In this section, we try to control the robot manipulators using Takagi–Sugeno fuzzy models.

Although Takagi–Sugeno fuzzy rules are believed to have stronger mapping ability than the Mamdani fuzzy rules, they are much more complex, especially when the input dimension is high. For example, for a rigid robot with N degrees of freedom, there are $3N$ input parameters (link position, velocity, and acceleration) and if each input is divided into six fuzzy sub-spaces, there will be 6^{3N} fuzzy rules in total. Suppose first-order Takagi–Sugeno rules are adopted. Then each rule consists of $N \times (3N + 1)$ consequent parameters and consequently there are $N(3N + 1)6^{3N}$ parameters to be estimated. If $N = 6$, the number is about 1.16×10^{16}, which is very huge and makes it impossible for real-time implementation. It is therefore sensible to decouple the robot dynamics before we apply the Takagi–Sugeno fuzzy model to robot control. In the decoupled robot dynamics, there are only two input variables, namely, position and velocity, for each link. In this case, the number of parameters to be estimated will be greatly reduced. In the preceding example, only $N \times 6^2$ fuzzy rules are needed, each consisting of three consequent parameters. When N is 6, the total number of parameters is only 648.

A. Decoupling of the Robot Dynamics

Two main approaches are used by most researchers to derive the dynamics of robot manipulators, namely, Lagrange–Euler and Newton–Euler formulations. From the control point of view, the Lagrange–Euler formulation is very desirable. For an N-degree-of-freedom rigid robot, the Lagrange equation of motion is as follows:

$$\tau = H(q)\ddot{q} + M(q, \dot{q}) + G(q), \tag{34}$$

where τ is the N-dimensional torque vector, $H(q)$ is the $N \times N$ inertia matrix, $M(q, \dot{q})$ is the N-dimensional coriolis and centrifugal force vector, $G(q)$ is the N-dimensional gravity vector, and q, $\dot{q}$, and $\ddot{q}$ are N-dimensional angular, velocity, and acceleration, respectively. Let $x_i = q_i$, $x_{N+i} = \dot{q}_i$, ($i = 1, 2, \ldots, N$), then the robot dynamics can be written

$$\dot{X} = A(X) + B(X)U, \tag{35}$$

$$Y = C(X), \tag{36}$$

where $X = [X_1 X_2]^T = [x_1, \ldots, x_N, x_{N+1}, \ldots, x_{2N}]^T$, $U = [\tau_1, \ldots, \tau_N]^T$, $C(X) = X_1$, and

$$A(X) = \begin{bmatrix} X_2 \\ -H^{-1}(M + G) \end{bmatrix}, \quad B(X) = \begin{bmatrix} 0 \\ -H^{-1} \end{bmatrix}. \tag{37}$$

Define the following operator [16]:

$$N_A^j C_i(X) = \left[\frac{\partial}{\partial X} N_A^{j-1} C_i(X) \right] A(X), \quad j = 1, 2, \ldots, N - 1, \tag{38}$$

$$N_A^0 C_i(X) = C_i(X), \tag{39}$$

where $C_i(X)$ is the ith row of $C(X)$. Define the relative degree of the system:

$$d_i = \min_j \left\{ \left[\frac{\partial}{\partial X} N_A^{j-1} C_i(X) \right] B(X) \neq 0 \right\}, \qquad j = 1, 2, \ldots, N. \quad (40)$$

Then we have the following control that decouples the robot dynamics:

$$U = F(X) + G(X)V, \quad (41)$$

where V is the new control vector of the decoupled linear system, and

$$F(X) = -(D^*)^{-1}(X)(F_1^*(X) + F_2^*(X)), \quad (42)$$

$$G(X) = -(D^*)^{-1}(X)\Lambda, \quad (43)$$

$$D_i^*(X) = \frac{\partial}{\partial X}\left[N_A^{d_i} C_i(X) \right] B(X), \quad (44)$$

$$F_{1i}^*(X) = N_A^{d_i} C_i(X), \quad (45)$$

$$F_{2i}^*(X) = \sum_{k=1}^{d_i-1} a_{k,i} N_A^k C_i(X), \quad (46)$$

$$\Lambda = \mathrm{diag}[\lambda_1, \ldots, \lambda_N], \quad (47)$$

where $D_i^*(X)$, $F_{1i}^*(X)$, and $F_{2i}^*(X)$ are the ith row of matrix $D^*(X)$, $F_1^*(X)$, and $F_2^*(X)$, respectively, and $a_{k,i}$ are some constants to be determined. For the robot system given in Eq. (35), because

$$\frac{\partial}{\partial X}\left[N_A^0 C_i(X) \right] B(X) = 0 \quad (48)$$

and

$$\frac{\partial}{\partial X}\left[N_A^1 C_i(X) \right] B(X) \neq 0, \quad (49)$$

the relative degree of the system is

$$d_i = 2, \qquad i = 1, 2, \ldots, N. \quad (50)$$

Thus we have the following decoupled linear model for the robot system:

$$\begin{bmatrix} \dot{x}_i \\ \vdots \\ \dot{x}_N \\ \dot{x}_{N+1} \\ \vdots \\ \dot{x}_{2N} \end{bmatrix} = \begin{bmatrix} x_{N+1} \\ \vdots \\ x_{2N} \\ -a_{0,1}x_1 - a_{1,1}x_N \\ \vdots \\ -a_{0,N}x_N - a_{1,N}x_{2N} \end{bmatrix} + \begin{bmatrix} & & 0 \\ \lambda_1 & & \\ & \ddots & \\ & & \lambda_N \end{bmatrix} \begin{bmatrix} v_1 \\ \vdots \\ v_N \end{bmatrix}. \quad (51)$$

It is straightforward that the subsystem for each link is a time-constant two-input single-output linear system. The parameters of the linear system should be chosen in such a way that the linear subsystems are stable.

As mentioned previously, the decoupled linear systems are time-constant if the dynamics of the robot is exactly known. In this case, a conventional PD controller will perform successfully. However, there are always parameter errors in real robotic systems. Moreover, it is difficult to model such dynamics as nonlinear friction, backlash, and other uncertainties in robot systems. Therefore, adaptive fuzzy controllers are necessary to deal with the uncertainties.

B. Simulation Study

For the sake of simplicity, a two-degree-of-freedom rigid manipulator is studied in this simulation. The dynamics of the system is expressed as follows:

$$
\begin{bmatrix} \tau_1 \\ \tau_2 \end{bmatrix} = \begin{bmatrix} H_{11} & H_{12} \\ H_{21} & H_{22} \end{bmatrix} \begin{bmatrix} \ddot{q}_1 \\ \ddot{q}_2 \end{bmatrix} + \begin{bmatrix} M_1 \\ M_2 \end{bmatrix} + \begin{bmatrix} G_1 \\ G_2 \end{bmatrix},
\tag{52}
$$

where

$$
H_{11} = (m_1 + m_2)l_1^2 + m_2 l_2^2 + 2m_2 l_1 l_2 \cos(q_2),
\tag{53}
$$

$$
H_{12} = H_{21} = m_2 l_2^2 + m_2 l_1 l_2 \cos(q_2),
\tag{54}
$$

$$
H_{22} = m_2 l_2^2,
\tag{55}
$$

$$
M_1 = -2m_2 l_1 l_2 \sin(q_2)\dot{q}_1\dot{q}_2 - m_2 l_1 l_2 \sin(q_2)\dot{q}_2^2,
\tag{56}
$$

$$
M_2 = m_2 l_1 l_2 \sin(q_2)\dot{q}_1^2,
\tag{57}
$$

$$
G_1 = (m_1 + m_2)gl_1 \cos(q_1) + m_2 gl_2 \cos(q_1 + q_2),
\tag{58}
$$

$$
G_2 = m_2 gl_2 \cos(q_1 + q_2),
\tag{59}
$$

where m_1 and m_2 are the mass of the two links, l_1 and l_2 are the link lengths, and g is the gravity. Without loss of generality, the parameters of the linearized model are chosen as follows:

$$
0.1\ddot{q}_i + \dot{q}_i = V_i, \qquad i = 1, 2.
\tag{60}
$$

The two inputs are $x_1^i = \dot{q}_i$ and $x_2^i = q_i^d - q_i$, and the output is $y_i = v_i$ $(i = 1, 2)$. The diagram of the control system is shown in Fig. 4. The desired trajectories for the two links are

$$
q_1^d(t) = \exp(0.5t)(\text{rad}),
\tag{61}
$$

$$
q_2^d(t) = 0.5 + \exp(0.4t)(\text{rad}).
\tag{62}
$$

To observe how the controller behaves in the presence of various uncertainties, two cases of uncertainty, namely, parameter variation and unmodeled friction, are considered.

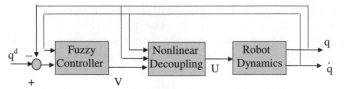

FIGURE 4 Diagram of the robot control system.

1. Parameter Variations

We first suppose both the mass and the length of the two links have an error of 10%. After 10 iterations of learning, the tracking errors of the two links are acceptable (see Fig. 5), noticing that there exist initial position errors of 0.2 (rad) and 0.1 (rad). The membership functions of link 1 and link 2 are provided in Figs. 6 and 7, where the membership functions in (a) are for input x_1^i and those in (b) are for x_2^i. It is seen that all the fuzzy partitions are complete and fairly distinguishable thanks to the measures that are taken to preserve the interpretability.

2. Unmodeled Friction

The Lagrange model of robot dynamics usually does not consider the nonlinear friction. To investigate the performance of the controller in the presence of unmodeled nonlinear friction, the following nonlinear friction is

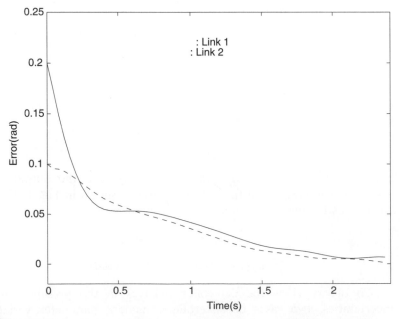

FIGURE 5 Position tracking errors in the presence of parameter uncertainties.

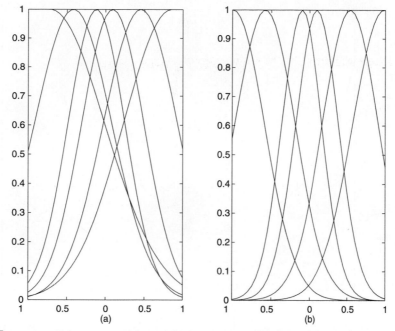

FIGURE 6   Membership functions of the fuzzy controller for link 1.

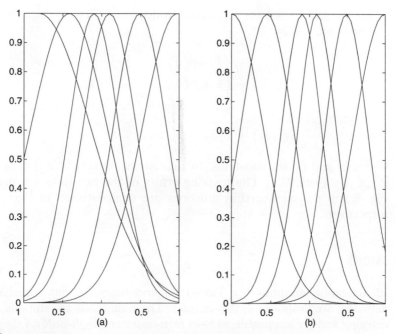

FIGURE 7   Membership functions of the fuzzy controller for link 2.

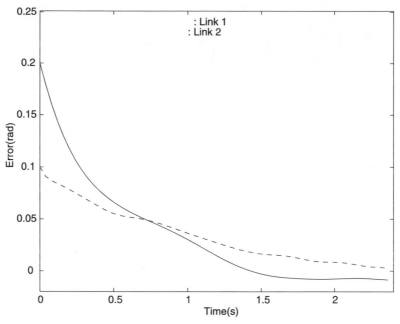

FIGURE 8 Position tracking errors in the presence of unmodeled friction.

added in simulation:

$$f_i = f_{q_i}(\dot{q}_i, \tau_i) + f_{v_i}(\dot{q}_i), \qquad i = 1, 2, \tag{63}$$

where f_{q_i} and f_{v_i} are the Columbus and viscous friction, respectively, which can be expressed by

$$f_{q_i}(\dot{q}_i, \tau_i) = \begin{cases} k_i \operatorname{sgn}(\dot{q}), & |\dot{q}_i| > 0, \\ k_i \operatorname{sgn}(\tau_i), & |\dot{q}_i| = 0, |\tau_i| > k_i, \\ \tau_i, & |\dot{q}_i| = 0, |\tau_i| < k_i, \end{cases} \tag{64}$$

$$f_{v_i}(\dot{q}_i) = C_i \dot{q}_i, \tag{65}$$

where k_i and C_i are constants. In simulation, we set $[k_1, k_2] = [0.2, 0.5]$ and $[C_1, C_2] = [0.05, 0.01]$. The tracking errors of the two links are presented in Fig. 8 and the membership functions are demonstrated in Figs. 9 and 10, respectively.

VI. CONCLUSIONS

Identification methods for Takagi–Sugeno fuzzy systems based on neural networks are proposed in this chapter. These methods are suitable for online learning and are applicable to both zero-order and high-order Takagi–Sugeno fuzzy systems. To maintain the interpretability of fuzzy systems, some mea-

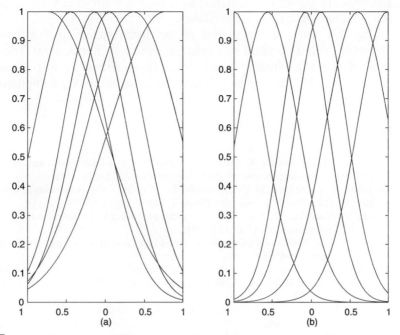

FIGURE 9   Membership functions of the fuzzy controller for link 1.

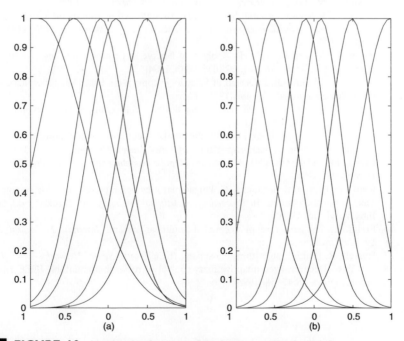

FIGURE 10   Membership functions of the fuzzy controller for link 2.

sures are taken in the process of learning. The effectiveness of the proposed methods are shown by simulation studies of control of robot manipulators in the presence of various uncertainties.

In this work, the structure of the fuzzy system is fixed beforehand. However, such a standard rule structure is usually not optimal. The optimization of the rule structure can be realized either by introducing neural network pruning techniques [17] or by using evolutionary computation [18].

REFERENCES

1. Sugeno, M. and Tanaka, K. Successive identification of a fuzzy model and its application to prediction of a complex system. *Fuzzy Sets Systems* 42:315–324, 1991.
2. Takagi, T. and Sugeno, M. Fuzzy identification of systems and its application to modeling and control. *IEEE Trans. Systems Man Cubernet.* 14:349–355, 1985.
3. Lin, C. T. and Lee, C. S. G. Neural network based fuzzy logic control and decision systems. *IEEE Trans. Comput.* 40:1320–1336, 1991.
4. Jang, J.-S. R. ANFIS: Adaptive-network-based fuzzy inference systems. *IEEE Trans. Systems Man Cybernet.* 23:665–685, 1993.
5. Jin, Y. C., Jian, J. P., and Zhu, J. Neural network based fuzzy identification and its application to modeling and control of complex systems. *IEEE Trans. Systems Man Cybernet.* 25:990–997, 1995.
6. Hunt, K. J., Haas, R., and Brown, M. On the functional equivalence of fuzzy inference systems and spline-based networks. *Internat. J. Neural Systems* 6:171–184, 1995.
7. Berenji, H. R. and Khedkar, P. Learning and tuning fuzzy controllers through reinforcement. *IEEE Trans. Neural Networks* 3:724–739, 1992.
8. Jin. Y. C., v. Seelen, W., and Sendhoff, B. Generating FC^3 fuzzy rule system from data using evolutionary strategies. *IEEE Trans. Systems Man Cybernet.* 29, 1999.
9. Jin, Y. C., v. Seelen, W., and Sendhoff, B. An approach to rule-based knowledge extraction. In *Proceedings of IEEE International Conference on Fuzzy Systems*, pp. 1188–1193. Anchorage, Alaska, 1998.
10. Yager, R. R. and Feliv, D. P. Analysis of flexible structured fuzzy logic controllers. *IEEE Trans. Systems Man Cybernet.* 24:1035–1043, 1994.
11. Gupta, M. M. and Qi, J. Design of fuzzy logic controllers based on generalized *t*-operators. *Fuzzy Sets Systems* 40:473–489, 1991.
12. Jin, Y. C. and v. Seelen, W. Evaluating flexible fuzzy controllers via evolution strategies. *Fuzzy Sets Systems*, 1998. To appear.
13. Durbin, R. and Rumelhart, D. E. Product units: A computationally powerful and biologically plausible extension to backpropagation networks. *Neural Computation* 1:133–142, 1989.
14. Brown, M. and Harris, C. J. *Neurofuzzy Adaptive Modeling and Control.* Prentice Hall, Hemel Hempstead, 1994.
15. Runkler, T. A. and Bezdek, J. C. Polynomial membership functions for smooth first order Takagi–Sugeno systems. In *Proceedings in Artificial Intelligence* (C. Freksa, Ed.), pp. 382–387. Infix, Augustin, 1996.
16. Freud, E. The structure of decoupled nonlinear system. *Internat. J. Control* 21:443–450, 1975.
17. Russell, R. Pruning algorithms—a survey. *IEEE Trans. Neural Networks* 4:740–747, 1993.
18. Jin, Y. C. Decentralized adaptive fuzzy control of robot manipulators. *IEEE Trans. Systems Man Cybernet.* 28:47–58, 1998.

6
FUZZY CONTROL WITH REFERENCE MODEL-FOLLOWING RESPONSE

C. M. LIAW

Department of Electrical Engineering, National Tsing Hua University, Hsinchu, Taiwan, Republic of China

Y. S. KUNG

Department of Electrical Engineering, Nan-Tai Institute of Technology, Tainan, Taiwan, Republic of China

I. INTRODUCTION 130
II. STATEMENT OF THE PROBLEM 131
III. THE INDIRECT FIELD-ORIENTED INDUCTION MOTOR DRIVE 133
IV. OUTPUT FEEDBACK LINEAR MODEL-FOLLOWING CONTROLLER 136
  A. Controller Design 136
  B. Reference Model Selection 137
V. CONVENTIONAL FUZZY CONTROLLER 139
VI. FUZZY CONTROLLER WITH REFERENCE MODEL-FOLLOWING RESPONSE 145
VII. DESIGN OF THE PROPOSED CONTROLLER FOR INDUCTION MOTOR DRIVE 149
  A. Design of the Linear Model-Following Controller 149
  B. Design of the Fuzzy Controller 151
VIII. CONCLUSIONS 155
  REFERENCES 158

Fuzzy control possesses some remarkable merits and has been successfully applied in many kinds of plants. However, the dynamic response trajectory of the plant controlled by a conventional fuzzy controller generally may deviate significantly due to changes in the system operating conditions and parameters. In this chapter, the systematic design of a conventional fuzzy controller and its major problems are first described. Then a fuzzy model-following controller (FMFC) combining the advantages of a model-following controller and a fuzzy controller is introduced. In the design of this hybrid controller, a reference model is selected and a linear model-following controller (LMFC) is designed based on the roughly estimated plant model at the nominal operating point. As changes occur in the operating conditions, to let the response of the controlled plant still closely follow the output generated by the reference model, a fuzzy adaptation mechanism driven by the model-following error signal is constructed. For illustration, the proposed fuzzy model-following controller is applied to the speed control of an indirect field-oriented induction motor drive. Dynamic signal analysis of the model-following behavior is made; accordingly, the procedure for constructing the control algorithms is introduced in detail. Some simulation and measured results are provided to demonstrate the performance of the motor drive controlled by the fuzzy model-following controller.

I. INTRODUCTION

Fuzzy logic theory has been applied to many fields [1–12], such as industrial process control, motor drive control, temperature control, steam engine control, communication, signal processing, and manufacturing. In the control applications, similar to the adaptive controllers, the fuzzy controllers can also be classified as fuzzy signal tuning controllers [5] and fuzzy parameter tuning controllers [6]. The former are based on the direct fuzzy tuning of the control law. In the latter, incremental fuzzy tuning of controller parameters is adopted. Generally speaking, a fuzzy controller has the following features: (a) Rather than using mathematical derivations, its control algorithms are built up based on intuition and experience of the plant to be controlled. As a result, the plant dynamic model is not needed in the controller design stage. This makes it suited to control systems with unknown or unmodeled dynamics. (b) It possesses some adaptive capability. However, although barely satisfactory dynamic performance of the controlled plant can easily be obtained through roughly designing the fuzzy control mechanism, good response generally cannot be achieved without tediously adjusting the key components of the fuzzy controller, such as the input and output weighting factors, the quantization levels in fuzzification, the shape of the membership function, and the algorithm of defuzzification. Moreover, the dynamic response trajectory of the plant controlled by the conventional fuzzy controller may deviate significantly due to the changes in system operating conditions and parameters.

It is known that by applying the model-following controller [13, 14], plant output can be controlled to follow the desired dynamic response generated by the reference model. Hence, in this chapter, a hybrid controller combining the advantages of model-following controllers and fuzzy controllers is introduced. In this hybrid controller, an output linear model-following controller [13, 14] is first designed according to the roughly estimated plant model at nominal operating point and the chosen reference model. Then an adaptation signal is yielded by a fuzzy controller, which is driven by the model-following error and its change. In the nominal case, the model following is perfect; no control signal is contributed by the fuzzy controller. But when the operating condition and parameter changes occur, an adaptation signal will be generated automatically by the fuzzy controller to let the desired model-following control performance be preserved. The proposed controller can be applied to the control of many kinds of plants. For ease of explanation, we look at the speed control of an induction motor drive by the proposed controller.

Compared with DC motors, squirrel-cage induction motors possess many structural advantages (e.g., low cost, small size, high reliability, and minimum maintenance). However, control of an induction motor is considerably more complex and difficult than control of DC motor, because it is a highly nonlinear and time-varying multivariable plant [15]. In the past decades, there has been great progress made in solid-state induction motor drives, and various control methods have been developed to give the induction motor drives improved performance. Furthermore, with the advent of indirect field-oriented (IFO) control [15], an induction motor can be operated like a separately excited DC motor and can be employed for high-performance applications. However, the major drawback of the IFO induction motor drive is that its performance depends heavily on rotor parameter variations, particularly the rotor resistance. To obtain better performance, it must be controlled using sophisticated control techniques. As a result, the IFO induction motor drive is considered to be a good study example for developing advanced and practical control techniques. In the past decades, many types of controller (see, e.g., PI controller [15, 16], two-degree-of freedom controller [17], optimal controller [18], variable-structure system controller [19], adaptive controller [13, 14, 20], and robust controller [15, 16]) have been developed to improve the performance of induction motor drives. Each approach has its advantages and disadvantages in practical realization. However, all these controllers must be designed mathematically based on an accurate plant model. So if the structure and the parameters of the plant model are not accurately known, good control performance will not be achieved. This common drawback can be overcome using fuzzy control techniques [7, 8].

Dynamic signal analysis for model-following response is performed in applying the fuzzy model-following controller to speed control of an indirect field-oriented induction motor drive, for easy incorporation of experience into the control algorithms. Accordingly, the fuzzy control algorithm is systematically constructed. Because in a practical controller minimization of control energy consumption must be considered, the strategy for avoiding the inherent chattering problem [7] of fuzzy systems is also proposed here. In addition to tracking response, good speed regulating response can be obtained simultaneously by this fuzzy model-following controller.

II. STATEMENT OF THE PROBLEM

The block diagram of a fuzzy model-following controller for an indirect field-oriented induction motor drive is shown in Fig. 1 [7], consisting of an output feedback linear model-following controller (LMFC) [13, 14] and a fuzzy controller. The composite control signal contributed by these two controllers is represented by $u_p(= i_{qs}^*) = u_{pl} + u_{pf}(e_o)$. In the design of this controller, the parameters K_x and K_u of the LMFC are first calculated

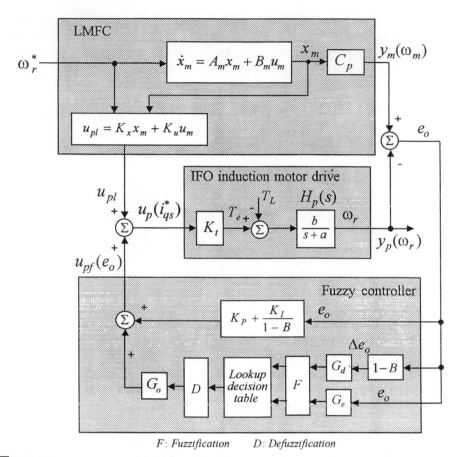

FIGURE 1 System configuration of the fuzzy model-following controller. Reprinted with permission from Y. S. Kung and C. M. Liaw, *IEEE Trans. Fuzzy Systems* 2:194–202 (© 1994 IEEE).

according to a chosen reference model and the roughly estimated dynamic motor drive model at nominal operating point. Then the model-following error is used to drive the fuzzy controller, and its output signal $u_p(e_o)$ is applied to let the model-following performance be insensitive to the system operating condition and parameter variations. It should be mentioned that the LMFC shown in Fig. 1 can be neglected. In this case, because all of the control signal will be contributed by the fuzzy controller, adjusting it to achieve good control performance will be rather difficult. However, by the addition of LMFC, the control effort produced by the fuzzy controller will be much reduced. It follows that tuning the fuzzy controller becomes easier and the operating stability of the whole drive system is greatly improved. The control system shown in Fig. 1 using model-following error as the input of a fuzzy controller can also be regarded as a fuzzy controller improving a LMFC [7], or it can also be said that a LMFC is augmented with a fuzzy adaptation mechanism.

III. THE INDIRECT FIELD-ORIENTED INDUCTION MOTOR DRIVE

Generally, the dynamic model of an induction motor is derived using reference-frame theory and linearization techniques. According to the detailed derivations made in [15], the small-signal voltage equations and torque equation in a synchronously rotating frame can be expressed as

$$
\begin{bmatrix} v_{qs} \\ v_{ds} \\ 0 \\ 0 \end{bmatrix} = \begin{bmatrix} R_s + sL_s & \omega_e L_s & sL_m & \omega_e L_m \\ -\omega_e L_s & R_s + sL_s & -\omega_s L_m & sL_m \\ sL_m & \omega_{sl} L_m & R_r + sL_r & \omega_{sl} L_r \\ -\omega_{sl} L_m & sL_m & -\omega_{sl} L_r & R_s + sL_r \end{bmatrix} \begin{bmatrix} i_{qs} \\ i_{ds} \\ i_{qr} \\ i_{dr} \end{bmatrix}, \tag{1}
$$

$$
T_e = \frac{3P}{4} L_m \left(i_{qs} i_{dr} - i_{ds} i_{qr} \right) = T_L + B\omega_r + J\frac{d\omega_r}{dt}, \tag{2}
$$

where

L_m = magnetizing inductance per phase
R_s = stator resistance per phase
L_s = stator inductance per phase
R_r = rotor resistance per phase referred to stator side
L_r = rotor inductance per phase referred to stator side
P = number of poles
v_{qs} (v_{ds}) = q-axis (d-axis) stator voltage
i_{qs} (i_{ds}) = q-axis (d-axis) stator current
i_{qr} (i_{dr}) = q-axis (d-axis) rotor current referred to stator side
ω_e = electrical angular speed
ω_{sl} = electrical slip angular speed
J = total mechanical inertia
B = total damping ratio
T_L = load torque

Equations (1) and (2) can be rearranged and written as

$$
\frac{d}{dt}\begin{bmatrix} i_{ds} \\ i_{qs} \\ \lambda_{dr} \\ \lambda_{qr} \end{bmatrix} = \begin{bmatrix} -\dfrac{R_s}{\sigma L_s} - \dfrac{R_r(1-\sigma)}{\sigma L_r} & \omega_e & \dfrac{L_m R_r}{\sigma L_s L_r^2} & \dfrac{P\omega_r L_m}{2\sigma L_s L_r} \\ -\omega_e & \dfrac{-R_s}{\sigma L_s} - \dfrac{R_r(1-\delta)}{\sigma L_r} & \dfrac{-P\omega_r L_m}{2\sigma L_s L_r} & \dfrac{L_m R_r}{\sigma L_s L_r^2} \\ \dfrac{L_m R_r}{L_r} & 0 & -\dfrac{R_r}{L_r} & \omega_{sl} \\ 0 & \dfrac{L_m R_r}{L_r} & -\omega_{sl} & -\dfrac{R_r}{L_r} \end{bmatrix}
$$

$$\times \begin{bmatrix} i_{ds} \\ i_{qs} \\ \lambda_{dr} \\ \lambda_{qr} \end{bmatrix} + \frac{1}{\sigma L_s} \begin{bmatrix} v_{ds} \\ v_{qs} \\ 0 \\ 0 \end{bmatrix}, \tag{3}$$

$$T_e = \frac{3}{2} \frac{P}{2} \frac{L_m}{L_r} (i_{qs} \lambda_{dr} - i_{ds} \lambda_{qr}) = T_L + B\omega_r + J\frac{d\omega_r}{dt}, \tag{4}$$

where

$\sigma = 1 - L_m^2 / L_{sr}$
$\lambda_{qr} = L_m i_{qs} + L_r i_{qr} = q$-axis rotor flux linkage referred to stator side
$\lambda_{dr} = L_m i_{ds} + L_r i_{dr} = d$-axis rotor flux linkage referred to stator side

For ideally decoupling the induction motor, it is desired that the d-axis of a synchronously rotating frame align with the rotor flux vector λ_r, which is equivalently represented by the complex phasor $\lambda_r = \lambda_{qr} - j\lambda_{dr}$, that is,

$$\lambda_{qr} = 0, \qquad \frac{d\lambda_{qr}}{dt} = 0. \tag{5}$$

In this case, if the electrical time constant (L_r/R_r) is neglected compared with the mechanical time constant and using Eq. (3), the rotor flux linkage $\lambda_r = \lambda_r^*$ can be set directly by the rated stator flux component current i_{ds}^* as follows:

$$\lambda_{dr} = \lambda_r^* = \frac{L_m i_{ds}^*}{1 + s(L_r/R_r)} \cong L_m i_{ds}^*, \tag{6}$$

where s denotes Laplace operator. From Eqs. (5), (6), and (3), one can derive the predicted slip angular speed ω_{sl} command for obtaining the ideal field orientation:

$$\omega_{sl} = \omega_{sl}^* = \frac{R_r i_{qs}^*}{L_r i_{ds}^*}, \tag{7}$$

where i_{ds}^* and i_{qs}^* denote the stator flux and torque component current commands generated from the outer speed control loop. The expression of developed torque T_e in Eq. (4) can be simplified to

$$T_e = \frac{3P}{4} \frac{L_m^2}{L_r} i_{ds}^* i_{qs}^* \triangleq k_t' i_{ds}^* i_{qs}^* \triangleq k_t i_{qs}^* = T_L + B\omega_r + J\frac{d\omega_r}{dt}, \tag{8}$$

where k_t' is the torque generating constant and $k_t \triangleq k_t' i_{ds}^*$ for constant current command i_{ds}^*.

The block diagram of a typical indirect field-oriented induction motor drive is drawn in Fig. 2a. It mainly consists of an induction motor, a dynamic load (a DC generator with switched resistors), a current-controlled pulse-width-modulated (PWM) inverter, a slip angular speed estimator, a coordi

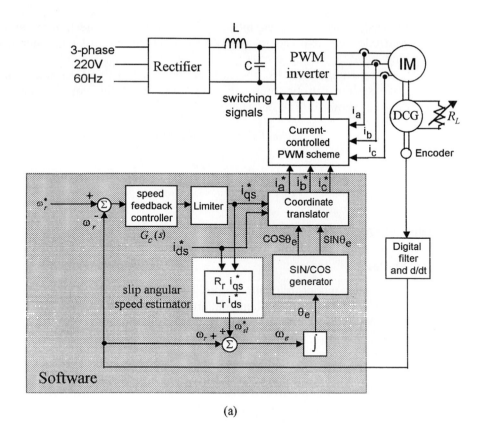

(a)

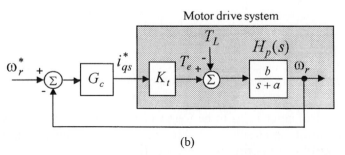

(b)

FIGURE 2 An indirect field-oriented induction motor drive: (a) system configuration; (b) control system block diagram.

nate translator, and an outer speed feedback control loop. The induction motor used here is a three-phase Δ-connected four-pole 1-hp, 220-V, 60 Hz motor. The torque component current command i_{qs}^* is generated from the speed error between the command and the measured rotor speed through the torque controller G_c. According to Eq. (7), the estimate of slip angular speed ω_{sl}^* is obtained using the regulated stator torque component current i_{qs}, the preset stator flux component current i_{ds}^*, and the rotor time constant L_r/R_r. Although the current-controlled pulse-width-modulated switching scheme is realized using analog circuits, the field-orientation mechanism and speed controller are realized in a PC/AT 486-DX66 control computer using C-language.

According to Eq. (8), the dynamic behavior of this IFO induction motor drive system can be reasonably represented by the control system block diagram shown in Fig. 2b, where

$$\omega_r(s) = H_p(s)(T_e(s) - T_L(s)), \tag{9}$$

$$H_p(s) = \frac{1/J}{s + B/J} \triangleq \frac{b}{s + a}. \tag{10}$$

IV. OUTPUT FEEDBACK LINEAR MODEL-FOLLOWING CONTROLLER

A. Controller Design

The linear model-following controller based on output feedback has the advantage of ease of implementation. A theoretical basis of this LMFC has been derived in detail in [14]; only a brief description is given here. Suppose that the roughly estimated motor drive model at nominal case and the chosen reference model are expressed by

$$\dot{x}_p = A_p x_p + B_p u_p, \tag{11}$$

$$y_p = C_p x_p, \tag{12}$$

$$\dot{x}_m = A_m x_m + B_m u_m, \tag{13}$$

$$y_m = C_p x_m, \tag{14}$$

where $x_p \in R^n$, $x_m \in R^n$, $y_p \in R^q$, $y_m \in R^q$, $u_m \in R^p$, and $u_p \in R^p$, and A_p, B_p, C_p, A_m, and B_m are constant matrices of appropriate dimensions. The pair (A_p, B_p) is assumed to be stabilizable and A_m is a stable matrix. The control input u_p is to be found such that the plant states and hence the plant output can closely track those of reference model. To achieve this goal, the control input is proposed to be [14]

$$u_p = K_x x_m + K_u u_m + K_e e_o, \tag{15}$$

where e_o is the error between the plant output and the model output, and K_x, K_u, and K_e are constant gain matrices of appropriate dimensions. Define the state error vector as

$$e \triangleq x_m - x_p. \tag{16}$$

Then, from Eqs. (11)–(16), one can obtain the following error dynamic equation:

$$\dot{e} = (A_p + B_p K_e C_p)e + (A_m - A_p - B_p K_x)x_m + (B_m - B_p K_u)u_m. \tag{17}$$

It is seen from Eq. (17) that if A_m, B_m, K_x, K_u, and K_e are chosen to let $(A_p - B_p K_e C_p)$ be a Hurwitz matrix and

$$K_x = B_p^+ (A_m - A_p), \qquad K_u = B_p^+ B_m, \qquad B_p^+ = \left(B_p^T B_p \right)^{-1} B_p^T, \quad (18)$$

then the error vector in (17) will decay to zero asymptotically in the presence of the state mismatch, and the output of the controlled plant will follow the output of reference model. In Eq. (18), B_p^+ is the left Penrose pseudo inverse of B_p [13].

It is known that the performance of LMFC is sensitive to system parameter variations, and experience shows that the stability, control effort, and model-following characteristics of the LMFC are much affected by the control signal contributed from the model-following error, that is, the term $K_e e_o$ in Eq. (15). Particularly for the LMFC designed based on reduced-order reference model, because the unmodeled system dynamics may cause the result system to exhibit significant performance degradation and even instability.

To solve this problem, the control law of the proposed fuzzy model-following controller is proposed to be

$$u_p = K_x x_m + K_u u_m + u_p(e_o), \qquad (19)$$

where the parameters of K_x and K_u are calculated using Eq. (18), and the error-driven control signal $u_p(e_o)$ is generated by the fuzzy controller. The latter is added to reduce the model-following error due to the inaccurate plant dynamic model and the unmodeled dynamics. The synthesis of $u_p(e_o)$ is made by applying fuzzy theory, which will be introduced in the following section.

B. Reference Model Selection

The reference model in a model-following controller defines the desired tracking response of the controlled plant. Two suggestions for choosing the reference model are given as follows.

1. Method A

According to plant dynamic behavior as well as the desired and achievable tracking performance, a transfer function of proper order is directly chosen. Then its state-space realization expressed as Eqs. (13) and (14) is made.

2. Method B

For the speed control of a motor drive, it is usually more convenient to specify the desired control performance with tracking and regulation time response specifications. To incorporate these specifications into the reference model, the motor drive in Fig. 2 can be first controlled by a two-degree-of-freedom controller (2DOFC) as shown in Fig. 3a [14] at nominal case. In this control system, $\bar{a}$ and $\bar{b}$ are the nominal plant model parameters, and G_c and

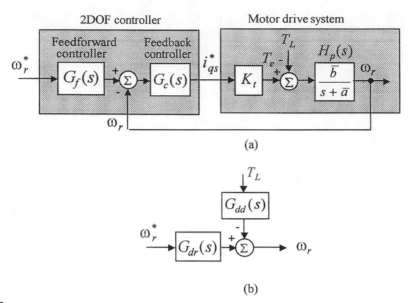

FIGURE 3 A two-degree-of-freedom controller for motor drive: (a) system configuration; (b) closed-loop transfer function block diagram.

G_f denote the feedback and feedforward controllers, respectively. Although G_c is emphasized in regulation response, G_f is designed to meet the desired tracking response.

In the design of G_c and G_f, the following drive specifications are prescribed: (1) the steady-state errors of the speed responses due to step command and load torque changes are zero; (2) the response time of the step speed tracking response is given; (3) the overshoot of the step speed tracking response is zero; and (4) the maximum speed due to step load torque change is given. According to the motor drive model at nominal case and the required performance, the parameters of G_c and G_f can be found following a systematic procedure derived in [14]. Having finished the design, the resulting motor drive can be represented by the block diagram shown in Fig. 3b with closed-loop tracking and regulation transfer functions

$$G_{dr}(s) \triangleq \frac{\omega_r(s)}{\omega_r^*(s)}\bigg|_{T_L(s)=0}, \qquad G_{dd}(s) \triangleq \frac{\omega_r(s)}{T_L(s)}\bigg|_{\omega_r^*(s)=0},$$

where $G_{dr}(s)$ and $G_{dd}(s)$ represent the desired tracking and regulation time responses, respectively. Thus $G_{dr}(s)$ can be chosen as a reference model and its realization with the expression of Eqs. (13) and (14) can be used for the design of the LMFC described previously. After applying the model-following fuzzy controller, the tracking speed response of the controlled motor drive can also closely follow that of $G_{dr}(s)$ as the operating conditions are changed, and the regulation response can also be further improved although it has not been quantitatively treated by the fuzzy controller.

V. CONVENTIONAL FUZZY CONTROLLER

As mentioned previously, fuzzy control is very well suited for controlling plants whose dynamic models are not accurately known. Figure 4a shows the basic configuration of a fuzzy controller [3, 4], the basic functions of its components are briefly described next.

Fuzzification Interface. The measured input variables are suitably scaled, quantized, and converted into linguistic variables, which are defined by the chosen membership functions. Generally, the input variables to a conventional fuzzy speed controller are defined as

$$e(k) \triangleq \omega_r^*(k) - \omega_r(k), \tag{20}$$

$$\Delta e(k) \triangleq e(k) - e(k-1), \tag{21}$$

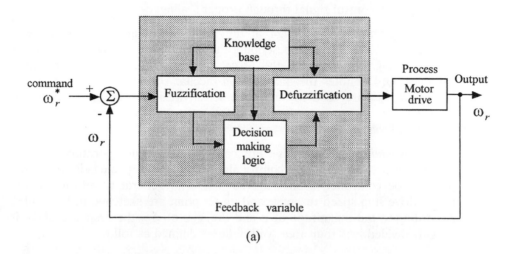

(a)

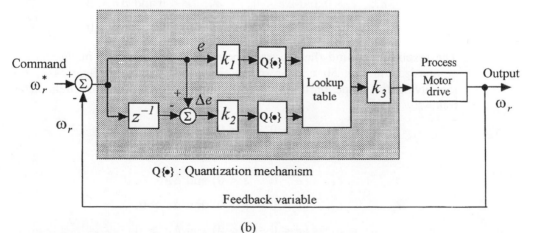

(b)

FIGURE 4 A conventional fuzzy control system: (a) basic configuration; (b) lookup-table-based fuzzy controller.

where

$\omega_r(k)$ = speed command at the kth sampling interval
$\omega_r(k)$ = rotor speed response at the kth sampling interval
$e(k)$ = speed error at the kth sampling interval
$\Delta e(k)$ = speed error change at the kth sampling interval

Knowledge Base. This consists of a data base and a linguistic control rule base. The former provides necessary definitions to define the linguistic control rules and fuzzy data manipulation. The latter characterizes the control goals and control policy of the domain by using a set linguistic control rules defined by experts.

Decision-Making Logic. This has the capability of simulating human decision making based on fuzzy concepts and the fuzzy set operations.

Defuzzification Interface. This converts the fuzzy output control variable into nonfuzzy control signal through proper scaling.

Having finished the design of the fuzzy controller, a lookup table is builtup. The block diagram describing the execution of a fuzzy controller is shown in Fig. 4b, in which the error e and error change Δ_e are input variables, $Q\{\cdot\}$ denotes the quantization mechanism, and the scaling factors k_1, k_2, and k_3 are used to adjust the sensitivities of the input and output variables. The design procedure for constructing the lookup table is summarized as follows.

1. *Dynamic signal analysis*—For convenience in incorporating intuition and experience into the fuzzy control algorithms, the dynamic behavior of the plant to be controlled is investigated first. The general waveforms of the motor drive step speed response and its set point are sketched in Fig. 5a [5]. According to the magnitude of e and the sign of Δe, the response plane is roughly divided into four areas with indexes defined as follows:

$$a_1: \quad e > 0 \text{ and } \Delta e < 0, \qquad a_2: \quad e < 0 \text{ and } \Delta e < 0,$$
$$a_3: \quad e < 0 \text{ and } \Delta e > 0, \qquad a_4: \quad e > 0 \text{ and } \Delta e > 0. \tag{22}$$

The responses around the crossover points and the extremes in Fig. 5a are emphasized in Fig. 5b and 5c, respectively. The crossover indexes c_i for identifying the slope of the response across the set point are defined as follows:

$$c_1: \quad (e > 0 \rightarrow e < 0) \text{ and } \Delta e \lll 0,$$
$$c_2: \quad (e > 0 \rightarrow e < 0) \text{ and } \Delta e \ll 0,$$
$$c_3: \quad (e > 0 \rightarrow e < 0) \text{ and } \Delta e < 0,$$
$$c_4: \quad (e < 0 \rightarrow e > 0) \text{ and } \Delta e > 0, \tag{23}$$
$$c_5: \quad (e < 0 \rightarrow e > 0) \text{ and } \Delta e \gg 0,$$
$$c_6: \quad (e < 0 \rightarrow e > 0) \text{ and } \Delta e \ggg 0.$$

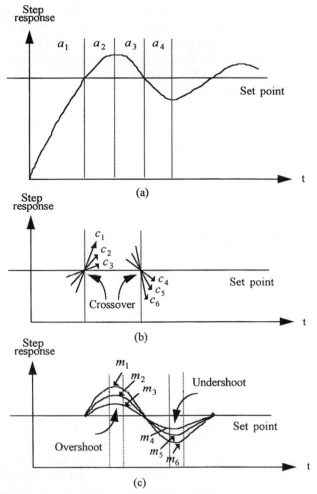

FIGURE 5 Dynamic signal analysis of a conventional fuzzy controller: (a) typical step response around the set point; (b) responses around the crossover points; (b) responses around the extremes.

The magnitude indexes for representing the extent of overshoot and undershoot are defined as follows:

$$m_1: \quad \Delta e \approx 0 \text{ and } e \lll 0, \qquad m_2: \quad \Delta e \approx 0 \text{ and } e \ll 0,$$

$$m_3: \quad \Delta e \approx 0 \text{ and } e < 0, \qquad m_4: \quad \Delta e \approx 0 \text{ and } e > 0, \qquad (24)$$

$$m_5: \quad \Delta e \approx 0 \text{ and } e \gg 0, \qquad m_6: \quad \Delta e \approx 0 \text{ and } e \ggg 0.$$

2. *Fuzzification*—The control variables are quantized and mapped into suitable fuzzy variables through the chosen membership functions. The commonly used linguistic set is {PB, PM, PS, ZE, NS, NM, NB}, where P, N, B, M, S, and ZE denote Positive, Negative, Big, Medium, Small, and Zero, respectively. Depending on the applications, many types of membership functions can be defined. The commonly used membership functions are

triangular shape, bell shape, and trapezoidal shape. The choice of membership functions depends on the user's preference and the particular applications. Commonly used trapezoidal-shaped membership functions are shown in Fig. 6 [5], where the input variable is quantized into 13 levels with universe of discourse $(-6, -5, \ldots, 5, 6)$. The membership function corresponding to ZE can be mathematically expressed as

$$\text{ZE:} \quad f(x) = \begin{cases} 0, & 6 \geq x > 2, \\ -2/3(x - 0.5) + 1, & 2 \geq x > 0.5, \\ 1, & 0.5 \geq x > -0.5, \\ 2/3(x + 0.5) + 1, & -0.5 \geq x > -2, \\ 0, & -2 \geq x > -6. \end{cases} \quad (25)$$

The expressions of other membership functions in Fig. 6 can be obtained from Eq. (25) by suitable modification of the argument.

3. *Construction of linguistic control rules*—According to the dynamic behavior of the plant and the chosen fuzzy numbers, the linguistic control rules can be defined. The typical form of linguistic control rule is

$$\text{IF } e \text{ is PB and } \Delta e \text{ is NB THEN the control input is PB.} \quad (26)$$

4. *Defuzzification*—Defuzzification strategy is aimed at producing a nonfuzzy control action that best represents the possibility distribution of an inferred fuzzy control action. The commonly used strategies are the max criterion, the mean of maximum (MOM), and the center of area (COA) [4]. The max criterion is seldom used. Although the MOM yields better transient performance, the COA is emphasized in steady-state behavior.

5. *Construction of decision lookup table*—With the defined linguistic control rules and the chosen membership functions, many sets of control inputs may exist for a pair of error and error change. The final control inputs can be synthesized using the COA defuzzification strategy. Then a decision lookup table is constructed. Finally, the analog control system is produced through proper scaling. The performance of a fuzzy controller is highly dependent on the chosen scaling factors k_1, k_2, and k_3. Generally, it is suggested that the output scaling factor k_3 is first decided according to the control input characteristic of the process to be controlled. Then the input scaling factors

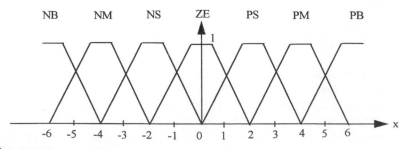

FIGURE 6 Trapezoidal-shaped membership functions.

k_1 and k_2 are adjusted by trial and error, such that the phase portrait is properly allocated as shown in Fig. 7. For a detailed adjustment procedure, refer to [9]. For analyzing the properties of the fuzzy controller in detail, it can be approximately equivalent to a conventional PID controller by using the product–sum–gravity method and the simplified fuzzy reasoning method [10–12], through which the adjustment of scaling factors is equivalent to modifying the location of the pole and zero of the closed-loop system in the sense of a linear control system. On the other hand, the input and output scaling factors can also be set to be some special functions to possess adaptive capability.

Basically, there are two types of fuzzy controllers, which are shown in Fig. 8a and 8b namely the PD-type fuzzy controller and the PI-type fuzzy controller. The control input u of the former type is constructed using error e and error change Δe. It has good transient response, but its steady-state error cannot be completely eliminated. As to the PI-type fuzzy controller, the incremental control input Δu is directly constructed from e and Δe. This type of controller possesses better static response without steady-state error.

Generally, the fuzzy controller, which is realized based on the decision lookup table, cannot lead to good performance in both the transient and

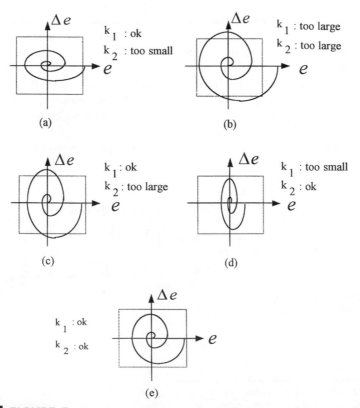

FIGURE 7 The phase portraits corresponding to various chosen input scaling factors.

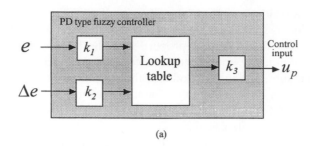

(a)

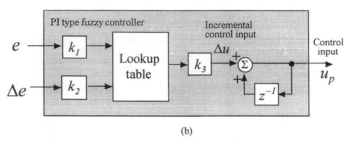

(b)

FIGURE 8 Two types of fundamental fuzzy controllers: (a) PD type; (b) PI type.

static periods. Because the quantization is too coarse, overshoot and hunting around the set point in the static period may result. To solve this problem, the fine decision lookup tables are usually used to replace the coarse table when the error falls within a preset limit. However, this increases the complexity of implementation and the hunting problem still exists. A fuzzy controller with a simple but practical approach to solving this problem is shown in Fig. 9, in which an integral controller is employed to replace the fuzzy controller as the error falls within a specified region ($|e| < \varepsilon$, ε is a small positive number). Within this region, the fuzzy controller is disconnected with its final control signal output being memorized by an integrator,

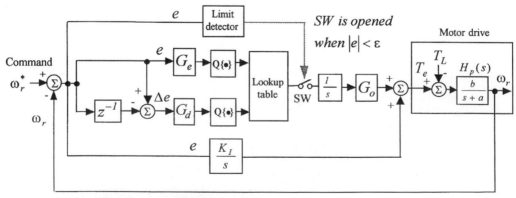

$Q\{\bullet\}$: Quantization mechanism

FIGURE 9 A conventional fuzzy controller for induction motor drive.

and the integral controller remains to eliminate the error in steady-state. In the fuzzy control system shown in Fig. 9, because the fuzzy controller and the integral controller are emphasized in the transient and the static responses, respectively, the integral gain and output gain are set to have the following error adaptation characteristics [5]:

$$K_I(e) = K_{Is} - K_{Ip}|e|, \qquad G_o(e) = K_{os} + K_{op}|e|. \tag{27}$$

VI. FUZZY CONTROLLER WITH REFERENCE MODEL-FOLLOWING RESPONSE

The major drawback of the conventional fuzzy controller introduced in the previous section is that the dynamic response trajectory of the controlled plant may deviate significantly due to changes in system operating conditions and parameters. The fuzzy model-following controller shown in Fig. 1 [7] can be used to solve this problem. The design of this fuzzy controller is briefly introduced as follows:

1. *Fuzzification*—The process variable to be observed and the control variable of the motor drive are chosen to be the rotor speed ω_r and the torque current i_{qs}^*. In this fuzzy controller, the system variables are defined as the rotor speed model-following error e_o (the error between the rotor speed and the reference-model output) and error change Δe_o [7]:

$$e_o(k) \triangleq \omega_m(k) - \omega_r(k), \tag{28}$$

$$\Delta e_o(k) \triangleq e_o(k) - e_o(k-1), \tag{29}$$

where

$\omega_m(k)$ = the response generated by the reference model at the kth sampling instant

$\omega_r(k)$ = the rotor speed response of the motor drive at the kth sampling instant

The error, error change, and control variable are all quantized the corresponding universe of discourse. In this chapter, the quantizations of e_o and Δe_o are shown in Table 1, where the scaling factor is set as 3.77 V to 1800 rpm. Having made the quantizations of system variables, the quantized input data are then converted into suitable linguistic variables which may be viewed as labels of fuzzy sets. Here the linguistic set is also chosen to be {PB, PM, PS, ZE, NS, NM, NB}, and the trapezoidal-shaped membership function shown in Fig. 5 is also employed.

2. *Dynamic signal analysis*—For convenience in incorporating the intuition and experience into the fuzzy control algorithms, the dynamic behavior of the model-following error signal is investigated first. For the proposed fuzzy model-following controller, the desired rotor speed step response trajectory will be deviated around the response produced by a reference model. For convenience of analysis, the general rotor speed response is drawn in Fig. 10a

TABLE I   Quantized Error and Error Change[a]

Error e_o (mV)	Error change Δe_o (mV)	Quantized level
-1600	-1600	-6
-800	-800	-5
-400	-400	-4
-200	-200	-3
-100	-100	-2
-50	-50	-1
0	0	0
50	50	1
100	100	2
200	200	3
400	400	4
800	800	5
1600	1600	6

[a] Reprinted with permission from Y. S. Kung and C. M. Liaw, *IEEE Trans. Fuzzy Systems* 2:194–202 (© 1994 IEEE).

[7], in which, $c_1, c_2, c_3, \ldots$ and $m_1, m_2, m_3, \ldots$ denote the crossover points and the extreme points. The properties at these points are as follows:

$$
\begin{aligned}
c_1: &\quad (e_o < 0 \rightarrow e_o > 0) \text{ and } \Delta e_o \ggg 0, \\
c_2: &\quad (e_o > 0 \rightarrow e_o < 0) \text{ and } \Delta e_o \lll 0, \\
c_3: &\quad (e_0 < 0 \rightarrow e_o > 0) \text{ and } \Delta e_o \gg 0, \\
c_4: &\quad (e_o > 0 \rightarrow e_o < 0) \text{ and } \Delta e_o \ll 0, \\
c_5: &\quad (e_0 < 0 \rightarrow e_o > 0) \text{ and } \Delta e_o > 0, \\
c_6: &\quad (e_o > 0 \rightarrow e_o < 0) \text{ and } \Delta e_o < 0;
\end{aligned}
\tag{30}
$$

$$
\begin{aligned}
m_1: &\ \Delta e_o \approx 0 \text{ and } e_o \ggg 0, \quad & m_2: &\ \Delta e_o \approx 0 \text{ and } e_o \lll 0, \\
m_3: &\ \Delta e_o \approx 0 \text{ and } e_o \gg 0, \quad & m_4: &\ \Delta e_o \approx 0 \text{ and } e_o \ll 0, \\
m_5: &\ \Delta e_o \approx 0 \text{ and } e_o > 0, \quad & m_6: &\ \Delta e_o \approx 0 \text{ and } e_o < 0.
\end{aligned}
\tag{31}
$$

On the other hand, $A_1, A_2, A_3, \ldots, A_{12}$ in Fig. 10a denote the reference ranges. The polarities of e_o and Δe_o at each range are also indicated in Fig. 10a. For easily observing the model-following error convergence characteristics and justifying the linguistic rule, the phase-plane trajectory corresponding to Fig. 10a is shown in Fig. 10b. The crossover points, the extreme points, and the reference ranges are combined and shown in Table 2 for later reference.

3. *Derivation of fuzzy control rules* [7]—The fuzzy control rules are developed based on expert experience and control engineering knowledge. In order to obtain good model-following characteristics, according to the dynamic signal analysis made in Fig. 10a and experience, the fuzzy control rules decided at the crossover points and extreme points are listed in Table 3. These control rules are also listed in the linguistic-rule table of Table 4 (the

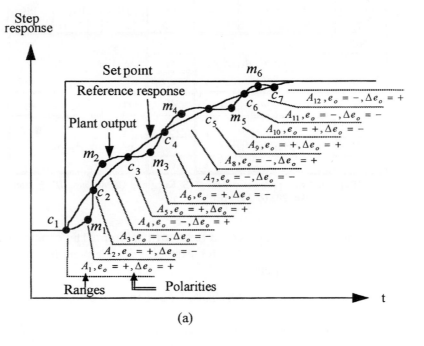

(a)

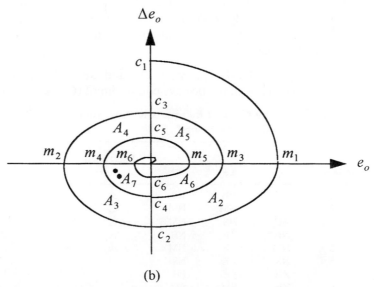

(b)

FIGURE 10 (a) Dynamic signal analysis of the fuzzy model-following controller; (b) the phase-plane trajectory corresponding to (a). Reprinted with permission from Y. S. Kung and C. M. Liaw, *IEEE Trans. Fuzzy Systems* 2:194–202 (© 1994 IEEE).

TABLE 2 The Indices in the State Plane[a]

Δe_o \ e_o	NB	NM	NS	ZE	PS	PM	PB
NB				c_2			
NM	A_3, A_7, A_{11}			c_4	A_2, A_6, A_{10}		
NS				c_6			
ZE	m_2	m_4	m_6	ZE	m_5	m_3	m_1
PS				c_5			
PM	A_4, A_8, A_{12}			c_3	A_1, A_5, A_9		
PB				c_1			

[a]Reprinted with permission from Y. S. Kung and C. M. Liaw, *IEEE Trans. Fuzzy Systems* 2:194–202 (© 1994 IEEE).

elements in the fourth row and the fourth column). The conditional rules are implied in these two tables; for example, the element in the third row of Table 3 implies at point c_6

$$\text{If } (e_o \text{ is ZE and } \Delta e_o \text{ is NS) Then (the change of control input (CI) is NS)}$$
$$(32)$$

The rule justification at the reference ranges can be made according to the dynamic signal analysis shown in Fig. 10a and the phase-plane trajectory shown in Fig. 10b. Some observations used for determining control rules are made as follows:

a. At areas A_1, A_5, A_9—$e_o = $ "+" and $\Delta e_o = $ "+", the error is positive and increased, the positive control input (CI) is set to reduce the error.

TABLE 3 Fuzzy Control Rules at the Reference Crossover and Extreme Points[a]

Rule no.	e_o	Δe_o	CI	Reference points
1	ZE	NB	NB	c_2
2	ZE	NM	NM	c_4
3	ZE	NS	NS	c_6
4	NB	ZE	NB	m_2
5	NM	ZE	NM	m_4
6	NS	ZE	NS	m_6
7	ZE	PB	PB	c_1
8	ZE	PM	PM	c_3
9	ZE	PS	PS	c_5
10	PB	ZE	PB	m_1
11	PM	ZE	PM	m_3
12	PS	ZE	PS	m_5
13	ZE	ZE	ZE	Set point

[a]Reprinted with permission from Y. S. Kung and C. M. Liaw, *IEEE Trans. Fuzzy Systems* 2:194–202 (© 1994 IEEE).

TABLE 4   The Linguistic Rule Table[a]

CI e_o Δe_o	NB	NM	NS	ZE	PS	PM	PB
NB	NB	NB	NB	NB	ZE	ZE	ZE
NM	NB	NB	NM	NM	ZE	ZE	ZE
NS	NB	NB	NS	NS	PS	PS	PM
ZE	NB	NM	NS	ZE	PS	PM	PB
PS	NM	NS	NS	PS	PS	PB	PB
PM	ZE	ZE	ZE	PM	PM	PB	PB
PB	ZE	ZE	ZE	PB	PB	PB	PB

[a]Reprinted with permission from Y. S. Kung and C. M. Liaw, *IEEE Trans. Fuzzy Systems* 2:194–202 (© 1994 IEEE).

  b. At areas A_2, A_6, A_{10}—e_o = "+" and Δe_o = "−", the error is still positive, but it is decreased gradually, the control input (CI) is set to be small.

The observation about the areas A_3, A_7, A_{11} and A_4, A_8, A_{12} are dual to that listed in items a and b, respectively. According to the preceding observations, the linguistic rules at the reference areas are selected and listed in Table 4.

  4. *Defuzzification strategies*—Basically, defuzzification is a mapping from a space of fuzzy control actions defined over an output universe of discourse into a space of nonfuzzy (crisp) control actions. The defuzzification strategy is aimed at producing a nonfuzzy control action that best represents the possibility distribution of an inferred fuzzy control action. As mentioned previously, many strategies can be used for performing the defuzzification. The center-of-gravity method is adopted here. Based on the linguistic control rules shown in Table 4, a decision lookup table constructed using the center-of-gravity method is listed in Table 5. By proper scaling, the real control input (torque component current i_{qs}^*) can be generated according to the contents of the decision lookup table.

VII. DESIGN OF THE PROPOSED CONTROLLER FOR INDUCTION MOTOR DRIVE

A. Design of the Linear Model-Following Controller

  Because the dynamic model of the drive including the mechanical load is not easy to obtain using the physical modeling technique introduced in Section III, the stochastic technique proposed in [21] is used as an alternative here to find the motor drive model. The parameters of this motor drive model shown in Fig. 2b at a typical operating point (ω_r = 1000 rpm, R_L = 160 Ω) are

TABLE 5   The Decision Look Table[a]

CI e_o / Δe_o	-6	-5	-4	-3	-2	-1	0	1	2	3	4	5	6
-6	-6	-6	-6	-6	-6	-6	-6	-3	0	0	0	0	0
-5	-6	-6	-6	-6	-5	-5	-5	-2	0	0	0	0	0
-4	-6	-6	-6	-5	-4	-4	-4	-2	0	0	0	0	0
-3	-6	-6	-6	-4	-3	-3	-3	-1	1	1	1	2	2
-2	-6	-6	-6	-4	-2	-2	-2	0	2	2	2	3	4
-1	-6	-6	-5	-4	-2	-2	-1	0	2	2	3	4	5
0	-6	-5	-4	-3	-2	-1	0	1	2	3	4	5	6
1	-5	-4	-3	-2	-2	0	1	2	2	4	5	6	6
2	-4	-3	-2	-2	-2	0	2	2	2	4	6	6	6
3	-2	-2	-1	-1	-1	1	3	3	3	4	6	6	6
4	0	0	0	0	0	2	4	4	4	5	6	6	6
5	0	0	0	0	0	2	5	5	5	6	6	6	6
6	0	0	0	0	0	3	6	6	6	6	6	6	6

[a] Reprinted with permission from Y. S. Kung and C. M. Liaw, *IEEE Trans. Fuzzy Systems* 2:194–202 (© 1994 IEEE).

estimated to be

$$K_t = 1.479, \qquad H_p(s) = \frac{b}{s+a}, \qquad a = 0.103, \qquad b = 0.4275; \quad (33)$$

the plant model used for the design of the linear model-following controller is

$$H_p(s) = K_t \frac{b}{s+a} = \frac{0.6323}{s+0.103}. \quad (34)$$

The method of reference model selection described in Section IV.A.1 is used here. According to the plant model of (34) and experience about the performance achievable by the drive, a first-order reference model is chosen as

$$H_m(s) = \frac{8}{s+8}. \quad (35)$$

The state-space models corresponding to $H_p^*(s)$ and $H_m(s)$ are expressed by Eqs. (11)–(14) with the following parameters:

$$A_p = -0.103, \qquad B_p = 1.0, \qquad C_p = 0.6323, \quad (36)$$

$$A_m = -8.0, \qquad B_m = 12.652, \qquad C_m = 0.6323. \quad (37)$$

Based on (36) and (37), the parameters of LMFC are calculated using (18):

$$K_x = -7.8974, \qquad K_u = 12.652. \tag{38}$$

B. Design of the Fuzzy Controller

The main purpose of the fuzzy controller in the proposed control scheme is to synthesize an augmented control signal from the model-following error to improve the model-following performance. The decision lookup table has been designed and is shown in Table 5. To enhance the error-driven adaptive capability of the fuzzy controller, the input and output sensitivity adjustment gains are chosen to be error dependent as shown in Fig. 11, which can also be expressed by [7]

$$G_e = \begin{cases} K_1, & a_0 \geq |e_o|, \\ K_1 - \dfrac{K_1 - K_2}{a_1 - a_0}(|e_o| - a_0), & a_1 > |e_o| > a_0, \\ K_2, & |e_o| \geq a_1, \end{cases} \tag{39}$$

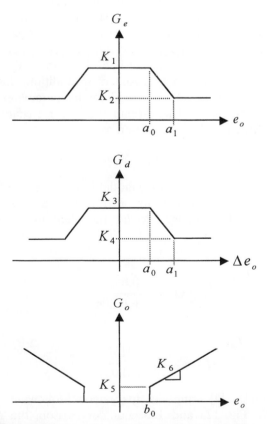

FIGURE 11 The functions of the input and output scaling factors. Reprinted with permission from Y. S. Kung and C. M. Liaw, *IEEE Trans. Fuzzy Systems* 2:194–202 (© 1994 IEEE).

$$G_d = \begin{cases} K_3, & a_0 \geq |e_o|, \\ K_3 - \dfrac{K_3 - K_4}{a_1 - a_0}(|e_o| - a_0), & a_1 > |e_o| > a_0, \\ K_4, & |e_o| \geq a_1, \end{cases} \tag{40}$$

$$G_o = \begin{cases} 0, & b_0 \geq |e_o|, \\ K_5 + K_6(|e_o| - b_0), & |e_o| \geq b_0, \end{cases} \tag{41}$$

with

$$a_0 = 0.03, \quad a_1 = 0.05,$$

$$K_1 = 60, \quad K_2 = 20, \quad K_3 = 30, \quad K_4 = 10,$$

$$b_0 = 0.001, \quad K_5 = 2, \quad K_6 = 30$$

The functions of G_e and G_d indicate that the sensitivities of error and error change will be increased linearly as their magnitudes reduce gradually. Just as the variable structure system (VSS) controller, the output of a fuzzy controller generally may exhibit hunting around the set point when the error is approaching zero [2, 5]. To solve this problem, the function G_o of (41) is emphasized in improving the transient response speed and to let the output from the fuzzy controller be zero when the error becomes smaller than a preset limit. At this instant, only the PI controller exists to eliminate the steady-state error, so the inherent chattering behavior around the set point possessed by the fuzzy controller can be avoided. In addition, this PI controller provides an additional degree of freedom; its parameters can be chosen to emphasize obtaining good load-regulating control characteristics. The parameters of the PI controller are chosen as

$$K_P = 32, \quad K_I = 96.6. \tag{42}$$

Before implementing the proposed fuzzy model-following controller, some simulation results are provided to show its performance. Suppose that the mechanical inertia constant is changed to let the plant model $H_p(s)$ be changed from that of Eq. (34) to

- Case 1,

$$H_{p1}(s) = \frac{0.855}{s + 0.206} \tag{43}$$

- Case 2,

$$H_{p2}(s) = \frac{0.21375}{s + 0.0515} \tag{44}$$

The rotor speed responses [7] using the fuzzy model-following controller (FMFC) are plotted in Fig. 12a and 12b. For comparison, the responses obtained using only the LMFC with $K_e = 10$ are shown in Fig. 12c and 12d. One can observe from the results shown in Fig. 12a–d that the proposed

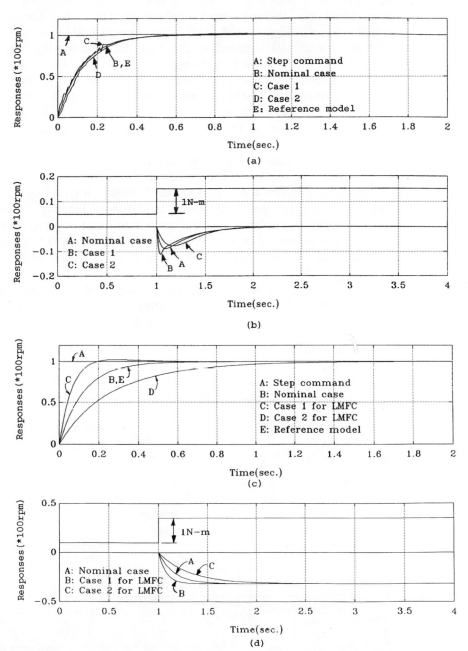

FIGURE 12 Simulated rotor speed responses: (a) due to unit-step command change by the FMFC; (b) due to step load torque (1 N-m) change by the FMFC; (c) due to unit-step command change by the LMFC; (d) due to step load torque (1 N-m) change by the LMFC. Reprinted with permission from Y. S. Kung and C. M. Liaw, *IEEE Trans. Fuzzy Systems* 2:194–202 (© 1994 IEEE).

controller gives better model-following and load-regulating responses. In addition, the responses of the same motor drive obtained using the conventional fuzzy signal tuning controller and the fuzzy parameter tuning controller [22] are shown in Figs. 13 and 14 [7]. The results indicate that the rotor speed response trajectories obtained by the fuzzy model-following controller are rather insensitive to parameter changes.

Having confirmed the effectiveness of the FMFG by simulations, the implementation of the FMFC is carried out. Generally, there are two possibilities for implementing fuzzy controllers: (1) using the design procedure described in Section V, the lookup tables are built up offline, then the controller evaluates the lookup table during run-time; (2) by run-time fuzzy inference is performed using a specialized CAD package and/or specialized VLSI circuits or processors. The former approach is adopted here because of its simplicity in circuit configuration and flexibility in changing control algorithms. The control algorithms are written using C-language. Some experimental results [7] are given here for further testing the performance of the drive system controlled by the FMFC. Figure 15a shows the dynamic speed responses obtained by the FMFC due to step speed command change of 100 rpm applied when the motor was running at (ω_{r0} = 1000 rpm, R_L = 160 Ω) and (ω_{r0} = 1000 rpm, R_L = 33 Ω), respectively. The results show that good model-following characteristics are obtained. As to the load regulating characteristics, Fig. 15b shows the dynamic speed responses due to step load changes of (ω_{r0} = 1000 rpm, R_L = 160 $\Omega \rightarrow$ 68 Ω) and (ω_{r0} = 1000 rpm,

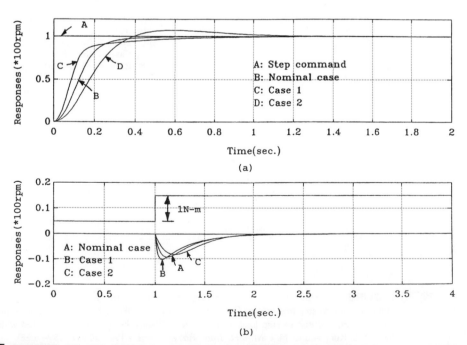

FIGURE 13 The simulated rotor speed responses by the conventional fuzzy signal tuning controller: (a) due to unit-step command change; (b) due to step load torque (1 N-m) change. Reprinted with permission from Y. S. Kung and C. M. Liaw, *IEEE Trans. Fuzzy Systems* 2:194–202 (© 1994 IEEE).

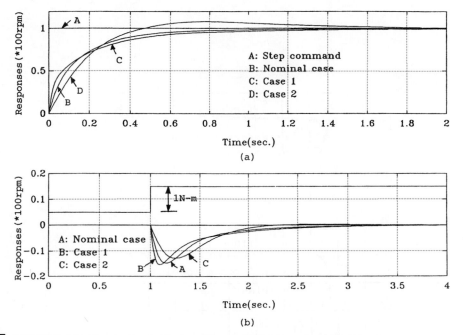

FIGURE 14 The simulated rotor speed responses by the conventional fuzzy parameter tuning controller: (a) due to unit-step command change; (b) due to step load torque (1 N-m) change. Reprinted with permission from Y. S. Kung and C. M. Liaw, *IEEE Trans. Fuzzy Systems* 2:194–202 (© 1994 IEEE).

$R_L = 160\ \Omega \to 33\ \Omega$). Good load regulating performances can also be observed from the results shown in Fig. 15b. To further observe the effect of operating condition changes on the control performance, Fig. 16a and 16b [7] shows the dynamic speed responses due to step speed command change and step load changes when the rotor was running at $\omega_{ro} = 500$ rpm. The results shown in Figs. 15 and 16 indicate that control performance of the fuzzy model-following controller is rather insensitive to operating condition changes.

VIII. CONCLUSIONS

In this chapter, the system configuration, the dynamic signal analysis, and the design and implementation of a conventional fuzzy controller for speed control of an induction motor drive are introduced. It has been shown that the dynamic response trajectory of the conventional fuzzy controller may deviate significantly due to changes of system operating condition and parameters. To let the dynamic response trajectory of a fuzzy controller be quantitatively controlled, a fuzzy model-following controller and its application to the speed control of induction motor drive are then presented. In that, an output feedback linear model-following controller is first designed according to a roughly estimated drive model and a chosen reference model. Then a fuzzy controller, which is driven by the model-following error, is designed to reduce the effects of parameter variations and unmodeled dynamics on

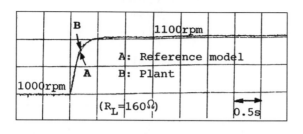

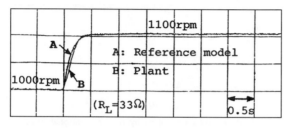

(a)

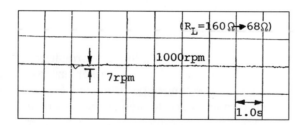

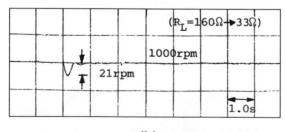

(b)

FIGURE 15 The experimental rotor speed responses by the FMFC (ω_{r0} = 1000 rpm): (a) due to step command change ($\Delta \omega_r$ = 100 rpm); (b) due to step load resistance change. Reprinted with permission from Y. S. Kung and C. M. Liaw, *IEEE Trans. Fuzzy Systems* 2:194–202 (© 1994 IEEE).

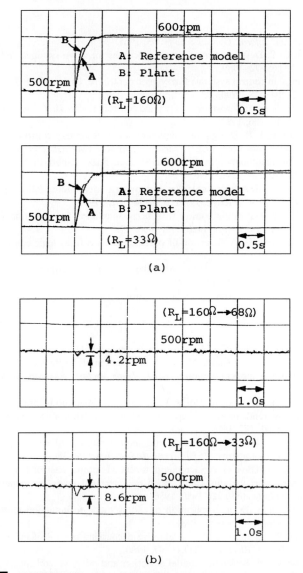

FIGURE 16 The experimental rotor speed responses by the FMFC ($\omega_{r0} = 500$ rpm): (a) due to step command change ($\Delta \omega_r = 100$ rpm); (b) due to step load resistance change. Reprinted with permission from Y. S. Kung and C. M. Liaw, *IEEE Trans. Fuzzy Systems* 2:194–202 (© 1994 IEEE).

control performances. Reference model selection and systematic design of the fuzzy model-following controller are described in detail. Having tested the effectiveness of the designed FMFC by computer simulations, the realization of the designed controller is performed. The simulated and experimental results have confirmed the validity of this controller. Finally, it can be concluded from the material presented here that the hybrid controller, by suitably combining the traditional controller and the fuzzy controller, sometimes may result in improved control characteristics.

REFERENCES

1. Zadeh, L. A. Fuzzy set. *Inform. Control* 8:338–353, 1965.
2. Li, Y. F. and Lau, C. C. Development of fuzzy algorithms for servo systems. *IEEE Control Systems Magazine* 9(1):65–72, 1989.
3. Lee, C. C. Fuzzy logic in control systems: Fuzzy logic controller. Part I. *IEEE Trans. Systems Man Cybernet.* 20(2):404–418, 1990.
4. Lee, C. C Fuzzy logic in control systems: Fuzzy logic controller. Part II. *IEEE Trans. Systems Man Cybernet.* 20(2):419–436, 1990.
5. Liaw, C. M. and Wang, J. B. Design and implementation of a fuzzy controller for a high performance induction motor drive. *IEEE Trans. Systems Man Cybernet.* 21(4):921–929, 1991.
6. Tzafestas, S. and Papanikolopopoulos, N. P. Incremental Fuzzy Expert PID control. *IEEE Trans. Indust. Appl.* 37(5):365–371, 1990.
7. Kung, Y. S. and Liaw, C. M. A fuzzy controller improving a linear model following controller for motor drives. *IEEE Trans. Fuzzy Systems* 2(3):194–202, 1994.
8. Liaw, C. M. and Lin, F. J. Position control with fuzzy adaptation for induction servomotor drive. *IEE Proc. Electric Power Appl.* 142(6):397–404, 1995.
9. Braae, M. and Rutherford, D. A. Selection of parameters for a fuzzy logic controller. *Fuzzy Sets Systems* 2:185–199, 1979.
10. Moon, B. S. Equivalence between fuzzy controllers and PI controllers for single input systems. *Fuzzy Sets Systems* 69:105–113, 1995.
11. Galichet, S. and Foulloy, L. Fuzzy controllers: Synthesis and equivalencies. *IEEE Trans. Fuzzy Systems* 3(2):140–148, 1195.
12. Mizumoto, M. Realization of PID controls by fuzzy control methods. Fuzzy Sets Systems 70:171–182, 1995.
13. Landau, I. D. *Adaptive Control: The Model Reference Approach.* Dekker, New York, 1982.
14. Liaw, C. M., Pan, C. T., and Chen, Y. C. Design and implementation of an adaptive controller for current-fed induction motor. *IEEE Trans. Indust. Electron.* 35(3):393–401, 1988.
15. Bose, B. K. *Power Electronics and AC Drives.* Prentice Hall, Englewood Cliffs, 1986.
16. Bose, B. K. Technology trends in microcomputer control of electrical machines. *IEEE Trans. Indust. Electron.* 35(1):160–177, 1988.
17. Liaw, C. M., Kung, and Wu, C. M. Design and implementation of a high performance field-oriented induction motor drive. *IEEE Trans. Indust. Electron.* 38(4):275–282, 1991.
18. Bellini, A., Figalli, G., and Ulivi, G. A microcomputer based optimal control system to reduce the effect of the parameter variations and speed measurement errors in induction motor drives. *IEEE Trans. Indust. Appl.* **22**:42–50, 1986.
19. Sabanoic, A. and Izosimov, D. B. Application of sliding mode to induction motor control. *IEEE Trans. Indust. Appl.* 17(1):41–49, 1981.
20. Chan, C. C., Leung, W. S., and Ng, C. W. Adaptive decoupling control of induction motor drives. *IEEE Trans. Indust. Electron.* 37(1):41–47, 1990.
21. Lin, F. J. and Liaw, C. M. Reference model selection and adaptive control for induction motor drives. *IEEE Trans. Automat. Control* 38(10):1594–1600, 1993.
22. Liaw, C. M., Liu, T. S., Liu, A. H., Chen, Y. T., and Lin, C. J. Parameter estimation of excitation systems from sampled data. *IEEE Trans Automat. Control* 37(5):663–666, 1992.
23. Liaw, C. M. Design and implementation of fuzzy controllers for field-oriented induction motor drives. Project report, 1991. (Supported by National Science Council, R.O.C., NSC-80-0404-E007-09.

7
FUZZY-SET-BASED MODELS OF NEURONS AND KNOWLEDGE-BASED NETWORKS: TECHNIQUES AND APPLICATIONS

WITOLD PEDRYCZ

Department of Electrical and Computer Engineering, University of Alberta, Edmonton, Alberta, Canada T6G 2G7

I. INTRODUCTION 160
II. LOGIC-BASED NEURONS 161
  A. Aggregative OR and AND Logic Neurons 161
  B. OR / AND Neurons 163
  C. Conceptual and Computational Augmentations of Fuzzy Neurons 164
III. LOGIC NEURONS AND FUZZY NEURAL NETWORKS WITH FEEDBACK 166
IV. REFERENTIAL LOGIC-BASED NEURONS 169
V. ARCHITECTURES OF FUZZY NEURAL NETWORKS 171
VI. LEARNING IN FUZZY NEURAL NETWORKS 172
  A. Learning Policies for Parametric Learning in Fuzzy Neural Networks 173
  B. Performance Index 173
VII. INTERPRETATION OF FUZZY NEURAL NETWORKS 174
VIII. CASE STUDIES 176
  A. Logic Filtering 176
  B. Minimization of Multiple-Output Two-Valued Combinational Systems 177
  C. Sensor Fusion via Fuzzy Neurons 182
  D. Logic Signal Compression 183
  E. Revealing Logical Descriptors of Data: An Application of Fuzzy Neurocomputing to Data Mining 184
IX. CONCLUSIONS 186
  REFERENCES 186

This chapter is devoted to a class of hybrid neuro-fuzzy structures and their applications. We introduce basic processing units that originate from the theory of fuzzy sets (by modeling basic logic operations) and exhibit substantial parametric flexibility supporting learning mechanisms inherent to neural networks. Transparency and plasticity are the two most essential features characterizing this class of fuzzy neural networks. Their importance is exemplified through a number of detailed illustrative examples.

I. INTRODUCTION

Here we look at an important class of fuzzy neural networks. Such networks have highly heterogeneous neural architectures, seemingly combining the learning abilities residing within neural networks while enjoying an explicit format of knowledge representation that is an inherent functional component originating from the theory of fuzzy sets. The need for this type of interaction between computational and learning abilities has been expressed in many different ways and in different contexts. It is perhaps instructive to quote von Neumann [1] who put the quest in a lucid way many years before symbiotic relationships between fuzzy sets and neural networks occurred:

> [W]e have seen how this leads to a lower level of arithmetical precision but to a higher level of logical reliability: a deterioration in arithmetic has been traded for an improvement in logic.

This observation makes a point of emphasizing the need for some trade-offs and cooperation—fuzzy sets just attempt to fill this important computational niche. We are concerned with fuzzy neural networks: a conceptual and computational vehicle combining the technology of fuzzy sets and neural networks. The term itself carries a number of representations and, in fact, materializes into a series of diverse topologies (and ensuing learning algorithms). Since the inception of fuzzy sets there have been interesting approaches resulting in the form of neuro-fuzzy structures. The reader may refer to Lee and Lee [2] as the first hybrid approach to the design of neural networks. Some other research was reported by Buckley and Hayashi [3] and Keller and Hunt [4]. It is instructive to place all the neuro-fuzzy (or fuzzy-neuro) architectures in a two-dimensional plane of learning abilities—knowledge transparency (Fig. 1). Transparency usually implies more efficient

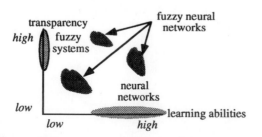

FIGURE I   Fuzzy systems and neural networks: a synergy.

training; we avoid learning from scratch by being navigated by some initial domain knowledge (which itself could be qualitative). Some relationships could be defined in advance (at least at the linguistic level). Some topologies can be also predetermined and fixed by mapping (implanting) existing domain knowledge. As the resulting network is transparent (as opposed to opaque numeric neural networks), this process is highly simplified and significantly augmented. Referring again to Fig. 1, the points situated along the first axis represent neural networks of different intensity of learning abilities. The elements distributed along the second axis concern fuzzy systems (models). The clusters of elements in this space depict an entire spectrum of hybrid systems; again some of them lean more toward fuzzy structures with some touch of learning abilities whereas some others are very much neural-network-dominant.

The content of this chapter is organized as follows: We start with an introduction to the basic computational elements—fuzzy neurons. This material is covered in Sections II and III. It starts with AND and OR neurons as the two basic processing elements. In the sequel, we discuss referential fuzzy neurons (Section IV) and proceed with referential neurons. Section V deals with specific architectures and learning abilities of the networks constructed with such processing elements. As opposed to standard numerically dominated neurons, the networks built from fuzzy neurons are inherently heterogeneous. This calls for more specialized learning mechanisms. Section VIII deals with selected case studies.

II. LOGIC-BASED NEURONS

We commence by introducing and analyzing basic properties of fuzzy neurons developed with the aid of logic operations [5–7]. Let us consider a collection of inputs $x_i, i = 1, 2, \ldots, n$, and arrange them in a vector form $(\mathbf{x})$. They are subsequently viewed as the elements of a unit hypercube, $\mathbf{x} \in [0\,1]^n$. The connections of the neurons distributed again in the unit hypercube are denoted by $\mathbf{w}, \mathbf{v}, \ldots$ The first class of neurons (called, aggregative logic neurons) realizes an aggregation of the input signals; the second is oriented toward some type of referential processing.

A. Aggregative OR and AND Logic Neurons

The OR neuron realizes a mapping $[0, 1]^n \rightarrow [0, 1]$ and is described as

$$y = \text{OR}(\mathbf{x}; \mathbf{w}).$$

Its coordinatewise description is then given by

$$y = \text{OR}[x_1 \text{ AND } w_1, x_2 \text{ AND } w_2, \ldots, x_n \text{ AND } w_n],$$

where $\mathbf{w} = [w_1, w_2, \ldots, w_n] \in [0, 1]^n$ summarizes a collection of the connections (weights) of the neuron.

The standard implementation of fuzzy set connectives involves triangular norms, which means that the OR and AND operators are realized by some s and t norms, respectively. This produces the following expression:

$$y = \overset{n}{\underset{i=1}{S}} \left[x_i \, t \, w_i \right]$$

In the AND neuron, which is somewhat a dual structure, the OR and AND operators are placed in a reversed order, producing the following expression:

$$y = \text{AND}(\mathbf{x}; \mathbf{w}),$$

which, using the notation of triangular norms is expressed as

$$y = \overset{n}{\underset{i=1}{T}} \left[x_i \, s \, w_i \right].$$

The AND and OR neurons realize "pure" logic operations on the input values. The role of the connections is to differentiate between particular levels of impact that the individual inputs may have on the result of aggregation. In particular, let us consider all the connections of the neuron equal to 0 or 1. For the OR neuron with $\mathbf{w} = \mathbf{1}$, its formula reduces to the form

$$y = x_1 \text{ OR } x_2 \text{ OR } \cdots \text{ OR } x_n$$

whereas the connections $\mathbf{w} = \mathbf{0}$ yield $y = 0$. For the AND neuron one gets

$$y = x_1 \text{ AND } x_2 \text{ AND } \cdots \text{ AND } x_n \qquad \text{for } \mathbf{w} = \mathbf{0},$$

$$y = 1 \qquad \text{for } \mathbf{w} = \mathbf{1}.$$

In general, despite the form of the triangular norms used in these constructs, the neurons coincide at the vertices of the unit hypercube and produce logical values that are consistent with the two-valued OR or AND connective.

Due to the boundary conditions of the triangular norms, we conclude that higher values of the connections in the OR neuron exercise a stronger influence than the corresponding inputs on the output of the neuron. The opposite weighting (ranking) effect takes place in the case of the AND neuron: here the values of w_i close to one make the influence of x_i almost negligible.

The reader can also observe an interesting generalization furnished by these logic neurons, as outlined in Fig. 2. The highest level with the underlying constructs of multivalued logic leads to intermediate versions of the world in which modeling activities take place, namely,

- Boolean (0–1) world, fuzzy (analog) connections
- Boolean (0–1) connections, continuous world

At the lowest level emerge two-valued logic constructs of standard OR and AND digital gates.

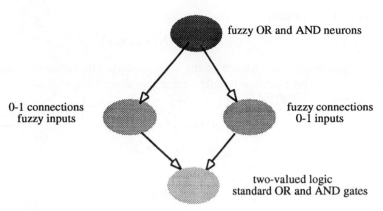

fuzzy OR and AND neurons

0-1 connections
fuzzy inputs

fuzzy connections
0-1 inputs

two-valued logic
standard OR and AND gates

FIGURE 2   A stream of a specialization of logic neurons.

B. OR / AND Neurons

As a straightforward extension of the two aggregative neurons discussed so far, we introduce a neuron with intermediate logical characteristics. The OR/AND neuron [8], is constructed by bundling several AND and OR neurons into a single two-layer structure as shown in Fig. 3.

The main motivation behind combining several neurons and considering them as a single computational entity (which, in fact, constitutes a small fuzzy neural network), lies in an ability of this neuron to synthesize intermediate logical characteristics. These characteristics can be located anywhere between the "pure" OR and AND characteristics generated by the previous neurons. The influence coming from the OR (AND) part of the neuron can be properly balanced by selecting suitable values of the connections v_1 and v_2 during this neuron's learning. In the limit, when $v_1 = 1$ and $v_2 = 0$, the OR/AND neuron operates as a pure AND neuron. In the second extremal situation (for which $v_1 = 0$ and $v_2 = 1$), the structure functions as a pure OR

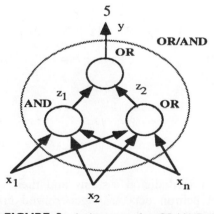

FIGURE 3   Architecture of an OR / AND neuron.

neuron. The notation

$$y = \mathrm{OR}/\mathrm{AND}(\mathbf{x}; \mathbf{w}, \mathbf{v})$$

clearly underlines the nature of the intermediate characteristics produced by the neuron. The relevant detailed formulas describing this architecture read as follows:

$$y = \mathrm{OR}([\mathbf{z}_1 \quad \mathbf{z}_2]; \mathbf{v}),$$

$$z_1 = \mathrm{AND}(\mathbf{x}; \mathbf{w}_1),$$

and

$$z_2 = \mathrm{OR}(\mathbf{x}; \mathbf{w}_2)$$

with $\mathbf{v} = [v_1 \, v_2]$ and $\mathbf{w}_i = [w_{i1} \, w_{i2} \, \cdots \, w_{in}]$, $i = 1, 2$, being the connections of the corresponding neurons. We can encapsulate the preceding expressions into a single formula by writing

$$y = \mathrm{OR}/\mathrm{AND}(\mathbf{x}; \mathbf{connections}),$$

where the **connections** summarize all the connections of the network.

C. Conceptual and Computational Augmentations of Fuzzy Neurons

Two further enhancements of fuzzy neurons can be anticipated. They are aimed at making these logical processing units more flexible, from a conceptual as well as computational point of view.

Representing Inhibitory Information

As the coding range being commonly encountered in fuzzy sets constitutes the unit interval, the inhibitory effect to be conveyed by some variables can be achieved by including their complements instead of the direct variables themselves, say $\bar{x}_i = 1 - x_i$. Hence, the higher the value of x_i, the lower the activation level associated with it. Thus the original input space $[0, 1]^n$ is augmented and the neurons are now described as follows:

- OR neurons,

$$y = \mathrm{OR}\big([x_1 \quad x_2 \quad \cdots \quad x_n \quad \bar{x}_1 \quad \bar{x}_2 \quad \cdots \quad x_n];$$
$$[w_1 \quad w_2 \quad \cdots \quad w_n \quad w_{n+1} \quad w_{n+2} \quad \cdots \quad w_{2n}]\big)$$

- AND neuron,

$$y = \mathrm{AND}\big([x_1 \quad x_2 \quad \cdots \quad x_n \quad \bar{x}_1 \quad \bar{x}_2 \quad \cdots \quad x_n];$$
$$[w_1 \quad w_2 \quad \cdots \quad w_n \quad w_{n+1} \quad w_{n+2} \quad \cdots \quad w_{2n}]\big)$$

The reader familiar with two-valued digital systems and their design can easily recognize that any OR neuron acts as a generalized maxterm [9] summarizing x_i and their complements whereas the AND neurons can be

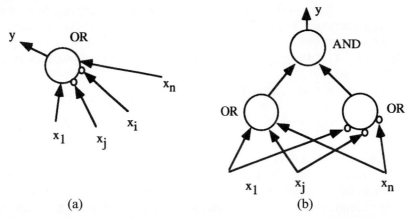

(a)

(b)

FIGURE 4 Representing inhibitory and excitatory information in logic neurons.

viewed as the generalization of miniterms (product terms) encountered in digital circuits. Symbolically, the complemented variable (input) is denoted by a small dot as visualized in Fig. 4a.

There is also another way of representing inhibitory information in the neuron (or the network). Instead of combining all the inputs (both inhibitory and excitatory), we split these inputs and expand the architecture by adding few extra neurons. The concept is illustrated in Fig. 4b. The inhibitory inputs are first aggregated in the second neuron. The outputs of the neurons are combined AND-wise by the third neuron, situated in the outer layer. Consider that all the inhibitory inputs are equal to 1. Then the output z_2 is equal to 0. Assuming that the connections of the AND neuron are equal to zero, this draws the value of the output (y) down to zero.

Computational Enhancements of the Neurons

Despite the well-defined semantics of these neurons, the main concern one may eventually raise about these constructs happens to be on a numerical side. Once the connections (weights) are set (after learning) each neuron realizes an *in* (rather than an *on*) mapping between the unit hypercubes, that means that the values of the output y for all possible inputs cover a subset of the unit interval but not necessarily the entire interval $[0, 1]$. More specifically, for the OR neuron the values of y are included in $[0, S_{i=1}^{n} w_i]$ whereas the accessible range of the output values of the AND neuron is limited to $[T_{i=1}^{n} w_i, 1]$.

This observed shortcoming could be alleviated by augmenting the neuron by a nonlinear element placed in series with the previous logical component. The neurons obtained in this manner are formalized accordingly:

$$y = \psi \left(\underset{i=1}{\overset{n}{S}} \left(x_i \, t \, w_i \right) \right) = \psi(u),$$

$$y = \psi \left(\underset{i=1}{\overset{n}{T}} \left(x_i \, s \, w_i \right) \right) = \psi(u),$$

where $\psi: [0, 1] \rightarrow [0, 1]$ is a nonlinear monotonic mapping. In general, we can even introduce mappings whose monotonicity is restricted to some regions of the unit interval.

A useful two-parameter-family of sigmoidal nonlinearides is specified in the form

$$y = \psi(u) = \frac{1}{1 + \exp[-(u - m)\sigma]},$$

$u, m \in [0, 1]$, $\sigma \in \mathbf{R}$.

By adjusting the parameters of the function (i.e., m and σ), various forms of the nonlinear characteristics of the element can be easily obtained. Especially, the values of σ determine either an increasing or decreasing type of the characteristics of the obtained neuron, whereas the second parameter (m) shifts the all characteristics along the unit interval.

III. LOGIC NEURONS AND FUZZY NEURAL NETWORKS WITH FEEDBACK

The logic neurons studied so far realize a static memoryless nonlinear mapping in which the output depends solely upon the inputs of the neuron. In this form, the neurons are not capable of handling dynamical (memory-based) relationships between the inputs and outputs. This aspect might be, however, essential in a proper problem description.

As an example, consider the following diagnostic problem. A decision about system's failure should be issued while one of the system's sensors provides information about an abnormal (elevated) temperature. Observe that the duration of the phenomenon has a primordial impact on expressing confidence about failure. If the elevation of the temperature is prolonged, confidence about failure rises. On the other hand, some short temporary temperature elevations (spikes) reported by the sensor might be almost ignored (filtered out) and need not have any impact on the decision about failure. To capture this dynamical effect properly, one has to equip the basic logic neuron with a certain feedback link. A straightforward extension of this nature is schematically illustrated in Fig. 5.

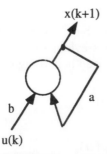

FIGURE 5   A logical neuron with feedback.

The type of neuron with feedback is described accordingly:

$$x(k + 1) = [b \text{ OR } u(k)] \text{ AND } [a \text{ OR } x(k)]$$

The dynamics of the neuron obtained in this fashion is uniquely defined by the feedback connection (a) as it determines a speed of evidence accumulation $x(k)$. The initial condition $x(0)$ expresses a priori confidence associated with x. After a sufficiently long period of time $x(k + 1)$ could take on higher values compared with the level of the original evidence present at the input.

High-order dynamical dependencies to be accommodated by the network, if necessary, have to be taken care of via a feedback loop consolidating several pieces of a temporal information, for example,

$$x(k + 2) = [b \text{ OR } u(k)] \text{ AND } [a_1 \text{ OR } x(k)] \text{ AND } [a_2 \text{ OR } x(k + 1)].$$

One can also refer to the two preceding relationships as examples of *fuzzy difference equations*.

The limit analysis of the neuron (or network) with feedback allows us to reveal some general dynamical properties of the system. As an example let us study the neuron described by the relationship

$$x(k + 1) = \bar{x}(k) \text{ OR } a,$$

where a characterizes the feedback loop; $a \in [0, 1]$. Additionally, let the s norm be given as the probabilistic sum. This yields

$$x(k + 1) = a + (1 - x(k)) - a(1 - x(k)).$$

Iterating from $x(0)$, that is, computing $x(1), x(2), \ldots$, we can unveil the steady state behavior of the system (assuming that it does exist). This example is simple enough to analyze in detail. Let us denote the steady state by $x(\infty)$. Then we get

$$x(\infty) = a + \bar{x}(\infty) - a\bar{x}(\infty).$$

Rearranging the terms one obtains

$$x(\infty)(2 - a) = 1$$

and finally

$$x(\infty) = \frac{1}{2 - a}.$$

As the simulations reveals (Fig. 6), for the connection in $(0, 1)$ the neuron does not converge to any specific value but oscillates continuously assuming states between 0 and 1. In fact, the value $x(\infty)$ becomes an average of the states the neuron can occupy. The same observation holds for the remaining values of the feedback parameter except that lower values of a yield a more profound amplitude of the oscillations.

Now let us study the neuron with feedback governed by the expression

$$x(k + 1) = x(k) \text{ OR } a.$$

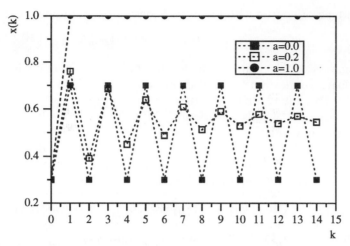

FIGURE 6 States of a fuzzy neuron $x(k + 1) = a$ OR $\bar{x}(k)$ for several values of the feedback connection; initial condition $x(0) = 0.3$.

Confining ourselves to the probabilistic sum we derive

$$x(k + 1) = a + x(k) - ax(k).$$

Steady-state analysis leads to the following expression:

$$x(\infty) = 1.$$

In this case the neuron converges to $x(\infty)$ as confirmed by the simulation experiments, (Fig. 7). Here the rate of convergence depends on the feedback connection.

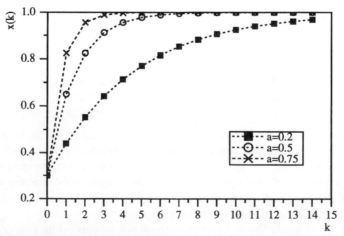

FIGURE 7 States of a fuzzy neuron $x(k + 1) = a$ OR $x(k)$ for several values of the feedback connection; initial condition $x(0) = 0.3$.

IV. REFERENTIAL LOGIC-BASED NEURONS

Compared with the AND, OR, and OR/AND neurons realizing aggregative operations, the class of neurons discussed now is useful in realizing reference computations. The main idea behind this structure is that the input signals are not directly aggregated as in the aggregative neuron but rather are analyzed first (e.g., compared) with respect to the given reference point. The results of this analysis (involving such operations as matching, inclusion, difference, and dominance) are afterward summarized in the aggregative part of the neuron as previously described. In general one can describe the reference neuron as

$$y = \text{OR}(\text{REF}(\mathbf{x}; \textbf{reference\_point}), \mathbf{w})$$

(a disjunctive form of aggregation) or

$$y = \text{AND}(\text{REF}(\mathbf{x}; \textbf{reference\_point}), \mathbf{w})$$

(a conjunctive form of aggregation), where the term REF($\cdot$) stands for the reference operation carried out with respect to the provided point of reference.

In spite of the specific form of the reference operation, the functional behavior of the neuron is described accordingly (the following formulas all pertain to the disjunctive form of aggregation):

i. The MATCH neuron is

$$y = \text{MATCH}(\mathbf{x}; \mathbf{r}, \mathbf{w})$$

or, equivalently,

$$y = \overset{n}{\underset{i=1}{\text{S}}} \left[w_i \, \text{t} \, (x_i \equiv r_i) \right]$$

where $\mathbf{r} \in [0, 1]^n$ stands for a reference point defined in the unit hypercube. The equality operation [10] is defined as follows:

$$x \equiv y = \tfrac{1}{2}\left[(x \to y) \wedge (y \to x) + (\bar{x} \to \bar{y}) \wedge (\bar{y} \to \bar{x}) \right]$$

The implication operation in this expression is associated with a certain t norm [11] that is,

$$x \to y = \sup\{c \in [0, 1] | x \, \text{t} \, c \le y\}.$$

Moreover, the complement is defined in the usual way,

$$\bar{x} = 1 - x.$$

To emphasize the referential character of this processing carried out by the neuron one can rewrite the basic formula in the equivalent form

$$y = \text{OR}(\mathbf{x} \equiv \mathbf{r}; \mathbf{w}).$$

The use of the OR neuron indicates an "optimistic" (disjunctive) character of the final aggregation. The pessimistic form of this aggregation is produced by using the AND operation.

    ii. The difference neuron combines degrees to which $\mathbf{x}$ is different from the given reference point $\mathbf{g} = [g_1, g_2, \ldots, g_n]$. The output is interpreted as a global level of difference observed between the input $\mathbf{x}$ and this reference point,

$$y = \mathrm{DIFFER}(\mathbf{x}; \mathbf{w}, \mathbf{g}),$$

that is,

$$y = \mathop{S}_{i=1}^{n} \left[w_i \, \mathbf{t} \left(x_i \equiv\!| \, g_i \right) \right],$$

where the difference operator is defined as a complement of the equality index,

$$a \equiv\!| \, b = 1 - (a \equiv b).$$

As before, the referential character of processing is emphasized by noting that

$$\mathrm{DIFFER}(\mathbf{x}; \mathbf{w}, \mathbf{g}) = \mathrm{OR}(\mathbf{x} \equiv\!| \, \mathbf{g}; \mathbf{w})$$

    iii. The inclusion neuron summarizes the degrees of inclusion to which $\mathbf{x}$ is included in the reference point $\mathbf{f}$,

$$y = \mathrm{INCL}(\mathbf{x}; \mathbf{w}, \mathbf{f}),$$

$$y = \mathop{S}_{i=1}^{n} \left[w_i \, \mathbf{t} \left(x_i \rightarrow f_i \right) \right].$$

    iv. The dominance neuron expresses a relationship dual to that carried out by the inclusion neuron

$$y = \mathrm{DOM}(\mathbf{x}; \mathbf{w}, \mathbf{h}),$$

where $\mathbf{h}$ is a reference point. In other words, the dominance relationship generates a degrees to which $\mathbf{x}$ dominates $\mathbf{h}$ (or, equivalently, $\mathbf{h}$ is dominated by $\mathbf{x}$). The coordinatewise notation of the neuron reads

$$y = \mathop{S}_{i=1}^{n} \left[w_i \, \mathbf{t} \left(h_i \rightarrow x_i \right) \right].$$

    To model complex situations, the referential neurons can be encapsulated in a form of neural network. An example is a tolerance neuron that consists of DOMINANCE and INCLUSION neurons placed in the hidden layer and a single AND neuron in the output layer (Fig. 8). The neuron discussed generates a tolerance region as shown in Fig. 9.

    The ordinal sum of t norms can be used in another model of a fuzzy neuron with more diversified processing occurring at the level of synaptic processing. Assume that the number of t norms as well as their type have

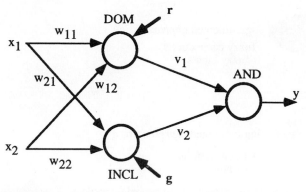

FIGURE 8 Tolerance neuron.

been fixed. We leave, however, the ranges of the individual t norms variable and regard them as adjustable parameters of the neuron. In a concise form this reads

$$y = \text{OR}(\mathbf{x}; \mathbf{w}, \mathbf{a}, \mathbf{b}),$$

where $\mathbf{a}$ and $\mathbf{b}$ are the vectors of the ranges of the contributing t norms.

V. ARCHITECTURES OF FUZZY NEURAL NETWORKS

Both aggregative and referential neural networks form basic building blocks of larger architectures of fuzzy neural networks. The networks are knowledge-oriented as their topologies are very much induced by the specificity of the problem at hand. In this sense the available pieces of domain knowledge are directly ported onto the structure. As the section on case studies reveals, the resulting networks are highly heterogeneous and the diversity depends

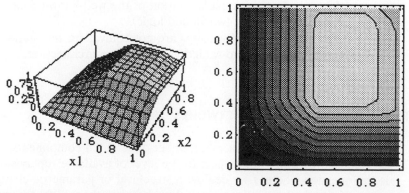

FIGURE 9 2D and 3D characteristics of a tolerance neuron: AND neuron, min operator; INCL and DOM neurons, $a \rightarrow b = \min(1, b/a)$, $a, b \in [0, 1]$; $w_{ij} = 0.05$, $v_i = 0.0$; reference points, (INCL neuron) $\mathbf{r} = [0.8\ 0.9]$ and (DOM neuron) $\mathbf{g} = [0.5\ 0.4]$.

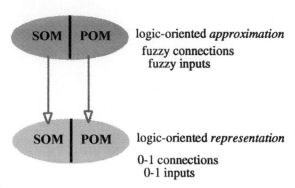

logic-oriented *approximation*
fuzzy connections
fuzzy inputs

logic-oriented *representation*

0-1 connections
0-1 inputs

FIGURE 10   Logic processors: multivalued and two-valued versions.

upon the form of the problem to be tackled. Interestingly enough, there is a broad category of *logic-inclined approximation* problems. The form of the approximation is driven by a logical expression—a fuzzy function [12]. The form of the function calls for a fuzzy neural network—a so-called logic processor consisting of three layers. Two alternatives are sought:

- A sum-of-products version (The SOM type of logic processor involves a hidden layer formed by the AND neurons. The output layer comprises OR neurons.)
- A product-of-sums version (The POM type of logic processor has a hidden layer formed by the OR neurons. This layer is followed by the output layer consisting of the AND neuron.)

These two networks approximate experimental data by a logical relationship —a fuzzy function. It is essential to note that what is produced as a result of the approximation is easily interpretable as a collection of rules. The relationship between the logic processor (LP) and the reduced two-valued version (Fig. 10) underlines two important features:

1. The LP *approximates* data; the two-valued counterpart *represents* data (this is, in fact, an implementation of the well-known Shannon theorem encountered in two-valued logic).
2. The POM and SOM versions are not equivalent; this stands in sharp contrast with the two-valued version of the architecture.

VI. LEARNING IN FUZZY NEURAL NETWORKS

The issue of learning embraces a number of essential components. First is the question of learning with respect to the level of available supervision. Second is the question whether we deal with structural or parametric changes of the network or whether we can work with both of them. We commence with these issues and then focus on a number of case studies that illustrate several learning procedures.

A. Learning Policies for Parametric Learning in Fuzzy Neural Networks

Learning in the fuzzy neural network can vary from case to case and usually depends heavily on the initial information available about the problem that can be accommodated immediately by the network. For instance, in many situations it is obvious in advance that some connections will be weak or even nonexistent. This allows us to build an initial configuration of the network very divergent from a fully connected network. This initial knowledge tangibly enhances the learning procedure, eliminating the need to modify all the connections of the network and thus preventing us from proceeding with learning from scratch. On the other hand, if the initial domain knowledge about the problem (network) is not sufficient, then a fully connected structures yielding higher values of its entropy function [13, 14] would be strongly recommended.

In many cases the role of the individual layers is also obvious so that one can project the behavior of the network (and evaluate its learning capabilities). The following two general strategies of learning are worth pursuing:

1. In the *successive reductions* strategy, one starts with a large and eventually excessive neural network (containing many elements in the hidden layer), analyzes the results of learning and, if possible, resumes the size of the network. These reductions are carried out as far as they do not drastically affect the quality of learning (by slowing it down significantly and/or elevating the values of the minimized performance index). The main advantage of this strategy lies in a fast learning. This is achieved due the "underconstraint" nature of the successive networks. A certain shortcoming is that the network constructed in this way can be fairly "overdistributed".

2. In the *successive expansions* strategy, the starting point is a small neural network that afterward expanded successively based on the values of the performance index obtained. Too high values of the index may suggest further expansions. The network derived in this way could be made compact; nevertheless, under some circumstances a total computational overhead (many unsuccessfully extended structures of the neural networks) may not be acceptable and could make this approach computationally quite costly.

B. Performance Index

One can easily envision a number of different ways in which the similarity (distance) between the outputs of the neural network and the target fuzzy sets can be expressed. Although the Euclidean distance is common, we confine ourselves to the equality index. There are two reasons behind studying this option. First, the equality index retains a clear logic set-oriented interpretation that could be of particular interest in the setting of fuzzy neural networks. Second, with a suitable selection of the underlying t norms, it may assure some robustness features.

Let x and y denote two membership values, $x, y \in [0, 1]$. We consider the equality index already defined. Owing to the robustness aspects we are interested in, it is instructive to concentrate on the Lukasiewicz implication as being a part of this quality index,

$$x \to y = \begin{cases} 1, & \text{if } x \leq y, \\ 1 - x + y, & \text{if } x > y. \end{cases}$$

After simple calculations we derive the following expression:

$$x \equiv y = \begin{cases} 1 - x + y, & \text{if } x > y, \\ 1, & \text{if } x = y, \\ 1 - y + x, & \text{if } x < y. \end{cases}$$

The equality index becomes a piecewise-linear function of the arguments. Evidently,

$$x \equiv x = 1.$$

The equality index can be rewritten in a slightly different manner by observing that both $1 - x + y$ and $1 - y + x$ can be combined into a single expression $1 - |x - y|$ that is nothing but a complement of the Hamming distance between x and y. Thus

$$x \equiv y = 1 - |x - y|.$$

It is also evident that the Lukasiewicz-based equality index based on the L_1 metric promotes significant robustness properties of the fuzzy neural network obtained.

VII. INTERPRETATION OF FUZZY NEURAL NETWORKS

In contrast to the way neural networks are interpreted, an interpretation of fuzzy neural networks (FNNs) is far easier and can be done in a more comprehensive and exhaustive fashion. This stems from the fact that FNNs are highly heterogeneous structures in which each processing element exhibits well-defined functional characteristics. In spite of the flexibility of the network at the parametric level, each element retains its underlying characteristics. What is more important is that the interpretation of the network is not carried out in the standard input—output black-box fashion but rather by gaining some direct insight into the structure of the network.

As an example consider the so-called LP structure shown in Fig. 11. The hidden layer is composed of AND neurons and the output layer is built up with the aid of OR neurons. The connections of the neurons obtained after learning are visualized in Fig. 11.

The following formulas are inferred directly from the structure (we have ignored the values of the weakest connections; more detailed explanation is provided later):

- If (x_1 and not(x_2)) or x_3 then y_1
- If (x_1 and x_5) or x_2 then y_2

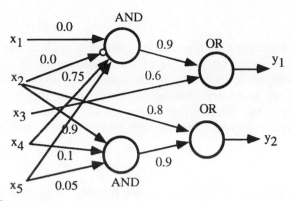

FIGURE 11 Four-input, two-output fuzzy neural network.

and can be regarded as two rules induced (conveyed) by the network. The black-box approach is not capable of revealing this type of dependence; in particular, the internal relationships are not tackled at all.

The interpretation can be made even more transparent by dropping the least relevant connections of the neurons. Depending upon the form of the neuron, two reduction schemes apply:

- In OR neurons reduce (drop) the connections with *low* value.
- In AND neurons ignore the connections with *high* values (values that are close to 1).

These two simple guidelines follow an analysis of the boundary conditions of the t and s norms. We get

- If $w \approx 0$ then $w\,t\,x \approx 0\,t\,x = 0$
- If $w \approx 1$ then $w\,t\,x \approx 1\,t\,x = 1$

meaning that the value of the second argument (x) becomes completely irrelevant if the connections are too low or too high.

The reduction procedure requires a specification of two thresholds situated in the unit interval. Denote them by μ and λ; depending on the form of neuron, the connections below or above these thresholds become pruned. More specifically,

- OR neuron, if $w < \lambda$ then eliminate connection
- OR neuron, if $w < \lambda$ then eliminate connection

Continuing with the interpretation example, let us set the thresholds as equal: $\mu = 0.5$, $\lambda = 0.5$. This leads to the already shown interpretation of the network coming in the form of two rules.

In general, by changing the thresholds the pruning can be made more radical. For detailed studies on the selection of these parameters refer to [12].

VIII. CASE STUDIES

In this section we highlight a number of representative applications of FNNs. In particular, we elaborate on the role of the transparency of a networks' architecture with respect to its ability to accommodate domain knowledge as well as carry out all necessary learning activities. It is also important to underline that due to the prudent representation of domain knowledge, the learning itself is quite supported and may not be carried out from scratch.

A. Logic Filtering

One of the most popular methods of signal filtering deals with a sort of averaging (Fig. 12).

Considering that the values of $x(k)$ are situated in the unit interval, we propose a discrete-time, logic-based filter of the form

$$y(k + 1) = \mathrm{OR}([x(k + 1), y(k)], [w\, 1 - w])$$

where

$x(k + 1)$ is the input signal
$y(k)$ is the output signal
w is the weight factor in the range of $[0, 1]$

One can rewrite the filter expression in a compact vector format,

$$y(k + 1) = \mathrm{OR}(\mathbf{z}, \mathbf{v}),$$

where now $\mathbf{z} = [y(k)\, x(k + 1)]$, $\mathbf{v} = [w\, 1 - w])$. That is,

$$y(k + 1) = [y(k)\, \mathbf{t}\, w]\, \mathbf{s}\, [x(k + 1)\, \mathbf{t}\, (1 - w)].$$

The weight factor is used to achieve required filtering properties. Note that high values of w promote strong filtering [$y(k)$ depends heavily on the previous output of the filter]. For $w \approx 1$, we get $y(k + l) \approx x(k + 1)$ and filtering gets very much limited. Let us realize the t norm and s norm as the product and the probabilistic sum, respectively. Consider that the input $x(k)$ is a mixture of a constant signal with some superimposed noise component

$$x(k) = 05 + 0.20 * z(k),$$

where $z(k)$ comes from a random variable of a uniform distribution in $[0, 1]$. The role of the filtering parameter (w) is clearly visible from Fig. 13. Higher values of w imply a more profound averaging phenomenon.

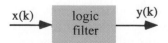

FIGURE 12 Logic filter as an input–output system.

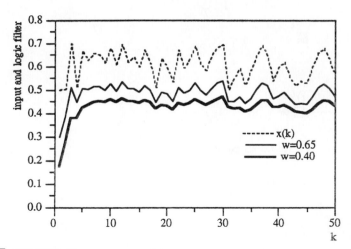

FIGURE 13 Input signal $x(k)$ and response of the filter for two selected values of the weight factor.

B. Minimization of Multiple-Output Two-Valued Combinational Systems

The problem of minimization of Boolean functions has been extensively studied in the literature on digital systems. This, in fact, constitutes a cornerstone of any design process of combinational as well as sequential systems. For a small number of independent variables (say, up to 5–6), this optimization can be carried out manually and it usually uses Karnaugh maps (K maps). The problem becomes more challenging when one wants to minimize several Boolean functions one at a time. The minimization should lead to the most compact circuit that uses a minimal amount of hardware. To illustrate the way in which this task could be handled through computing with fuzzy neural networks, let us consider several functions given in a canonical format of a sum of minterms. We start with a simple illustrative problem involving only a single output.

1. We consider a well-known exclusive-OR (XOR) problem that is commonly viewed as a testbed for analyzing various learning algorithms for neural networks. In the simplest scenario the training set consists of four two-dimensional patterns distributed at the vertices of the unit square (Fig. 14).

The logical expression describing these patterns reads as

$$y = (x_1 \text{ AND } \bar{x}_2) \text{ OR } (\bar{x}_1 \text{ AND } x_2).$$

The training is completed using the network with two hidden AND units; see Fig. 15. The learning rate is 0.15. The standard MSE performance index recorded over the course of learning is shown in Fig. 16. The resulting connections are given next; the first matrix summarizes the connections between the inputs and the hidden layer; the second one concerns the

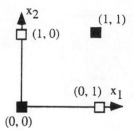

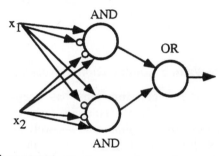

FIGURE 14   Two-dimensional exclusive OR problem.

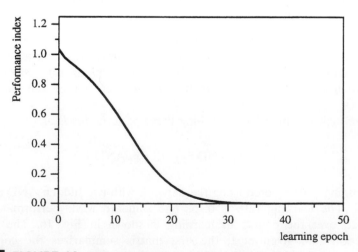

FIGURE 15   Fuzzy neural network supporting XOR problem.

FIGURE 16   Performance index in successive learning epochs.

connections of the OR neuron in the output layer:

$$\begin{bmatrix} 0.0000 & 0.9997 & 1.0000 & 0.0000 \\ 1.0000 & 0.0004 & 0.0005 & 1.0000 \end{bmatrix},$$

$$[1.0000 \quad 0.9998].$$

Even without any pruning one can conclude that the neural network fully complies with the logical expression of the XOR problem.

2. The data (given next) involve two boolean functions with three arguments (variables);

$$\begin{array}{ccccc} 0 & 0 & 0 & 1 & 0 \\ 0 & 0 & 1 & 1 & 1 \\ 0 & 1 & 1 & 0 & 0 \\ 0 & 1 & 0 & 1 & 0 \\ 1 & 0 & 0 & 0 & 1 \\ 1 & 0 & 1 & 1 & 1 \\ 1 & 1 & 1 & 0 & 0 \\ 1 & 1 & 0 & 0 & 0 \end{array}$$

The learning was carried out for several dimensions of the hidden layer (h); h was varied from 2, 3, and 4. The learning rate was 0.05. The results are displayed in Figs. 17 and 18. The learning led to successful results (zero performance index) or got stuck at some level of the performance index due to an insufficient topology of the network (too small size of the hidden layer). This effect happens in the case of $h = 2$. The training was successful both for $h = 3$ and $h = 3$. The results (connections of the network) are as follows:

• for $h = 2$,

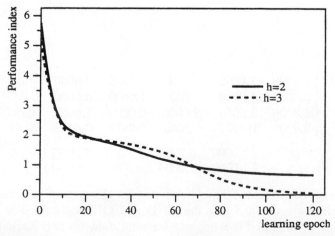

FIGURE 17 Performance index in successive learning epochs for $h = 2$ and $h = 3$.

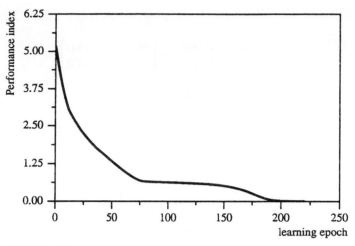

FIGURE 18 Performance index in successive learning epochs for $h = 4$.

input—hidden layer (the successive columns are denoted by x_1, x_2, and x_3, followed by their complements),

$$\begin{bmatrix} 0.9675 & 1.0000 & 0.3592 & 1.0000 & 0.0000 & 1.0000 \\ 1.0000 & 1.0000 & 1.0000 & 0.0000 & 1.0000 & 0.0060 \end{bmatrix}$$

output—hidden layer,

$$\begin{matrix} 0.9635 & 1.0000 \\ 1.0000 & 0.0000 \end{matrix}$$

- for $h = 3$,

$$\begin{bmatrix} 1.0000 & 1.0000 & 1.0000 & 0.0000 & 1.0000 & 0.0000 \\ 0.0000 & 1.0000 & 1.0000 & 1.0000 & 0.0000 & 0.0707 \\ 1.0000 & 1.0000 & 0.0000 & 1.0000 & 0.0000 & 1.0000 \end{bmatrix},$$

$$\begin{bmatrix} 1.0000 & 0.0000 & 1.0000 \\ 0.0000 & 1.0000 & 1.0000 \end{bmatrix}$$

- for $h = 4$,

$$\begin{bmatrix} 1.0000 & 1.0000 & 0.0000 & 1.0000 & 0.0000 & 1.0000 \\ 0.0000 & 1.0000 & 1.0000 & 1.0000 & 0.0000 & 0.06759 \\ 1.0000 & 1.0000 & 1.0000 & 0.0000 & 1.0000 & 0.0000 \\ 0.0000 & 0.4327 & 0.1666 & 0.0817 & 0.4682 & 0.7320 \end{bmatrix},$$

$$\begin{bmatrix} 1.0000 & 0.0000 & 1.0000 & 0.3077 \\ 1.0000 & 1.0000 & 0.0000 & 0.5494 \end{bmatrix}$$

All the networks obtained have been subject to some slight pruning with threshold levels set to 0.5 (for the AND and OR neurons). For comparative reasons we mapped the original learning data on two Karnaugh maps (Fig. 19).

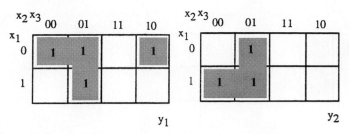

FIGURE 19 Karnaugh maps (K-maps) of the training data.

For $h = 2$ we derive the following boolean expression:

$$y_1 = \bar{x}_2 \text{ OR } (\bar{x}_1 \text{ AND } \bar{x}_3),$$

$$y_2 = \bar{x}_2,$$

which when mapped onto the K maps do not coincide with the original Boolean functions, Fig. 20. In fact, the network has introduced some additional entries whose existence is visible by comparing these two maps with the K maps of the original functions.

For $h = 3$ we obtain

$$y_2 = (\bar{x}_2 \text{ AND } \bar{x}_3) \text{ OR } (x_3 \text{ AND } \bar{x}_2),$$

$$y_2 = (x_1 \text{ AND } \bar{x}_2 \text{ AND } \bar{x}_3) \text{ OR } (x_3 \text{ AND } \bar{x}_2).$$

The fuzzy neural networks discovered the same product terms as those used for building the original functions (see Fig. 21).

The network with $h = 4$ (Fig. 22) becomes excessively large and this produces the following results

$$y_1 = (\bar{x}_1 \text{ AND } \bar{x}_3) \text{ OR } (x_3 \text{ AND } \bar{x}_2),$$

$$y_2 = (x_1 \text{ AND } \bar{x}_2) \text{ OR } (x_3 \text{ AND } \bar{x}_2)$$

$$\text{OR } (x_1 \text{ AND } x_2 \text{ AND } x_3 \text{ AND } \bar{x}_1 \text{ AND } \bar{x}_2).$$

The network is excessively large—which becomes fully reflected in the second expression, whose last product does not make any sense (as it cancels

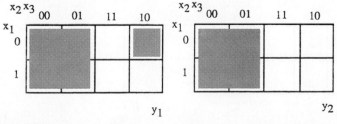

FIGURE 20 K-map derived from the fuzzy neural network for $h = 2$.

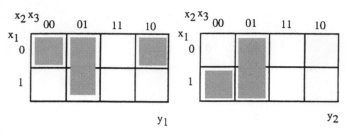

FIGURE 21 K-map derived from the fuzzy neural network for $h = 3$.

out automatically). An interesting phenomenon happens for the first function; it becomes more compact by taking advantage of a reduced product term. In fact, these terms can be obtained by "eyeballing" the Karnaugh maps of the relationships. Interestingly enough, the network was capable of minimizing these logical expressions as a by-product of the minimization of the performance index.

C. Sensor Fusion via Fuzzy Neurons

A broad category of problems of sensor fusion can be formulated as follows. Consider a collection of n senors that provide signals about a possible fault in a system. Each sensor has its own characteristics. For the purposes of this study we can assume that the higher the signal of the sensor, the more profound the level of anticipated fault. As the importance of each sensor could be different, we associate with the indication of the sensor a certain level of confidence. Taking this into account, a simple model of sensor fusion can be realized in the form of a single AND neuron in which the connections of the neuron are utilized to model the credibility levels associated with the corresponding sensors. More interesting and far more realistic scenarios emerge when we are interested in incorporating eventual relationships between the indications of the sensors. For instance, consider three sensors whose indications x_1, x_2, and x_3 are fused to infer information about the level of fault (f) of the system. Furthermore, we know that there is some relationship between the first and second sensors in the sense that usually the indications of the first are lower than that reported by the second. This constraint of inclusion relationship has to be accommodated as a part of the

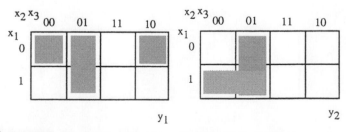

FIGURE 22 K-map derived from the fuzzy neural network for $h = 4$.

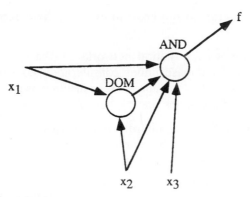

FIGURE 23 Fuzzy neural network in modeling senor fusion with dominance relationship.

model. Note that when x_1, is higher than x_2 even though both of them are high and should suggest a substantial level of failure, the violation of the satisfaction of the dependency between the evidence furnished by the sensors should lead to a much lower level of the failure—simply put, we cannot accept their signals as fully credible. This logic specifications can be easily added into the model of sensor fusion by adding an extra dominance neuron whose output serves as an additional input to the AND neuron (see Fig. 23).

D. Logic Signal Compression

The idea of logic-based signal compression can be realized in an architecture shown in Fig. 24. Essentially, it is composed of two logic processors. The first plays the role of decoder while the other encodes the original signal. The number of nodes in the output layer determines the level of information compression one can gain. By changing this size one can control compression rate versus the quality of compression itself. Considering a certain data set $\{x(k)\}$, $k = 1, 2, \ldots, N$, to be compressed, the training is completed in a supervised mode: for each $x(k)$ the target is the same pattern to be reproduced at the second logic processor (encoder level). Note that this type of data compression (or more precisely, transformation) is quite common in digital systems. For instance, the decoding scheme may produce an internal binary representation of the inputs. Then the encoder delivers an n-to-1 encoding pattern.

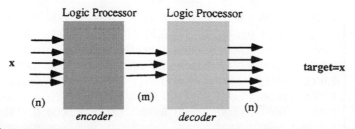

FIGURE 24 Decoding–encoding scheme using logic processors.

E. Revealing Logical Descriptors of Data: An Application of Fuzzy Neurocomputing to Data Mining

In this example we show how fuzzy neural networks can be used to reveal rules out of data. The small database under consideration concerns fitting contact lenses [15] and consist of 24 instances. Each instance is characterized by 4 nominal attributes and falls under one of the following classes:

1. The patient should be fitted with hard contact lenses.
2. The patient should be fitted with soft contact lenses.
3. The patient should not be fitted with contact lenses.

The attributes along with their values are as follows:

1. Age of the patient—(1) young, (2) pre-presbyopic, (3) presbyopic
2. Spectacle prescription—(1) myope, (2) hypermetrope
3. Astigmatic—(1) no, (2) yes
4. Tear production rate—(1) reduced, (2) normal

The encoding used in the design of the network uses 1-of-n encoding, meaning that a single bit is assigned to the description of the attribute. For instance, as the first attribute (age) assumes three values, we consider 3-bit encoding where the assignment reads

001, young
010, pre-presbyotic
100, presbyotic

Altogether, we end up with $3 + 2 + 2 + 2$ bits, that is, 9. Similarly, the network has three outputs, each for the individual class (again using the 3-out-of-1 encoding scheme). The computations are carried out starting from the smallest architecture of the network and successively expanding its hidden layer so that the minimal topology of the network capable of solving the problem becomes determined. It could obviously happen that the learning is not successful and one has to resort to a larger size of the hidden layer. Subsequently, some connections need to be pruned (eliminated) at the very end of the learning procedure. In the series of experiments conducted here, the point at which the performance index drops to zero is obtained for the hidden layer composed of nine neurons. The performance index achieved for successive sizes of the hidden layer is included in Fig. 25.

The matrices of the connections for this size of the hidden layer are

$$
\text{hidden layer}
\begin{bmatrix}
1.0000 & 1.0000 & 1.0000 & 0.0000 & 1.0000 & 1.0000 & 0.0000 & 0.0000 & 1.0000 \\
1.0000 & 1.0000 & 0.0000 & 1.0000 & 0.7653 & 1.0000 & 0.0000 & 0.0000 & 1.0000 \\
1.0000 & 1.0000 & 0.0000 & 1.0000 & 1.0000 & 0.0000 & 1.0000 & 0.0000 & 1.0000 \\
0.0000 & 1.0000 & 1.0000 & 0.0000 & 1.0000 & 0.0000 & 1.0000 & 0.6023 & 1.0000 \\
1.0000 & 0.0000 & 1.0000 & 1.0000 & 0.4915 & 1.0000 & 0.0000 & 0.0000 & 1.0000 \\
1.0000 & 1.0000 & 1.0000 & 1.0000 & 1.0000 & 1.0000 & 1.0000 & 1.0000 & 0.0000 \\
0.0000 & 1.0000 & 1.0000 & 1.0000 & 0.0000 & 1.0000 & 0.0000 & 0.9364 & 1.0000 \\
1.0000 & 1.0000 & 1.0000 & 1.0000 & 0.0000 & 0.0000 & 1.0000 & 0.0000 & 1.0000 \\
1.0000 & 0.0000 & 1.0000 & 0.0000 & 1.0000 & 0.0000 & 1.0000 & 0.0461 & 1.0000
\end{bmatrix},
$$

Input layer

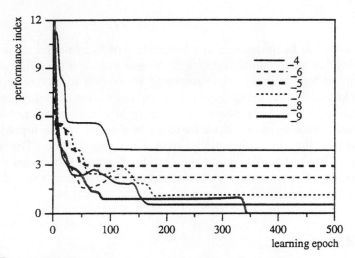

FIGURE 25 Training the neural network: a performance index in successive learning epochs.

hidden layer

$$\text{output} \atop \text{layer} \begin{bmatrix} 0.0000 & 0.0000 & 1.0000 & 0.0000 & 0.0000 & 0.0000 & 0.0000 & 1.0000 & 0.0000 \\ 1.0000 & 1.0000 & 0.0000 & 0.0000 & 1.0000 & 0.0000 & 0.0000 & 0.0000 & 0.0000 \\ 0.0000 & 0.0000 & 0.0000 & 1.0000 & 0.0000 & 1.0000 & 1.0000 & 0.0000 & 1.0000 \end{bmatrix}.$$

The induced rules read as follows (note that some weak connections have been pruned already:

- Class, hard contact lenses

  age = young *and* astigmatic = yes and tear production = normal
  or
  spectacle = myope *and* astigmatic = yes *and* tear production = normal

- Class, soft contact lenses

  Spectacle = hypermetrope *and* astigmatic = no *and* tear production = normal
  or
  age = young *and* astigmatic = no and tear production = normal
  or
  age = pre-presbyotic *and* spectacle = myope *and* astigmatic = no *and* tear production = normal

- Class, no contact lenses

  age = young *and* spectacle = myope *and* astigmatic = yes
  or
  tear production = reduced
  or
  age = young *and* spectacle = myope *and* astigmatic = no
  or
  age = pre-presbyotic *and* spectacle = myope *and* astigmatic = yes *and* tear production = normal

IX. CONCLUSIONS

This chapter has been devoted to a broad class of fuzzy neural networks. The underlying processing units embrace high computational flexibility characteristics for artificial neurons while retaining an explicit format of represented domain knowledge (as being driven by triangular norms of highly pronounced logical nature). We have discussed learning in fuzzy neural networks and emphasized the role of domain knowledge in shaping up the topology of the architectures thereby substantially reducing training time. The examples originating from several areas of application illustrate the design procedure and quantify the advantages of the assumed structures over their pure fuzzy-set-based or neural counterparts.

ACKNOWLEDGMENT

Support from the Natural Sciences and Engineering Research Council (NSERC) is gratefully appreciated.

REFERENCES

1. von Neumann, J. In *The Computer and the Brain*, pp. 66–82. Yale Univ. Press, New Haven, CT, 1958.
2. Lee, S. C. and Lee, E. T. Fuzzy neural networks. *Math. Biosci.* 23:151–177, 1975.
3. Buckley, J. and Hayashi, Y. Fuzzy neural networks: A survey. *Fuzzy Sets Systems* 66:1–14, 1994.
4. Keller, J. M. and Hunt, D. J. Incorporating fuzzy membership functions into the perceptron algorithm. *IEEE Trans. Pattern Anal. Machine Intelligence* PAMI-7:693–699, 1985.
5. Pedrycz, W. Neurocomputations in relational systems. *IEEE Trans. Pattern Anal. Machine Intelligence* 13:289–296, 1991.
6. Pedrycz, W. Fuzzy neural networks and neurocomputations. *Fuzzy Sets Systems* 56:1–28, 1993.
7. Pedrycz, W. and Rocha, A. F. Fuzzy-set based models of neurons and knowledge-based networks, *IEEE Trans. Fuzzy Systems* 1:254–266, 1993.
8. Hirota, K. and Pedrycz, W. OR/AND neuron in modeling fuzzy set connectives, *IEEE Trans. Fuzzy Systems* 2:151–161, 1994.
9. Schneeweiss, W. G. *Boolean Functions with Engineering Applications*. Springer-Verlag, Berlin, 1989.
10. Pedrycz, W. Direct and inverse problem in comparison of fuzzy data. *Fuzzy Sets Systems* 34:223–236, 1990.
11. Butnariu, D. and Klement, E. P. *Triangular Norm–Based Measures and Games with Fuzzy Coalitions*. Kluwer Academic, Dordrecht/Boston, 1993.
12. Pedrycz, W. *Fuzzy Sets Engineering*. CRC Press, Boca Raton, FL, 1995.
13. Machado, R. J. and Rocha, A. F. The combinatorial neural network: a connectionist model for knowledge based systems. In *3rd International Conference on Information Processing and Management of Uncertainty in Knowledge-Based Systems*, Paris, France, July 2–6, 1990, pp. 9–11.
14. Rocha, A. F. Neural nets: A theory for brain and machine, *Lecture Notes in Artificial Intelligence*, Vol. 638, Springer-Verlag, Berlin, 1992.
15. Cendrowska, J. PRISM: An algorithm for inducing modular rules. *Internat. J. Man–Machine Studies* 27:349–370, 1987.

8
IDENTIFYING FUZZY RULE-BASED MODELS USING ORTHOGONAL TRANSFORMATION AND BACKPROPAGATION

LIANG WANG

Staff Scientist, Casa Incorporated, Los Alamos, New Mexico 87544

REZA LANGARI

Department of Mechanical Engineering, Texas A & M University, College Station, Texas 77843

JOHN YEN

Department of Computer Science, Texas A & M University, College Station, Texas 77843

I. INTRODUCTION 188
II. FUZZY MODELS 189
III. METHOD 190
    A. Backpropagation Algorithm for Parameter Estimation 191
    B. Singular Value Decomposition and Fuzzy Rule Selection 192
    C. Computational Steps 193
IV. RESULTS 193
V. CONCLUSION 201
    APPENDIX: FACTORIZABLE PROPERTY OF GAUSSIAN MEMBERSHIP FUNCTIONS 201
    REFERENCES 202

An important issue in fuzzy rule-based modeling is how to select a set of important fuzzy rules from a given rule base. Even though it is conceivable that removal of redundant or less important fuzzy rules from the rule base can result in a compact fuzzy model with better generalizing ability, the decision as to which rules are redundant or less important is not an easy exercise. In this paper we introduce an orthogonal-transformation-based method for rule selection. This method is used in combination with the well-known backpropagation (BP) learning algorithm in neural-network modeling. Initially, an oversized fuzzy model that consists of a large number of fuzzy rules is trained using the BP algorithm. Then a singular value decomposition (SVD) and QR with column-pivoting algorithm is used to identify and remove redundant fuzzy rules. This results in a reduced fuzzy model that has

Fuzzy Theory Systems: Techniques and Applications, Vol. 1
Copyright © 1999 by Academic Press. All rights of reproduction in any form reserved.

a better generalizing ability. The reduced model is trained using the BP algorithm with initial parameter values determined in the oversized model. The approach is demonstrated using two nonlinear system modeling examples.

I. INTRODUCTION

Fuzzy rule-based modeling has become an active research field because of its unique merits in solving complex nonlinear system identification and control problems. Primary advantages of this approach include the facility for explicit knowledge representation in the form of if–then rules, the mechanism of reasoning in human-understandable terms, the capacity of taking linguistic information from human experts and combining it with numerical information, and the ability of approximating complicated nonlinear functions with simpler models. Unlike conventional modeling where a single model is used to describe the global behavior of a system, fuzzy rule-based modeling is essentially a *multimodel* approach in which individual rules (where each rule acts like a "local model") are combined to describe the global behavior of the system.

When using a fuzzy model to approximate an unknown system, it is desired that the model include as many rules as possible so that it can approximate the training data as closely as possible; yet it is also desired that the model include as few rules as possible because the generalizing ability of the model decreases as the number of rules increases. When we speak here of *generalization* we are referring to the system's mean performance in terms of approximation accuracy evaluated over some independent test data set. The trade-off between goodness of fit and simplicity is a fundamental principle underlying various general theories of statistical modeling and inductive inference [1, 2].

Several research efforts have been made in the fuzzy logic community to strike a balance between reducing the fitting error and increasing the model complexity. For example, an *entropy criterion* was proposed by Yager and Filev [3] to find a simple fuzzy-model structure by minimizing the rate of interaction between fuzzy rules. The number of fuzzy rules in this approach was determined using an *unbiasedness criterion* suggested by Sugeno and Kang [4] (see also [5]). Berenji and Khedkar [6], proposed a *pruning and merging* strategy to eliminate redundant fuzzy rules in a given rule base. Yen and Wang [7] suggested several modified *statistical information criteria* to determine the optimal number of fuzzy rules constituting the underlying model. Methods based on *genetic algorithms* have also been used for extracting fuzzy rules for control and classification problems [8–12].

In this chapter we introduce a method based on *orthogonal transformation*, for selecting important fuzzy rules from a given rule base. In particular, we use *singular value decomposition* (SVD) to determine the number of important fuzzy rules constituting the underlying model. The position of these important fuzzy rules in the rule base is identified by a subset selection method known as *SVD-QR with column-pivoting algorithm* [13]. After the less important fuzzy rules are removed, we retrain the antecedent parameters and consequent parameters of the model using the *backpropagation* (BP) learning algorithm. We illustrate the practical applicability of the approach using two nonlinear system modeling examples.

II. FUZZY MODELS

A fuzzy model is a set of fuzzy if–then rules that maps inputs to outputs. The antecedents of fuzzy rules partition the input space into a number of local fuzzy regions, and the consequents describe the system behavior within a given region via various constituents. The consequent constituent could be a membership function, a constant, or a linear equation. Different consequent constituents result in different fuzzy models, but their antecedents are always the same.

We consider fuzzy models that have the following form:

$$R_i: \quad \text{if } x_1 \text{ is } A_{i1} \text{ and } \cdots \text{ and } x_p \text{ is } A_{ip}$$

$$\text{then } y = B_i, \qquad i = 1, 2, \ldots, M, \tag{1}$$

where p and M are the number of input variables and rules, respectively; x_i and y are the input and output variables, respectively; A_{ij} and B_i are fuzzy membership functions.

This is a multi-input and single-output fuzzy model. Using the *center average defuzzifier*, the total output of the model can be computed as [14]

$$y = \frac{\sum_{i=1}^M w_i c_i}{\sum_{i=1}^M w_i} \equiv \sum_{i=1}^M v_i c_i, \tag{2}$$

where c_i is the *center* of B_i, w_i denotes the *firing strength* of the ith rule, which is defined by

$$w_i = A_{i1}(x_1) \times A_{i2}(x_2) \times \cdots \times A_{ip}(x_p), \tag{3}$$

and v_i is the *normalized* firing strength of the ith rule, which is defined by

$$v_i = \frac{w_i}{\sum_{i=1}^M w_i}. \tag{4}$$

The membership functions in model (1) can be triangles, trapezoids, bell curves, or Gaussian functions. In this paper we have adopted Gaussian functions as membership functions. A unique property of Gaussian functions is that they are *factorizable* (see Appendix for a brief illustration of this property). Given such a property, we may synthesize a multi-dimensional membership function as the product of one-dimensional membership functions. Also, we may obtain the parameters of one-dimensional membership functions by first determining the parameters of the multi-dimensional membership function. In this case the model (1) can be identically expressed as

$$R_i: \quad \text{if } \mathbf{x} \text{ is } \mathbf{A_i} \text{ then } y \text{ is } B_i$$

$$i = 1, 2, \ldots, M. \tag{5}$$

where $\mathbf{x}$ is a p-dimensional input vector, and $\mathbf{A_i}$ is a p-dimensional Gaussian membership function.

Substituting A_{ij} as in Eq. (3) with Gaussian membership functions, Eq. (4) can be written as

$$v_i = \frac{\Pi_{j=1}^{p} \exp\left(-(x_j - m_{ij})^2/2\sigma_{ij}^2\right)}{\Sigma_{i=1}^{M}\Pi_{j=1}^{p} \exp\left(-(x_j - m_{ij})^2/2\sigma_{ij}^2\right)}, \tag{6}$$

where m_{ij} and σ_{ij} denote the *center* and *width* of the Gaussian functions, respectively.

Equation (2) can be viewed as a special case of the linear regression model [14]

$$y(k) = \sum_{i=1}^{M} p_i(k)\theta_i + e(k) \tag{7}$$

with $p_i(k)$ and θ_i given by

$$p_i(k) \equiv v_i, \qquad \theta_i \equiv c_i, \tag{8}$$

where $p_i(k)$ are known as the *regressors*, θ_i are the parameters, and $e(k)$ is an error signal that is assumed to be uncorrelated with the regressors $p_i(k)$. Given N input–output pairs $\{\mathbf{x}(k), y(k)\}$, $k = 1, 2, \ldots, N$, where $\mathbf{x}(k) = (x_1(k), x_2(k), \ldots, x_p(k))$, it is convenient to express Eq. (7) in the matrix form

$$\mathbf{y} = P\theta + \mathbf{e}, \tag{9}$$

where $\mathbf{y} = [y(1), y(2), \ldots, y(N)]^T \in \Re^N$, $ = [\mathbf{p}_1, \mathbf{p}_2, \ldots, \mathbf{p}_M] \in \Re^{N \times M}$ with $\mathbf{p}_i = [p_i(1), p_i(2), \ldots, p_i(N)]^T \in \Re^N$, $\theta = [\theta_1, \theta_2, \ldots, \theta_M]^T \in \Re^M$, and $\mathbf{e} = [e(1), e(2), \ldots, e(N)]^T \in \Re^N$. Note that each column of P corresponds to one of the fuzzy rules in the rule base. We will call P the *firing strength matrix* and $P\theta$ the *predictor* throughout this chapter, for notational simplicity. In building a fuzzy model, the number of available training data points is usually larger than the number of fuzzy rules in the rule base. This implies that the row dimension of the matrix P is larger than its column dimension, that is, $N > M$.

III. METHOD

The proposed method is a combination of orthogonal transformation and backpropagation, which consists of four main steps. First, a candidate fuzzy model is considered, which may be exhaustive and oversized (i.e., have a large rule base) but not undersized. The parameters of the model are trained using the BP algorithm. Second, the number of important fuzzy rules in the rule base is specified using SVD. Third, the position of these fuzzy rules in the rule base is identified using a subset selection algorithm, known as SVD-QR with column-pivoting algorithm. Finally, the important fuzzy rules are retained, whereas the less important or redundant fuzzy rules are removed from the rule base. A compact fuzzy model is then constructed using the retained fuzzy rules, whose parameters are retrained using BP. Next we provide a detailed description of the proposed method.

A. Backpropagation Algorithm for Parameter Estimation

Backpropagation algorithm is a learning algorithm for adjusting the parameters in a neural network to minimize the error function $J(k)$ [15]:

$$J(k) = e^2(k) \equiv \tfrac{1}{2}[d(k) - y(k)]^2, \tag{10}$$

where $d(k)$ is the output of the real system and $y(k)$ is the output of the identified model. This algorithm adjusts the parameters of the neural network such that they move along the gradient of $J(k)$. Therefore, BP can essentially be viewed as a gradient-descent algorithm.

The backpropagation algorithm has also been widely used in the fuzzy logic community to train fuzzy models [16–19]. In this chapter we want to use this algorithm to determine the centers m_{ij} and the widths σ_{ij} in the antecedent membership functions A_{ij} as well as the centers c_i in the consequent membership functions B_i of the model (1). For this purpose we substitute $y(k)$ in Eq. (10) with the output of the fuzzy model defined by Eq. (2) and get

$$J(k) = e^2(k) \equiv \tfrac{1}{2}\left[d(k) - \sum_{i=1}^{M} v_i(k)c_i(k)\right]^2. \tag{11}$$

Equation (11) represents the instantaneous error squares for the kth observation. Differentiating $J(k)$ with respect to c_i, m_{ij}, and σ_{ij}, respectively, and using the chain rule, we have the following BP algorithm for parameter estimates of the fuzzy model:

$$c_i(k) = c_i(k-1) + \alpha_1 e(k)v_i(k), \qquad i = 1,2,\ldots,M, \tag{12}$$

$$m_{ij}(k) = m_{ij}(k-1) + \alpha_2 e(k)v_i(k)$$

$$\times \left[c_i(k) - \sum_{l=1}^{M} v_l(k)c_l(k)\right]\left[\frac{x_j(k) - m_{ij}(k-1)}{\sigma_{ij}^2(k-1)}\right],$$

$$i = 1,2,\ldots,M, j = 1,2,\ldots,p, \tag{13}$$

$$\sigma_{ij}(k) = \sigma_{ij}(k-1) + \alpha_3 e(k)v_i(k)$$

$$\times \left[c_i(k) - \sum_{l=1}^{M} v_l(k)c_l(k)\right]\left[\frac{(x_j(k) - m_{ij}(k-1))^2}{\sigma_{ij}^3(k-1)}\right]$$

$$i = 1,2,\ldots,M, j = 1,2,\ldots,p, \tag{14}$$

where α_1, α_2, and α_3 are learning parameters.

B. Singular Value Decomposition and Fuzzy Rule Selection

The singular value decomposition (SVD) of a matrix is a factorization of the matrix into a product of three matrices. For the firing strength matrix P, the decomposition can be written as

$$P = U\Sigma V^T, \tag{15}$$

where $U \in \Re^{N \times N}$ and $V \in \Re^{M \times M}$ are orthogonal matrices, $\Sigma = \text{diag}(\sigma_1, \sigma_2, \ldots, \sigma_M) \in \Re^{N \times M}$ is a diagonal matrix with $\sigma_1 \geq \sigma_2 \geq \cdots \geq \sigma_M \geq 0$. The diagonal elements of Σ are called the *singular values* of P.

An important property of SVD is that it reveals the *rank* of P. From Eq. (15) it follows that $\text{rank}(P) = \text{rank}(\Sigma)$. Consequently, the number of nonzero singular values indicates the rank of the matrix P. Let $s = \text{rank}(P)$. Then Eq. (15) can be rewritten in an alternative form,

$$P = \sum_{i=1}^{s} \sigma_i u_i v_i^T, \tag{16}$$

where $\sigma_1, \sigma_2, \ldots, \sigma_s$ are the s nonzero singular values of P, and u_i and v_i are the ith column of U and V, respectively.

In practice, the matrix P can usually be approximated by

$$P' = \sum_{i=1}^{r} \sigma_i u_i v_i^T, \tag{17}$$

where $r \leq s$ is some numerically determined estimate of s. It can be proved [20] that P' is the closest matrix to P that has rank r.

Replacing P by P' amounts to filtering the small singular values and can make a great deal of sense in those situations where P is derived from noise data. In our application, however, the existence of small singular values implies the presence of *redundant* or *less important rules* among the rules that compose the underlying model [21, 22]. In this case, we are not interested in a predictor such as $P'\theta'$ that involves all M rules. Instead, a predictor $P\theta$ should be sought, where θ has at most r nonzero components. The position of the nonzero entries determines which columns of P, that is, which rules in the rule base, are to be used in constructing the model and in approximating the observation vector y. How to pick these columns is the problem of *rule selection*.

Mouzouris and Mendel [21], an SVD-QR with column-pivoting algorithm was used to solve this problem. This algorithm was originally proposed by Golub *et al.* [13] to solve the subset selection problem in regression analysis and used by Kanjilal and Banerjee [23] to select hidden nodes in a feed-forward neural network. A comparison of this algorithm with other orthogonal-transformation-based methods was given in [22]. This algorithm is summarized as follows:

SVD-QR WITH COLUMN-PIVOTING ALGORITHM FOR RULE SELECTION.

1. Compute the SVD of P using Eq. (15) and save Σ and V.

2. Check the singular values in $\Sigma = \text{diag}(\sigma_1, \sigma_2, \ldots, \sigma_M)$, and determine the number of fuzzy rules that are to be used to construct the model as r, where $r \leq \text{rank}(P)$.

3. Partition V as

$$V = \begin{bmatrix} V_{11} & V_{12} \\ V_{21} & V_{22} \end{bmatrix}, \qquad \text{where } V_{11} \in \Re^{r \times r}, V_{21} \in \Re^{r \times (M-r)};$$

form $\overline{V}^T = [V_{11}^T \ V_{21}^T]$.

4. Apply *QR with column-pivoting algorithm* [20] to $\overline{V}$ and get the *permutation matrix*

$$\Pi \in \Re^{M \times M}: \quad Q^T \begin{bmatrix} V_{11}^T & V_{21}^T \end{bmatrix} \Pi = \begin{bmatrix} R_{11} & R_{12} \end{bmatrix},$$

where $Q \in \Re^{r \times r}$ is orthogonal and $R_{11} = \Re^{r \times r}$ is upper triangular. The position of the entries 1s in the first r columns of Π indicates the position of the r most important fuzzy rules in the rule base.

C. Computational Steps

Up to this point we have introduced two key components of the proposed method: BP for parameter estimates and orthogonal transformation for rule selection. For clarity, we now summarize the complete computational steps of the method as follows:

Step 1.   Consider a fuzzy model that includes a large number of fuzzy rules; train its parameters using the BP algorithm.

Step 2.   Form the firing strength matrix P; compute its SVD.

Step 3.   Determine the number of fuzzy rules that are to be retained based on the distribution of the singular values; use SVD-QR with column-pivoting algorithm to identify the position of these retained rules in the rule base.

Step 4.   Build a reduced fuzzy model using these retained rules; train its parameters using the BP algorithm. The initial parameter values are taken as the parameter values of these rules in the original model constructed in Step 1.

IV. RESULTS

In this section we present two examples to illustrate the proposed method. The first example is a simple nonlinear function approximation problem that is taken from Chiu [24], and the second example is a nonlinear plant modeling problem that is borrowed from Narendra and Parthasarathy [25]. In both examples, the learning parameters α_1, α_2, and α_3 for the BP algorithm are set to 0.25, that is, $\alpha_1 = \alpha_2 = \alpha_3 = 0.25$.

EXAMPLE 1. Consider the nonlinear function [24]

$$y = \frac{\sin(x)}{x}. \tag{18}$$

We want to approximate this function using the fuzzy model (1). For this purpose, we generated 100 training data points using equally spaced x values in the interval $[-10, 10]$. We first built a fuzzy model with 30 rules whose antecedent and consequent parameters were trained using the BP algorithm with the initial parameter values $c_i(0) \in [-1, 1]$, $m_{ij}(0) \in [-10, 10]$, and $\sigma_{ij}(0) \in [1, 2]$. These fuzzy rules were labeled as $1, 2, \ldots, 30$ to indicate their position in the rule base. For each of the 100 input data points $x(k)$, $k = 1, 2, \ldots, 100$, of the model, we computed the normalized firing strengths of the 30 rules using Eq. (4). A 100-by-30 firing strength matrix P was then formed. Applying SVD to P, the resulting singular values are shown in Fig. 1. Based on the distribution of the singular values, we retained 12 rules that were used to construct a reduced fuzzy model. The position of these retained rules in the rule base was identified using the SVD-QR with column-pivoting algorithm as $3, 5, 30, 16, 1, 13, 8, 24, 19, 29, 23, 14$. The order of the position also indicated the importance of the associated fuzzy rules in the rule base. The antecedent and consequent parameters of the reduced 12-rule fuzzy model were retrained using the BP algorithm with the initial parameter

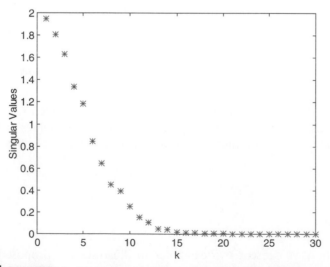

FIGURE 1 Distribution of singular values of the 100-by-30 firing-strength matrix.

values determined in the 30-rule fuzzy model. Figure 2 shows the log(MSE) (mean squared error) curves of the 30-rule fuzzy model and the 12-rule fuzzy model over 1000 epochs. As a comparison, Fig. 2 also shows a 12-rule fuzzy model trained using the BP algorithm with random initial parameter values $c_i(0) \in [-1, 1]$, $m_{ij}(0) \in [-10, 10]$, and $\sigma_{ij}(0) \in [1, 2]$. Table 1 gives the MSEs of the three fuzzy models after 1000 epochs. From Fig. 2 and Table 1 we know that the reduced 12-rule model with predetermined initial parameter values has achieved its feasible parameter sets much faster than the one with random initial parameter values. This is because the BP algorithm is a nonlinear optimization algorithm and its convergence depends to a great extent on the choice of initial parameter values. The original 30-rule fuzzy model gives a small MSE compared with the two reduced 12-rule fuzzy models. This is not surprising because the former contains a larger number of fuzzy rules. However, good fit on training data points does not always assure good performance on future data points. A fuzzy model with a large number of rules may have a high risk of overfitting: capable of fitting training data points completely, but incapable of generalizing to untrained data points satisfactorily. We will see this point in the following example. Figure 3 shows the real output of the nonlinear function and the outputs of 30-rule fuzzy model and 12-rule fuzzy model with predetermined initial parameter values. Figure 4 shows the membership functions of the input variable x in the 30-rule fuzzy model and in the 12-rule fuzzy model with predetermined initial parameter values.

EXAMPLE 2. Consider the second-order nonlinear plant [25]

$$y(k) = f(y(k-1), y(k-2)) + u(k), \tag{19}$$

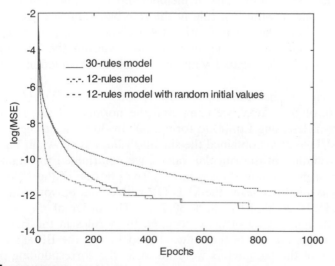

FIGURE 2 Log(MSE) curves for three fuzzy models.

TABLE I   MSEs of Three Fuzzy Models after 1000 Epochs

Fuzzy model	MSE
Original model (30 rules)	2.7902e − 6
Reduced model (12 rules, with predetermined initial values)	2.8342e − 6
Reduced model (12 rules, with random initial values)	6.8712e − 6

where

$$f(y(k-1), y(k-2)) = \frac{y(k-1)y(k-2)[y(k-1)+2.5]}{1+y(k-1)^2+y(k-2)^2}. \quad (20)$$

Our goal was to approximate the nonlinear component f using the fuzzy model (1). For this purpose, 700 simulated data points were generated from the plant model (19). The first 500 data points were obtained by assuming a random input signal $u(k)$ uniformly distributed in the interval $[-2, 2]$, and the last 200 data were obtained by using a sinusoidal input signal $u(k) = \sin(2\pi k/25)$. The 700 simulated data points are shown in Fig. 5.

We used the first 500 data points to construct the model with $y(k-1)$ and $y(k-2)$ as the input variables. The performance of the resulting model was tested using the remaining 200 data points. We first constructed a 30-rule fuzzy model whose antecedent and consequent parameters were trained using the BP algorithm with the initial parameter values $c_i(0) \in [-2, 2]$, $m_{ij}(0) \in [-4, 4]$, and $\sigma_{ij}(0) \in [1, 2]$. These 30 rules were labeled by the number $1, 2, \ldots, 30$, which indicated the position of the rules in the rule base as well as the associated combinations of membership functions. For example, the number "1" indicates the first rule in the rule base, which is associated with the membership function combination $\{A_{11}(x_1), A_{12}(x_2)\}$ [where $x_1 \equiv y(k-1)$, $x_2 \equiv y(k-2)$], whereas the number "30" indicates the 30th rule in the rule base, which is associated with the membership function combination $\{A_{30,1}(x_1), A_{30,2}(x_2)\}$.

For each of the 500 input data points $\{y(k-1), y(k-2)\}$, $k = 1, 2, \ldots, 500$, of the model, we computed the normalized firing strengths of the 30 fuzzy rules using Eq. (4) to form a 500-by-30 firing strength matrix P. Applying SVD to P, we obtained the singular values as shown in Fig. 6. Based on the distribution of the singular values, we determined to retain 15 fuzzy rules to construct a reduced fuzzy model. The position of the 15 rules in the rule base was identified using the SVD-QR with column-pivoting algorithm as $10, 23, 1, 6, 15, 14, 26, 29, 16, 11, 8, 20, 30, 2, 18$. The order of the position also indicated the importance of the associated fuzzy rules in the rule base. We retrained the parameters of the reduced model using the BP algorithm. The initial values of the parameters were taken as the corresponding parameter values in the original 30-rule fuzzy model. Figure 7 shows the log(MSE) curves of the 30-rule fuzzy model and the 15-rule fuzzy model over 1000

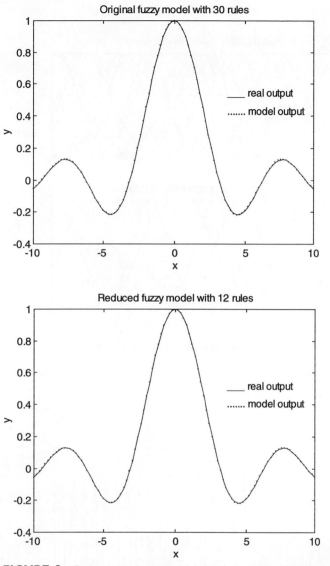

FIGURE 3   Comparison of outputs of real system and identified fuzzy models.

epochs. As a comparison, Fig. 7 also shows a 15-rule fuzzy model trained using the BP algorithm with random initial parameter values $c_i(0) \in [-2, 2]$, $m_{ij}(0) \in [-4, 4]$, and $\sigma_{ij}(0) \in [1, 2]$. It can be seen that the 15-rule model with predetermined initial parameter values shows a much faster convergence than the 15-rule model with random initial parameter values. Table 2 shows the MSEs of the three fuzzy models in both the training stage and the testing stage. As expected, the 30-rule fuzzy model gives the smallest MSE among the three fuzzy models in the training stage. However, the 15-rule fuzzy model with predetermined initial parameter values shows the best performance in the testing stage. This indicates that a well-trained simple fuzzy

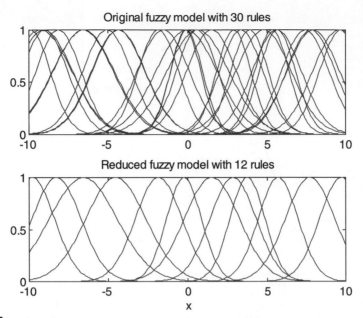

FIGURE 4 Membership functions of input variable *x* in the original and the reduced fuzzy models.

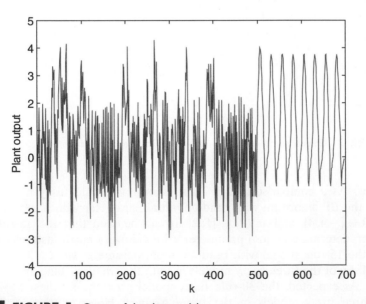

FIGURE 5 Output of the plant model.

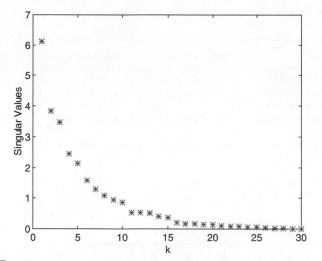

FIGURE 6   Distribution of singular values of the 500-by-30 firing-strength matrix.

model has a better generalizing ability than a complex fuzzy model. Figure 8 shows the outputs of the real plant, the 30-rule fuzzy model and the 15-rule fuzzy model with predetermined initial parameter values for the sinusoid input signal $u(k) = \sin(2\pi k/25)$ in the testing stage. Figure 9 shows the centers m_{ij} of the membership functions in the 30-rule fuzzy model and in the 15-rule fuzzy model with predetermined initial parameter values.

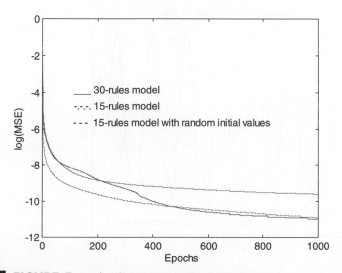

FIGURE 7   Log(MSE) curves for three fuzzy models.

TABLE 2 MSEs of Three Fuzzy Models for Training and Testing Data

Model	MSE (training)	MSE (testing)
Original model (30 rules)	1.7448e − 5	2.3648e − 5
Reduced model (15 rules, with predetermined initial values)	1.9375e − 5	1.8914e − 5
Reduced model (15 rules, with random initial values)	4.9907e − 5	5.4542e − 5

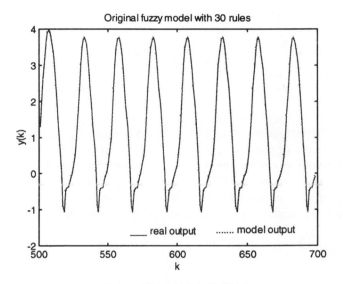

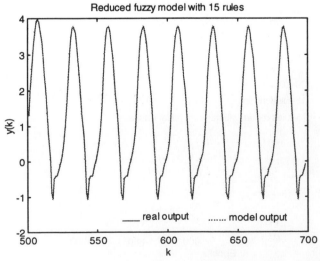

FIGURE 8 Outputs of the real plant and the identified fuzzy models for $u(k) = \sin(2\pi k / 25)$.

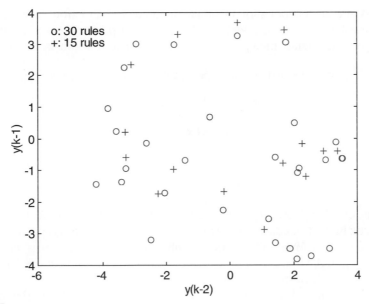

FIGURE 9   The centers m_{ij} of the Gaussian membership functions in the 30-rule and 15-rule fuzzy models.

V. CONCLUSION

In this paper we combine orthogonal transformation and backpropagation learning algorithms to the construction of fuzzy rule-based models. Unlike conventional methods where multiple iterations are usually required to find the "optimal" number of fuzzy rules, the proposed method is a noniterative procedure. It starts with an oversized rule base, and then removes redundant or less important fuzzy rules through a "one-pass" operation. Therefore, this method is computationally less expensive compared to the conventional methods, especially when the number of input variables is large. Moreover, this method is extremely robust numerically and can be implemented using some well-established numerical computation packages.

This method is capable of identifying and eliminating less important and redundant fuzzy rules in a large rule base, producing models that generalize better. Simulations show that this method is promising and should become a valuable addition to the fuzzy modeling workers' tool kit.

APPENDIX: FACTORIZABLE PROPERTY OF GAUSSIAN MEMBERSHIP FUNCTIONS

A Gaussian membership function is specified by two parameters $\{m, \sigma\}$ as follows:

$$\text{gauss}(x; m, \sigma) = \exp\left(-\frac{(x - m)^2}{2\sigma^2}\right), \qquad (A1)$$

where m and σ denote the center and width of the function, respectively.

A unique property of Gaussian membership functions is that they are factorizable [26]. To demonstrate this important property, consider the simple case of a Gaussian membership function with unit width in two dimensions

$$
\begin{aligned}
\mathrm{gauss}(\mathbf{x}; \mathbf{m}) &= \mathrm{gauss}(\|\mathbf{x} - \mathbf{m}\|^2) \\
&= \exp\left(-\tfrac{1}{2}\|\mathbf{x} - \mathbf{m}\|^2\right) \\
&= \exp\left(-\tfrac{1}{2}(x_1 - m_1)^2 - \tfrac{1}{2}(x_2 - m_2)^2\right) \\
&= \exp\left(-\tfrac{1}{2}(x_1 - m_1)^2\right)\exp\left(-\tfrac{1}{2}(x_2 - m_2)^2\right) \\
&= \mathrm{gauss}(x_1; m_1)\,\mathrm{gauss}(x_2; m_2).
\end{aligned} \tag{A2}
$$

Equation (A2) shows that a two-dimensional Gaussian membership function $\mathrm{gauss}(\mathbf{x}; \mathbf{m})$ with a center $\mathbf{m} = [m_1, m_2]^T$ is equivalent to the product of a pair of one-dimensional Gaussian membership functions $\mathrm{gauss}(x_1; m_1)$ and $\mathrm{gauss}(x_2; m_2)$, where x_1 and x_2 are the elements of the vector $\mathbf{x}$, and m_1 and m_2 are the elements of the center $\mathbf{m}$.

The result of Eq. (A2) may be readily generalized for the case of a multivariate Gaussian membership function that computes the weighted norm, assuming that weighting matrix Λ is a diagonal matrix. To be specific, suppose that we have

$$
\Lambda = \mathrm{diag}(\sigma_1, \sigma_2, \ldots, \sigma_p). \tag{A3}
$$

We may then decompose $\mathrm{gauss}(\mathbf{x}; \mathbf{m}, \Lambda)$ as follows:

$$
\begin{aligned}
\mathrm{gauss}(\mathbf{x}; \mathbf{m}, \Lambda) &= \mathrm{gauss}(\|\mathbf{x} - \mathbf{m}\|_\Lambda^2) \\
&= \prod_{k=1}^{p} \exp\left(-\frac{(x_k - m_k)^2}{2\sigma_k^2}\right) \\
&= \mathrm{gauss}(x_1; m_1, \sigma_1) \cdots \mathrm{gauss}(x_p; m_p, \sigma_p),
\end{aligned} \tag{A4}
$$

where m_k denotes the kth element of the center $\mathbf{m}$.

Given such a property, we may synthesize a multi-dimensional membership function as the product of one-dimensional membership functions. Also, we may obtain the parameters of one-dimensional membership functions by first determining the parameters of the multi-dimensional membership function. More important, because the model can be expressed in a compact form (in a multi-dimensional membership function), the "curse of dimensionality" caused by the exponential increase in the number of rules resulting from the exhaustive combination of membership functions of separable variables can be avoided.

REFERENCES

1. Akaike, H. A new look at statistical model identification. *IEEE Trans. Automat. Control* 19:716–723, 1974.

2. Angluin, D. and Smith, C. Inductive inference: Theory and methods. *ACM Comput. Surveys* 15:237–269, 1984.

3. Yager, R. R. and Filev, D. P. Unified structure and parameter identification of fuzzy models. *IEEE Trans. Systems Man Cybernet.* 23:1198–1205, 1993.

4. Sugeno, M. and Kang, G. T. Structure identification of fuzzy model. *Fuzzy Sets Systems* 28:15–33, 1988.

5. Sugeno, M. and Yasukawa, T. (1993). A fuzzy logic based approach to qualitative modeling. *IEEE Trans. Fuzzy Systems* 1:7–31, 1993.

6. Berenji, H. R. and Khedkar, P. S. Clustering in product space for fuzzy inference. Unpublished.

7. Yen, J. and Wang, L. Application of statistical information criteria for optimal fuzzy model construction. *IEEE Trans. Fuzzy Systems*, to appear.

8. Homaifar, A. and McCormick, E. Simultaneous design of membership functions and rule sets for fuzzy controllers using genetic algorithms. *IEEE Trans. Fuzzy Systems* 3:129–139, 1995.

9. Ishibuchi, H., Nozaki, K., Yamamoto, N., and Tanaka, H. Selecting fuzzy if–then rules for classification problems using genetic algorithms. *IEEE Trans. Fuzzy Systems* 3:260–270, 1995.

10. Karr, C. L. Applying genetics to fuzzy logic. *AI Expert* 6:38–43, 1991.

11. Lee, M. A. and Takagi, H. Integrating design stages of fuzzy systems using genetic algorithms. In *Proceedings of the Second IEEE International Conference on Fuzzy Systems*, San Francisco, 1993, pp. 612–617.

12. Thrift, P. Fuzzy logic synthesis with genetic algorithms. In *Proceedings of the Fourth International Conference Genetic Algorithms*, San Diego, 1991, pp. 509–513.

13. Golub, G. H., Klema, V., and Stewart, G. W. Rank degeneracy and least squares problems. Technical Report TR-456, Department of Computer Science, University of Maryland, College Park, 1975.

14. Wang, L. X. (1994). *Adaptive Fuzzy Systems and Control: Design and Stability Analysis*. Prentice-Hall, Englewood Cliffs, NJ, 1994.

15. Rumelhart, D. E., Hinton, G. E., and Williams, R. J. Learning internal representations by error propagation. In *Parallel Distributed Processing: Explorations in the Microstructure of Cognition* (D. E. Rumelhart and J. L. McCelland, Eds.), Vol. I, Chap. 8, MIT Press, Cambridge, MA, 1986.

16. Jang, J.-S. R. ANFIS: Adaptive-network-based fuzzy inference system. *IEEE Trans. Systems Man Cybernet.* 23:665–685, 1993.

17. Lin, C.-T. and Lee, C. S. G. Neural-network-based fuzzy logic control and decision system. *IEEE Trans. Comput.* 40:1320–1336, 1991.

18. Nomura, H., Hayashi, I., and Wakami, N. A learning method of fuzzy inference rules by descent method. In *Proceedings of the First International Conference on Fuzzy Systems*, San Diego, 1992, pp. 203–210.

19. Wang, L. X. and Mendel, J. M. Backpropagation fuzzy systems as nonlinear dynamic system identifiers. In *Proceedings of the First IEEE International Conference Fuzzy Systems*, San Diego, 1992, pp. 1409–1418.

20. Golub, G. H. and Van Loan, C. F. *Matrix Computations*, 2nd ed. Johns Hopkins Press, Baltimore, 1989.

21. Mouzouris, G. C. and Mendel, J. M. Designing fuzzy logic systems for uncertain environments using a singular-value-QR decomposition method. In *Proceedings of the Fifth IEEE International Conference on Fuzzy Systems*, New Orleans, 1996, pp. 295–301.

22. Yen, J. and Wang, L. Simplification of fuzzy rule based systems using orthogonal transformation. In *Proceedings of the Sixth IEEE International Conference on Fuzzy Systems*, Barcelona, Spain, 1997, pp. 253–258.

23. Kanjilal, P. P. and Banerjee, D. N. On the application of orthogonal transformation for the design and analysis of feedforward networks. *IEEE Trans. Neural Networks* 6:1061–1070, 1995.

24. Chiu, S. L. Fuzzy model identification based on cluster estimation. *J. Intelligent Fuzzy Systems* 2:267–278, 1994.

25. Narendra, K. S. and Parthasarathy, K. Identification and control of dynamical systems using neural networks. *IEEE Trans. Neural Networks* 1:4–26, 1990.
26. Haykin, S. *Neural Networks: A Comprehensive Foundation.* Macmillan College Co., New York, 1994.
27. Chao, C. T., Chen, Y. J., and Teng, C. C. Simplification of fuzzy–neural systems using similarity analysis. *IEEE Trans. Systems Man Cybernet.* 26(B):344–354, 1996.
28. Lin, C.-T. and Lee, C. S. G. *Neural Fuzzy Systems: A Neural–Fuzzy Synergism to Intelligent Systems.* Prentice Hall, Upper Saddle River/New York, 1996.
29. Sun, C.-T. Rule-base structure identification in an adaptive-network-based fuzzy inference system. *IEEE Trans. Fuzzy Systems* 2:64–73, 1994.

9
EVOLUTIONARY NEURO-FUZZY MODELING

CHUEN-TSAI SUN

HAO-JAN CHIU

Department of Computer and Information Science, National Chiao Tung University, 1001 Ta Hsuch Rd., Hsinchu, Taiwan 30050, Republic of China

I. INTRODUCTION   205
II. ANFIS: ADAPTIVE-NETWORK-BASED FUZZY
INFERENCE SYSTEM   206
III. RBFN: RADIAL BASIS FUNCTION NETWORK   208
IV. NICHE: DIVERSITY OF SPECIES   209
V. THE EVOLUTIONARY MODEL   210
   A. Skeleton of the Model   210
   B. Design Criteria   213
VI. EXPERIMENTAL RESULTS   216
   A. Experiment 1: Recognizing the Iris Data Set   216
   B. Experiment 2: Recognizing the Breast Cancer Data Set   218
VII. CONCLUSION   221
   REFERENCES   221

Neuro-fuzzy systems [1–4] are powerful hybridized systems within artificial neural network (ANN) and fuzzy systems. Identifying a neuro-fuzzy system is a relevant issue that has received extensive attention. In this chapter, we present an evolutionary model that fulfills the two phases of identifying a system: simultaneously identifying the structure and the parameter. The proposed model facilitates construction of a neuro-fuzzy system.

I. INTRODUCTION

Designing a neuro-fuzzy system hinges on identifying the system, which consists of identifying the structure and the parameter. Structure identification aims at the rough designation of the membership functions. It determines the number of membership functions, linguistic variables, and linguistic values associated with the linguistic variables. Parameter identification entails fine-tuning the parameters related to membership functions, which largely determines the shape of the membership functions.

Originally, the system identification of a fuzzy system must be manually tuned by experts who are familiar with the specific problem. However, when a certain problem is extremely complex or no expertise is available, manually tuning the fuzzy system would be inappropriate. Therefore, structure identification is frequently implemented by some partitioning or clustering approaches, for example, fuzzy $k-d$ tree partitioning, fuzzy grid partitioning, and k-mean clustering. These approaches, although enhancing the design phase of structure identification, have certain limitations. Consider the k-mean clustering approach. If the initial positions of the k seeds in the problem domain are not adequately assigned, the k seeds do not spread out and are not representative of the problem domain. As generally known, artificial neural network (ANN) is quite powerful with respect to its learning ability. Applying the backpropagation (BP) algorithm to identify the parameter of a fuzzy system facilitates fine-tuning the parameters of the membership functions [5] and the weights of the ANN structure. However, a local optimum may prevent BP from fine-tuning the parameters.

Evolutionary computation has received extensive interest in diverse research fields. Here we apply evolutionary computation to identify the system of a neuro-fuzzy system. An attempt is also made to resolve the data classification problems that have been thoroughly described [6–13]. Herein, we assign the membership functions to the predators and the training data to the prey. The interactions between predators and the prey [14] may eventually reach a stable state, implying that the membership functions have accurately classified the problem domain. The evolutionary model proposed herein simultaneously identifies the structure and the parameter while averting the constraints of BP.

The adaptive-network-based fuzzy inference system (ANFIS) [1, 5, 15] is a renowned neuro-fuzzy system that is functionally equivalent [15] to the radial basis function network (RBFN) [16]. Therefore, we construct a neuro-fuzzy system on RBFN. The rest of this chapter is organized as follows. Section II briefly introduces ANFIS. Section III describes the structures and functions of RBFN. Section IV discusses diversity within population in the evolutionary process, which is achieved via niches. Section V presents the proposed evolutionary model. Section VI summarizes the experimental results of the classification of the iris data set and the breast cancer data set, respectively. Section VII gives concluding remarks regarding the evolutionary model proposed herein.

II. ANFIS: ADAPTIVE-NETWORK-BASED FUZZY INFERENCE SYSTEM

An adaptive-network-based fuzzy inference system (ANFIS) is a neuro-fuzzy system whose structure is a multilayered ANN. ANFIS embeds fuzzy rules with the ANN and use a BP-like algorithm to fine-tune the parameters. Figure 1 illustrates a simple example of an ANFIS. The two rules of the example are as follows:

Rule 1.   If X is A1 and Y is B1, then $z1 = p1X + q1Y + r1$.

Rule 2.   If X is A2 and Y is B2, then $z2 = p2X + q2Y + r2$.

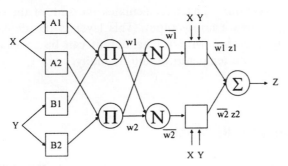

FIGURE 1 Diagram of ANFIS.

In this example, X and Y are both linguistic variables. A1 and A2 are linguistic values of X. B1 and B2 are linguistic values of Y. Assume that X denotes temperature and Y represents humidity. A1 and A2 represent hot and cold, respectively. B1 and B2 are dry and humid, respectively. Therefore, Rule 1 considers whether the weather is both hot and dry, whereas Rule 2 examines whether it is cold and humid. In what follows, we decipher the functions of each layer of ANFIS on the basis of the preceding example.

Layer 1 (Fuzzification). This layer is used to fuzzify the values of the input variables. Each node in this layer is a membership function, and its activation function is a membership function. Consider Fig. 1, in which A1 and A2 estimate the memberships of the input temperature as hot and cold, respectively. B1 and B2 measure the memberships of the input humidity with respect to dry and humid, respectively.

The clustering techniques are employed during the structure identification phase to conventionally determine the number of nodes and the rough structure of the membership functions. The parameters are fine-tuned by a BP-like algorithm while identifying the parameter. The parameters in this layer are generally referred to as *premise parameters*.

Layer 2 (Fuzzy AND). This layer attempts to calculate each rule's firing strength. Each node performs a fuzzy AND operation on the premise part of the fuzzy rules. As Fig. 1 depicts, the firing strength of Rule 1 is calculated via the fuzzy AND of the memberships of hot and dry. This strength closely resembles the estimation of the firing strength of Rule 2.

Layer 3 (Normalization). In this layer, we normalize the firing strength of each rule by calculating the ratio of each rule's firing strength to the sum of that of each rule. According to Fig. 1, wi denotes the firing strength of Rule i.

Layer 4 (Fuzzy Inference). Layer 4 estimates each rule's output. The outputs are calculated by the functions present in the then part of the rules. Figure 1 indicates that each rule's output is a crisp value. The parameters, (pi, qi, and ri) are adjusted during the phase of parameter identification. These parameters are frequently referred to as the *consequence parameters*.

Layer 5 (Defuzzification). Layer 5 calculates the sum of the outputs of all the rules and produces a crisp output. This layer is the defuzzification process of the fuzzy inference system. The output can be considered as a corresponding action to the temperature and humidity.

III. RBFN: RADIAL BASIS FUNCTION NETWORK

A radial basis function network (RBFN) is used to extract the input–output mappings. Each node's activation function in the hidden layer is a radial basis function (RBF). The input–output mapping is achieved by clustering the input values via the RBFs. Therefore, adjusting the RBFs allows the RBFN to precisely cluster the training data.

Figure 2 illustrates the structure of a RBFN. The number of input features and the number of classes of the data set determine the number of nodes in the first and the third layer, respectively. The number of nodes in the hidden layer and the adjustments of the RBFs are the critical design issues in constructing a RBFN. Several attempts have been made to construct an effective RBFN [16]. Theoretically, for a more complex problem domain, more RBFs are necessary. Meanwhile, for an easier one, less RBFs are involved.

Similar to ANFIS, RBFN is also a neuro-fuzzy system. According to a previous investigation, RBFN and ANFIS are functionally equivalent to each other [15]. Jang and Sun [15] indicated that these two architectural configurations are iso-functional. RBFN and ANFIS differ only in that the former uses the RBFs to cluster the data set while the latter uses rules to describe each cluster's location.

Next, the functions of RBFN are briefly explained layer by layer.

Layer 1 (Input). Layer 1 reads inputs from the problem domain. The number of features of the problem domain determines the number of nodes in this layer. Layers 1 and 2 are fully connected and the weights of the links are fixed to 1.

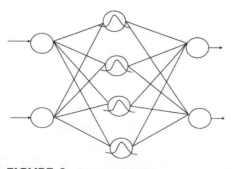

FIGURE 2 Diagram of RBFN.

Layer 2 (Clustering).   Each node of the hidden layer adopts a RBF as its activation function; each RBF is functions as a rule to classify the problem domain. As assumed herein, the RBF having the largest output responding to the input data represents the input data. The closer the input is to the RBF, the closer the output of the RBF is to 1.

Gaussian functions are frequently applied to the RBFs. Equation 1 provides the activation function of a Gaussian function. In practice, the centers of the RBFs are determined during the structure identification. The width is adjusted in the parameter identification phase. The dimensions of the center and the input of a RBF are the same.

Layer 3 (Output).   Layer 3 summarizes the classification results of the input data. Weights are located between Layer 2 and Layer 3, which are adjusted via BP in the learning phase.

RBFN is a two-phase approach: clustering phase and learning phase. During the clustering phase, the RBFs' locations in the problem domain are determined. Adjusting the weights and fine tuning the RBFs are fulfilled in the learning phase via BP,

$$f(x) = \exp\left(-\frac{\|\bar{x} - \bar{c}\|^2}{\sigma^2} \right). \tag{1}$$

IV. NICHE: DIVERSITY OF SPECIES

Many resources and living species abound in a natural environment and each living species lives on different kinds of resources. This biological discovery clarifies the concept of niche, that is, each species tends to live on the resource for which no other species competes. This competition subsequently results in diversity of species in the environment. Competition only occurs between species that live on the same resource. Therefore, in a stable state of the distribution of the living species and the resources, each species has its own resource to live on; each resource has its own predators as well. This competition averts a circumstance in which all species contend for the same resource, possibly leading to a situation in which only a few species survive; meanwhile, other species starve to death due to the failure in resource competition.

The diversity within a population must also be maintained in evolutionary computation. Without diversity, all agents of a generation tend to compete for the solution that has the greatest feedback to the winner. To this end, the population converges to a certain answer. However, addressing the multiobject optimization problem requires different optimal answers for different problems.

The diversity of a generation can be achieved in many ways. A conventional means, as introduced by Goldberg [17], obtains the diversity of a generation by computing the Hamming distance between the agents of a generation. The resource possessed by an agent must be shared with the other agents via a sharing function. Equation (2) displays the rate of sharing that agent i must share with agent j, where $d(x, y)$ is the Hamming distance between x and y, and k is the effective range of sharing. The closer an agent is to the specific agent the greater share it obtains. The resource possessed by an agent after sharing with other agents is calculated by dividing the raw resource by the amount shared with all other agents in the generation. Equation (3) displays the resource left after sharing. Results obtained from Goldberg indicate that the environmental resources have their own predators; in addition, the number of predators is proportional to the amount of resource. This observation ensures not only the diversity within a population, but also that each agent has its own niche to live in,

$$\text{sharing}_j = 1 - \frac{1}{k} \times d(\text{agent}_i, \text{agent}_j), \tag{2}$$

$$\text{resource\_after\_sharing}_i = \frac{\text{resource}_i}{\Sigma_{j \ne i} \text{sharing}_j}. \tag{3}$$

V. THE EVOLUTIONARY MODEL

Identifying a neuro-fuzzy system is a critical design issue. Many approaches have been proposed to enhance identification of the system [18–20]. A successful neuro-fuzzy system hinges on the accuracy of the fuzzy membership functions [21]. Therefore, in addition to determining the rough structure of the membership functions, their fine-tuning is critical. In this work, we present an evolutionary model to fulfill the identification of a neuro-fuzzy system.

An RBFN is an adaptive neuro-fuzzy model that is functionally equivalent with ANFIS. Therefore, this work also constructs a RBFN via the proposed evolutionary computation model. Notably, constructing a RBFN involves determining the number of nodes in the hidden layer, as well as the parameters of their activation functions. Herein, we choose the Gaussian function as the radial basis function (RBF) of the hidden layer.

A. Skeleton of the Model

This work identifies the structure and parameter of a RBFN via the evolutionary computation. The genetic algorithm (GA) and several other biological concepts are applied in the evolutionary model proposed herein. In practice, the system is identified by first identifying the structure and, then, the parameter. However, in the proposed model, the structure and parameter are simultaneously identified.

The proposed model also applies the concept of diversity (as mentioned in Section IV) to the evolutionary process of the RBFs. Restated, each RBF only covers a class of the data set. Moreover, as generally assumed, the RBFs that encompass more than one class of data in their coverage attain low fitness in the evaluation phase of the GA. By utilizing the diversity of the RBFs, the evolutionary process tends to evolve the RBFs that only encompass a certain class of data. The diversity is maintained via the concept of niche.

Each RBF evolves to encompass a pure portion of the data set, as well as to adopt the pure coverage as its niche. When the purity within the coverage of a RBF is 1.0, it is assumed to be the fittest one; it attains its niche in the evolutionary process as well.

According to the evolutionary explanations of the proposed model, each RBF is viewed as a predator in a biological environment; in addition, the data set is viewed as prey in the environment. Each predator only feeds on a certain kind of prey. Moreover, the classes of prey are determined in line with the classes of the data in the data set. When eating a prey that is not supposed to be devoured by itself, the predator gets a negative feedback to its fitness. In contrast, the predator gains in its fitness in line with the amount of resource it receives. Although the prey resources are initially set to be identical, the resource of the prey increases when either the prey causes a negative feedback to the fitness of a predator or no predator feeds on it. The interactions between the RBFs and the data set forces our model to create the most appropriate RBFs to encompass the data set.

The biological implication of our interpretation in the previous paragraph is that an animal may become ill when it consumes something that is harmful to itself. In contrast, devouring something that is good provides nutrition. From the perspective of food in an environment, the fact that some foods cause a predator to become ill implies that they are more apt to survive in the environment. Therefore, those food resources multiply. Those food resources that are never devoured by any predators increase as well. In addition, the interactions between the predators and the prey in the environment tend to converge to a stable state. A stable state implies that all predators have the highest fitness and all of the prey have the initial amount of resource. This observation suggests that no predator devours the wrong classes of prey.

Figure 3 depicts the skeleton of our model. A population of RBFs initially exists, with each chromosome of the population encoding the detailed parameters of the RBFs, the center and the width σ. Then, each RBF is applied to the data set to estimate its fitness. The fittest RBF is the one that

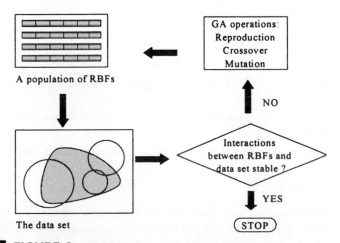

FIGURE 3 Flowchart of the evolutionary model.

encompasses only a certain class of data in the data set, that is, its purity equals 1.0. When all the RBFs have a purity of 1.0 and all data in the data set are correctly covered, a stable state between the RBFs and the data set is reached. This state is also the stopping point of the evolutionary process. The evolved RBFs are the desired ones for the RBFN when each RBF represents a certain class of the data set. The weights between a BRF and the output layer are all set to 0, except for the output node that has the same class as the representative class of the RBF. The weight of the link is set to 1. Further adjusting the weights between the hidden layer and the output layer of the RBFN is unnecessary. Therefore, the evolutionary process not only identifies the structure of a RBFN, but also completely identifies the parameters of these RBFs. We present an algorithm to explain the mechanism of the model proposed herein:

Step 1 (Initialization).   Initialize a population of the membership functions. Each RBF is encoded as a chromosome in the population. The chromosome encodes the center and the width of a RBF.

Step 2 (Measurements).

a. Determine the class of data that each RBF represents.
b. Assign the amount of resources that each data contains in the data set.
c. Assign the fitness of each RBF with respect to the purity of its coverage.

Step 3 (Stopping criteria).

a. Does the amount of average resource that the data set possess equal the initial value, say, 1.0?
b. Does the average fitness of the RBFs equal the fittest value, say, 1.0?
c. Is the variance of the resource between two generations of the data set equal to 0.0?
d. If all stopping criteria are adhered to, GOTO Step 5; otherwise, GOTO Step 4.

Step 4 (GA operations).

a. Reproduction: reproduce the RBFs that have a fitness equal to 1.0.
b. Selection: select RBFs from the population in proportion to their fitness.
c. Crossover: produce the offspring via the recombination of the parameters of the parents.
d. Mutation: mutate the parameter of an offspring with a certain probability.
e. GOTO Step 2.

Step 5 (Stop evolution).   The evolutionary process of the RBFN is fulfilled, in which all the RBFs in the population are correctly representative of the data set; the learning process with respect to the data set is completed as well. A testing data set can be used to evaluate the performance of the RBFN.

B. Design Criteria

Section V.A introduced the operational mechanism of our evolutionary model via an algorithm. In this section, we discuss the critical implementation criteria of the proposed model. Section V.B.1 explains how to determine the representative class of each RBFs. Section V.B.2 introduces the estimation of the fitness of each RBF and the resource of the data set. Finally, Section V.B.2 clarifies the interactions between the RBFs and the data via the notion of diversity in the evolutionary process.

1. Determination of the Representative Class of a RBF via Its Purity

A Gaussian function is a RBF that is frequently used as a fuzzy membership function. With the overlapping receptive fields of many RBFs over the data set, the representative RBF for each datum in the overlapping area can be determined by selecting the RBF that has the strongest firing strength fired by the datum. Therefore, using RBFs to cluster a problem domain would not make it necessary to insist on crisp boundaries between the RBFs. Figure 4 illustrates the notion of applying RBFs to cluster a one dimensional data set. All of the RBFs do not have explicit boundaries between each other. As long as the RBF with the strongest firing strength has the same representative class as that of the data that it encompasses, the RBF is assumed to correctly cluster the data. As mentioned in Section II, a RBF has infinite boundaries. Although approaching zero as they stretch out, the boundaries may never be equal to zero. In the proposed model, we constrain the effective range of a RBF to three times its width, σ.

$$\text{class} = \{c(M) | f(M) = \max\{f(x)\}, M, x \subseteq \text{dataset}\}. \tag{4}$$

The datum having the greatest membership determines the class that a RBF represents. Equation (4) provides the estimation of the representative class of a RBF, where M and x denote the data in the data set, $c(M)$ represents the class of M, and $f(x)$ is the membership of x in the RBF. The class of datum

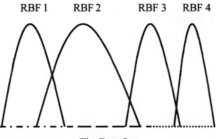

The Data Set

FIGURE 4   Use RBFs to classify the one-dimensional data set.

having the greatest membership is selected to allow each RBF to focus on the data that are in close proximity to its center during the evolutionary process. This focusing ability averts the possibility that some impurities are still near the center of a RBF, perhaps preventing other RBFs from representing those impurities. Therefore, concentrating the RBFs on the data near their centers during the evolutionary process is highly desired. Experimental results in this chapter confirm that such an approach is an effective means of clustering the data set via RBFs. Figure 5 presents the concept of selecting the class of the datum that is nearest the center of a RBF as the representative class of the RBF. As the evolutionary process proceeds, the RBF tends to minimize its range to concentrate on the data near its center, while leaving the other data to be covered by other RBFs.

2. Interactions between the RBFs and the Data Set

Interactions between the RBFs and the data set occur during every generation of the RBFs. During such interactions, the fitness of the RBFs and the resources of the data set are determined. Interactions between both sides attempt to maintain an evolutionary pressure to push the evolution of the RBFs rapidly and robustly. When some data are misclassified by some RBFs or they are not covered by any RBFs, their resources increase. This mechanism pushes the RBFs to encompass the data set correctly. The fact that (a) GA tends to evolve toward the good solutions and (b) those with higher resources are assumed to be good solutions to the evolution of RBFs, by assigning higher resources to those that are not correctly classified, accounts for why the evolutionary model tends to evolve RBFs to encompass them. On the other hand, the RBFs also have their fitness assigned with respect to the purity of their coverage. Higher purity of a RBF implies a higher likelihood that the RBF will survive to the next generation. As the evolutionary process proceeds, the average resource of the data set approaches the initial value and the average fitness of the RBFs approaches the fittest value.

Interactions between the RBFs and the data set form an evolutionary pressure to force the RBFs to correctly encompass the data set. Therefore, during the evolutionary process, there are increasingly fewer misclassified or uncovered data and, eventually, the RBFs evolve to classify the data set

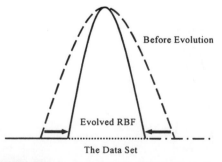

FIGURE 5 Evolution of a RBF to correctly cover a pure portion of the data set.

correctly. When the evolutionary process evolves to a stable state, that is, all data are correctly covered, the average resource of the data set equals their initial value and the average fitness of the RBFs equals the fittest value. Then the evolutionary process is complete. Figure 6 depicts both the stable and unstable states of the classification of a one-dimensional data set. At the unstable state, the purity of the RBFs is low, implying that some data are still incorrectly classified by the RBFs. At the stable state, all the data are classified correctly by the RBFs. The overlap between the RBFs representing different classes is allowed at the stable state as long as the representative RBF has a higher membership than other RBFs of its representative data.

At the beginning of each generation, the resources of the data set are equally set to an initial value, except for those that are not covered at all in the last generation. Both the uncovered and misclassified data have their resources increased by a certain value. Furthermore, the misclassified data have their resources reset to the initial value at the beginning of each generation. Meanwhile, those uncovered data maintain the updated value regardless of the coming of a new generation. The updated value of the resource of the uncovered data is not reset until they are covered by any RBFs. This phenomenon is because the misclassified data give a direct feedback to the fitness of the RBFs in the generation; however, those that are not covered will not. Therefore, the updated values of the uncovered data are maintained until they are covered by any RBFs. At that time, they promulgate the fitness of the RBFs that encompass them.

The resources of the data are calculated in line with the condition of how the data are covered. The representative RBF of a datum is the one that has the highest membership. If the class of the RBF does not correlate with that of the datum, it is assumed to be misclassified. Then the resource of the datum is increased. When a datum is not covered by any RBFs, its resource is also increased. However, when a datum is correctly covered, its resource remains unchanged.

The fitness of a RBFs is determined according to the resource of the data within its coverage. A datum, if correctly classified by the RBF, contributes a positive gain to the fitness of the RBF. In contrast, if a datum is misclassified by the RBF, it contributes a negative gain to the fitness of the RBF. Equation (5) displays the calculation of the fitness of a RBF, where RBF k represents

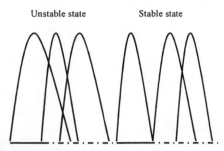

FIGURE 6 The classification of the data set under stable and unstable states.

data i, and j denotes the misclassified data. If Equation (5) does not consider the resources, it simply computes the purity of RBF k. Consider the resources of the data set in the computation of the fitness of a RBF that strengthens the effects provided by the data set. When a RBF encompasses an incorrect datum, the negative gain to the fitness of the RBF is in proportion to its resource. This phenomenon is contrary to the situation in which a RBF encompasses a correct datum. This results in a portion of the interactions between the RBFs and the data set. The interactions force the RBFs to correctly encompass all the data in the data set,

$$\text{fitness}_k = \frac{\Sigma_i f(i) \times \text{resource}_i}{\Sigma_i f(i) \times \text{resource}_i + \Sigma_j f(j) \times \text{resource}_j}. \quad (5)$$

At a desired stable state, the structure of the RBFs and the detailed parameters of the RBFs are fine-tuned by the evolutionary computation. The RBFs can be directly applied to the RBFN, and the weights between the hidden layer and the output layer are set to 1.0, if it is the weight between the RBF and its representative class in the output layer. The remaining weights are all set to 0.0. Therefore, in the proposed model, only an evolutionary process is necessary to identify the structure and parameter of the RBFN without the need to further apply the fine-tuning algorithm to tune the parameters of the RBFN. Moreover, the model proposed herein can rapidly and effectively construct a neuro-fuzzy system like RBFN.

VI. EXPERIMENTAL RESULTS

This section describes experimental results of the classification of both the iris data set and the breast cancer data set. In both experiments, the settings of the parameters of our model are set equal to each other. The initial resource of each datum is set to be 1.0. When the data are misclassified or uncovered at all, 1.0 is added to their resources. The fittest value of the fitness of the RBFs is 1.0. The mutation rate of the GA applied to the evolution of the RBFs is 40%. The dimension of the iris data set is 4, and its population size in the GA is 50. The dimension of the breast cancer data set is 9, that is, much more complex than that of the iris data. Therefore, we choose 2000 chromosomes as the population size of the GA.

A. Experiment 1: Recognizing the Iris Data Set

The iris data set, as created by R. A. Fisher [22], is one of the best known databases for data recognition. This data set has 3 classes, and each plant has 4 features. In the database, there are 50 instances for each of the 3 classes of the iris plant. Among the 3 classes of the iris plant, one of them is linearly separable from the others; meanwhile, the other 2 classes are not linear separable. Table 1 lists the detailed features, classes, and the relative range of each feature.

TABLE I   **Features and Classes of the Iris Data Set**

Features of iris data set	Range of feature	Classes of the iris data set
Sepal length in cm	Min = 4.3 and Max = 7.9	Iris setosa
Sepal width in cm	Min = 2.0 and Max = 4.4	Iris versicolour
Petal length in cm	Min = 1.0 and Max = 6.9	Iris virginica
Petal width in cm	Min = 0.1 and Max = 2.5	

In our experiment with the iris data set, we attempt to predict the class of a certain iris plant from its four features. Among the 150 instances of the database, we separate them into three distinct data subsets. Two of the data subsets are selected as the training data and third as the testing data in our experiment. Therefore, there are three experiments in the recognition of the iris data. Table 1 reveals that the ranges of the four features are not the same. Therefore, before our experiments, the ranges of the four features are rescaled within the range 1–10.

Figure 7 presents the average resource of the data set. The initial value for the resource is 1.0 and, at the beginning of the experiment, the average resource is high. This finding suggests that the data set is not correctly covered by the RBFs. As the evolutionary process proceeds with, the average resource approaches 1.0. This finding implies that the RBFs evolves to correctly encompass all the data in the data set. In the three subexperiments, the final average values are all 1.0, indicating that all the three subexperiments successfully learn from the training data.

Figure 8 displays the average fitness of the RBFs, which is the complementary information for the evolution of Fig. 7. The higher the average fitness of the RBFs is, the more data are correctly covered by the RBFs. According to Figs. 7 and 8, the evolution process tends to converge to a stable state, where both the resource possessed by the data set and the fitness of the RBFs are all 1.0. In the iris data set, owing to its simplicity, only 50 RBFs are used herein in each generation. Experimental results further indicate that 50 RBFs would be sufficient to recognize the iris data set.

Figure 9 illustrates the number of RBFs that can correctly encompass a subset of the data set. As the GA evolves new generations of RBFs, an increasing number of RBFs are qualified to represent a certain portion of the

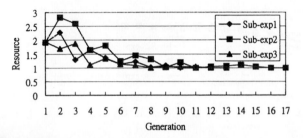

FIGURE 7   Average fitness of the data set.

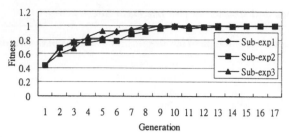

FIGURE 8 Average fitness of the RBFs.

data set. Figure 10 displays the variance of the sum of the resource of the data set between the generations. When approaching the stable state, the variance converges to zero. A situation in which the variance does not converge to zero implies that the evolutionary process is still attempting to evolve the appropriate RBFs for the training set. Experimental results of the recognition rate of the three subexperiments are 94%, 96%, and 98%, respectively. The average recognition rate is 96%.

B. Experiment 2: Recognizing the Breast Cancer Data Set

The data set of breast cancer, as created by Dr. William H. Wolberg [23], aims at medically predicting the likelihood of breast cancer. There are nine features and two classes of each instance in the data set. The nine features describe the characteristics of the cell nuclei in the digitized images that are obtained via the fine needle aspirate (FNA) of the breast mass. The two classes are "benign" and "malign." The data set contains 699 instances, among which, 16 of the instances have missing data. In our experiment, we neglect those 16 instances. In the proportion of the instances with the incomplete data included, there are 458 benign instances out of the 699 instances and 241 malign instances out of the 699 instances. Table 2 thoroughly describes the 9 features and their ranges. According to this table, all the ranges of the 9 features are the same. Therefore, the values of the features do not need to be rescaled. We take 33% of the data set as the testing data, and the remaining instances of the data set are taken as the training data.

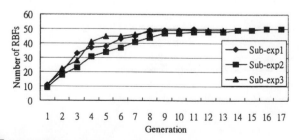

FIGURE 9 Number of RBFs evolved successfully.

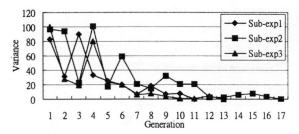

FIGURE 10   Variance between data set and RBFs.

Experimental results of the breast cancer data set resemble those of the iris data set. However, the dimension of the data set is nine, which is much complex than the four-dimensional iris data set. Figure 11 depicts the convergence of the source of the data set, and Fig. 12 displays the fitness of the RBFs evolving to 1.0 as the evolutionary progress proceeds. When both the data set's average resource and the RBFs' average fitness approach to 1.0, it is the stable state of the evolutionary process. However, this figure reveals that when the average fitness of the RBFs equals 1.0, the average resource of the data set still has not reached 1.0 yet. This occurrence may be attributed to two factors. First, the number of RBFs may not be sufficient to encompass the data set. Second, due to the fitness of all RBFs is 1.0, the RBFs cannot evolve to encompass the uncovered data set. This phenomenon allows GA to reproduce all RBFs to the next generation, thereby preventing the evolution of new RBFs.

Figure 13 displays the number of the RBFs qualified to represent a certain portion of the data set within the evolutionary process. Owing to the high dimension of the problem domain, selecting 1000 RBFs in a population of the GA performs the best recognition rate.

Figure 14 shows the variance of the sum of the resource of the data set between two generations. The variance approaches zero around the stable state. Moreover, the recognition rate of the breast cancer data set is 95%.

TABLE 2   Features and Classes of the Breast Cancer Data Set

Features	Range
Clump thickness	Min = 1.0 and Max = 10.0
Uniformity of cell size	Min = 1.0 and Max = 10.0
Uniformity of cell shape	Min = 1.0 and Max = 10.0
Marginal adhesion	Min = 1.0 and Max = 10.0
Single epithelial cell size	Min = 1.0 and Max = 10.0
Bare nuclei	Min = 1.0 and Max = 10.0
Bland chromation	Min = 1.0 and Max = 10.0
Normal nucleoli	Min = 1.0 and Max = 10.0
Mitoses	Min = 1.0 and Max = 10.0
Catalogues	
Benign	
Malign	

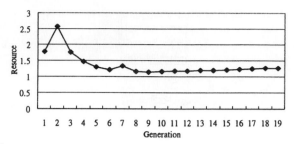

FIGURE 11   Average fitness of the data set.

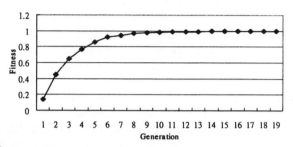

FIGURE 12   Average fitness of the RBFs.

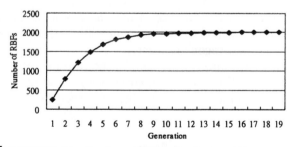

FIGURE 13   Number of RBFs evolved successfully.

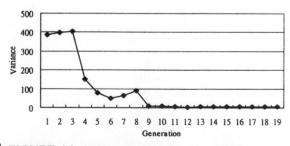

FIGURE 14   Variance between data set and RBFs.

VII. CONCLUSION

Expertise is indispensable in structure identification in conventional approaches to construct a neuro-fuzzy system. In constructing such a system, BP is frequently applied to deal with the parameter identification. However, expertise is not always available to data classification problems; in addition, the local optimum possibly traps BP. Therefore, in this work, we apply evolutionary concepts to construct the neuro-fuzzy system. Experimental results demonstrate the effectiveness of the proposed model to identify the structure and the parameter via the evolutionary process. The model proposed herein averts the problems frequently confronted in the system identification of a neuro-fuzzy system.

The proposed model adopts as a data set the prey in the environment. Each prey possesses a certain amount of resources. Herein, the RBFs are viewed as predators living on the resources provided by the prey. The optimal solution lies in the stable state of the interactions between the RBFs and the data set. When the stable state is reached, the RBFs can be used to construct a RBFN that can classify the data set. The structures of the RBFs and their detailed parameters are fine-tuned at the stable state, thereby averting the necessity to further fine-tune algorithms for parameter identification. Each RBF represents only a certain class of the data set, allowing us to directly determine the weights between the hidden layer and the output layer of the RBFN. Therefore, further weight adjustments are unnecessary. The proposed model effectively and rapidly identifies a neuro-fuzzy system.

An ANN is severely limited by the unavailability of the knowledge it has attained [24]. However, in the proposed model, after the evolutionary process, each RBF can be viewed as a certain rule to classify the data set. In addition, the center and width of each rule explicitly gives readable knowledge to the users. This feature allows the user to extract useful knowledge attained by the ANN.

Future works should focus on more complex problems, where explicit features are unknown. Therefore, feature extraction should be initially applied to the problem and, in doing so, the evolutionary model proposed herein can be applied to classify the problem domain. In addition, the proposed model can be enhanced by closely examining the competition within a certain species. More stress on the competition within a species helps reduce the redundant RBFs in each generation of GA. Furthermore, the proposed model can be used to construct other neuro-fuzzy systems, in which the ANN parts are not attached to RBFN.

REFERENCES

1. Jang, J. S. and Sun, C. T. Neuro-fuzzy modeling and control. *Proc. IEEE* 83(3):378–406, 1995.
2. Joshi, A., Ramakrishman, N., Houstis, E. N., and Rice, J. R. On neurobiological, neuro-fuzzy, machine learning, and statistical pattern recognition techniques. *IEEE Trans. Neural Networks* 8(1):18–31, 1997.

3. Lin, C. T. and Juang, C. F. An adaptive neural fuzzy filter and its applications. *IEEE Trans. Systems Man Cybernet.* 27(4):635–656, 1997.

4. Lin, C. T. and Lu, Y. C. A neural fuzzy system with fuzzy supervised learning. *IEEE Trans. Systems Man Cybernet.* 26(5):744–763, 1996.

5. Jang, J. S. ANFIS: Adaptive-network-based fuzzy inference system. *IEEE Trans. Systems Man Cybernet.* 23(03):665–685, 1993.

6. Cho, S. B. Neural-network classifiers for recognizing totally unconstrained handwritten numerals. *IEEE Trans. Neural Networks* 8(1):43–53, 1997.

7. Chen, C. L. P. and Lu, Y. FUZZ: A fuzzy-based concept formation system that integrates human categorization and numerical clustering. *IEEE Trans. Systems Man Cybernet.* 27(1):79–94, 1997.

8. Gader, P. D., Mohamed, M., and Chiang J. H. Handwritten word recognition with character and intercharacter neural networks. *IEEE Trans. Systems Man Cybernet.* 27(1):158–164, 1997.

9. Li, Q. and Tufts, D. W. Principal feature classification. *IEEE Trans. Neural Networks* 8(1):155–160, 1997.

10. Maniati, E. V., Kurz, L., and Kowalski, J. M. A neural-network approach to nonparametric and robust classification procedures. *IEEE Trans. Neural Networks* 8(2):288–298, 1997.

11. Pedrycz, W. and Waletzky, J. Fuzzy clustering with partial supervision. *IEEE Trans. Systems Man Cybernet.* 27(5):787–795, 1997.

12. Ridella, S., Rovetta, S., and Zunino, R. Circular backpropagation networks for classification. *IEEE Trans. Neural Networks* 8(1):84–97, 1997.

13. Sun, C. T. and Jang, J. S. Fuzzy classification based on adaptive networks and genetic algorithms. In *Genetic Algorithms and Fuzzy Logic Systems: Soft Computing Perspectives* (E. Sanchez, T. Shibata, and L. A. Zadeh, Eds.), World Scientific, Singapore, 1996.

14. Rosin, C. D. and Belew, R. K. New methods for competitive coevolution. *Evolutionary Comput.* 5:1–29, 1997.

15. Jang, J. S. and Sun, C. T. Functional equivalence between radial basis function networks and fuzzy inference system. *IEEE Trans. Systems Man Cybernet.* 23(03):665–685, 1993.

16. Hwang, Y. S. and Bang, S. Y. An effective method to construct a radial basis function neural network classifier. *Neural Networks* 10(8):1495–1503, 1997.

17. Goldberg, D. E. Genetic algorithms. In *Search, optimization, and machine learning*, pp. 185–197, Addison Wesley, Reading, MA, 1989.

18. Lee, K. M., Kwak, D. H., and Kwang, H. L. Fuzzy inference neural network for fuzzy model tuning. *IEEE Trans. Systems Man Cybernet.* 26(4):637–645, 1996.

19. Luciano, A. M. and Savastano, M. Fuzzy identification of systems with unsupervised learning. *IEEE Trans. Systems Man Cybernet.* 27(1):138–141, 1997.

20. Torbaghan, M. F. and Hildebrand, L. Model-free optimization of fuzzy rulebased systems using evolution strategies. *IEEE Trans. Systems Man Cybernet.* 27:270–277, 1997.

21. Zeng, X. J. and Singh, M. G. A relationship between membership functions and approximation accuracy in fuzzy systems. *IEEE Trans. Systems Man Cybernet.* 26:176–180, 1996.

22. Fisher, R. A. The use of multiple measurements in taxonomic problems. *Ann. Eugenics* 7(II):179–188, 1936.

23. Wolberg, W. H. and Mangasarian, O. L. Multisurface method of pattern separation for medical diagnosis applied to breast cytology. *Proc. Nat. Acad. Sci. U.S.A.* 87:9193–9196, 1990.

24. Narazaki, H., Watanabe, T., and Yamamoto, M. Recognizing knowledge in neural networks: An explanatory mechanism for neural networks in data classification problems. *IEEE Trans. Systems Man Cybernet.* 26:107–117, 1996.

10
TECHNIQUES AND APPLICATIONS OF FUZZY THEORY IN QUANTIFYING RISK LEVELS IN OCCUPATIONAL INJURIES AND ILLNESSES

PAMELA R. McCAULEY-BELL

Department of Industrial Engineering and Management Systems,
University of Central Florida, Orlando, Florida 32816

LESIA L. CRUMPTON-YOUNG

Department of Industrial Engineering, Mississippi State University,
Mississippi State, Mississippi 39762

ADEDEJI BODUNDE BADIRU

School of Industrial Engineering, University of Oklahoma,
Norman, Oklahoma 73019

I. INTRODUCTION   224
   A. Objective   224
   B. Current State of Occupational Illnesses in American Industry   225
   C. Occupational Injuries and Illnesses   226
II. PREDICTIVE MODELS FOR OCCUPATIONAL INJURIES   228
III. FUZZY MODELING   230
   A. Numeric Basis of FST   231
   B. Membership Functions   232
IV. REVIEW OF FUZZY APPLICATIONS TO MODELING OCCUPATIONAL INJURIES   237
   A. Predictive Models for CTDs   238
V. PRINCIPLES FOR APPLYING FUZZY MODELING TO MODELING CTDs   239
   A. Nature of the Injury   239
   B. Presence of Data   239
   C. Knowledge Elicitation   240
   D. Application of Methodology   242
VI. CONCLUSION   259
   REFERENCES   260

I. INTRODUCTION

Predictive models are needed to identify the risk of developing occupational injuries and illnesses in today's workplace. Prediction models will aid significantly in preventing and controlling the development of occupational injuries and illnesses, thereby minimizing the frequency and severity of these problems. The creation of predictive models for occupational injuries and illnesses is often hampered by the variability associated with human anthropometry and performance. Traditional modeling techniques often experience difficulties when trying to identify the onset of injuries and illnesses, particularly with cumulative trauma disorders (CTDs). Cumulative trauma disorders (i.e., carpal tunnel syndrome, tendonitis, tenosynvitis, trigger finger, etc.) are occupational illnesses that develop gradually as a result of repeated stress and strain being placed on the musculoskeletal system. The dichotomous nature of conventional logic is inadequate in representing the stages that an individual may undergo in the transition from the condition of "no injury" to the condition of "injury." However, fuzzy set theory (FST) provides a tool to address this variability. Occupational injuries that are most compatible with the principles of fuzzy modeling are those that occur over an extended period or work-related musculoskeletal disorders. Cumulative trauma disorders affecting the upper extremities of the body (i.e., hand, wrist, forearm, arm) are plaguing virtually every aspect of hand-intensive industries. In order to treat them, the various contributing factors of the problem must be assessed and analyzed. One of the problems with understanding CTDs is defining the specific response of the musculoskeletal system resulting from exposure to the occupational risk factors. The use of fuzzy modeling permits representation of these risk factors to analyze the resulting damage to the musculoskeletal system.

A. Objective

The objective of this chapter is to provide an overview of the applicability of fuzzy set theory to modeling risks of occupational injuries and illnesses. The modeling of occupational injuries is inherently vague due to the variation in work environments, individual differences, and application techniques. In spite of these difficulties, it is essential to represent the risk of injuries to reduce the likelihood of occurrence. This is even more difficult with the proliferation of cumulative trauma disorders because these injuries usually occur gradually over a period of time and exposure. Due to the gradual onset of these injuries and the absence of specific measures with respect to the dose−response relationship for suspected risk factors, alternatives to traditional probabilistic techniques are required. Fuzzy modeling is a very appropriate tool to address this area due to the ability to accommodate the variability, use linguistic variables, and provide quantitative outputs.

B. Current State of Occupational Illnesses in American Industry

The number of occupational injuries and illnesses is growing at an alarming rate. The occurrence of these incidents has a direct impact on the cost associated with service and products for the consumer. The Occupational Safety and Health Administration (OSHA) defines an occupational injury as an injury such as a cut, fracture, or sprain, that results from an exposure involving a single incident in the work environment. OSHA defines an occupational illness as any abnormal condition or disorder, other than one resulting from an injury, caused by exposure to environmental factors associated with employment. Thus, cumulative trauma disorders should be classified as occupational illnesses. There is a great demand for prediction models to assist in the evaluation of CTDs.

Fuzzy modeling is the use of fuzzy set theory to represent the status of an entity, or its parameters, in a particular situation or environment. A fuzzy set A is characterized by a membership function $\mu_A(x)$. Unlike conventional (crisp set theory), where objects are either in or out of the set, fuzzy theory allows objects to have partial membership and this progression aptly represents the manner in which cumulative trauma disorders develop. Fuzzy modeling is more compatible than the randomness of probabilistic modeling when representing the occurrence of CTD risk factors or the development of injury. For example, a case of carpal tunnel syndrome (CTS) should be detected in the initial stage to prevent development into a permanent disability. Fuzzy modeling allows detection of CTS at the preliminary stages by permitting partial membership in the set, incidence of occurrence, and CTS can thus be recognized sooner. Even though the use of subjective probability is one way to describe event ambiguity, fuzzy set theory is an alternative that is easier to apply as well as offering the option to represent linguistic variables. Further, the flexibility of fuzzy set theory permits a more realistic approach to evaluating the risks of musculoskeletal injuries through the linguistic representation of risk level and gradual set memberships.

Traditional regression analysis is useful in assessing the impact of various independent risk factors on a given dependent variable. In the case of CTDs the various risk factors, or independent variables, are used to predict the outcome of the dependent variable, CTD injury. Thus it is necessary to consider a multitude of risk factors for a single task. Naturally, tasks or operations that comprise actions containing multiple CTD risk factors are expected to have a greater chance of causing CTDs. However, risk-factor presence alone is not sufficient; the true impact will be dependent on the relative degree of severity of each factor's existence. An alternative to traditional regression is fuzzy regression. Fuzzy regression gives rise to a possibility distribution that accounts for the imprecise nature in the development of CTD. This technique can be used at varying levels to model risk of CTDs [1]. Therefore in some CTD analyses fuzzy regression is more appropriate than traditional regression for analyzing ambiguous processes. Also, because the amount of exposure to a risk factor or potential synergistic impact of multiple risk factors is not well documented, fuzzy set theory can be used to accommodate this shortcoming by permitting linguistic representation of the impact to produce a transition to a fuzzy membership function. These tools allow fuzzy modeling to produce appropriate representation of factor level and an effective means to aggregate model inputs.

C. Occupational Injuries and Illnesses

Since the early 1980s, the U.S. Bureau of Labor Statistics (BLS) has reported a steady increase in the number of reported cases of cumulative trauma disorders from 26,700 in 1983 to 332,000 in 1994 as shown in Fig. 1 [2]. However, in 1995 and 1996, there were slight decreases in the number of reported cases from 308,000 in 1995 (down 9% from 1994) to 281,000 in 1996 (down 15% from 1994) [3, 2]. The 281,000 reported cases of CTDs in 1996 accounted for 64% of all occupational injuries and illnesses.

In 1991, more than 100,000 operations were performed for CTD-related disorders, with an average cost of $4,000 per operation [4]. Additionally, CTDs have a long recovery period, which can add to this cost by more than $29,000 [4]. The costs associated with CTDs are more than 50% higher than any other work-related injury or illness.

Back injuries, one of the most common CTDs, accounted for 40% of all worker-compensation costs in 1991, while worker compensation costs for CTDs of the upper extremities accounted for 3% [5, 4].

Total worker compensation costs in 1995 for CTDs were greater than $20 billion, illustrating how expensive CTDs can be to an organization in worker-compensation costs alone [6]. In addition, CTDs accounted for approximately 37,500 lost work days in 1995, with the average number of days missed being 30 days [3].

Table 1 includes a listing of the several types of CTDs and the number of cases resulting in lost work days [3].

Given the costs as well as the problems of lost work days, the occurrence of CTDs is a major problem for certain types of industries. CTD cases reported in various industries for 1996 and 1997 are shown in Table 2 [2].

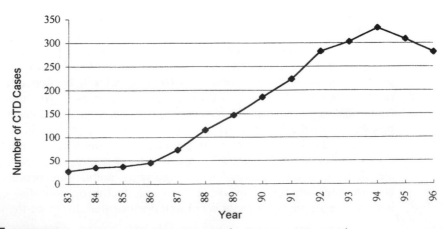

FIGURE I   Number of reported CTD cases (numbers are in thousands).

TABLE 1   Nature of Illness for CTD Cases That Resulted in Lost Work Days

Nature of Illness	No. of Cases	Percentage of Cases
CTS	31,154	37.7%
Tendonitis	12,324	14.9%
Back Pain	1,023	1.2%

Research conducted over the past few decades has identified several occupational factors such as repetitive movements, forceful exertions, and sustained or constrained postures that contribute to the development of CTDs. Also, it is hypothesized that nonoccupational factors such as age, gender, smoking, alcoholism, use of oral contraceptives or hormone replacement medication, obesity, prior hand or wrist problems, arthritis, and diabetes may contribute to the occurrence of CTDs by making some employees more susceptible to its development. For example, the percentage of CTD cases that resulted in lost work days is higher for females [3].

Additional research strides are needed to develop predictive models for the development of CTDs based on the presence of occupational and nonoccupational risk factors. The aforementioned statistics demonstrate why CTDs continue to be a top concern of employers. Companies are concerned with the cost associated with CTDs; therefore, they are eagerly pursuing ways to identify, prevent, and control CTDs before they become problematic.

The impact that CTDs are having on industry includes financial burdens, inhibition of quality, loss of productivity, and reduction in worker morale [1]. These costly burdens are felt across American industry and in many countries around the world. The significance of the problem of CTS is enhanced upon realizing that as many as 25% of employees in hand-intensive industries are being afflicted with injuries serious enough to require medical attention [7]. Perhaps one of the more serious areas of concern is video display terminal (VDT) tasks. In some cases, these jobs are considered a breeding ground for

TABLE 2   CTD Cases By Industry

Industry	Cases in 1995[a]	Cases in 1996[a]
Motor vehicle	49.5	44.4
Meat products	36.7	30.2
Hospitals	6.1	7.3
Grocery stores	5.1	6.2
Aircraft	6.4	5.5
Men's, boy's furnishings	6.2	5.1
Misc. plastic products	5.9	5.0
Metal forging, stampings	5.2	4.6
Electronic components	4.3	4.0
Telephone communications	4.6	3.6

[a] Rate per 10,000 workers.

occupational CTDs. The World Health Organization concluded that "musculoskeletal discomfort was commonplace during work with VDTs" and that "injury from repeated stress...is possible" [8]. Understanding the global market associated with current organizations, it is imperative to acknowledge and address this problem on an international level. An epidemic of CTDs affecting VDT operators swept Australia in the last decade. In fact, prevalence rates of almost 35% were recorded in some Australian organizations [8]. The high-tech, and often admired, Japanese industry is not exempt from the occurrence of CTDs. Japanese organizations have reported surges in the number of occupational-related musculoskeletal injuries [8].

II. PREDICTIVE MODELS FOR OCCUPATIONAL INJURIES

Mathematical and graphical modeling have also become increasingly popular as a means to assess a job and evaluate risk factors associated with CTDs. McCauley-Bell and Crumpton [9] developed a fuzzy linguistic model designed to predict the risk of carpal tunnel syndrome (CTS) in an occupational setting. The methodology of this model involves the acquisition of knowledge to identify and categorize a holistic set of risk factors that include task-related, personal, and organizational categories. Analytic hierarchy analysis (AHP) was used to determine relative factor importance. Fuzzy set theory was used to quantify linguistic input parameters and to develop a mathematical representation of the CTS risk. The results of an evaluation of the model including determination of sensitivity and specificity indicate the fuzzy model has potential for accurately predicting risks of injury for the identified risk factors.

Crumpton and Congleton [10] generated a model describing median nerve conduction using factors commonly associated with CTS to determine the best subset of variables to use as predictors of median nerve conduction time. Questionnaires were used to collect information on the occupation, work history, medical history, and leisure activities. The Nerve Pace Electroneurometer was used to collect nerve conduction time measurements of the median nerve. Multiple linear regression techniques were used to develop a model for predicting median nerve conduction time values. Results of this study suggest the possibility of using personal attributes as well as occupational risk factors to approximate median nerve conduction values. In addition, findings of this study suggest that regression techniques are powerful tools for analyzing factors and the combined effects of these factors on median nerve conduction. The researchers suggest inferences can be drawn about factors that may influence the development of CTS through the application of these principles.

Karwowski *et al.* [11] developed a three-dimensional (3D) graphical model to allow for better understanding of the relationship between several risk factors of lower-back disorders and the related lower-back injury risk potential. One of the variables is suspended in space with respect to the other two, and the surface-based algorithm allowed for the creation of the iso-potential surfaces. The low-risk jobs were those with at least three years showing no injuries and no turnover. High-risk jobs were those with at least 12 injuries per 200,000 hours of exposure. In the model, the risk space for occupationally related lower-back disorders was associated with a combination of five factors representing the workplace and the trunk motion factors, including load moment, lifting frequency, and three trunk motion factors. The graphic modeling was found to allow for better understanding in the risk behavior in terms of 3D space. In addition, results of this study revealed the dynamic systems approach and 3D graphic visualization models are powerful tools for analyzing complex interrelationships between critical risk factors for lower-back disorders. The risk behavior in relationship to the injury potential can also be examined; however, the relationships need to be carefully examined to improve understanding of lower-back disorders in the workplace.

Schoenmarklin and Marras [12] developed a 3D biomechanical model of the wrist to estimate the compression and shear forces on the wrist during isokinetic exertions. The model did not optimize muscle or reaction forces on the wrist joint. Instead it was data assisted in that wrist loads were calculated from estimates of forearm muscles as measured by electromyography (EMG). Fine-wire EMG electrodes were inserted into seven muscle bellies by a physician. Two subjects gripped a custom-built handle exerting flexion torque at constant velocities. Each trial began with an extension posture of greater than 60° and ended at a full-flexion angle greater than 60°. There were two torque levels, 4 and 8 N-m, and two grip orientations, power and pinch. Three repetitions were completed within each experimental condition for a total of 24 trials. Final gains ranged from 20 to 60 N/per square centimeter of muscle area. Torque predicted under pinch conditions agreed with external torque. Absolute compression was found to be a peak of 500 N on the wrist joint. Compression normalized to external torque was approximately 50–60 N per Newton-meter of external torque. In the radial–ulnar and flexion–extension planes, shear forces normalized to external torque were about 10 and 5 N per Newton-meter of torque. Results of this study suggest that it is possible to quantitatively estimate the loadings on the wrist joint during dynamic exertions. Matching the predicted external torque with the actual values validated the model. In addition, results of this study suggest this model could be used to explain in part why wrist motions associated with high risk of CTDs are injurious to workers.

Grant and Galinsky [13] examined the extent to which physical differences in grocery checkout employees and merchandise clerks can be reflected in actigraph data. Ten grocery cashiers and four merchandise clerks were observed working and were categorized as data was obtained from actigraphs affixed to the wrists and ankles. The subjects were monitored for certain lengths of time and then the actigraphs were removed and the data downloaded to a microcomputer. The cashiers' records revealed high levels of wrist activity and lower levels of ankle activity, while the clerks' records revealed moderate levels of wrist activity and higher levels of ankle activity. This difference could possibly have been due to the clerks additional walking tasks. The wrist activity was far greater during the work period than during periods of rest, and right-hand activity exceeded left-hand activity during work, but not necessarily during rest. Wrist activity was far greater among cashiers than clerks during work, whereas ankle activity was higher for clerks than cashiers during work. Even when the task varied (tendering money, bagging groceries, scanning groceries) the activity was greater for the dominant wrist (right). The results of this study indicate that actigraphy provides some advantages over traditional methods; however, observations are still an important part of the analysis. Combining the sampled data with actigraph data would help researchers be able to derive a fairly comprehensive index describing the activities of workers performing different jobs.

Ayoub and Hsiang [14] conducted a study to estimate the kinematics and kinetics of lifting movement under given conditions using simulation models. The generation of the possible postures composed by the joints at each point during the lift must first be done, and then the kinematics and kinetics of the lift can be evaluated. After this, the optimal principles to improve the lifting task can be reviewed, keeping in mind the possible designs of the workstation, the job design, and the individual's training. The goal was to maximize output while minimizing fatigue, stress, strain, and impacting forces on the body. A lifting task was performed, lifting from the floor to shoulder height and from the floor to knuckle height. Five subjects were used using two container sizes while being photographed with the Motion Analysis System. The model required inputs of the first and last body posture, the weight of the containers, and physical constraints. The predicted joint motions were then compared with the actual motion patterns. Based on the output, it was determined that the human body required the optimal pattern for the critical tasks with large loads or heavy containers, but that noncritical tasks can be performed when mechanical stress levels are low. There are currently very few models predicting the optimum pattern a worker should take. However, this model was found to provide a feasible alternative to obtaining knowledge and optimization of human motion.

III. FUZZY MODELING

Fuzzy set theory (FST) has provided a consistent and proven means to model many real-world environments [15, 16]. FST does not sharply define sets as traditionally done in ordinary set theory. Ordinary set theory is governed by binary principles such that a variable either belongs to a set, which would indicate a membership value of 1, or it does not belong to the set and maintains a membership value of 0. Conversely, FST does not restrict variable conditions to these rigid guidelines. Instead, FST deals with the imprecision associated with many variables by permitting a grade of membership to be defined over the interval [0, 1]. The grade of membership expresses the degree of strength with which a particular element belongs to a fuzzy set.

The nature of FST makes it useful for handling a variety of imprecise or inexact conditions. "Inexactness" in cognitive information may arise due to a number of situations. To differentiate between these situations, or classes of problems associated with the use of FST, Kandel (in [15]) has subdivided the categories within which most problems assessed by FST fall; these include generality, ambiguity, and vagueness.

1. Generality is the use of FST for specifying a general condition that can apply to a number of different states. A variety of situations can be characterized in this manner where the defined universe is not just a point [16]. For example, FST was used in a previous study with nurses to specify the level of fatigue associated with performing specific job demands [17].
2. Ambiguity is the use of FST to describe a condition where more than one distinguishable subconcept can simultaneously exist. For example, FST was used in a research study in the construction industry to describe the overall resulting awkward posture of a worker's hand, wrist, and arm while performing job tasks.
3. Vagueness is the use of FST to present those cases whose precise boundaries are not well defined. The boundaries in this case may be described as nonprecise or noncrisp. This is particularly relevant when attempting to define linguistic variables. For example, when evaluating the impact of a risk factor such as awkward joint posture on injury development, the exact degree of deviation that produces a particular risk level is not clearly defined and is thus considered vague.

All of these types of fuzziness (i.e., generality, ambiguity, and vagueness) are present in real-world applications and can be represented mathematically by a fuzzy set. This chapter focuses on the application of FST to the evaluation and measurement of occupational-illness risks through the use of FST and natural language. The measurement of cumulative trauma risks is challenging due to the cadre of risk factors as well as intrinsic human differences. Such a real-world situation is appropriately represented with fuzzy linguistic values and membership functions. Specifically, FST has been used to translate linguistic terms into numeric values that can then be used to get aggregate measures given several input factors.

A. Numeric Basis of FST

A fuzzy set is a class of objects with a continuum of grades of membership defined for a given interval. Such a set is characterized by a membership function that assigns a degree of membership ranging between zero and one to each object. To understand the mathematical definition of fuzzy sets, consider a finite set of objects X.

DEFINITION 1. Define the finite set as

$$X = x_1, x_2, \ldots, x_n, \tag{1}$$

where x_i are elements in the set X. Each element, x_i has a particular membership value μ_i, which represents its grade of membership in a fuzzy set [18]. The set of membership values μ_i associated with the fuzzy set occur along the continuum $[0, 1]$. A fuzzy set A can thus be represented as a linear combination of the following form:

$$A = \mu_1(x_2), \mu_2(x_2), \ldots, \mu_n(x_n). \tag{2}$$

A fuzzy set could also be expressed as a vector, a table, or a standard function whose parameters can be adjusted to fit a given system. The interval over which a fuzzy set applies, known as a universe of discourse U, is thus characterized by a membership function that associates each element x_i of X with a degree of membership μ_i.

B. Membership Functions

Several geometric mapping functions have been developed, including S-, π-, trapezoidal-, and triangular-shaped functions. All of these functions have utility in characterizing particular environments [19]. In ordinary (i.e., crisp) subsets, a phenomenon is represented by a characteristic function. The characteristic function is associated with a set, S, which is represented as a binary mapping function

$$\mu_s: X \rightarrow [0,1] \tag{3}$$

such that for any element x in the universe,

$$\mu_s(x) = \begin{cases} 1, \text{if } x \text{ is a member of } S \\ 0, \text{if } x \text{ is not a member of } S. \end{cases}$$

Figure 2 illustrates the mapping of a characteristic function for a crisp set. The process of translating a crisp set to a fuzzy set is known as fuzzification. To be "fuzzified," the real-world characteristics of a phenomenon must be mapped to a fuzzy function. The goal of this function is to map a subjective and ambiguous real-world phenomenon X into a membership domain, for example, $[0,1]$. This mapping function is a graphical representation of an element as it passes throughout a continuum (i.e., nonbinary) set of potential membership values. In other words, the mapping function provides a means to view the progression of the changes in the state of a given variable. Thus, this representation is referred to as a *membership function* for the fuzzy set

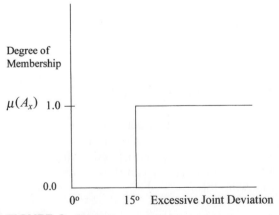

FIGURE 2 Crisp set.

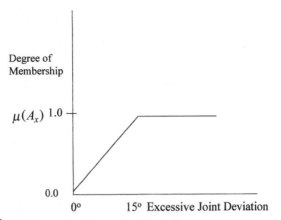

FIGURE 3 Basic membership function.

(Fig. 3). The term "membership function" emphasizes the previously stated premise of fuzzy sets: for a fuzzy set A, each x value within the set has an associated $\mu(x)$ value that indicates the degree to which x is a member of the set A.

Membership functions are a characteristic of the data set under analysis and can take on many forms. Given the primary geometric mapping functions of S-, π-, trapezoidal-, and triangular-shaped functions, the sinusoidal mapping functions, which include the S and π shapes [16], are the most frequently implemented. For all S and π mapping functions discussed here, consider the fuzzy phenomenon X, defined over a real (i.e., nonnegative) interval $[x_m, x_M]$, where x_m and x_M correspond to the lower and upper bounds of the set X, respectively.

I. S-Shaped (Sigmoid–Logistic) Mapping Functions

These mapping functions are termed "S" because they are shaped similar to the letter S. The curves comprised by the S mapping functions may be referred to as *growth* and *decline* curves [19]. The growth S-curve set moves from no membership at its extreme left-hand side, to complete membership at its extreme right-hand side. The decline S curve behaves in just the opposite manner, beginning with complete membership at its extreme left-hand side and progressing to zero membership at its extreme right hand side. Two common S-shaped curves are the S_1 and S_2 membership functions. Each of these functions has utility in the representation of fuzzy elements. An illustration and example of each of these functions are provided in Figure 4.

S_1 *Mapping Functions.* The mapping function S_1 maps x_i ($x_m \leq x_i \leq x_M$) into a non-symmetric sinusoidal membership function (see Fig. 3).

S_2 *Mapping Functions.* For the S_2 sinusoidal mapping function a symmetrical crossover point x_c is defined as follows:

$$x_c = \tfrac{1}{2}(x_m + x_M) \tag{4}$$

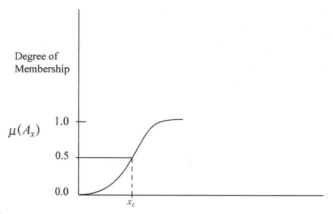

FIGURE 4   S-shaped membership function.

(see Fig. 3). The mapping function S_2 assigns low fuzzy set membership values to points below the crossover point ($x_i \leq x_c$; $0.0 \leq x_i \leq 0.5$) and higher membership values to points above the crossover point ($x_i \geq x_c$, i.e., $0.5 \leq x_i \leq 1.0$]. In other words, this crossover value divides the function into two portions and is located at the center point of the curve.

2. π Mapping Functions

The π mapping functions are so named because they approximately simulate the shape of the Greek letter π. Figure 5 illustrates a symmetrical π mapping function for a fuzzy phenomenon X over the interval $[x_m, x_M]$, where x_m, represents the minimum value for the curve and x_M represents the maximum value in the given curve. For this mapping function, the symmetrical point x_s is defined as the mid-point of x_m and x_M,

$$x_s = \tfrac{1}{2}(x_m + x_M) \tag{5}$$

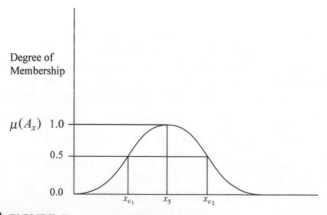

FIGURE 5 π-shaped membership function.

and the lower and upper crossover points are defined as

$$x_{c1} = \tfrac{1}{2}(x_m + x_s),$$ (6)

$$x_{c2} = \tfrac{1}{2}(x_s + x_M).$$ (7)

The π mapping function is a convex function that increases monotonically from 0 to 1 over the interval $[x_m, x_s]$ and decreases monotonically from 1 to 0 over the interval $[x_S, x_M]$. At the crossover points x_{c1} and x_{c2}, the value of the function is 0.5.

The π-shaped mapping function is often the default method of representing a fuzzy variable [19] because this method of representation allows a gradual descent from complete membership for a number in both directions, thus representing the concept of approximation. The symmetric π curve is centered on a single value and as the curve moves away from the ideal value (value with complete membership), the degrees of membership begin to taper off until the curve reaches a point of no membership, where $\mu_s = 0$.

Linear Membership Functions

In cases were the universe of discourse X is a real line, the fuzzy set can be expressed as a line (Fig. 6) or as some functional form. Two primary types of linear membership functions include the triangular and trapezoidal membership functions.

The linear membership function is perhaps the simplest membership function and is often used as a starting point when mapping a function. After initial representation using a basic linear membership function, set refinement often leads to more sophisticated linear and nonlinear membership functions. Triangular-shaped membership functions are used to represent relationships that are expected to be linear with a suspected optimal point or value and symmetry about this optimal point. This membership function is constructed under the same premise as the π-shaped membership function: as the value moves bilaterally away from the suspected optimal point, the

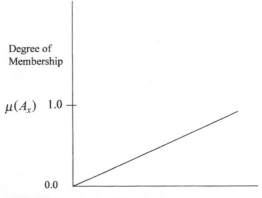

FIGURE 6 Linear membership function.

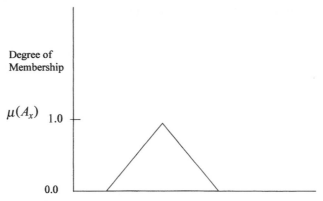

FIGURE 7 Triangular-shaped membership function.

degrees of membership begin to decrease until the value of no membership ($\mu_s = 0$) is reached at each end of the function (Fig. 7).

As with the triangular-shaped membership functions, the trapezoidal membership function is used to represent a set that is expected to exhibit a linear relationship. In this instance, there is not an optimal point or value that has complete membership. Rather, there is a range of values that have complete membership in the set (Fig. 8) and the descent from this complete membership completes the trapezoid. This approach was utilized in previous modeling of occupational injuries by (McCauley-Bell and Badiru [20, 21].

Fuzzy sets may also take on a combination of triangular and trapezoidal membership functions. For instance, in process control systems variables are decomposed into overlapping arrays of triangular-shaped membership functions. The endpoints of these variables represent regions that begin and end in complete membership for a given set. These outer membership functions are often expressed as "shouldered" sets and they appear as bisected trapezoids (Fig. 9).

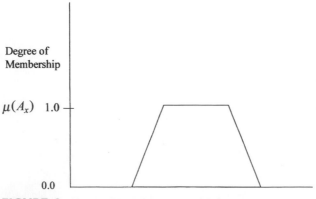

FIGURE 8 Trapezoidal-shaped membership function.

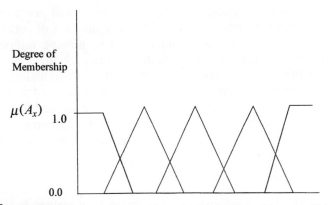

Degree of
Membership

$\mu(A_x)$ 1.0

0.0

FIGURE 9 Triangular-shaped membership function with trapezoidal shoulders.

The development of an appropriate membership function is critical for effective representation and modeling of a fuzzy set. It is generally possible to represent virtually any domain through a membership function because such functions may take on a variety of different shapes and forms to accommodate a given data set. Irregular and uniquely shaped membership functions can also be developed to represent a fuzzy set in unusual cases. The previously mentioned membership functions, however, will be useful at graphically representing most fuzzy sets.

IV. REVIEW OF FUZZY APPLICATIONS TO MODELING OCCUPATIONAL INJURIES

Although FST and occupational injury representation are compatible in many respects, the application of this technology to ergonomics and modeling injury risk has been limited. Karwowski and Mital [22] produced a volume designed to highlight fuzzy applications to human factors and ergonomics [22]. This book is entitled *Applications of Fuzzy Set Theory in Human Factors* and hosts work that applied fuzzy modeling to the representation of occupational back-injury risk levels and material-handling tasks as well as other topics. The following are some of the topics included in this book: "Towards an algorithmic/procedural human consistency of decision support systems: A fuzzy logic approach"; "A fuzzy method for the modeling of HCI in information retrieval tasks"; "Techniques for fitting fuzzy connective and logical operators to human judgement data in design and evaluation of man–machine systems"; and "Dealing with the vagueness of natural languages in man–machine communication." Although the projects presented in this volume were useful and of sound methodology, they did not produce widespread application of fuzzy modeling to human-factors and ergonomic problems.

Additional applications of fuzzy modeling to ergonomics were explored by the authors [23, 24, 9, 1, 25]. In 1996, Crumpton *et al.* [25] applied fuzzy modeling techniques to develop a model for evaluating fatigue using data from several estimators of fatigue. Some of the fatigue estimators included in this model were choice reaction time, heart rate, perceived exertion, and skin temperature. These research projects included application of fuzzy modeling to various occupational settings including risks of nursing injuries and an array of cumulative trauma disorders. Sufficient time has not lapsed to judge the impact of this work on the proliferation of fuzzy modeling in human factors and ergonomics. Nonetheless, there still exists a need to provide the occupational-injury, human-factors, and ergonomics communities with practical, comprehensive methodologies to produce reliable and quantitative models of occupational risks using the techniques associated with fuzzy set theory.

A. Predictive Models for CTDs

To investigate the relationships among a group of variables it is useful to create a model for those variables. One of the most widely used processes of finding a mathematical model that best fits the data is known as regression analysis. The traditional multiple-regression techniques and logistic regression have been applied in the areas of ergonomics and human factors engineering. Each of these techniques is described in the following sections.

I. Traditional Multiple Regression Analysis

In traditional multiple regression, the dependent variable y is a function of k independent variables, $x_1, x_2, \ldots, x_k$. The random error term is added to make the model probabilistic rather than deterministic, indicating the amount of variance in the dependent variable not accounted for by the linear combination of the k independent variables. The value of the coefficient β_i determines the contribution of the independent variable x_i [26]. Traditional regression is useful in identifying the contribution of multiple independent variables to a given dependent variable when ample data is available.

2. Logistic Regression Analysis

The logistic-regression approach is an approach that fits a model specially designed for a dichotomous, dependent, qualitative variable at two levels [26]. Applications occur frequently in the social sciences where people or actions of people often fall into one of two categories. The general logistic model is not a linear function of the β parameters as in traditional regression. Silverstein [27] used logistic regression to determine whether there were associations between CTDs and job attributes of force and repetitiveness. To test the hypotheses of no association between exposure categories and CTDs, student t tests were calculated to determine if there was a statistically significant difference in the overall means between low- and high-force jobs. Logistic regression techniques were used to evaluate the association between exposure variables and personal characteristics (i.e., sex, age, years on the job, etc.) in the development of CTDs of the hand.

3. Fuzzy Linear Regression Analysis

The ambiguity or fuzziness of human judgment is influential yet difficult to address in predictive modeling. Therefore, effective modeling of these systems is challenging [28]. Alternatively, the concept of fuzzy set theory seems to be applicable for modeling such systems [28]. Fuzzy linear regression models are introduced via the concept of possibility. In fuzzy linear regression models, deviations between the observed values and the estimated values are assumed to be dependent on the fuzziness of the parameters of the systems, in contrast to traditional linear regression analysis, where deviations are attributed to observation errors [28]. Because the parameters of the fuzzy linear regression model are fuzzy numbers, the model can accommodate the distortion introduced by the linearization. Fuzzy linear regression has as its core the principles of fuzzy set theory coupled with linear programming or least squares regression models. This technique employs the characteristics of regression analysis while using the strengths of fuzzy modeling in the management of uncertainty.

V. PRINCIPLES FOR APPLYING FUZZY MODELING TO MODELING CTDs

The application of fuzzy set theory to modeling cumulative trauma disorder risks can be accomplished through an efficient methodology. The stages in the application of this technique to occupational injuries is as follows:

- Determination of the nature of the injury (are injuries acute, cumulative, combination)
- Evaluation of the types of data available for methodological development
- Presence of experts or expertise
- Application of Methodology

Each of these stages is discussed.

A. Nature of the Injury

In the analysis of injury risks, the characteristics that define the occurrence of the injury should be understood. For example, if the injury is the result of an acute incident then the tools associated with FST may not provide improvement in prediction over existing techniques (i.e., NIOSH lifting guide). However, if the injury is produced over a sustained period of exposure then the techniques of fuzzy set theory may prove to be very useful.

B. Presence of Data

Various data sources are generally available in most organizations for use in the analysis of tasks. Obtaining access and most effectively utilizing these data sources is essential in model development. Archival data are particularly useful because they are readily available, but there are disadvantages associated with these types of data [29]. Thus it is necessary also to evaluate the existing state of the occupational environment. Some of the data sources to analyze include [29] the following:

Fixed data—Examples of fixed data include facility layout drawings, production rates, location of facilities, and production layouts.

Production records—Examples of production records include order execution, routing slips, manual material handling (MMH) details, and material requirements planning (MRP) details.

Industrial engineering (IE) *records*—The relationship of IE to methods and time study means that often records are kept on the breakdown and analysis of tasks. This is generally a good source of information on the details of task performance.

Personnel data records—A considerable amount of information is maintained in personnel records. The most pertinent information available through personal records with respect to model development is fixed data on job description. This information should be available for all jobs and will generally provide the basic elements of task performance. Although all organizations may not maintain these types of records, the establishment of the Americans with Disabilities Act (ADA) has been an impetus for promoting the maintenance of detailed records of task requirements. Additional sources of data contained in personnel records include accident and injury data. These data sources may provide insight into the types of risks that are present in an area. Examples of the type of information that may be gained from these records include an increased likelihood of injury during a particular work shift, history of injury for a particular type of tool, and environmental issues that appear to impact the risk of injury (i.e., oil on a floor). Finally, individual employee data is also available in the personnel records. Although this information has been used in determining risk of injury [21, 22], caution must be exercised when using this information. This is primarily due to the fragility of using personal data as a viable means in determining suitability for task performance. This runs counter to the principles of ergonomics, as the ergonomist is tasked with designing the task to fit the person. Thus this information though important, and at times relevant to the likelihood of injury development, must be handled very delicately.

C. Knowledge Elicitation

Upon analysis of fixed data sources additional sources of expertise may be needed. This expertise is generally obtained through interaction with individuals or observation in the workplace. Some of the individuals that may be useful in obtaining this information include production managers, safety personnel, ergonomic experts, and experienced employees. The interaction with individuals to obtain relevant task information will entail some type of knowledge elicitation. In selecting individuals to interview it is important to get a holistic perspective (i.e., interview a foreman and a construction engineer) of task performance as well as access to expertise regarding task-performance capabilities (i.e., ergonomist). Knowledge elicitation or acquisition is required in the development of knowledge-based systems and the most commonly used techniques are discussed later. These areas overlap and it is useful to use a combination of methods to solicit knowledge [30]. These methods are as follows:

1. Text analysis (analysis of existing or archival data)
2. Interview analysis

3. Behavior analysis
4. Machine induction
5. Concept mapping

Each of these methods can be used individually, or they can be synthesized such that a knowledge engineer can capitalize on the benefits of each approach while minimizing the impact of their respective shortcomings.

1. Text Analysis

Text analysis consists of knowledge acquisition without recourse to an expert, but through the use of printed or archived material. In the evaluation of occupational injuries, the primary resources for obtaining knowledge through text analysis are covered in the previous section on the use of archival data sources. Text analysis should not be used as a sole method of knowledge acquisition, particularly when the knowledge engineer is unfamiliar with the subject matter. If the knowledge engineer is unfamiliar with the topic and attempts to make inferences based on printed material alone, he is expected to behave as an expert and is generally not qualified to do so. One domain where text analysis has proven useful is the development of systems that contain rules and regulations, such as legal advice systems [30]. However the use of text analysis as a sole source of knowledge elicitation is not encouraged.

2. Interview Analysis

Interview analysis involves direct interaction with the expert. The choice of interview style will be a function of the desired goals or constraints of a project, including access to individuals, volume of input sought, and financial constraints. Included in the interview analysis class of knowledge elicitation are questionnaires, structured interviews and unstructured interviews [31], and one-on-one interviews such as problem discussion, problem description, and problem analysis [30].

Questionnaires are the simplest method of interview analysis. However, to be successful, the interviewer must know which questions to ask and be careful not to be too restrictive in the answers.

Interviews, in one form or another, are the most widely used form of knowledge elicitation. The advantages and disadvantages associated with the varying types of interviews will determine the mode used to obtain the information needed for the model. Guidelines for effectively using structured and unstructured interviews in ergonomic-related problems are available in Shadbott and Burton [31]. Three important things that an expert can do to provide insight include problem discussion, description, and analysis.

Problem discussion involves asking the expert to explore the kinds of data, knowledge, and procedures needed to solve specific problems. Problem description involves a sort of backward methodology. To accomplish this, the knowledge engineer asks the expert to describe a representative problem for each category of answers in the domain.

Problem analysis involves presenting the expert with a series of related case studies to solve aloud. While the expert is solving the problems, the knowledge engineer is probing the rationale behind the reasoning steps.

3. Behavior Analysis

Behavior analysis involves observing the expert as he performs a given analysis. Although this is the most comprehensive technique of knowledge acquisition, it is also the most time consuming and can also be intrusive if conducted in a real-world setting.

4. Machine Induction

The goal of machine induction is to induce a general rule that covers all cases from a sample of example cases. The premise of this approach is that many expert tasks or subtasks come down to a classification problem. An advantage of machine induction is that it offers the potential to deduce new knowledge. For example, it may be possible to list all the factors that influence a decision without understanding their impacts, and to induce a rule that works successfully [30]. This methodology would prove very beneficial in an environment where there is plentiful test data and a well-defined classification problem to be solved.

5. Concept Mapping

Concept mapping is a knowledge-acquisition tool that has been designed to capture and graphically represent the relationships that exist between concepts in the domain expert's understanding of the problem space, and the solutions that the expert applies to the problem [32]. Human expertise is often composed of procedural and perceptual knowledge. Thus, it is sometimes difficult for experts to recall specific aspects of knowledge or to verbalize these factors discretely. Concept mapping minimizes the likelihood that important knowledge is discarded by providing a user-centered knowledge-acquisition methodology that captures the entire solution domain. This methodology supports both the expert and the knowledge engineer as knowledge users because they are both actively involved in the mapping process [33], using the concept map as a shared medium for communication. Concept mapping also allows multiple inputs (i.e., experts) and indexing of the knowledge base to access specific information readily.

D. Application of Methodology

Prior to discussing the example, the basic format for the application of these techniques to determining risks of occupational injury will be presented. The following outline may be used a guide in applying these techniques:

- Problem identification
- Knowledge elicitation
- Determination of *dependent variables* (e.g., task duration, repetition)
- Determination of *independent variables* (e.g., risk of injury, fatigue index)
- Data collection (Data should be collected on relevant task components.)
  - Task Analysis

- Development of universe of discourse and linguistic levels for variables
- Weighting of variables
- Output interpretation
- Model validation–verification

Although listed sequentially, some of the previously mentioned activities may be performed simultaneously.

I. Problem Identification

Problem identification consists of narrowing the scope of the project and identifying the specific objectives of the proposed model. For example, in modeling CTDs the problem is the absence of a quantitative measure of risk as well as the need to integrate the impact of numerous quantitative and qualitative variables. The motivation for desiring a quantitative measure is the ability that numeric measures provide for considering aggregate inputs and more thoroughly evaluating the sensitivity of the model. Thus the objective in modeling CTDs is to produce a numeric measure that accommodates a comprehensive set of risk factors.

2. Knowledge Elicitation

Knowledge elicitation for occupational injury analysis should consist of interaction with a cross section of individuals who are knowledgable about the subject matter. This approach should include text analysis of organizational documents that contain specifics about task performance, occupational safety guidelines, and any relevant regulations that impact the work environment or task performance expectations. Upon compilation of this information, additional sources of information should be collected from subject-matter experts. These subject-matter experts may include plant ergonomists, task operators, safety personnel, or medical professionals.

3. Determination of Dependent and Independent Variables

In trying to model a particular problem, a suspected portion of independent variables generally exist. Although this prior knowledge may be useful in initially developing a model, it is essential not to introduce bias into the model based on the developers opinions. The text analysis and knowledge elicitation should be the foundational material for the establishment of the independent variables, or input parameters. The dependent variable should be determined as a function of the problem identification and objective. In other words, the goal of the system should be to produce a measure that meets the objectives of the system.

4. Data Collection

In this stage the researchers are tasked with collecting relevant information on the identified independent variables. It is essential that data-collection techniques be consist and accurate as this information will directly impact the quality of the model. The collection of the data often includes task analysis to obtain information such as task duration or joint deviation or questionnaire analysis to solicit larger categories of information.

5. Development of Universe of Discourse

To develop the universe of discourse three things must be understood:

- Definition of the set (e.g., degree of repetition)
- Range of existence for a variable (e.g., repetition exists from 0 to 100 units per hour),
- The units of measurement for the variable (each time a unit is assembled this represents one repetition)
- Interpretation of the degrees membership with respect to the units (e.g., how much repetition has a membership of 0.5)
- (optional) Development of associated linguistic risk levels for categories of the universe of discourse (e.g., low, medium, or high repetition)

This is more directly accomplished if finite numeric limits exist but often in the modeling of occupational injuries the variables are linguistic and not easily quantified. Consider the risk factor repetition, which may be quantified if the number of repetitive exertions is known or available through production records. However, attention must be given to understanding exactly what is being measured by the figure representing repetition. For example, in the evaluation of an assembly plant the factor repetition may be considered. The amount of repetition associated with attaching a bracket to a unit (e.g., to assemble a hard drive for a computer) is a concern. The issue becomes whether or not to measure the number of hard drives assembled or the number of times the power driver is used to torque a screw in securing the subassembly. Although the outcome associated with risk will not be impacted solely by the choice of measurement, the interpretation of the results can be directly impacted. For instance, if an operator is expected to assemble 30 units (hard drives) within an hour and each hard drive has four screws to be attached, then the measures for the drive will be 30 units per hour of repetition but the repetition for the action will be 120 per hour. If the factor differentials are understood by those compiling and utilizing the results, then this will not impact model usability. However, to support model generality and prevent misinterpretations based on measurement differentials, the researchers encourage the use of "actions" or motions in measuring repetition rather than the use of units. Often measures are linguistic and in these cases, again, it is important to understand the limits and assign the degrees of existence of categories. The results of the knowledge-elicitation stage should be useful in assigning linguistic measures an associated degree of membership.

6. Weighting of Variables

Various approaches exist to obtaining weighting of different factors. These approaches to obtain a determination of factor input to a given outcome utilize techniques such as rating scales, pairwise comparisions, and other subjective approaches. An approach commonly used is analytic hierarchy processing (AHP), which takes all known factors into consideration. Although this approach has been widely utilized, a criticism of AHP is the validity of the relative weights upon identification of additional inputs. This technique is discussed more in Example 1.

7. Output Interpretation

Prior to model development the researchers should understand how the results will be aggregated and interpreted. This involves understanding what the numeric results will be used to infer as well as how they will be used. For example, if a numeric output is obtained will this be the information that is imported into another system for further analysis or will these results be assigned a linguistic meaning? Other alternatives include designing the system such that a predefined interpretation result is attached to a meaningful interpretation for the end user.

8. Model Validation

Upon completion of the model the results should be evaluated to determine utility and applicability. Validation is generally performed immediately upon completion of the model with an alternative set of data; however, interim and follow-up evaluations should be performed to determine long term model validity.

EXAMPLE 1. (Application of methodology). (This text closely follows [23] and [24]).

To illustrate these steps further, a study conducted by McCauley-Bell and Badiru [23, 24] will be presented. The stages in the methodology for developing the linguistic risk level and associated membership functions included the following:

1. Knowledge elicitation
   Development of dependent variables
   Development of independent variables
2. In-depth analysis
3. Defining universe of discourse
   Development of membership functions,
   Development of fuzzy linguistic variables
4. Weighting of variables—AHP analysis.
5. Output interpretation

These steps are discussed in more detail next.

Knowledge Elicitation. The knowledge elicitation process consisted of a hybrid of knowledge acquisition methodologies. The three primary components included

1. Preliminary analysis
2. Text analysis
3. Traditional interview analysis and concept-mapping interviews

These aspects of knowledge acquisition were performed sequentially to build a framework for the classification of the knowledge.

TABLE 3   Risk Factors Identified in Preliminary Analysis

Task-Related risk factors	Personal risk factors	Organizational risk factors
Force	Previous CTD	No ergonomics program in a facility
Repetition	Years on the job	High CTD statistics
Awkward posture	Obesity	No training for hand-intensive tasks
Excessive use of grip or pinch force	Age	
Workshifts that exceed eight hours	Gender	
Extreme temperatures in the work environment	Health-related problems that inhibit circulation	

In-Depth Text Analysis.   Prior to the formal knowledge-elicitation stage, a preliminary analysis of the literature was performed to get an outline of the problem area (Table 3). The formal knowledge acquisition began with an in-depth text analysis to identify risk factors associated with CTDs of the forearm and hand. The results of this analysis yielded risk factors that were very similar to, but more extensive than, those gathered in the preliminary knowledge-acquisition sessions. The literature contained an abundance of information about the potential causes of upper extremity CTDs [34–40] (see the end of our References section for an extended bibliographical listing of citations). There was consistency across the literature, particularly with respect to task-related risk factors. A significant outcome of the knowledge sessions was the suggestion that there is a core of task-related risk factors that promote CTDs, and within that core are force, repetition, and awkward joint posture. These factors consistently surface. The results of the text analysis were reviewed and each of the factors was placed in the appropriate category. The results of the formal text analysis for the task-, personal-, and organization-related characteristics are listed in Tables 4, 5, and 6, respectively.

TABLE 4   Text Analysis Results: Task-Related Risk Factors

Repeated or sustained exertions	Hand tools
Awkward postures	Suspension of forearm
Mechanical stresses	Flexion of the wrist
Vibration	Extension of the wrist
Low temperatures	Ulnar and radial deviation
Gloves	Shoulder abduction
Vibrating hand tools	Forceful pinching
Repetitive motions	Force
Wrist deviation	Muscular fatigue, repetition induced
Joint angle	Forceful grasping

▪ **TABLE 5** **Text Analysis Results: Personal Risk Factors**

Diabetes	Hand length
Hypertension	Hand breadth
Oral contraceptives	Race
Obesity	Regular exercise
Gender	Age
Years on the job	Hand dominance
Smoking	Pregnancy
Lack of physical activity	Oral contraceptives
High blood pressure	Weight lifting
Strength and flexibility	Thyroid disorders
Chronic musculoskeletal problems	Inflammatory arthritis of the wrist
Circulatory problems	Excessive alcohol consumption
Extracurricular activities	

Knowledge Elicitation: *Interview Analysis*. The experts were selected based upon their expertise, years of experience, and accessibility. A total of four subject-matter experts (SME) were initially chosen. The intent in SME selection was to obtain individuals with a well-rounded understanding of CTDs yet a strength in one of the three modules identified in the preliminary analysis. Thus the system utilized the following experts: two orthopedic hand surgeons, who were considered to have primary strengths in the personal characteristics category; one occupational ergonomist whose expertise was in the task characteristics category; and, finally, one safety engineer–human-resources professional whose strength was considered to be in the occupational characteristics category. Unfortunately, one of the orthopedic surgeons was unable to complete the series of knowledge acquisition sessions. Therefore, the information contained in his initial knowledge-acquisition sessions was not used. The first interview session utilized interview analysis and problem analysis. This session lasted approximately 1.5 hours. The interview analysis consisted of a question-and-answer session and verbal problem solving. Observation analysis was also conducted by observing the expert as she/he performed routine evaluations of individuals suspected of having CTDs. The second session utilized concept mapping. This session took place approximately two weeks after the first session. The second session lasted for 2 days and consisted of 4 hours of knowledge acquisition. Each of the two 4-hour sessions was broken into two 2-hour intervals to minimize the likelihood of expert's block [18]. Although more information was obtained in the

▪ **TABLE 6** **Text Analysis Results: Organizational Risk Factors**

Decreasing work capacity	Peer influence
Management philosophy	Job security
Training	Ergonomic program
Treatment–rehabilitation for CTD sufferers	Production rate
Workplace design	Level of automation

TABLE 7   Knowledge Elicitation: Comparison of Risk Factors

	Text analysis results	Interview analysis results
Task-Related factors	Excessive force	Excessive force
	Repetition	Repetition
	Awkward joint posture	Awkward joint posture
	Extreme temperatures	Cold temperatures
	Extended work shifts	
	Gloves	
Personal factors	Wrist intensive hobbies or sports	Circulatory problems
	Excessive alcohol	Pregnancy
	Pregnancy	Oral contraceptives
	Oral contraceptives	Hormonal imbalances
		Knitting
		Racquetball
		Bowling
Organizational factors	On-site ergonomics program	Unions
		On-site ergonomics program

second session, when these results were compared with the knowledge obtained from the first session, there was consistency in the rankings of factors between the various knowledge acquisition sessions. Again, the purpose of using a hybrid of knowledge-acquisition methodologies was to ensure thorough coverage of the knowledge necessary to identify the risk factors. Table 7 presents an abbreviated list illustrating a comparison of the text analysis and interview analysis results.

Knowledge Elicitation: Concept Mapping.   Concept mapping is a knowledge acquisition tool that is designed to capture and graphically represent the relationships that exist between concepts in the domain expert's understanding of the problem space. The solutions that the expert applies to the problems are also captured in this methodology [32]. Human expertise is often composed of procedural and perceptual knowledge, making it difficult for experts to recall specific aspects of knowledge or to verbalize these factors directly. Concept mapping minimizes the likelihood that important knowledge is discarded by providing a user-centered knowledge-acquisition methodology that captures the entire solution domain. This methodology supports both the expert and the knowledge engineer as knowledge users because they are both actively involved in the mapping process, using the concept map as a shared medium for communication [33]. Concept mapping also allows multiple inputs (i.e., several experts) and indexing of the knowledge base to allow ready access to specific information. This tool may be used as a means to represent various components of a problem and the relationships that exist between these components. In addition, the graphical representation of the knowledge during knowledge acquisition serves as useful medium for tracking the flow of knowledge and solution methodologies.

The goal of each concept-mapping session was to obtain detailed representations of the knowledge necessary to address the problems and to understand the relationships that exist among the various concepts. The specific steps of this methodology were as follows:

1. Explain the elements of concept mapping and how the information that the expert provides will be represented.
2. Present an example of concept mapping.
3. Ask the expert if there was a need for clarification of the approach.
4. Conduct the mapping session.
5. Review the map with the expert.
6. Obtain relative factor significance through the analytic hierarchy process (AHP): expert performs pairwise comparisons of all of the elements at the various levels of the map.

More information was obtained in the concept-mapping sessions than in the traditional interview analysis. This may be likely due to the open-ended nature of the methodology. This information was then compared and contrasted with the knowledge obtained from the first interview session. Throughout the mapping session, the expert's conceptualization of the identification and analysis of potential risk factors was captured in the concept map. As the expert presented information, the concept map was drawn on a white, dry-marker board in front of him/her so that both he/she and the interviewer could easily see and discuss the concepts that were being represented. The fact that the map was drawn in front of the domain expert so that he/she could read it as it emerged is a particularly important aspect of the concept-mapping process. The expert was encouraged to interact with the map by reading, editing, and correcting it throughout the concept-mapping session. Also, upon completion of the concept map the expert was asked to re-evaluate the links that had been drawn to verify the accuracy of the transcription.

At appropriate times throughout the mapping session, the expert was probed with questions that asked for clarification or additional information on concepts that were being discussed. The intent of the probing was to stimulate additional discussion or to provide consistency within the given knowledge-acquisition session. No external information was used (i.e., information from the literature review or other expert interaction) to lead the expert.

Upon completion of each interview, an extensive review of the maps began. Once the review was completed, the interview map was transcribed and redrawn using Concept Interpreter. Concept Interpreter is an Apple Macintosh−based knowledge-acquisition tool that was developed by researchers at the Armstrong Laboratories at Wright Patterson AFB [41]. Upon completion of the concept maps, the user is given the opportunity to generate a database from the concept, review concept nodes, and interact with the information stored in other concept maps. This tool is particularly useful when combining input from several experts. These inclusive databases contain information from all of the experts and form the foundation for the development of the fuzzy rule-based model in phase 2 of the research.

Knowledge Acquisition Results.   Again, it should be emphasized that the concept mapping provided an illustration of the *relationship* between risk factors and the onset of injury and this information was crucial in the establishment of the knowledge base. Concept maps were generated from each of the knowledge-acquisition sessions.

After the concept maps were created the experts were asked to review them for accuracy. This accuracy review consisted of the expert analyzing the map and determining if the map actually represented valid relationships between the risk factors and injuries. At the same time, the expert had the opportunity to make any changes that he/she felt necessary in the information that was initially presented. The databases were created after the maps were complete and accepted by the experts. The results of the text analysis and concept-mapping knowledge acquisition led to the identification of more risk factors than those identified in the preliminary analysis.

Due to the volume of information collected via the various knowledge-acquisition methodologies, there was a need to identify the most significant risk factors in an effort to reduce the number of variables to be considered in the model. The risk factors that were consistently identified through the text analysis, traditional interview analysis, and concept mapping were considered to be the most significant in the development of injury. Thus they were determined to be the best candidates for inclusion in the risk-assessment model. The experts were also asked to rank the top risk factors. Again, this stage of the research resulted in consistency across experts. In cases where there was not strong consistency, the results from the expert within whose area of expertise the risk factor was contained received preference.

Ranking of Factors.   The results of the knowledge-acquisition and AHP analysis activities led to the generation of factors for representing the primary risk factors within each category. Table 8 lists the resulting risk factors deemed to be most significant by the experts and the literature within the task, personal, and organizational categories.

Development of Fuzzy Linguistic Variables.   At this point, risk factors and their associated levels of occurrence were qualified and quantified. This aspect of the project is particularly important due to the absence of standard-

TABLE 8   Final Risk Factors

Task-Related characteristics	Personal characteristics	Organizational characteristics
Force	Previous CTD	Equipment
Repetition	Habits and hobbies	Production rate
Awkward joint posture	Diabetes	CTD statistics
Hand tools	Thyroid problems	Peer influence
Length of work shift	Age	Training
Low-frequency vibration tools	Arthritis and/or degenerative joint Disease (DJD)	Ergonomics program and awareness

ized risk levels for the factors in current CTD risk evaluations. The goal of fuzzy linguistic variables is to represent the condition of an attribute over a given interval. In FST several intervals are defined and assigned linguistic variables over the continuum of a given variable. In the continuum of risk exposure as it relates to injury development, this linguistic variable can be considered to represent a range where the suspected risk factors produce the prescribed risk of injury for a given variable. The values obtained in the development of fuzzy linguistic variables are considered fuzzy measures. These values can then become the criteria for measuring attributes of objects [42] or, in this case, risks.

The methodology to derive the linguistic risk levels consisted of a re-evaluation of the information obtained in the initial literature search and a second stage of knowledge acquisition. The objective of the second stage of knowledge acquisition was to identify "ranges" where the experts could assign a particular linguistic value to an interval. To accomplish this goal, the experts were asked, first, to consider what the levels or approximate break-points were for the classification of the risk factors. They were then asked to partition the entire interval of factor existence into as many levels as they felt were necessary. Second, they were asked to classify the extent of risk associated with each level of the risk factors. This analysis indicated that four to five levels of linguistic variables would be useful for each of the risk factors. Again this was based on the text analysis and the experts' perspectives. The responses yielded the following categories of risk levels for each factor; each category was assigned an associated linguistic risk level (LRL):

1. LRL1, little or no risk: very little risk for the particular level of the given risk factor
2. LRL2, mild risk: some risk for the particular level of the given risk factor,
3. LRL3, moderate risk: average risk for the particular level of the given risk factor,
4. LRL4, strong risk: greater than average or considerable risk for the particular level of the given risk factor

and in some cases,

5. LRL5, very strong risk: definite risk for the particular level of the given risk factor

Using these variables, the levels for qualifying risks associated with each factor were established. The experts agreed that these variables could be used to represent the categories. However, the suggested boundaries for many of the linguistic variables differed among experts. These differences were addressed in the development of the membership functions.

Development of Membership Functions. As with most fuzzy modeling, the development of the membership functions was the most challenging portion of this project. Due to the inherent subjectivity that accompanies the construction of membership functions, an effort was made to follow a standard methodology with each expert and for each risk factor. Two membership functions were defined for each variable. The first membership function was designed to graphically represent the progression of a factor through the LRL. This membership function was constructed utilizing the linguistic categories identified in the knowledge-acquisition session and consisted of a set of overlapping π-shaped curves.

At this point, the SMEs were asked to keep in mind the linguistic variable that they had assigned to the particular range of interest in the development of the membership functions. They were asked to help rate the progression of each linguistic variable throughout a universe of discourse. Each of the linguistic variables (i.e., mild risk or strong risk) within the factor possesses an individual and overlapping π-shaped membership curve that travels throughout the entire interval $[0, 1]$ (Fig. 10). Overlapping functions were used because the experts and the literature concurred that, in the analysis of the risks associated with a factor, the risk levels may have "gray" or ill-defined boundaries. The results of the knowledge acquisition served to re-emphasize this point.

Linguistic Variables Membership Function. The first set of membership functions represents each risk factor's various LRLs. The results of the second stage of knowledge acquisition revealed ambiguity across and within the expert's analysis of the cutoff points for identifying LRL. The experts were asked to consider each factor and the five linguistic variables determined for each factor. For each LRL the expert was asked to identify three things: (1) the absolute minimum value for inclusion in the set; (2) the value that was considered to have complete membership in the set $[\mu(x) = 1]$; and (3) the value that was considered to be the absolute maximum point of membership in the set. Table 9 illustrates the results when considering the factor *force* (amount of exertion required by the hand), at the LRL of moderate risk. The five LRLs for this factor can be abbreviated as minimal risk, some risk, moderate risk, somewhat high risk, and high risk.

The determination of the upper endpoints for the individual membership curve was accomplished by taking the union of the three values. In FST the

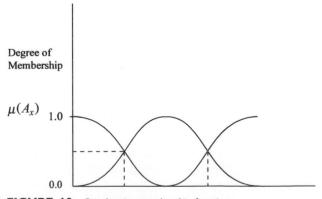

FIGURE 10   Overlapping membership function.

TABLE 9 Development of Membership Functions: Force

Source	Minimum (kg)	Average (kg)	Maximum (kg)
Expert 1	2.0	2.5	3.0
Expert 2	2.2	2.6	3.3
Expert 3	2.5	2.8	3.25

[a] Risk factor, force; value of LRL; moderate risk.

union operator ($\cup$) can be used to determine the maximum degree of membership associated with a number of fuzzy values. Conversely, the intersection operator ($\cap$) can be used to determine the minimal degree of membership associated with a number elements. This was used to determine lower endpoints. This basically meant that the most extreme value from the experts was used to represent the endpoint in the membership of the set. For example, when considering the minimum for the linguistic variable moderate force, the union of the values 2.0, 2.2, and 2.5 was taken. The objective of this union operation is to take the value that encompasses the other two values. Thus the value 2.0 is selected. Likewise, the maximum was obtained by considering the values, 3.0, 3.2, and 3.25. The most extreme value in this case is 3.25. Thus, the minimum and maximum values for this membership curve are 2.0 3.25 kg, respectively.

The determination of the value(s) with complete membership in the set was accomplished by taking a weighted average of the values. This was considered a variation of the approach used by Cox [19] to establish symmetric π-shaped membership functions. The purpose for weighting the averages is to give the expert into whose primary area of expertise the factor falls the largest amount of input in the development of the membership function. For example, the orthopedic surgeon holds primary expertise in the personal factor category due to his medical education, experience, and expertise. Likewise, the ergonomist is the primary expert in the task-related category. The weighted mean was determined by assigning a weight of 0.4 to the primary expert and 0.3 to each of the two secondary experts. Continuing with the factor force, given that expert #1 is the primary expert for task characteristics, the weighted mean would produce the following as the point of complete membership:

$$y = 0.4(2.5) + 0.3(2.6) + 0.3(2.8) = 2.62. \tag{8}$$

After the mean was determined the final membership function was developed. The minimum, mean, and maximum were used to construct the skeleton of the membership function. The goal was to allow a factor to have dual membership the neighboring membership function after the crossover point or midpoint in the given portion of the π-shaped membership function. In other words, after approximately half of the distance between the midpoint and the lower extreme point had been reached, the membership function should cross the preceding membership curve. Symmetry was used to define the right crossover point x_{c2} to be on the declining side of the π-shaped curve at a location symmetrical to x_{c1}. This methodology was the guideline for establishing the membership functions for each of the quantitative variables.

Hazard Level Membership Function. The second membership function is a graphical representation of the degree of risk, or hazard (i.e., the risk associated with a given level of performance for the risk factor of interest). The hazard function is an S-shaped growth curve, which moves from no membership at its extreme left to complete membership at its extreme right. In this case, degree of hazard is defined as the amount of risk produced by a particular linguistic level of exposure for the given factor. The purpose of the hazard function was to quantify the degree of hazard associated with each LRL within a given factor. Again, this required further knowledge acquisition. The experts were asked the amount of risk exposure that a particular linguistic level of the factor posed for the development of injury. For factors such as risk repetition or force, the greater the condition, the greater the hazard associated with the factor. However, for other factors such as previous CTD, the values of the hazard function may be the same for various linguistic risk levels (LRL). For example the LRLs for the factor "Previous CTD" would be:

LRL1, no history of CTD of the forearm and/or hand

LRL2, minor musculoskeletal irritation of the forearm and/or hand

LRL3, irritation of the forearm and/or hand that required medical treatment,

LRL4, medically diagnosed case of upper extremity CTD

LRL5, severe history of CTD related problems requiring corrective surgery.

In this case, although there is a difference in LRL3 and LRL4, they would both yield a similar value on the hazard-level curve because this factor was determined to be so significant in the development of CTDs. In addition to providing a degree of hazard associated with the given LRL (see Fig. 11) the hazard function also standardized the various linguistic levels. These levels are standardized because the result places all of the potential linguistic

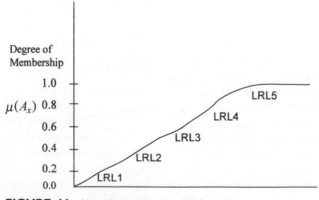

FIGURE 11   Hazard-curve membership function.

outputs on one membership curve that travels throughout the entire universe of discourse. This normalization results in an S-shaped membership function for representing the hazard throughout a continuum on the interval [0, 1] for the factor of interest. This membership function can thus be used to quantify the expected possibility of injury associated with each level of a risk factor. The linguistic levels within each parameter were then determined to have a certain rating on the hazard membership curve. However, the membership function for the factor *age* simulated a bathtub curve because the linguistic categories *young* and *old* had the highest hazard rating within the factor age.

At this point the fuzzy measures can be further classified as possibility measures. A possibility measure is considered a fuzzy measure if X is finite and if the possibility distribution is normal [43]. Consider the task-related risk factor *repetition* and the risk level associated with various levels of repetition in a hand-intensive task. If X represents the amount of repetition required for task performance and Π represents the possibility that the level of repetition will result in injury, the distribution in Table 10 may be used to represent the values associated with this hazard function. This possibilistic distribution can thus be used to represent the ranges that contain the LRL of the factor. The use of possibilistic functions was deemed appropriate because the type of uncertainty addressed in this situation is concerned with the possibility or likelihood of injury. Thus, this methodology was used to derive the standardized hazard membership functions for each risk factor.

At this point in model development each factor within the modules has been assigned a linguistic risk level, a membership function (universe of discourse), and a hazard function that indicates the degree to which the given level of the factor contributes to the possibility or risk of injury. The factors have experienced complete fuzzification by being transformed from linguistic terms or crisp values to fuzzy functions. The next step is to assign weights to the variables

Analytic Hierarchy Process (AHP). After the concept mapping knowledge acquisition sessions, each of the experts was asked to do pairwise comparisons involving all pairings of the risk factors within each of the three modules or categories. Each expert was asked to make comparisons in their area of expertise. However, to evaluate the feasibility of these ratings the experts were also asked to review the relative weights obtained through AHP for the other areas. Pairwise comparisons were used to determine the relative importance of the risk factors. The importance was quantified by using a scale that indicates the strength with which one risk factor dominated another. Comparisons were made within modules to determine the relationship be-

TABLE 10   Possibility Distribution for Repetition[a]

X	0	15	30	45	60	75	90	105	120	> 120
Π	0	0	0.1	0.2	0.5	0.6	0.7	0.8	1.0	1.0

[a]X = units produced per hour; Π = degree of membership in the possibility function representing level of risk.

tween the factors indicated by each expert as significant. The rating scale developed by Saaty [44] was utilized for all comparisons. In this scale, the degree of intensity or preference of the decision maker in the choice for each pairwise comparison is rated on a scale of 1 to 9. In AHP analysis the rating is used to define the degree of preference of one variable over another during the pairwise comparison. The lower end of the scale, 1, suggests equal importance of two variables x and y and is stated "x is equally important as y." The upper end of the scale is a rating of 9. The rating of 9 suggests that x is strongly more important than y. This scale permits the use of even numbers, and even numbers could also be used to represent compromise interpolation between the rating categories. The inverse of these values may also be used if the expert determines that an inverse relationship exists between the factors being evaluated. The quantities were then placed in a matrix of comparisons. Once a pairwise comparison matrix was constructed, the relative weights of the factors included in the matrix are obtained from the estimate of the maximum eigenvector of the matrix. The normalized average rating associated with each attribute provides an estimate for the maximum eigenvalue for the matrix of pairwise comparisons. This normalized average rating, sometimes referred to as priority weight, is used to depict the relative significance of each factor. In this case, the priority weights were used to represent relative significance in the mathematical model.

The six to seven items identified as risk factors for each of the three modules were evaluated for relative significance. The priority weights for the risk factors in the task-related and personal modules are listed in Tables 11 and Table 12, respectively.

In the evaluation the organizational risk factors, equipment was the most significant factor. In knowledge acquisition, the term "equipment" refers to the degree of automation for the machinery being used in the task under evaluation. The order of importance and priority weights for each of the risk factors are listed in Table 13. This module evaluated the impact of seven risk factors. However, upon further analysis and discussion, the awareness and ergonomics program categories were combined because, according to the experts and the literature, one of the goals of an ergonomics program is to provide awareness about the ergonomic risk factors present in a workplace.

After the factors within the categories (or modules) were compared, an AHP analysis was conducted to determine the relative significance of each of

TABLE 11   AHP Results: Task-Related Risk Factors

Ranking	Factor	Relative weight
1	Awkward joint posture	0.299
2	Repetition	0.189
3	Hand-tool use	0.180
4	Force	0.125
5	Task duration	0.124
6	Vibration	0.083

TABLE 12   AHP Results: Personal Risk Factors

Ranking	Factor	Relative weight
1	Previous CTD	0.383
2	Hobbies and habits	0.223
3	Diabetes	0.170
4	Thyroid problems	0.097
5	Age	0.039
6	Arthritis	0.088

the modules: task, personal, and organizational characteristics. The priority weights obtained for the task, personal, and organizational characteristics are listed in Table 14. The task characteristics module received a relative weight of 0.637. The personal characteristics module had a relative weight of 0.258, less than half of the relative weight of the task characteristics module. Finally, the organizational characteristics module received the smallest relative weight, 0.105.

Determination of Aggregate Risk Level. After the linguistic risk and the relative significance are generated an aggregated numeric value is obtainable. Equation (9) represents the model for the calculation of the numeric risk value for the task module. In Eq. (9), the w_i values represent the numeric values obtained from the user inputs for each of the six risk factors and the a_j values represent the relative significance or factor weight obtained from the AHP analysis. The numeric risk levels for the personal and organizational characteristics are represented by Eqs. (10) and (11), respectively. Likewise, the values of x_i and y_i represent the user inputs, whereas, the b_i and c_j values represent the AHP weights for the task and organizational characteristics, respectively. These linear equations are based on fuzzy quantification theory I [42]. The objective of theory I is to find the relationships between the qualitative descriptive variables and the numerical object variables in the fuzzy groups. An alternative to this approach is to use CTD epidemiological data to establish the regression weights rather than the relative weights that were derived from the AHP analysis with the experts. However, the lack of availability of comprehensive data for a regression model prevented the

TABLE 13   AHP Results: Organizational Risk Factors

Ranking	Factor	Relative weight
1	Equipment	0.346
2	Production rate and layout	0.249
3	Ergonomics program	0.183
4	Peer influence	0.065
5	Training	0.059
6	CTD level	0.053
7	Awareness	0.045

■ **TABLE 14   AHP Results: Module Risk Comparison**

Ranking	Module	Relative weight
1	Task	0.637
2	Personal	0.258
3	Organizational	0.105

application of regression analysis. The resulting equations represent the numeric risk levels for each category:

- Task-related risk,

$$R_1 = F(T) = a_1 w_1 + a_2 w_2 + a_3 w_3 + a_4 w_4 + a_5 w_5 + a_6 w_6 \quad (9)$$

- Personal risk,

$$R_2 = F(P) = b_1 x_1 + b_2 x_2 + b_3 x_3 + b_4 x_4 + b_5 x_5 + b_6 x_6 \quad (10)$$

- Organizational risk,

$$R_3 = F(O) = c_1 y_1 + c_2 y_2 + c_3 y_3 + c_4 y_4 + c_5 y_5 + c_6 y_6 \quad (11)$$

Interpretation of Results.   The numeric risk values obtained from each of the modules and the weights obtained from the AHP analysis were used to obtain the final crisp overall risk output. This output is a value that indicates the risk of subject injury for a specific task in a given workplace. The following equation was used to quantify the comprehensive risk of injury as a result of all three categories:

$$Z = d_1 R_1 + d_2 R_2 + d_3 R_3, \quad (12)$$

where,

Z = overall risk for the given situation
R_1 = the risk associated with the task characteristics
d_1 = weighting factor for the task characteristics
R_2 = the risk associated with the personal characteristics
d_2 = weighting factor for the personal characteristics
R_3 = the risk associated with the organizational characteristics
d_3 = weighting factor for the organizational characteristics

The weighting factors (d_1, d_2, d_3) represent the relative significance of the given risk-factor category's contribution to the likelihood of injury. These factors were determined through the AHP analysis. The numeric risk levels obtained from the previous equations exist on the interval $[0, 1]$. On this interval 0 means no risk of injury and 1 means extreme risk of injury. The partitioning of the categories is shown in Table 15.

TABLE 15 Categorization of aggregate numeric risk levels

Numeric risk level	Expected amount of risk associated with numeric value
0.00–0.20	Minimal risk: individual should not be experiencing any conditions that indicate musculoskeletal irritation
0.21–0.40	Less than average risk: very little risk; individual may experience irregular irritation but is not expected to experience regular musculoskeletal irritation
0.41–0.60	Average risk: individual may experience minor musculoskeletal irritation regularly
0.61–0.80	High risk: individual is expected to be experiencing regular minor or major musculoskeletal irritation
0.81–1.00	Very high risk: individual is expected to experience regular musculoskeletal irritation and/or medical correction for the condition

Model Validation. The model was validated by performing a series of evaluations with alternative sets of data. For ease of application the results were programmed into a rule-based system (for details see [21]).

Summary of Example. The results of this research yielded useful outcomes. First, a modular style for addressing the injuries associated with CTDs was produced that allows for categorical evaluation of risks. Second, the relative importance of the risk factors within the designated modules were determined; linguistic risk levels were determined for each of the identified risk factors; and, finally, a hazard membership function to was developed to represent the aggregate impact of the risk factors.

The research was successful in producing a quantifiable tool to predict therisk of injury. If utilized this approach may be useful in reducing the risk of CTDs of the upper extremity

VI. CONCLUSION

The techniques presented in this chapter represent an evolution in the modeling of occupational injuries. The ability to model these injuries in a manner that is commensurate with their occurrence is attainable with the tools of fuzzy set theory. The goal of the researchers is for others to be able to utilize this systematic approach in an effort to model and reduce the risk of occupational injuries. Future application and evaluation of these guidelines will be useful in determining the long term usefulness of this methodology.

REFERENCES

1. McCauley-Bell, P. R., Crumpton, L. L., and Wang, H. Measurement of cumulative trauma disorder risk in clerical tasks using fuzzy linear regression. *IEEE Trans. Systems Man and Cybernet.* 29C(1):1–14, 1999.
2. *CTDNews.* http://ctdnews.com/numbers.html, December 1997.
3. Bureau of Labor Statistics. Safety and health statistics—workplace injury and illness summary. http://www.bls.gov/news.release/soh.hws.htm, December 1997.
4. *Ergonomics and CTD's*, FOCUS, March 1995, http:// www.engest.unl.edu/ EngExt/ letter/ ergo.html, December 1997.
5. *CTDNews online.* CTD's in industry. http://ctdnews.com/business.html, December 1997.
6. *Ergonomic Solutions.* Cumulative trauma disorders. http://www.womensbusiness.com/CTD.htm, December 1997.
7. Armstrong, C. T., Fine, L. J., Goldstein, S. A., Lifshitz, Y. R., and Silverstein, B. A. Ergonomics considerations in hand and wrist tendinitis, Part 2. *J. Hand Surgery* 12-A(5):830–837, 1987.
8. Sauter, S., Schleifer, L., and Knutson, S. Work posture, work station design, and musculoskeletal discomfort in a VDT data entry task. *Human Factors* 33(2):151–167, 1991.
9. McCauley-Bell, P. and Crumpton, L. A fuzzy linguistic model for the prediction of carpal tunnel syndrome risks in an occupational environment. *Ergonomics* 40(8):790–799, 1997.
10. Crumpton, L. L. and Congleton, J. J. Use of risk factors commonly associated with carpal tunnel syndrome to model median nerve conduction. *Advances in Industrial Ergonomics and Safety VI* 511–514, 1994.
11. Karwowski, W., Gaddie, P., and Marras, W., S. A dynamic systems approach for analysis of the relationships between risk factors for low back disorders using the 3D graphical visualization models. *Advances in Industrial Ergonomics and Safety VI* 653–656, 1994.
12. Schoenmarklin, R. W. and Marras, W. S. An EMG assisted biomechanical model of the wrist joint. *Advances in Industrial Ergonomics and Safety IV* 777–781, 1992.
13. Grant, K. A. and Galinsky, T. L. Use of the actigraph for objective quantification of hand/wrist activity in repetitive work. In *Proceedings of the Human Factors and Ergonomics Society 37th Annual Meeting*, 1993, pp. 720–724.
14. Ayoub, M. M. and Hsiang, M. S. Biomechanical simulation of a lifting task. *Advances in Industrial Ergonomics and Safety IV* 831–838, 1992.
15. King, H.-H. An expert adaptive fuzzy logic control system, Louisiana Tech University, 1998, pp. 1–57. Unpublished dissertation.
16. Gupta, M., Knopf, G., and Nikiforuk, P. Sinusoidal-based cognitive mapping functions. *Fuzzy Logic in Knowledge-Based Systems*, pp. 69–92. Elsevier, Amsterdam, 1988.
17. Soh, T. N., Crumpton, L. L., and McCauley-Bell, P. The use of fuzzy logic to develop a mathematical model to quantify fatigue. *Advances in Occupational Ergonomics and Safety I* 123–128, 1996.
18. Badiru, A. B. In *Expert Systems Applications in Engineering and Manufacturing*, pp. 58–76, Prentice Hall, Englewood Cliffs, NJ, 1992.
19. Cox, E. *The Fuzzy Systems Handbook: A Practitioner's Guide to Building, Using, and Maintaining Fuzzy Systems*. Academic Press, San Diego, 1994.
20. McCauley-Bell, P. and Badiru, A. Fuzzy modeling and analytic hierarchy processing to quantify risk levels associated with occupational injuries. Part I. The development of fuzzy linguistic risk levels. *IEEE Trans. Fuzzy Systems* 4(2):124–131, 1996.
21. McCauley-Bell, P. and Badiru, A. Fuzzy modeling and analytic hierarchy processing as a means to quantify risk levels associated with occupational injuries. Part II. The development of a fuzzy rule-based model for the prediction of injury. *IEEE Trans. Fuzzy Systems* 4(2):132–138, 1996
22. Karwowski, W. and Mital, A. Fuzzy concepts in human factors/ergonomics research. *Applications of Fuzzy Set Theory in Human Factors*. Elsevier Science Publishers, B. V., Amsterdam, 1986.
23. McCauley-Bell, P. and Badiru, A. Fuzzy modeling and analytic hierarchy processing to quantify risk levels associated with occupational injuries part I: The development of fuzzy linguistic risk levels. *IEEE Trans. Fuzzy Systems* 4(2):124–131, 1996.

24. McCauley-Bell, P. and Badiru, A. Fuzzy modeling and analytic hierarchy processing as a means to quantify risk levels associated with occupational injuries part II: The development of a fuzzy rule-based model for the prediction of injury. *IEEE Trans. Fuzzy Systems* 4(2):132–138, 1996.

25. Crumpton, L., McCauley-Bell, P., and Soh, T. Modeling overall level of fatigue using linguistic modeling and fuzzy set theory. Submitted.

26. Mendenhall, W. and Sincich, T. *A Second Course in Business Statistics: Regression Analysis*, Macmillan Publishing Company, New York, 1993.

27. Silverstein, B. The prevalence of upper extremity cumulative trauma disorders in industry. Unpublished Ph.D. Thesis, The University of Michigan, Ann Arbor, 1985.

28. Sakawa, M. and Yano, H. Fuzzy linear regression and its applications. *Fuzzy Regression Analysis*. Omnitech Press, Warsaw and Physica-Verlag, Heidelberg, 1992, pp. 61–79.

29. Drury, C. G. Computerized data collection in ergonomics. In *Evaluation of Human Work: A Practical Ergonomics Methodology*. Taylor & Francis, London, 1995, ch. 9, 2nd ed., pp. 229–243.

30. Cleal, D. M. and Heaton, N. O. *Knowledge-Based Systems: Implications for Human-Computer Interface*. John Wiley & Sons, New York, 1988, pp. 1–129.

31. Turban, E. Knowledge acquisition and validation. In *Expert Systems and Applied Artificial Intelligence*. McMillan Publishing Company, New York, 1992.

32. McNeese, M. and Zaff, B. Knowledge as design: A methodology for overcoming knowledge acquisition bottlenecks in intelligent interface design. In *Proceedings of the 35th Annual Meeting of the Human Factors Society*, pp. 1181–1185. Human Factors Society, Santa Monica, CA, 1991.

33. Snyder, D., McNeese, M., Zaff, B., and Gomes, M. Knowledge acquisition of tactical air-to-ground mission information using concept mapping. Unpublished research report, Wright-Patterson Air Force Base, Air Force Institute of Technology, OH, 1992.

34. Armstrong, T. and Lifshits, Y. Evaluation and Design of Jobs for Control of Cumulative Trauma Disorders. The University of Michigan, Center for Ergonomics, 1988. Unpublished Research.

35. Burnette, J. CTD-123: A cumulative trauma disorder risk assessment model. Unpublished Ph.D. Thesis, Department of Industrial Engineering, North Carolina State University, 1989.

36. Dortch, H. and Trombly, C. The effects of education on hand use with industrial workers in repetitive jobs. *Amer. J. Occupational Therapy* 3:777–782, 1990.

37. Eastman Kodak. In *Ergonomic Design for People at Work*, Vol. 2, pp. 112–170. Van Nostrand–Reinhold, New York, 1986.

38. McKenzie, F., Storment, J., Van Hook, P., and Armstrong, T. T. A program for control of repetitive trauma disorders. *Amer. Industrial Hygiene Assoc. J.* 46(11):674–677, 1985.

39. Silverstein, B. The prevalence of upper extremity cumulative trauma disorders in industry, 1985, pp. 1–170. Unpublished Doctoral Dissertation. The University of Michigan, Ann Arbor.

40. Crumpton, L. An evaluation of risk factors commonly associated with carpal tunnel syndrome development. Published Doctoral Dissertation. Texas A & M University, College Station, 1993.

41. Snyder, D., McNeese, M., Zaff, B., and Gomes, M. Knowledge acquisition of tactical air-to-ground mission information using concept mapping. Wright-Patterson Air Force Base, Air Force Institute of Technology, Ohio, 1992. Unpublished Research Report.

42. Terano, T., Asai, K., and Sugeno, M. *Fuzzy Systems Theory and Its Applications*. Academic Press, San Diego, 1992, pp. 69–84.

43. Zimmerman, H. J. *Fuzzy Set Theory and Its Applications*. Kluwer Academic, Norwell, MA, 1991.

44. Saaty, T. *The Analytic Hierarchy Process*. McGraw-Hill, New York, 1980.

45. Adams, E. Using macro-ergonomics to design out cumulative trauma risks. *Occupational Health and Safety* 40, 1993.

46. Aptel, M. and Cail, F. An empirical index for evaluating the biomechanical wrist stress. *Advances in Industrial Ergonomics and Safety I* 2:421–426, 1996.

47. Aaras, A. and Ro, O. Back pain with sitting posture in the workplace. In *Proceedings of the 13th Triennial Congress of the International Ergonomics Association*, 1997, Vol. 4, pp. 13–15.

48. Armstrong, T., Buckle, P., Fine, L., Hagberg, M., Jonsson, B., Kilbom, A., Kuorinka, I., Silverstien, B., Sjogaard, G., and Viikari-Juntura, E. A conceptual model for work-related neck and upper-limb musculoskeletal disorders. *Scand. J. Work Environment and Health* 19(2):73–84, 1993.

49. Babski, K. and Crumpton, L. A review of the advancements in carpal tunnel syndrome research. *Advances in Industrial Ergonomics and Safety II* 399–402, 1997.

50. Bagchee, A., Bhattacharya, A., Succop, P. A., and Lai, C.-F. Development of a risk factor analysis model for predicting postural instability at workplace. *Advances in Occupational Biomechanics and Safety II*, 87–90, 1997.

51. Baker, E. and Ehrenberg, R. Preventing the work-related carpal tunnel syndrome: Physician reporting and diagnostic criteria, *Ann. Internal Medicine* 112(5):317–319, 1990.

52. Bonnisonne, P. A fuzzy sets based linguistic approach: Theory and Applications. In *Proceedings of the* 1980 *Winter Simulation Conference*, pp. 99–111.

53. Boughanim, D. and Gilad, I. A three dimensional lifting analysis. *Advances in Industrial Ergonomics and Safety IV*, 853–859, 1992.

54. Bruchal, L. C. Occupational knee disorders: An overview. *Advances in Industrial Ergonomics and Safety VII* 89–93, 1995.

55. Buckle, P. and Li, G. A practical approach to musculoskeletal risk assessment in the real workplace. In *Proceedings of the 13th Triennial Congress of the International Ergonomics Association*, 1997, pp. 138–140.

56. Buckle, P. Musculoskeletal injuries and their prevention—assessment of interventions. In *Proceedings of the 13th Triennial Congress of the International Ergonomics Association*, 1997, pp. 141–143.

57. Chaffin, D. and Andersson, G. *Occupational Biomechanics*. Wiley-Interscience, New York, 1975.

58. Chaffin, D. and Andersson, G. *Occupational Biomechanics*, 2nd ed., pp. 215–216, Wiley, New York, 1991.

59. Chang, T.-W., Liu, L., Chen, J.-C., and Jackson, A. S. Biomechanical analysis of torso lift and torso pull. *Advances in Industrial Ergonomics and Safety VII* 771–776, 1995.

60. Chung, M. K. *et al.* Biomechanical and postural analysis of machine repair tasks with relatively high complaints of low back pain. In *Proceedings of the 13th Triennial Congress of the International Ergonomics Association*, 1997, pp. 25–27.

61. Colombini, D. and Occhipinti, E. Proposal by the IEA TC for description and assessment of repetitive movements risks. In *Proceedings of the 13th Triennial Congress of the International Ergonomics Association*, 1997, pp. 34–36.

62. Crumpton, L. L. and Congleton, J. J. An evaluation of the relationship between subjective symptoms and objective testing used to assess carpal tunnel syndrome. *Advances in Industrial Ergonomics and Safety VI* 515–519, 1994.

63. Dellman, N. J. ISO/CEN standards on risk assessment for upper limb repetitive movements. In *Proceedings of the 13th Triennial Congress of the International Ergonomics Association*, 1997, pp. 37–39.

64. Dempsey, P. G. The study of work-related low-back disorders in industry: Opportunities and pitfalls. *Advances in Industrial Ergonomics and Safety II* 267–270, 1997.

65. Dempsey, P. G. and Westfall, P. H. Developing explicit risk models for predicting low-back disability: A statistical perspective. Unpublished.

66. Fathallah, F. A., Marras, W. S., and Parnianpour, M. Three dimensional spinal loading during complex lifting tasks. In *Proceedings of the Human Factors and Ergonomics Society 40th Annual Meeting*, 1996, pp. 661–665.

67. Fathallah, F. A. and Burdof, A. Exposure assessment methods of occupational low back disorders risk factors. In *Proceedings of the 13th Triennial Congress of the International Ergonomics Association*, 1997, pp.153–155.

68. Fernandez, J. E., Dahalan, J. B., Halpern, C. A., and Fredericks, T. K. The effect of deviated wrist posture on pinch strength for females. *Advances in Industrial Ergonomics and Safety IV* 693–700, 1992.

69. Ferguson, S. A., Marras, W. S., and Waters, T. R. Quantification of back motion during asymmetric lifting. *Ergonomics* 35(7/8):845–859, 1992

70. Ferrell, P. and Johnson, S. Exploration of causal relationships between poultry processing job characteristics and the development of carpal tunnel syndrome. *Advances in Industrial Ergonomics and Safety VI* 497–503, 1994.

71. Fleischer, L. B., Congleton, J. J., Foster, J. W., and Ghahramani, B. Development of an image analysis system for evaluation of a manual lift. *Advances in Industrial Ergonomics and Safety IV* 327–332, 1992.

72. Garg, A. and Moore, S. The strain index: A method to analyze jobs for risk of distal upper extremity disorders. In *Proceedings of the 13th Triennial Congress of the International Ergonomics Association*, 1997, pp. 40–42.

73. Granata, K. P., Marras, W. S. and Ferguson, S. A. Relation between biomechanical spinal load factors and risk of occupational low back disorders. In *Proceedings of the Human Factors and Ergonomics Society 40th Annual Meeting*, 1996, pp. 656–660.

74. Granata, K. P. and Marras, W. S. An EMG-assisted model of loads on the lumbar spine during asymmetric trunk extensions. *J. Biomechanics* 26(12):1429–1438, 1993.

75. Hagg, G. M. Forearm flexor and extensor muscles exertion during gripping—a short review. In *Proceedings of the 13th Triennial Congress of the International Ergonomics Association*, 1997, pp. 49–51.

76. Hallbeck, M. S., Sheeley, G. A., and Bishu, R. R. Wrist fatigue in pronation and supination for dynamic flexion and extension: A pilot study. *Advances in Industrial Ergonomics and Safety IV* 713–716, 1992.

77. Halpern, M., Skovron, M. L., and Nordin, M. Employee-rated job demands: Implications for prevention of occupational back injuries. *Advances in Industrial Ergonomics and Safety I* 1:275–280, 1996.

78. Hamrick, C. A., Gallagher, S., and Redfern, M. S. A biomechanical analysis of a bolter cable pulling task. *Advances in Industrial Ergonomics and Safety VI* 645–651, 1994.

79. Hignett, S. and Mcatamney, L. Rapid entire body assessment (REBA) in the health care industry. In *Proceedings of the 13th Triennial Congress of the International Ergonomics Association*, 1997, pp. 162–164.

80. Http://www.ufcw.org/idandtreat.html. Cumulative trauma disorders: Identification and treatment. 1997.

81. Http://mime1.marc.gatech.edu/mime/ergo/ansi/net/a2.html. ANSI Z–365. Working draft, 1997.

82. Http://www.hermanmiller.com/research/papers/support/chairs.html. Prevention of back discomfort. 1997.

83. Jegerlehner, J. Workers participation helps reduce CTD injuries. *Advances in Industrial Ergonomics and Safety VII* 339, 1997.

84. Kapitaniak, B. and Peninou, G. Software BIOVECT—Help for postural analysis. *Advances in Industrial Ergonomics and Safety I* 2:578–583, 1996.

85. Karlqvist, L. *et al.* Position of the computer mouse—a determinant of posture, muscular load and perceived exertion? In *Proceedings of the 13th Triennial Congress of the International Ergonomics Association*, 1997, pp. 61–63.

86. Karwowski, W., Caldwell, M., and Gaddie, P. Relationships between the NIOSH (1991) lifting index, compressive and shear forces of the lumbosacral joint, and low back injury incidence rate based on industrial field study. In *Proceedings of the Human Factors and Ergonomics Society, 38th Annual Meeting*, 1994, pp. 654–657.

87. Killough, M. K. and Crumpton, L. L. An investigation of cumulative trauma disorders in the construction industry. *Advances in Industrial Ergonomics and Safety VII* 81–87, 1995.

88. Laurie, N. E. Predicting tendon damage due to repetitive loading—A proposed model. *Advances in Industrial Ergonomics and Safety II* 387–394, 1997.

89. Leskinen, T. and Haijanen, J. Torque on the low back and the weight limits recommended by NIOSH in simulated lifts. In *Proceedings of the Fourth International Symposium on 3-D Analysis of Human Movement*, 1996.

90. Lin, C. J., Bernard, T. M., Macedo, A., and Macedo, J. A. Biomechanics of manual material handling through simulation. *Advances in Industrial Ergonomics and Safety VI* 635–640, 1994.

91. Mairiux, Ph., Dohogne, T., Laigle, F., Schleich, E., and Van Damme, J. Identification of occupational risk factors for low back disorders: The B.E.S. guide. *Advances in Industrial Ergonomics and Safety I* 2:388–393, 1996.

92. Marley, R. J. and Kumar, N. An improved musculaoskeletal discomfort assessment tool. *Advances in Industrial Ergonomics and Safety VI* 45–52, 1994.

93. Martin, C. *et al.* A project management approach to CTD prevention: case study in a furniture plant. In *Proceedings of the 13th Triennial Congress of the International Ergonomics Association*, 1997, pp. 309–311.

94. McCauley-Bell, P. A fuzzy linguistic artificial intelligence model for assessing risks of cumulative trauma disorders of the forearm and hand. Ph.D. Thesis, Industrial Engineering Department, University of Oklahoma, Norman, 1993.

95. McCauley-Bell, P. and Wang, H. Fuzzy linear regression models for assessing risks of cumulative trauma disorders. *Fuzzy Sets Systems* 92(3): 1997.

96. Meyer, J.-P. *et al.* Movement repetitivity assessment. In *Proceedings of the 13th Triennial Congress of the International Ergonomics Association*, 1997, pp. 201–203.

97. McGill, S. M. Using biomechanical models to reduce occupationally related low back injury. In *Proceedings of the 13th Triennial Congress of the International Ergonomics Association*, 1997, pp. 198–200.

98. Moore, A., Wells, R., and Raaney, D. Quantifying exposure in occupational manual tasks with cumulative trauma disorder potential. *Ergonomics* 34(12):1433–1453, 1991.

99. Neumann, P. *et al.* Comparison of four methods of determining peak spinal loading in a study of occupational low back pain. In *Proceedings of the 13th Triennial Congress of the International Ergonomics Association*, 1997, pp. 204–206.

100. Nussbaum, M. A. and Chaffin, D. B. Laboratory environment and biomechanical modeling tools for ergonomic assessment of material handling devices. *Advances in Occupational Ergonomics and Safety II* 233–236, 1997.

101. Occipinti, E. and Colombini, D. Proposal of a concise index for exposure assessment of repetitive movements of upper limbs (OCRA). In *Proceedings of the 13th Triennial Congress of the International Ergonomics Association*, 1997, pp. 90–92.

102. Palmerud, G. *et al.* Risk zone identification in work engaging the upper extremity—effects of arm position and external load on the intramuscular pressure in shoulder muscles. In *Proceedings of the 13th Triennial Congress of the International Ergonomics Association*, 1997, pp. 93–94.

103. Rempel, D. *et al.* Wrist postures while typing on a standard and split keyboard. *Advances in Industrial Ergonomics and Safety VII* 619, 1997.

104. Savic, D. A. and Pedrycz, W. Fuzzy linear regression models: Construction and evaluation. In *Fuzzy Regression Analysis*, pp. 91–100. Omnitech Press, Warsaw, and Physica-Verlag, Heidelberg, 1992.

105. Schoenmarklin, R. W., Marras, W. S., and Leurgans, S. E. Industrial wrist motions and incidence of hand/wrist cumulative trauma disorders. *Ergonomics* 37(9):1449–1459, 1994.

106. Silverstein, Barbara, Fine, L., and Armstrong, T. Occupational factors and carpal tunnel syndrome. *Amer. J. Industrial Medicine* 11:343–358, 1987.

107. Silverstein, B. The role of hazard surveillance in preventing work related musculoskeletal disorders. In *Proceedings of the 13th Triennial Congress of the International Ergonomics Association*, 1997, pp. 326–328.

108. Silverstein, B. The use of checklists for upper limb risk assessment. In *Proceedings of the 13th Triennial Congress of the International Ergonomics Association*, 1997, pp. 109–111.

109. Slappendel. A national injury prevention programme for occupational overuse syndrome: Concept, delivery, evaluation. In *Proceedings of the 13th Triennial Congress of the International Ergonomics Association*, 1997, pp. 329–331.

110. Stevenson J. M., Weber, C. L., Dumas, G. A., Smith, J. T., Albert, W. J., and Lapensee, M. The Queen's–DuPont longitudinal low back pain study: Initial examination of the physical measures. *Advances in Industrial Ergonomics and Safety II* 259–262, 1997.

111. Stewart, G. B. and Woldstad, J. C. A computer based method for recording three dimensional body postures. *Advances in Industrial Ergonomics and Safety IV* 587–592, 1992.

112. Tanaka, S. and McGlothin, J. D. A conceptual model to assess musculoskeletal stress of manual work for establishment of quantitative guidelines to prevent hand and wrist cumulative trauma disorders (CTDs). *Advances in Industrial Ergonomics and Safety I* 419–426, 1989.

113. United States Department of Labor Statistics. http://stats.bls.gov/oshhome.htm, January 1998.

114. Wick, J. L. Force and frequency—how much is too much? *Advances in Industrial Ergonomics and Safety VI* 521–525, 1994.

115. Wicks, J. and Weige, L. Ergonomics awareness training and ergonomics intervention. *Advances in Industrial Ergonomics and Safety VII*, 369, 1997.

116. Wilker, S. F. and Chen, A.-C. Accuracy and efficacy of using pictograms for self-reports of postures assumed when performing lifting tasks. *Advances in Industrial Ergonomics and Safety VI* 641–644, 1994.

117. Wood, D. D. *et al.* Minimizing fatigue during repetitive jobs: Optimal work–rest schedules. *J. Human Factors* 39(1), 1997.

118. Woodhouse, M., McCoy, R., Redondo, D., and Shall, L. Effects of back support on intra-abdominal pressure and lumbar kinetics during heavy lifting. *Human Factors* 582–590, 1995.

119. Yates, J. W. and Karwowski, W. An electromyographic analysis of seated and standing lifting tasks. *Ergonomics* 35(7/8):859–898, 1992.

120. Pazos, G. J., Rios, J., and Yagur. A methodology of linguistics evaluation in risk situations using fuzzy techniques. *Fuzzy Sets and Systems* 48:185–194, 1992.

121. Jetzer, T. Use of vibration testing in the early evaluation of workers with carpal tunnel syndrome. *J. Occupational Medicine*, 33(2):117–120, 1991.

122. Madri, A. The role of human factors in expert systems design and acceptance. *Human Factors* 30(4):395–414, 1988.

123. Meagher, S. Pre-employment anthropometric profiling. In *Advances in Industrial Ergonomics and Safety II*, pp. 375–379. North-Holland, Amsterdam, 1990.

124. Putz-Anderson, V. In *Cumulative Trauma Disorders: A Manual for Musculoskeletal Diseases of the Upper Limbs*. pp. 1–70. Taylor and Francis, London, 1988.

125. Smith, M. and Zehel, D. Cumulative trauma injuries control efforts in select Wisconsin manufacturing plants. In *Advances in Industrial Ergonomics and Safety II*, pp. 259–265. North-Holland, Amsterdam, 1990.

126. Steward, J. Personal Communication, Safety Engineer, Seagate Technology, 1991.

127. Stewart, M. Replacement testing and selection of workers: The industrial perspective, *Ergonomics*, 30(2):253–258, 1987.

128. Tabourn, S. Cumulative trauma disorders: An ergonomics intervention. In *Advances in Industrial Ergonomics and Safety II*, pp. 277–284. North-Holland, Amsterdam, 1990.

FUZZY INFERENCE WITH CONTROL SCHEMES FOR FUZZY CONSTRAINT PROPAGATION*

KIYOHIKO UEHARA

Communication and Information Systems Research Laboratories,
Research and Development Center, Toshiba Corporation,
1, Komukai Toshiba-cho, Saiwai-ku, Kawasaki 210-8582, Japan

KAORU HIROTA

Department of Computation Intelligence and Systems Science,
Tokyo Institute of Technology, 4259 Nagatsuta-cho,
Midori-ku, Yokohama 226-8502, Japan

I. INTRODUCTION   268
II. FUZZY INFERENCE BASED ON α-LEVEL SETS AND
   GENERALIZED MEANS   269
   A. α-Level Sets and Resolution Identity Theorem   269
   B. Generalized Mean   271
   C. Computational Steps for Fuzzy Inference Based on α-Level
      Sets and Generalized Means   272
III. EFFICIENT INFERENCE OPERATIONS   274
   A. Parallel Operations   274
   B. Membership-Function Simplification Based on α-Level
      Sets   274
IV. CONVEX AND RESOLUTIONALLY IDENTICAL FORMS IN
   DEDUCED CONSEQUENCES   276
V. CONTROL OF FUZZY CONSTRAINT PROPAGATION   277
   A. Fuzziness Propagation Control   277
   B. Specificity Propagation Control   279
VI. LEARNING ALGORITHM FOR FUZZY EXEMPLARS   280
VII. FUZZY CONSTRAINT PROPAGATION CONTROL
   REFLECTING FORMS OF GIVEN FACTS   281
   A. A Method for Control of Fuzzy Constraint Propagation
      Reflecting Forms of Given Facts   281
   B. Relations between Facts and Consequences in
      Fuzzy Constraints   281
VIII. SIMULATION STUDIES   282
IX. CONCLUSION   288
   REFERENCES   289

* Adapted from *Information Sciences*, Vol. 100, No. 1–4, K. Uehara and K. Hirota, "Parallel fuzzy inference based on α-level sets and generalized means," pp. 165–206, © 1997, with permission from Elsevier Science.

A parallel fuzzy inference is proposed in which the fuzzy constraint propagation from given facts to inference consequences can be controlled. The inference consequences are computed via α-level sets of fuzzy sets for then-parts by using generalized means. The proposed inference method proves that the consequences are obtained in the form of normal and convex fuzzy sets and therefore the consequences can be treated as fuzzy numbers. Moreover, this inference method can be performed for every α-level set independently, thus providing efficient inference computations by using hardware constructed in parallel.

In this chapter, first, the computational steps of the proposed inference method are explained, and efficient inference operations are then derived by exploiting the α-level-set based process of the proposed inference method. Second, the properties of the inference method are proved and the convexity in deduced consequences and the controllability of fuzzy constraint propagation are clarified. Third, a method for learning in fuzzy constraint propagation is shown. Moreover, a method is proposed for controlling the fuzzy constraint propagation so as to satisfy a set of the required properties of inference. Simulation studies show the feasibility of the proposed inference method in the control of fuzzy constraint propagation. Finally, this chapter concludes with a summary and a brief discussion.

I. INTRODUCTION

Parallel fuzzy inference has been applied in a wide variety of fields, including system control, system modeling, and decision support. In parallel fuzzy inference, conditional propositions are given as initial data sets and consequences are deduced as terminal data sets from the conditional propositions and given facts. In this inference, fuzzy sets function as fuzzy constraints on variables.

The rules for the inference process in fuzzy logic can be viewed as the rules governing fuzzy constraint propagation [1, 2]. In parallel fuzzy inference, fuzzy constraints in given facts are propagated to consequences. The fuzzy constraints can be characterized by their fuzziness [3, 4] and specificity [5]. Therefore, the fuzzy constraint propagation can be considered as the fuzziness and specificity propagation in parallel fuzzy inference.

As in conventional methods of parallel fuzzy inference, the fuzzy constraint propagation leads to larger fuzziness and smaller specificity in deduced consequences compared with those in the then-parts of conditional propositions. This fuzziness and specificity propagation mentioned previously can be seen as agreeable properties in parallel fuzzy inference, although this view depends on the background knowledge of conditional propositions in use. It means that this fuzziness and specificity propagation is required to be controlled according to the background knowledge of conditional propositions. This is because the degree to which the fuzziness and specificity of given facts are propagated to those of deduced consequences depends on the background knowledge of conditional propositions. When learning schemes are introduced into the parallel fuzzy inference, such knowledge is to be extracted from exemplar patterns. Namely, the learning extracts both the knowledge in the form of conditional propositions as initial data sets and the knowledge in the rules for governing the fuzzy constraint propagation in the inference process.

The conventional parallel fuzzy inference, however, does not have schemes for controlling the fuzziness and specificity propagation. In addition, conventional parallel fuzzy inference tends to cause excessive fuzziness increase and specificity decrease in deduced consequences, which also leads to fuzziness explosion and specificity dispersion in *multistage-parallel* fuzzy inference [6]. This arises from unifying inference consequences with disjunctive or conjunctive operators that preserve the positions of fuzzy sets for then-parts.

In this chapter, a parallel fuzzy-inference method is proposed to solve the problems mentioned previously. This method provides rules of inference for controlling the fuzzy constraint propagation on the basis of α-level sets and the generalized means [7]. For convenience in the following discussions, let the inference method proposed here be named α-*GEM* (an abbreviation for "α-level-set- and **g**eneralized-**m**ean-based inference"). The α-GEM method has the following advantages compared with conventional methods.

i. It can control the degree to which the fuzzy constraints of given facts are propagated to those of deduced consequences. It can also provide fuzzy constraint modification in deduced consequences by control schemes for the fuzzy constraint propagation.

ii. It proves that the consequences are obtained in the form of normal and convex fuzzy sets. Thus, the consequences can be treated as fuzzy numbers.

iii. It performs inference operations on every α-level set independently and then provides efficient inference computations by using hardware constructed in parallel.

In the following sections, we describe the α-GEM method and its important properties. Moreover, an algorithm is illustrated for learning in α-GEM from fuzzy exemplar patterns. A method is also proposed for controlling fuzzy constraint propagation so as to satisfy a set of inference properties. Simulation studies show the feasibility of the α-GEM method in the control of fuzzy constraint propagation. Finally, this chapter concludes with a summary and brief discussion of α-GEM.

II. FUZZY INFERENCE BASED ON α-LEVEL SETS AND GENERALIZED MEANS

A. α-Level Sets and Resolution Identity Theorem

A fuzzy set is originally defined by its membership function. The membership functions assign an element in the universe of discourse to the membership grade. A fuzzy set, however, can be defined by the family of its α-level sets equivalently to its membership function according to the *resolution identity theorem* [8], shown next.

THEOREM 1 (Resolution identity theorem; [8]).   *A fuzzy set A can be represented by the family of its α-level sets A_α ($0 < \alpha \leq 1$) as follows*:

$$A = \bigcup_{0 < \alpha \leq 1} \alpha A_\alpha \tag{1}$$

$$= \int_X \sup_{0 < \alpha \leq 1} \alpha \mu_{A_\alpha}(x)/x, \tag{2}$$

where $\mu_{A_\alpha}(x)$ is the membership function of A_α. The α-level set A_α is defined by

$$A_\alpha = \{x \mid \mu_A(x) \geq \alpha, 0 < \alpha \leq 1\},$$

where $\mu_A(x)$ is the membership function of A.

When a fuzzy set A is initially defined by a membership function, the membership function should be given by a single-valued function. If the membership function is not a single-valued function, the resolution identity theorem does not hold. Namely, the initially given membership function is not identical to the membership function reconstructed, by using Eq. (2), from the α-level sets that are obtained from the initially given membership function. On the other hand, when the fuzzy set A is initially defined by its α-level sets A_α, the following relation should be satisfied:

$$A_{\alpha'} \supseteq A_{\alpha''}, \qquad \alpha' < \alpha''. \tag{3}$$

If the α-level sets do not satisfy relation (3), the resolution identity theorem does not hold. Namely, the initially given α-level sets are not identical to the α-level sets obtained from the membership function that is reconstructed from the initially given α-level sets by using Eq. (2). The aforementioned points are important in proofs of Proposition 1 described later. When given membership functions are single-valued functions and given families of α-level sets satisfy relation (3); namely, membership functions and α-level sets are given so that the resolution identity theorem holds, we call them *resolutionally identical* in this chapter.

The next theorem is useful in the following discussions, together with Theorem 1.

THEOREM 2.   *If and only if a fuzzy set A is convex, its α-level set is convex.*

In this chapter, membership functions are defined by upper-semicontinuous functions in bounded and convex universes of discourse. In this case, when a fuzzy set A is a convex fuzzy set in the one-dimensional Euclidean universe of discourse, its α-level set A_α is represented by an interval according to Theorem 2.

B. Generalized Mean

The parallel fuzzy-inference method proposed in this chapter is based on the generalized mean defined next.

DEFINITION 1. The generalized mean of x_i, $M_\omega(x_i, p_i)$, is defined by

$$M_\omega(x_i, p_i) = \left[\frac{\Sigma_i p_i x_i^\omega}{\Sigma_i p_i}\right]^{1/\omega}, \qquad x_i > 0, \ p_i > 0 \ (i = 1, 2, \ldots, n). \quad (4)$$

The generalized mean $M_\omega(x_i, p_i)$ includes the typical mean operations as follows:

$$M_1(x_i, p_i) = \frac{\Sigma_i p_i x_i}{\Sigma_i p_i}, \qquad \text{weighted arithmetic mean;}$$

$$\lim_{\omega \to +0} M_\omega(x_i, p_i) = \prod_i x_i^{p_i / \Sigma p_i}, \qquad \text{weighted geometric mean;}$$

$$M_{-1}(x_i, p_i) = \left[\frac{\Sigma_i p_i x_i^{-1}}{\Sigma_i p_i}\right]^{-1}, \qquad \text{weighted harmonic mean.}$$

Moreover,

$$\lim_{\omega \to +\infty} M_\omega(x_i, p_i) = \max_i x_i, \qquad \text{maximum operation;}$$

$$\lim_{\omega \to -\infty} M_\omega(x_i, p_i) = \min_i x_i, \qquad \text{minimum operation.}$$

In this chapter, we treat $x_i \in (0, 1)$ and define the dual operation of $M_\omega(x_i, p_i)$, $\overline{M}_\omega(x_i, p_i)$, as follows:

DEFINITION 2. For $x_i \in (0, 1)$, the dual operation of $M_\omega(x_i, p_i)$, $\overline{M}_\omega(x_i, p_i)$, is defined by

$$\overline{M}_\omega(x_i, p_i) = 1 - M_\omega(1 - x_i, p_i). \quad (5)$$

The following theorems are important in investigating the properties of the parallel fuzzy-inference method proposed in this chapter.

THEOREM 3. *$M_\omega(x_i, p_i)$ is constant if and only if $x_1 = x_2 = \cdots = x_n$. Otherwise, $M_\omega(x_i, p_i)$ is an increasing function of ω. On the other hand, $\overline{M}_\omega(x_i, p_i)$ is constant if and only if $x_1 = x_2 = \cdots = x_n$. Otherwise, $\overline{M}_\omega(x_i, p_i)$ is a decreasing function of ω.*

THEOREM 4. *Both $M_\omega(x_i, p_i)$ and $\overline{M}_\omega(x_i, p_i)$ are nondecreasing functions of x_i.*

THEOREM 5. *The generalized mean operation $M_\omega(x_i, p_i)$ satisfies the following relation:*

$$\min_i x_i \leq M_\omega(x_i, p_i) \leq \max_i x_i.$$

C. Computational Steps for Fuzzy Inference Based on α-Level Sets and Generalized Means

In this chapter, we study the parallel fuzzy inference represented in the following scheme:

Conditional propositions:
$$\begin{bmatrix} \text{If } x \text{ is } P_1 \text{ then } y \text{ is } Q_1. \\ \text{If } x \text{ is } P_2 \text{ then } y \text{ is } Q_2. \\ \vdots \\ \text{If } x \text{ is } P_n \text{ then } y \text{ is } Q_n. \end{bmatrix}$$

Fact: x is $\tilde{P}$.

Consequence: y is $\tilde{Q}$.

where P, Q, $\tilde{P}$, and $\tilde{Q}$ are fuzzy sets. These fuzzy sets are often defined with membership functions, that is,

$$P_j = \int_X \mu_{P_j}(x)/x, \qquad Q_j = \int_Y \mu_{Q_j}(y)/y,$$

$$\tilde{P} = \int_X \mu_{\tilde{P}}(x)/x, \qquad \tilde{Q} = \int_Y \mu_{\tilde{Q}}(y)/y.$$

In this chapter, Q_j is defined by a normal and convex fuzzy set in $(0,1)$. Then, its α-level set $Q_{j\alpha}$ is given by a closed interval, namely,

$$Q_{j\alpha} = \left[y^l_{Q_{j\alpha}}, y^u_{Q_{j\alpha}} \right]. \tag{6}$$

In many applications, fuzzy numbers are often given by normal and convex fuzzy sets. Moreover, conditional propositions are often given by normal and convex fuzzy sets.

A parallel fuzzy-inference method is proposed next on the basis of α-level sets and the generalized means. For convenience in the following discussions, we name this proposed method α-GEM as mentioned in Section I. Its inference process consists of the following steps [7]:

i. Calculate the compatibility degree $\tilde{p}_j$ between $\tilde{P}$ and P_j ($j = 1, 2, \ldots, n$). The compatibility degree $\tilde{p}_j$ can be defined, for example, as follows:

$$\tilde{p}_j = \sup_x \left[\mu_{\tilde{P}}(x) \wedge \mu_{P_j}(x) \right], \tag{7}$$

$$\tilde{p}_j = \sup_x \left[\mu_{\tilde{P}}(x) \cdot \mu_{P_j}(x) \right], \tag{8}$$

where $\wedge$ denotes the minimum operation.

ii. Obtain the α-level sets of the consequence, $\tilde{Q}_\alpha$, using the following equations:

$$y^l_{\tilde{Q}_\alpha} = \overline{M}_{\omega(\alpha, x)}\left(y^l_{Q_{j\alpha}}, \tilde{p}_j \right), \qquad \omega(\alpha, x) \geq 1, \tag{9}$$

$$y^u_{\tilde{Q}_\alpha} = M_{\omega(\alpha, x)}\left(y^u_{Q_{j\alpha}}, \tilde{p}_j\right), \qquad \omega(\alpha, x) \geq 1, \qquad (10)$$

where

$$\tilde{Q}_\alpha = \left[y^l_{\tilde{Q}_\alpha}, y^u_{\tilde{Q}_\alpha}\right]$$

and $\omega(\alpha, x)$ is a nonincreasing function of α.

iii. As the need arises, obtain the membership function of $\tilde{Q}$, using the resolution identity theorem.

Figure 1 illustrates the process of α-GEM. In this figure, the symbols $\otimes$ and Σ denote multiplication and summation, respectively. The symbol $1/\Sigma$ denotes the inverse of $\Sigma_j \tilde{p}_j$. The symbols $(y)^\omega$, $(\bar{y})^\omega$, $(y)^{1/\omega}$, and $(\bar{y})^{1/\omega}$ denote $y^{\omega(\alpha, x)}$, $(1 - y)^{\omega(\alpha, x)}$, $y^{1/\omega(\alpha, x)}$, and $(1 - y)^{1/\omega(\alpha, x)}$, respectively. Although one-dimensional fuzzy sets are treated above, the α-GEM method can easily be extended for multidimensional fuzzy sets.

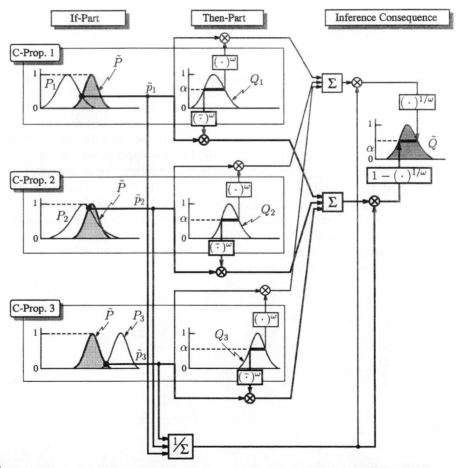

FIGURE I Parallel fuzzy inference based on α-level sets and generalized means. ("C-Prop." is an abbreviation for conditional proposition.)

III. EFFICIENT INFERENCE OPERATIONS

For efficient inference operations, parallel processing can be applied to α-GEM by taking advantage of its α-level-set based form. Moreover, the operations of α-GEM can be simplified by exploiting this advantage. This section describes the parallel processing and presents examples of the simplified operations for α-GEM.

A. Parallel Operations

The α-GEM method has operations that can be performed in parallel; the consequences can be calculated for every α-level set independently of each other. Therefore, to attain fast inference operations, parallel computation can be applied to α-GEM by hardware constructed in parallel or by multiprocessors. Each process for α-level sets is identical, and the same hardware components can be used. As the need arises, the number of quantized levels for α can be easily tuned by adding or removing identical hardware components.

B. Membership-Function Simplification Based on α-Level Sets

1. Trapezoidal Basis

By taking advantage of these α-level-set-based operations, the membership functions can be effectively simplified to attain efficient operations. The membership functions are often approximated by trapezoidal or triangular forms. The triangular forms can be seen as a special case of trapezoidal forms. Figure 2 shows an example of the approximation of membership functions on a trapezoidal basis. In such approximation, the parameters for representing membership functions are often the least upper and greatest lower bounds of α-level sets. According to the parameters in Fig. 2, $[y_1^l, y_1^u]$, $[y_{0.5}^l, y_{0.5}^u]$, and $[y_{\tilde{0}}^l, y_{\tilde{0}}^u]$ can be regarded as α-level sets. Here, $\tilde{0}$ means a number close to zero. Therefore, the inference operations are needed only for the α-level sets represented by these parameters. In this way, the inference operations can be efficiently performed, taking advantage of the α-level-set-based operations.

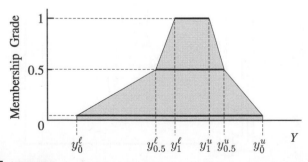

FIGURE 2 Approximation of a membership function with α-level-set-based parameters.

An α-level set can be viewed as a two-valued constraint at the level of α. This point shows that α-level sets have a suitable computational form for controlling the fuzzy constraints as exemplified in α-GEM. It is considered that membership functions can be approximated by the representatives of two-valued constraints given by α-level sets.

2. Gaussian Basis

For the operational simplification of α-GEM, the following radial basis function is used for a membership function:

$$g(x; x^\circ, \eta) = \exp\left[-\frac{(x - x^\circ)^2}{2\eta^2}\right]. \tag{11}$$

The membership function given by Eq. (11) is considered to be approximated by using α-level sets at the levels of 1 and α_* for α, where α_* is the value obtained so that the width of the α_*-level set of the fuzzy set defined by Eq. (11) is η.

Then the membership functions of P_j, Q_j, and $\tilde{P}$ are defined by

$$\mu_{P_j}(x) = g(x; x^\circ_{P_j}, \eta_{P_j}), \tag{12}$$

$$\mu_{Q_j}(y) = g(y; y^\circ_{Q_j}, \eta_{Q_j}), \tag{13}$$

$$\mu_{\tilde{P}}(x) = g(x; x^\circ_{\tilde{P}}, \eta_{\tilde{P}}). \tag{14}$$

As found in these equations, each of the membership functions is represented by two parameters. It reduces the complexity of operations in both the inference mode and learning mode described later.

In using Eqs. (12), (13), and (14), the function $\omega(\alpha, x)$ is simplified as follows:

$$\omega(\alpha, x) = \begin{cases} \omega_1(x), & \alpha = 1, \\ \omega_*(x), & \alpha = \alpha_*, \end{cases} \tag{15}$$

where $\omega_1(x) \leq \omega_*(x)$. In this case, the deduced consequences are often obtained in asymmetrical form. Then we define $\omega_{\tilde{Q}}(y)$ as follows:

$$\mu_{\tilde{Q}}(y) = \begin{cases} g(y; y^\circ_{\tilde{Q}}, \eta^l_{\tilde{Q}}), & y \leq y^\circ_{\tilde{Q}}, \\ g(y; y^\circ_{\tilde{Q}}, \eta^u_{\tilde{Q}}), & y > y^\circ_{\tilde{Q}}. \end{cases} \tag{16}$$

According to Eq. (16), $\tilde{Q}_{\alpha_*}$ is represented by

$$\tilde{Q}_{\alpha_*} = \left[y^\circ_{\tilde{Q}} - \tfrac{1}{2}\eta^l_{\tilde{Q}}, y^\circ_{\tilde{Q}} + \tfrac{1}{2}\eta^u_{\tilde{Q}}\right]. \tag{17}$$

From Eqs. (9), (10), and (17), we obtain

$$y_{\tilde{Q}}^{\circ} = M_{\omega_1(x)}\left(y_{Q_j}^{\circ}, \tilde{p}_j\right), \tag{18}$$

$$\eta_{\tilde{Q}}^{l} = 2\left[y_{\tilde{Q}}^{\circ} - \overline{M}_{\omega_*(x)}\left(y_{Q_j}^{\circ} - \tfrac{1}{2}\eta_{Q_j}, \tilde{p}_j\right)\right], \tag{19}$$

$$\eta_{\tilde{Q}}^{u} = 2\left[M_{\omega_*(x)}\left(y_{Q_j}^{\circ} + \tfrac{1}{2}\eta_{Q_j}, \tilde{p}_j\right) - y_{\tilde{Q}}^{\circ}\right]. \tag{20}$$

The radial basis functions mentioned make α-GEM differential. Therefore, the learning algorithm can easily be provided for α-GEM by using error backpropagation.

IV. CONVEX AND RESOLUTIONALLY IDENTICAL FORMS IN DEDUCED CONSEQUENCES

Fuzzy numbers as fuzzy constraints on numerical variables are important in many applications. If-parts and then-parts in conditional propositions are often given by fuzzy numbers. The fuzzy numbers are defined by normal and convex fuzzy sets in resolutionally identical forms. In this section, we study the convexity and resolutional identity of consequences deduced by using α-GEM. The α-GEM method proposed in this chapter satisfies the following proposition.

PROPOSITION 1 [7]. *If the then-parts in conditional propositions are all defined by normal and convex fuzzy sets, the consequence is obtained in the form of a normal and convex fuzzy set, in performing fuzzy inference based on the α-GEM method. Moreover, if $Q_{j\alpha}$ ($j = 1, 2, \ldots, n$) are all resolutionally identical, then it holds that $\tilde{Q}_\alpha$ is resolutionally identical.*

Proof. As described in Section II.C, each α-level set of the consequence $\tilde{Q}$ is obtained in the form of a single interval when Q_j is given by a normal and convex fuzzy set, that is, $Q_{j\alpha}$ is represented by a single interval at each level of α. According to Theorem 2, this proves that $\tilde{Q}$ is a convex fuzzy set. The normality of $\tilde{Q}$ can easily be proved from the inference process.

When $Q_{j\alpha}$ is resolutionally identical, the following relations are satisfied:

$$y_{Q_{j\alpha'}}^{l} \leq y_{Q_{j\alpha''}}^{l}, \qquad \alpha' < \alpha'', \tag{21}$$

$$y_{Q_{j\alpha'}}^{u} \geq y_{Q_{j\alpha''}}^{u}, \qquad \alpha' < \alpha''. \tag{22}$$

From Eqs. (21) and (22), the following relations are found to hold according to Theorems 3 and 4 together with the fact that $\omega(\alpha, x)$ is defined by a nonincreasing function:

$$y_{\tilde{Q}_{\alpha'}}^{l} \leq y_{\tilde{Q}_{\alpha''}}^{l}, \qquad \alpha' < \alpha'', \tag{23}$$

$$y_{\tilde{Q}_{\alpha'}}^{u} \geq y_{\tilde{Q}_{\alpha''}}^{u}, \qquad \alpha' < \alpha''. \tag{24}$$

Minkowski's inequality provides

$$\left[\sum_{j=1}^{n} (\alpha_j + b_j)^{\omega} \right]^{1/\omega} \leq \left[\sum_{j=1}^{n} a_j^{\omega} \right]^{1/\omega} + \left[\sum_{j=1}^{n} b_j^{\omega} \right]^{1/\omega},$$

$$a_j > 0, \, b_j > 0, \, \omega > 1. \quad (25)$$

From Eq. (25), we obtain

$$\overline{M}_{\omega}(x_i, p_i) \leq M_{\omega}(x_i, p_i), \qquad 0 < x_i \leq 1, \, p_i > 0. \quad (26)$$

Equation (26) and Theorem 4 yield

$$y_{\tilde{Q}_{\alpha}}^{l} \leq y_{\tilde{Q}_{\alpha}}^{u}. \quad (27)$$

Equations (23), (24), and (27) mean that the closed interval $[y_{\tilde{Q}_{\alpha}}^{l}, y_{\tilde{Q}_{\alpha}}^{u}]$, namely $\tilde{Q}_{\alpha}$ satisfies the following relation:

$$\tilde{Q}_{\alpha'} \supseteq \tilde{Q}_{\alpha''}, \qquad \alpha' < \alpha''. \quad (28)$$

Therefore, $\tilde{Q}_{\alpha}$ is resolutionally identical. ∎

Proposition 1 is effective in introducing learning algorithms to parallel fuzzy inference, in particular, when its exemplar patterns are given by normal and convex fuzzy sets for consequences.

V. CONTROL OF FUZZY CONSTRAINT PROPAGATION

In this section, we discuss control schemes for the fuzzy constraint propagation in α-GEM. Fuzzy constraint propagation is characterized by fuzziness and specificity propagation. The schemes for control are effective in extracting the knowledge in the rules for governing the fuzzy constraint propagation from fuzzy exemplar patterns with various levels of fuzziness and specificity. They provide useful ways for reflecting a set of given properties in the propagation to the rules. Namely, the modification of deduced consequences can be attained, reflecting the fuzziness and specificity in given facts.

A. Fuzziness Propagation Control

The α-GEM method has a control scheme for the degree to which the fuzziness of given facts is propagated to that of deduced consequences, by using $\omega(\alpha, x)$. In this chapter, fuzziness is defined by the following:

DEFINITION 3 [9]. When a fuzzy set A, in a universe of discourse given by a bounded set X, is normal and convex and is represented by the family of its α_i-level sets A_{α_i} $(i = 1, 2, \ldots, m)$, the fuzziness of A is defined as follows:

$$H_f(A) = \frac{2}{m\mathscr{R}(X)} \sum_{i=1}^{m} |w_{\alpha_i} - w^{near}|, \tag{29}$$

$$A_{\alpha_i} = \left[x^l_{\alpha_i}, x^u_{\alpha_i} \right], \qquad x^l_{\alpha_i} \in X, x^u_{\alpha_i} \in X,$$

$$A^{near} = \left[x^l_{\alpha^{near}}, x^u_{\alpha^{near}} \right], \qquad x^l_{\alpha^{near}} \in X, x^u_{\alpha^{near}} \in X,$$

$$w_{\alpha_i} = x^u_{\alpha_i} - x^l_{\alpha_i}, \qquad w^{near} = x^u_{\alpha^{near}} - x^l_{\alpha^{near}},$$

$$\mathscr{R}(X) = \sup x - \inf x, \qquad x \in X,$$

where if A_{α_i} is null, the value of w_{α_i} is regarded as zero for convenience. In the same way, if A^{near} is null, the value of w^{near} is regarded as zero. The symbol A^{near} denotes the crisp set nearest to A and is given by the α-level set of A at the level α of 0.5, α^{near}.

Note. The fuzziness defined in Definition 3 is based on Kaufmann's definition of fuzziness [3, 4]. Kaufmann's fuzziness is originally membership-function-based whereas the inference method studied in this chapter is α-level-set-based. Therefore, the membership-function-based definition was translated into its α-level-set-based definition.

From Definition 3 and Theorem 3, it can be found that the fuzziness becomes larger in deduced consequences as the value given by $H_{f\omega}$ defined next is larger:

$$H_{f\omega} = \sum_i |\omega(\alpha_i, x) - \omega_0|, \qquad \omega_0 = \omega(\alpha^{near}, x). \tag{30}$$

Figure 3a exemplifies $\omega(\alpha, x)$ for controlling fuzziness propagation. According to Eq. (30), $\omega''(\alpha, x)$ propagates the fuzziness of a given fact to that of the deduced consequence at a higher degree compared with $\omega'(\alpha, x)$. Namely, the fuzziness in consequences obtained by $\omega''(\alpha, x)$ is larger than that obtained by $\omega'(\alpha, x)$ even if the same compatibility degrees are given.

When $\omega(\alpha, x)$ is equal to 1 for all α, fuzzy constraint is not propagated from given facts to deduced consequences. In this case, the following relation can be derived [7, 9]:

$$H_f(\tilde{Q}) \leq \max_j H_f(Q_j). \tag{31}$$

Relation (31) shows that the fuzziness of deduced consequences is determined by that of the then-parts. In other words, the fuzziness of deduced consequences can be controlled only by that of the then-parts.

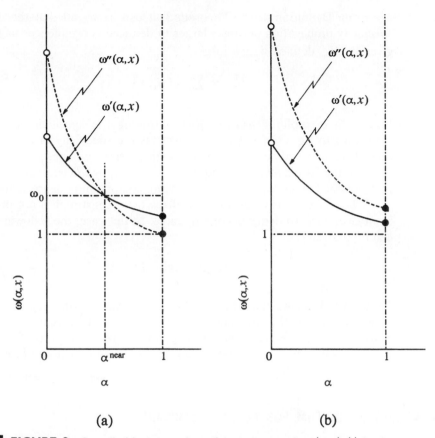

FIGURE 3 Control of fuzziness and specificity propagation by $\omega(\alpha, x)$: (a) fuzziness propagation control; (b) specificity propagation control.

B. Specificity Propagation Control

The α-GEM method has a control scheme for specificity propagation, by using $\omega(\alpha, x)$. In this chapter, specificity is defined by the following:

DEFINITION 4. When a fuzzy set A, in a universe of discourse given by a bounded set X, is normal and convex and is represented by the family of its α_i-level sets A_{α_i} $(i = 1, 2, \ldots, m)$, the specificity of A is defined by

$$H_s(A) = 1 - \frac{1}{m\mathscr{R}(X)} \sum_{i=1}^{m} \left(x_{\alpha_i}^u - x_{\alpha_i}^l \right), \tag{32}$$

where an α_i-level set of A, A_{α_i}, is given by

$$A_{\alpha_i} = \left[x_{\alpha_i}^l, x_{\alpha_i}^u \right].$$

Note. Specificity is another important factor related to the width of the fuzzy constraints. We define this specificity based on α-level sets according to [5].

From Definition 4 and Theorem 3, it can be found that the degree of specificity propagation becomes larger in deduced consequences as the value given by $\overline{H}_{s\omega}$ defined next is larger,

$$\overline{H}_{s\omega} = \sum_i |\omega(\alpha_i, x) - 1|. \tag{33}$$

Figure 3b exemplifies $\omega(\alpha, x)$ for controlling the specificity propagation. According to Eq. (33), $\omega''(\alpha, x)$ propagates the specificity of a given fact to that of the deduced consequence at a higher degree compared with $\omega'(\alpha, x)$. Namely, the specificity in consequences obtained by $\omega''(\alpha, x)$ is smaller than that obtained by $\omega'(\alpha, x)$ even if the same compatibility degrees are given.

When $\omega(\alpha, x)$ is equal to 1 for all α, fuzzy constraint is not propagated from given facts to deduced consequences. In this case, the following relation can be derived [7]:

$$H_s(\tilde{Q}) \geq \min_j H_s(Q_j). \tag{34}$$

Relation (34) shows that the specificity of deduced consequences is determined by that of the then-parts. In other words, the specificity of deduced consequences can be controlled only by that of the then-parts. This property is useful when the specificity of deduced consequences is to be proved within a certain value [10].

VI. LEARNING ALGORITHM FOR FUZZY EXEMPLARS

A learning algorithm for the α-GEM method is illustrated on the basis of error backpropagation [11]. As an example, the approximated inference operation, provided in Section III.B.2, is adopted here. Namely, the learning algorithm is derived for the inference operations with Eqs. (12)–(16). Moreover, we use $\omega_1(x) = 1$ in Eq. (15) for simplicity. Exemplar patterns are given by fuzzy-input/fuzzy-output pairs in the form of normal and convex fuzzy sets.

In deriving the learning algorithm, the compatibility degree $\tilde{p}_j$ is obtained by applying Eq. (8). In using Eqs. (13) and (16), we define the parameters as follows:

$$y_{Q_{\alpha_*}}^l = y_Q^\circ - \tfrac{1}{2}\eta_Q^l, \qquad y_{Q_{\alpha_*}}^u = y_Q^\circ + \tfrac{1}{2}\eta_Q^u.$$

The error E is defined by

$$E = \frac{1}{2n_t} \sum_c \left[\left(y_Q^\circ - \hat{y}_Q^\circ\right)^2 + \left(y_{Q_{\alpha_*}}^l - \hat{y}_{Q_{\alpha_*}}^l\right)^2 + \left(y_{Q_{\alpha_*}}^u - \hat{y}_{Q_{\alpha_*}}^u\right)^2 \right]$$

$$(c = 1, 2, \ldots, n_t), \quad (35)$$

where $\hat{y}_Q^\circ$, $\hat{y}_{Q_{\alpha_*}}^l$, and $\hat{y}_{Q_{\alpha_*}}^u$ denote the desired target data for y_Q°, $y_{Q_{\alpha_*}}^l$, and

$y^u_{\tilde{Q}_{\alpha_*}}$, respectively. The variable n_t gives the number of target data ($\hat{y}^{\circ}_{\tilde{Q}}$, $\hat{y}^l_{\tilde{Q}_{\alpha_*}}$, $\hat{y}^u_{\tilde{Q}_{\alpha_*}}$). The parameters $x^{\circ}_{P_j}$, η_{P_j}, $y^{\circ}_{Q_j}$, η_{Q_j}, and $\omega_*(x)$ can be tuned to minimize E by using the error backpropagation algorithm [7]. A learning algorithm for multistage-parallel inference based on α-GEM can be derived in the same way.

VII. FUZZY CONSTRAINT PROPAGATION CONTROL REFLECTING FORMS OF GIVEN FACTS

A. A Method for Control of Fuzzy Constraint Propagation Reflecting Forms of Given Facts

We propose a method for controlling the fuzzy constraint propagation in α-GEM so as to satisfy a set of required properties in relations between given facts and deduced consequences. This method is provided by the generation of $\omega(\alpha, x)$ as a function of fuzziness and specificity of $\tilde{P}$. For simplicity, let $\tilde{P}$ be a normal and convex fuzzy set that is defined by a symmetric triangular or trapezoidal membership function. Then let $\omega(\alpha, x)$ be given by the following equation so as to satisfy the set of required properties described in Section VII.B:

$$
\omega(\alpha, x) = \begin{cases} \varphi\big(1 - H_s(\tilde{P}) - H_f(\tilde{P})\big) + 1, & \alpha = 1, \\ \varphi\big(1 - H_s(\tilde{P})\big) + 1, & \alpha = \alpha_*(<1), \end{cases} \tag{36}
$$

where φ is a nondecreasing function such that $\varphi(0) = 0$. In using Eq. (36), the inference operations are performed for the α-level sets only at the levels α of 1 and α_*. This exploits the advantage that α-GEM can be performed for every α-level set independently. The membership functions of inference consequences are constructed by linear interpolation and extrapolation of the least upper and greatest lower bounds of the α-level sets deduced by the aforementioned process.

B. Relations between Facts and Consequences in Fuzzy Constraints

By using Eq. (36), fuzzy constraint propagation in α-GEM can be controlled as follows. The consequences deduced by using this control method are exemplified in Section VIII.

i. When a given fact is a singleton, the fuzzy constraint in the given fact is not propagated to the consequence. This also means that the fuzziness and specificity of the consequence can be controlled by those of then-parts [9, 10]. This property can be found from relations (31) and (34). These relations are derived from the equation $\omega(1, x) = \omega(\alpha_*, x) = 1$ obtained by Eq. (36).

ii. When a given fact is a closed interval, the specificity of the given fact is propagated to the consequence. The fuzziness of the consequence can be controlled by that of the then-parts because the functions given by Eqs. (4) and (5) are nondecreasing functions of x_i and the relation $\omega(1, x) = \omega(\alpha_*, x) > 1$ is derived from Eq. (36).

iii. When a given fact is a fuzzy set, both the fuzziness and specificity of the given fact are propagated to the consequence. This is because the functions given by Eqs. (4) and (5) are nondecreasing and nonincreasing functions of ω, respectively, and the relation $\omega(1, \omega) < \omega(\alpha_*, x)$ is derived from Eq. (36).

VIII. SIMULATION STUDIES

Some basic properties are illustrated using numerical examples. The following simulations were conducted on the basis of the simplified form of α-GEM described in Section III.B to grasp the basic functions studied in this chapter. The simulations were performed using Sun SPARCserver 1000 (CPU, Super SPARC; 60 MHz, RAM = 128 megabytes).

SIMULATION I. The first simulation was conducted to exemplify the effect of $\omega(\alpha, x)$. In this simulation, the simplified form of α-GEM described in Section III.B.2 was applied and compatibility degrees were calculated using Eq. (8). Eleven conditional propositions were used by setting their parameters of membership functions defined by Eqs. (12) and (13). The input–output relation of the inference by these conditional propositions is shown in Fig. 4,

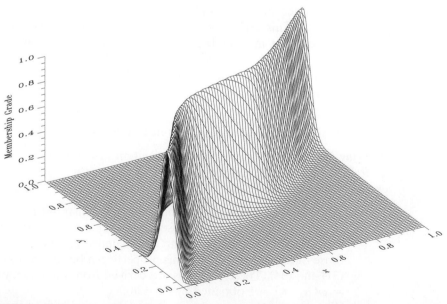

FIGURE 4 Input–output relation of fuzzy inference, which was used in Simulations I, II, and III; $\omega(\alpha, x) = 1$.

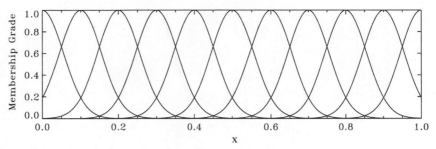

FIGURE 5 Membership functions of if-parts used in Simulations I–III.

in the case where $\omega(\alpha, x) = 1$. Figure 5 depicts the membership functions $\mu_{P_j}(x)$ for the if-parts.

Figure 6 shows the membership functions $\mu_{\bar{P}}(x)$ for given facts with various degrees of fuzziness and specificity. Figure 7 shows the consequences deduced from the given facts shown in Fig. 6. In deducing the consequences, Eqs. (15) and (16) were adopted, where $\omega(1, x) = 1$ for simplicity. When $\omega(1, x) = 1$ and $\omega(\alpha_*, x) = 1$, the fuzziness and specificity of the given facts are not propagated to the deduced consequences as shown in Fig. 7a. As a result, the fuzziness and specificity of the deduced consequences are determined by those of the then-parts. As can be seen in Fig. 7a–c, when the value of $\omega(\alpha_*, x)$ is larger, the fuzziness and specificity of the deduced consequences become larger and smaller, respectively, reflecting those of the given facts. This is because the degree of the fuzzy constraint propagation is larger as the value of $\omega(\alpha_*, x)$ is larger. This exemplifies the control of the fuzzy constraint propagation in α-GEM.

SIMULATION II. By adopting the simplified form of α-GEM with Eq. (8) and the conditional propositions used in Simulation I, some consequences were deduced from various given facts. This simulation was conducted at $\omega(1, x) = 1$ and $\omega(\alpha_*, x) = 3$. Figures 8 and 9 show the given facts and the consequences deduced from these given facts. As can be seen from the figures, deduced consequences were obtained that reflect the fuzziness and specificity of the given facts. In particular, Fig. 9 illustrates that the shapes of

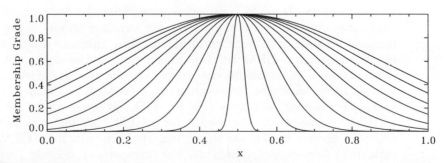

FIGURE 6 Membership functions of given facts in Simulation I.

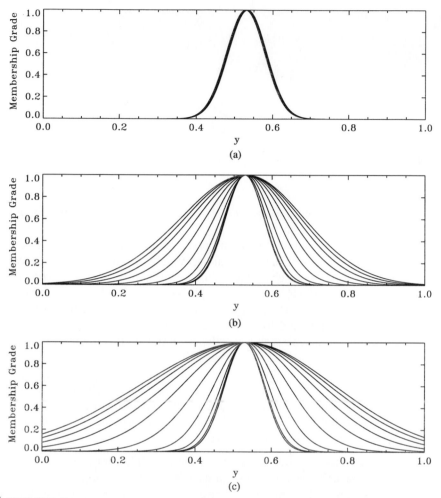

FIGURE 7 Membership functions of deduced consequences in Simulation I: (a) $\omega(1, x) = 1$, $\omega(\alpha_*, x) = 1$; (b) $\omega(1, x) = 1$, $\omega(\alpha_*, x) = 3$; (c) $\omega(1, x) = 1$, $\omega(\alpha_*, x) = 5$.

the membership functions for deduced consequences are reflected by the location of fuzzy sets in the conditional propositions.

SIMULATION III. This simulation exemplifies the learning of $\omega(\alpha, x)$ from fuzzy exemplar patterns by using the structure shown in Fig. 10. The function of α-GEM 1 is to perform the main inference by adopting the simplified form of α-GEM with Eq. (8) and the conditional propositions used in Simulation I. These conditional propositions are fixed in the simulation. To control fuzzy constraint propagation in α-GEM 1, α-GEM 2 generates $\omega(\alpha_*, x)$, assuming $\omega(1, x) = 1$ in Eq. (15) and using x_P° as the input of α-GEM 2 for simplicity. The if-parts for α-GEM 2 are the same as those for α-GEM 1 and are fixed in the simulation. The then-parts for α-GEM 2 are to be learned from exemplar patterns. The then-parts of α-GEM 2 are initialized with a uniformly random number in $(0, 1)$.

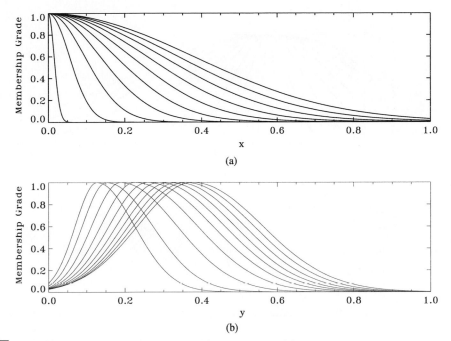

FIGURE 8 Membership functions of (a) given facts and (b) the consequences deduced from the facts in Simulation II [$\omega(1, x) = 1$, $\omega(\alpha_*, x) = 3$]; Example I.

Exemplar patterns are the pairs of the given facts and consequences. In generation of the exemplar patterns, the then-parts for α-GEM 2 were given by singletons at x of 6.5, 6, 5, 4, 3.5, 3, 3, 3, 3, 3, and 3 for the if-parts shown in Fig. 5 in order of increasing number of $x^\circ_{P_j}$, respectively. Namely, these numbers are to be learned from the randomly generated exemplar patterns. The values of $x^\circ_{\tilde{P}}$ were generated by a uniformly random number in $(0, 1)$. The values of $\eta_{\tilde{P}}$ were generated by a uniformly random number in the same range used in Simulation I. The total number of exemplar patterns was 100.

As a result of the simulation, the then-parts for α-GEM 2 were converged into the aforementioned numbers at the square error of 0.114×10^{-4}. Therefore, it can be found that $\omega(\alpha, x)$ was learned well from the given exemplar patterns.

SIMULATION IV. This simulation exemplifies the consequences deduced from some typical forms of fuzzy constraints in given facts and then-parts by using the method described in Section VII. For simplicity, let $\tilde{P}$ be a normal and convex fuzzy set that is defined by a symmetric triangular or trapezoidal membership function. The function $\omega(\alpha, x)$ was given by Eq. (36) and φ was defined by the following equation for simplicity:

$$\varphi(z) = c \cdot z, \tag{37}$$

where c is a constant value. In this simulation, the value of c was set to 50. Moreover, α_* in Eq. (36) was 0.5. The membership functions of conse-

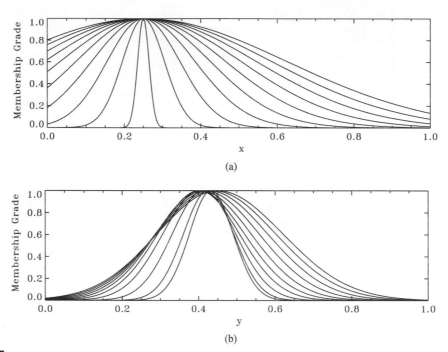

FIGURE 9 Membership functions of (a) given facts and (b) the consequences deduced from the facts in Simulation II [$\omega(1, x) = 1$, $\omega(\alpha_*, x) = 3$]; Example 2.

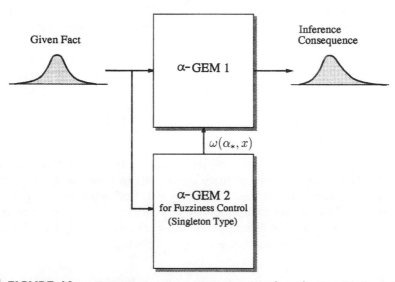

FIGURE 10 α-GEM with the inference for generating $\omega(\alpha_*, x)$, adopted in Simulation III: α-GEM 1 for main inference; α-GEM 2 for generating $\omega(\alpha_*, x)$.

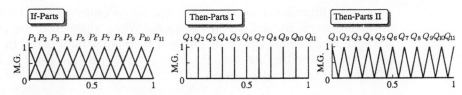

FIGURE 11 Membership functions for if-parts and then-parts used in Simulation IV. ("M.G." is an abbreviation for membership grade.)

quences were constructed by linear interpolation and extrapolation performed with $y_{Q_\alpha}^l$ and $y_{Q_\alpha}^u$ ($\alpha = 0.5, 1$). The compatibility degrees were calculated by using Eq. (7).

If-parts and given facts used in the simulations are shown in Figs. 11 and 12, respectively. By using Then-parts I and Then-parts II shown in Fig. 11, Consequences I and Consequences II in Fig. 12 were deduced, respectively. The simulation results shown in Fig. 12 exemplify the following properties. These properties are easily derived from (i), (ii), and (iii) in Section VII.B.

Consequences are obtained in crisp-set forms when both given facts and then-parts are defined by crisp sets. Then, the specificity in given facts is propagated to consequences. Even if then-parts are defined by crisp sets, consequences are obtained in fuzzy-set forms when facts are given by fuzzy sets. Namely, both the fuzziness and specificity in given facts are propagated to consequences.

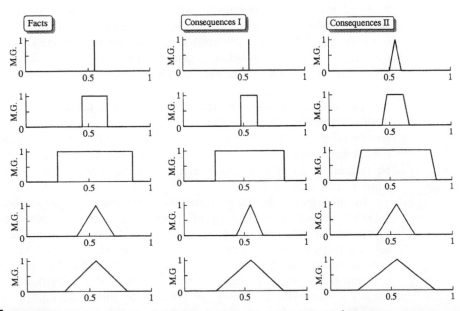

FIGURE 12 Given facts and deduced consequences in Simulation IV. ("M.G." is an abbreviation for membership grade.)

On the other hand, consequences are always obtained in fuzzy-set forms when then-parts are defined by fuzzy sets. When given facts are crisp sets, fuzziness of consequences is determined by that of then-parts. In the case where given facts are fuzzy sets, fuzziness of both given facts and then-parts is reflected to that of consequences.

IX. CONCLUSION

A parallel fuzzy-inference method was proposed in which fuzzy constraint propagation can be controlled. This method is based on α-level sets and generalized means. Namely, the least upper bounds and greatest lower bounds of the α-level sets are deduced on the basis of the generalized means of those of α-level sets defining then-parts. This proposed inference method was named α-GEM (an abbreviation for "α-level-set- and generalized-mean-based inference"). The α-GEM method includes, as a special case, the singleton-type fuzzy inference in which the then-parts are defined by singletons.

The control of the fuzzy constraint propagation in inference is important because how the fuzzy constraints are propagated from given facts to consequences is to be determined by not only the rules for the inference process but also the background knowledge in conditional propositions. The α-GEM method can reflect such background knowledge to the rule for governing the fuzzy constraint propagation. This fuzzy constraint propagation is characterized by fuzziness and specificity propagation. The conventional methods, however, do not have such a control scheme. Therefore, the α-GEM method is useful in representing background knowledge in conditional propositions, and thus can achieve more flexible inference compared with the conventional methods. In α-GEM, the conditional propositions can effectively be extracted from fuzzy exemplars, avoiding fuzziness explosion and specificity dispersion, particularly in multistage-parallel inference, because of the control schemes for fuzzy constraint propagation.

The α-GEM method deduces consequences in normal and convex forms if the then-parts of conditional propositions are defined by normal and convex fuzzy sets. The then-parts have often been represented by normal and convex fuzzy sets in many applications. Thus, the deduced consequences can be treated as fuzzy numbers which are often represented by normal and convex fuzzy sets.

In introducing learning algorithms into parallel fuzzy inference, the α-GEM method is particularly effective when fuzzy exemplar patterns are represented by normal and convex fuzzy sets. The normal and convex fuzzy sets are often given as the fuzzy exemplar patterns. The α-GEM method can provide the learning schemes for extracting the rules for governing fuzzy constraints propagation together with conditional propositions. Simulation studies showed the feasibility of α-GEM in this learning. In contrast to α-GEM, the introduction of such learning into conventional parallel fuzzy-inference methods is impeded because the methods cannot always deduce consequences in normal and convex forms and do not have control schemes for fuzzy constraint propagation.

The α-GEM method makes it possible to control the degree to which the fuzzy constraints in given facts are propagated to deduced consequences in various forms. Exploiting this advantage, a method was proposed for controlling fuzzy constraint propagation in α-GEM so as to provide a set of required properties in relations between given facts and deduced consequences. In this control method, fuzzy constraints in given facts are evaluated by fuzziness and specificity, and the results are reflected in the propagation control. As a result, the fuzziness and specificity of given facts can be propagated to consequences so as to satisfy the required properties of inference. Simulation studies demonstrated the feasibility of α-GEM in the control of fuzzy constraint propagation.

The α-level sets are effective in representing fuzzy constraints because each of the α-level sets can be represented by intervals. The intervals can easily be treated as constraints on variables because of their two-valued nature. The α-GEM method makes effective use of this property. The α-level sets have the potential to play an important role in the studies of fuzzy constraints and their propagation. Fuzzy inference with the compositional rule of inference (CRI) [8] was formulated into its α-level-set-based form [12, 13]. The multistage-parallel fuzzy inference with CRI was formulated into the form of linguistic-truth-value propagation and an α-level-set-based form [6, 13]. These formulations are considered to be effective in studies of the fuzzy constraint propagation in the fuzzy inference with CRI. Further studies are required to determine ways of governing constraint propagation in other types of inference, for which control schemes for fuzzy constraint propagation may need to be investigated.

REFERENCES

1. Zadeh, L. A. Fuzzy logic = computing with words. *IEEE Trans. Fuzzy Systems* 4(2):103–111, 1996.
2. Zadeh, L. A. Inference in fuzzy logic via generalized constraint propagation. In *Proceedings of the 26th International Symposium on Multi-Valued Logic* (*ISMVL'96*), 1996, pp. 192–195.
3. Kaufmann, A. *Introduction to the Theory of Fuzzy Subsets*, Vol. 1. Academic Press, New York, 1975.
4. Pal, N. R. and Bezdek, J. C. Measuring fuzzy uncertainty. *IEEE Trans. Fuzzy Systems* 2(2):107–118, 1994.
5. Yager, R. R. On the specificity of a possibility distribution. *Fuzzy Sets Systems*, 50(3):279–292, 1992.
6. Uehara, K. and Fujise, M. Multistage fuzzy inference formulated as linguistic-truth-value propagation and its learning algorithm based on back-propagating error information. *IEEE Trans. Fuzzy Systems* 1(3):205–221, 1993.
7. Uehara, K. and Hirota, K. Parallel fuzzy inference based on α-level sets and generalized means. *Inform. Sci.* 100(1–4):165–206, 1997.
8. Zadeh, L. A. The concept of a linguistic variable and its application to approximate reasoning, I, II, III. *Inform. Sci.* 8:199–249, 301–357; 9:43–80, 1975.
9. Uehara, K. Fuzzy inference based on a weighted average of fuzzy sets and its learning algorithm for fuzzy exemplars. In *Proceedings of the International Joint Conference of the Fourth IEEE International Conference on Fuzzy Systems and the Second International Fuzzy*

Engineering Symposium, FUZZ-IEEE/IFES'95, Yokohama, Japan, 1995, vol. IV, pp. 2253–2260.

10. Uehara, K. and Hirota, K. Fuzzy connection admission control for ATM networks based on possibility distribution of cell loss ratio. *IEEE J. Selected Areas Commun.* 15(2):179–190, 1997.

11. Rumelhart, D. E., Hinton, G. E., and Williams, R. J. Learning representations by back-propagating errors. *Nature* 323(6088):533–536, 1986.

12. Uehara, K. and Fujise, M. Fuzzy inference based on families of α-level sets. *IEEE Trans. Fuzzy Systems* 1(2):111–124, 1993.

13. Uehara, K. and Hirota, K. Parallel and multistage fuzzy inference based on families of α-level sets. *Inform. Sci.* 106(1–2):159–195, 1998.

14. Zadeh, L. A. Outline of a new approach to the analysis of complex systems and decision processes. *IEEE Trans. Systems Man Cybernet.* SMC-3(1):28–44, 1973.

15. Koczy, L. T. and Hirota, K. Approximate reasoning by linear rule interpolation and general approximation. *Internat. J. Approx. Reasoning* 9(3):197–225, 1993.

II

SYSTEM CONTROL METHODS AND APPLICATIONS

12

INTELLIGENT PROPORTIONAL, INTEGRAL, DERIVATIVE CONTROLLERS

KA CHING CHAN

School of Mechanical and Manufacturing Engineering,
University of New South Wales, Sydney, Australia

I. INTRODUCTION   293
II. INTELLIGENT APPROACHES TO PID
   CONTROLLER TUNING   294
   A. Feedback Control   296
   B. Process Description   296
III. NEURAL NETWORK-BASED TUNER   297
   A. Input–Output Space   298
   B. Input–Output Scaling   300
   C. Learning Strategy   300
   D. RMS Error   302
   E. Results and Discussion   303
IV. INTEGRATED FUZZY-NEURAL NETWORK TUNER   309
   A. A Fuzzy Rule Generation Scheme Using Networks   309
   B. Fuzzy-Neural Network Tuner   314
   C. Results and Discussion   315
V. ADAPTIVE VIRTUAL FUZZY TUNER   322
   A. Concept of a Virtual Fuzzy Set   324
   B. Virtual Fuzzy Rule Representation   325
   C. A Virtual Fuzzy Rule Generation Scheme   326
   D. Results and Discussion   332
VI. SUMMARY   338
   REFERENCES   342

I. INTRODUCTION

The conventional proportional, integral, and derivative (PID) feedback control is by far the most popular control action in most process industries. Compared to other control algorithms, feedback control is simple and easy to implement, and performs well for most applications. Because feedback control is purely error driven, the model of the process to be controlled is not needed.

Fuzzy Theory Systems: Techniques and Applications, Vol. 1
Copyright © 1999 by Academic Press. All rights of reproduction in any form reserved.

To design a feedback controller for a particular process, the parameters of the controller must be properly tuned to achieve reasonable performance. The Ziegler and Nichols tuning rules have been widely applied [1]. These rules may provide satisfactory results for initial trials in controller tuning, but fine tuning of the controller is often necessary and is left to the control engineer. During the tuning process, the parameters are adjusted iteratively according to the step response of the closed-loop controlled system. The control engineer recognizes the response pattern and adjusts the controller gains based on the past experience and some heuristic tuning rules (rules of thumb).

The manual tuning process is practical and has been very successful. In fact, such a tuning process can be considered as a form of adaptive control based on response pattern recognition, but without the need for an accurate mathematical model of the process to be controlled. Much research effort has been put into the automation of the tuning process by the use of artificial intelligence (AI) techniques such as expert systems, neural networks, and fuzzy systems. This chapter is intended to present the implementation of feedback controller tuners using a neural network, an integrated fuzzy-neural network system, and a virtual fuzzy system.

II. INTELLIGENT APPROACHES TO PID CONTROLLER TUNING

A few expert systems were developed in the past for controller tuning [2-5]. They were built by modeling the manual tuning procedure. Pattern recognition techniques were employed to recognize the response characteristics, and controller gains were adjusted based on a number of tuning rules captured from control engineers. Although successful results were reported in these approaches, the main drawback is that they did not have any learning mechanism and completely relied on human control experts to provide tuning rules.

More recently, neural networks were applied to the controller tuning problem. Ma and Loh [6] proposed the use of a back-propagation neural network to tune PID controllers. The network was trained to approximate the nonlinear relationship between certain steady-state variables of a specific process and the PID gain settings. Therefore, the neural network tuner was restricted to a specific process. The feasibility of the proposed method was demonstrated by simulations on a liquid-level control system. Another approach, proposed by Willis and Montague [7], was to build a dynamic model of the process with a neural network and to design a PI or PID controller based on the neural network model. The process model was used to simulate the response characteristics of the process for a variety of conditions. The integral squared error (ISE) was used and the controller settings were adjusted so that the cost function was minimized. A gradient free technique was chosen for the optimization.

The first part of this chapter describes the implementation of intelligent feedback controller tuners using neural networks. This approach differs significantly from previous attempts. First, unlike the expert system approaches, the neural network tuner extracts tuning knowledge automatically through the use of a representative process, and therefore, knowledge extraction from a human control expert is not required. Second, in contrast to the previous neural network approaches, this is a generic neural network tuner that can be used for arbitrary processes using step-response analysis. The neural network is developed to relate certain normalized response parameters to adjustments to the controller settings. Therefore, building a new neural network is unnecessary for a new process.

The system consists of a response recognition subsystem to identify open-loop responses and a tuning subsystem to make gain adjustment decisions. The tuning procedure includes an initial gain-setting procedure and a fine-tuning procedure. The initial gain settings are obtained by using the standard Ziegler–Nichols tuning rules, based on the open-loop step response of the process. The fine-tuning procedure is performed iteratively by using a neural network. The neural network suggests adjustments to the gain settings based on the closed-loop step response. Four parameters are defined to describe the response characteristics. They are the normalized peak rise time (R), normalized overshoot (O), normalized peak-to-peak height (H), and normalized final error (E). These four parameters are used as inputs to the neural network. The tuning knowledge of the neural network is extracted from the tuning of a representative process. Processes covering a wide range of dynamics are tested to demonstrate the excellent performance of the tuner.

Because of the inherent disadvantage of neural networks that it is difficult to incorporate structured knowledge, expressed as IF–THEN rules, a further attempt is made to develop an integrated neural-fuzzy system. This approach has significant improvement in the way that human tuning knowledge can be combined with automatically extracted numerical knowledge.

Although the neural-fuzzy approach has many advantages over the pure neural network approach, the long training time requirement of neural networks remains as the major bottleneck in the system. A logical extension is to develop a more efficient learning algorithm, replacing the neural network, for generating fuzzy rules from numerical data and combining linguistic rules into a common framework. Therefore, a new adaptive algorithm, based on a new concept, called the virtual fuzzy set, is developed.

The virtual fuzzy set concept provides a more accurate representation of the consequent part of a fuzzy production rule. A virtual fuzzy set, which consists of two consecutive fuzzy sets with different degrees of membership, is an imaginary fuzzy set located at a position most appropriate for a numerical training sample. The new concept is incorporated into the adaptive algorithm that is a one-pass build-up procedure, and time-consuming iterative training is not required. The proposed adaptive algorithm has been verified by its successful application to feedback controller tuning.

This chapter is organized as follows. The background of feedback control and process description will first be presented. Then the implementation of the three tuning approaches, neural network, integrated fuzzy-neural system, and virtual fuzzy system, will be provided.

A. Feedback Control

The objective of feedback control is to constrain a process response to follow input set points. A feedback controller may consist of proportional (P), integral (I), and derivative (D) actions. The task of a control engineer is to determine which of these control actions should be used, and in what proportion. The typical combinations of feedback control actions include P, PI, PD, and PID. The fuzzy tuner is developed to determine the proportions in which these control actions should be used by adjusting their corresponding gains.

In this study, experiments are conducted using computer digital simulation, and the following discrete-time expression will be used for the controller:

$$u(n) = K_p \left[e(n) + \frac{1}{T_i} \Delta t \sum_{k=0}^{n} e(k) + T_d \frac{e(n) - e(n-1)}{\Delta t} \right], \quad (1)$$

where K_p is the proportional controller gain, T_i is the integral or reset time, and T_d is the derivative time. $e(n)$ is the error or the input to the controller, and $u(n)$ is the controller output or the process input at the nth time interval, and Δt is the sampling period. The feedback control action is implemented in such a way that the controller output is compared to the set point to generate an error signal $e(n)$ together with its integral and derivative, which are computed and fed back to the controller to produce the control signal $u(n)$. For P, PI, or PD controllers, only the relevant term(s) of the complete PID expression will be included.

B. Process Description

The advantages of using feedback control are that detailed knowledge of the process to be controlled is not required and the entire control action is error driven. However, some crude idea about the process dynamic is necessary to pretune the tuner. The widely used Ziegler–Nichols tuning rules will be employed to set initial gain values according to the open-loop step response of the process. The intelligent tuners (the neural network, integrated fuzzy-neural system, or virtual fuzzy system) will then be used to fine tune the system until a satisfactory response is obtained.

The performance of Ziegler–Nichols tuned systems has been assessed by characterization of process dynamics using the normalized dead time or the normalized process gain [8–10]. The aim of the assessment was to identify the main areas of applications of different feedback control laws. The definition of normalized dead time is defined as the ratio of the apparent dead time to the apparent time constant:

$$\theta_1 = \frac{\tau_d}{\tau}. \quad (2)$$

The normalized dead time, θ_1, which can be estimated from the process reaction curve, provides a good measure of the difficulty of controlling a process. A small θ_1 indicates that the process is easy to control, whereas a large θ_1 indicates that the process is difficult to control.

The normalized process gain, K_1, is an alternative to θ_1 in assessing the difficulty of controlling a process. K_1 is inversely proportional to θ_1. An empirical relation between K_1 and θ_1 is given as

$$K_1 = 2 \times \frac{11\theta_1 + 13}{37\theta_1 - 4}. \tag{3}$$

The existence of such a relation means K_1 and θ_1 can be used interchangeably to assess process dynamics and, consequently, to predict the achievable performance of certain feedback control actions. It was shown that the prime application area of PI and PID control is when $0.15 < \theta_1 < 0.6$ or $2.25 < K_1 < 15$ [10]. Derivative action often improves performance significantly. Therefore, the development of the intelligent tuners will focus on these ranges.

III. NEURAL NETWORK-BASED TUNER

The neural network approach in implementing an automatic tuner is based upon the methodology a control engineer might use. The tuning procedure involves the following steps:

1. Obtain the process reaction curve.
2. Determine the process gain K, the time delay τ_d, and the apparent time constant τ.
3. Set initial controller gains, K_p and T_i, using the Ziegler–Nichols tuning rules.
4. Monitor the closed-loop controlled response to a unit step input.
5. Analyze the result, and based on the previous experience and knowledge, fine-tune K_p and T_i.
6. Repeat steps 4 and 5 until a satisfactory result is achieved.

This manual procedure is exactly the approach employed by the neural network-based tuner given in Fig. 1. The tuner collects information about the process response and recommends adjustments to be made to the controller gains. This is an iterative procedure that continues until the fastest possible critical damping for the controlled system is achieved.

The main components of the tuner include a response recognition system and an embedded neural network. The response recognition system is used to monitor the controlled response and gain extract knowledge about the performance of the current controller gain setting. The network will then suggest suitable changes to be made to the controller gains.

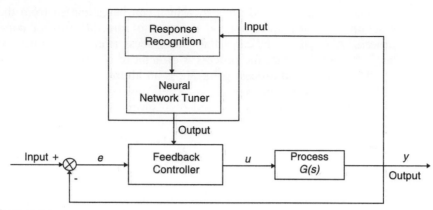

FIGURE I Neural network tuner structure.

A. Input–Output Space

Stable responses can be broadly divided into two categories: monotonic and oscillatory. Monotonic responses can be dealt with very easily. The required tuning action is either to increase K_p or to decrease T_i, or to take both actions. Therefore, in the implementation of the intelligent tuner, the problem of a monotonic response is solved by using an IF–THEN statement that increases K_p by 10% and decreases T_i by 10% simultaneously. But for oscillatory responses, the required tuning knowledge is not as simple. To capture the knowledge and represent it by using a neural network, the input and output of the neural network must be defined first.

Because a neural network is a numerical estimator, its inputs and outputs must be numeric. Therefore, the response has to be quantified before feeding into the neural network. Four variables, which can provide sufficient information regarding the shape and size of oscillatory responses, are chosen. These variables are peak rise time (R), overshoot (O), peak-to-peak height (H), and final error (E), and are depicted in Fig. 2a and b. Figure 2a shows the case where the first peak is above the set point, and Fig. 2b shows the case where the first peak is below the set point. However, the same set of variables can describe and differentiate them adequately. Because the tuner is intended to be used for a wide variety of processes, the actual values of these variables cannot be used directly—they must first be normalized. The normalized equations are given as follows:

$$R_n = \frac{R}{\tau_d}$$

$$O_n = \frac{O - SP}{SP}$$

$$(4)$$

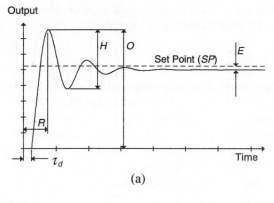

(a)

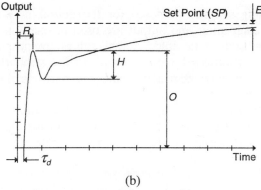

(b)

FIGURE 2 (a) Oscillatory response with overshoot above set point. (b) Oscillatory response with overshoot below set point. Reprinted from *Computers in Industry*, K. C. Chan, Development of a feedback controller tuner using virtual fuzzy sets, 28:219–232, © 1996, with permission from Elsevier Science.

$$H_n = \frac{H/2}{SP}$$

$$E_n = \frac{E}{SP},$$

where R_n, O_n, H_n, and E_n are the normalized rise time, normalized overshoot, normalized peak-to-peak height, and normalized final error, respectively. SP is the set-point value. These normalized values are used as inputs to the neural network tuner.

The purpose of the neural network tuner is to fine-tune the feedback controller based on the closed-loop step response performance. Rather than suggesting exact controller gains (as in the case of initial gain setting), the outputs of the neural network are multiplying factors of the current gains in logarithmic scale, $\log(mK_p)$ and $\log(mT_i)$.

For a proportional only feedback controller, the input to the neural network only includes R_n and H_n, because if the goal is to achieve the fastest critically damped response, the final error and overshoot are uncontrollable because of the limitations of pure proportional control. However, for PI, PD, and PID controllers, all four variables are required.

B. Input–Output Scaling

The tuner was implemented as a back-propagation neural network with one or more hidden layers. The sigmoid activation function was used for each node within the network. The cumulative delta rule was used for back-propagation error learning. However, a common cause of problems stems from presenting data to a back-propagation network as raw values, rather than in values that have been suitably scaled to the transfer function being used. Therefore, it is necessary to scale both the input and output values to suitable ranges.

Before the training samples are used for the neural network training stage, the ranges of each input and output are used to compute the proper scale and offset for each data field. Actual values are then scaled or preprocessed to network ranges for presentation to the network. After the network has produced a network-scaled result, the result is then descaled back to real-world units.

For a sigmoid transfer function, the input and output variables are generally scaled to be within the range $[0, 1]$ or a smaller range, such as $[0.2, 0.8]$. The four input variables are scaled to the range $[0, 1]$, and the following scaling equation is used:

$$\beta_n = \frac{\beta - \beta_{\min}}{\beta_{\max} - \beta_{\min}}, \tag{5}$$

where β can be any one of the input variables in the unscaled domain; β_n can be any one of the input variables in the scaled domain; $\beta_{\min}$ is the minimum value of β in the unscaled domain; and $\beta_{\max}$ is the maximum value of β in the unscaled domain.

The output variables are scaled to the range $[0.2, 0.8]$, and the following scaling equation is used:

$$\gamma_n = 0.6 \frac{\gamma - \gamma_{\min}}{\gamma_{\max} - \gamma_{\min}} + 0.2, \tag{6}$$

where γ can be any one of the output variables in the unscaled domain, γ_n can be any one of the output variables in the scaled domain, $\gamma_{\min}$ is the minimum value of γ in the unscaled domain, and $\gamma_{\max}$ is the maximum value of γ in the unscaled domain.

C. Learning Strategy

A rule of thumb for building neural networks is, if possible, to divide the problem into smaller parts, using a separate smaller neural network to handle each small domain, rather than to use a very large network to handle a very large problem. Therefore, different neural networks are employed for different categories of processes and feedback algorithms.

The neural network system requires sufficient representative numerical training samples for its learning cycles. The plants employed for knowledge extraction for different process groups are shown in Table 1. For stable processes, the plant used to obtain the training data set is a first-order plant plus a pure time delay. For unstable processes, the plant used to obtain the training data set is a process with integration plus a pure time delay.

To obtain tuning knowledge for different process groups, the pure time delay is adjusted so that different normalized dead times, θ_1 and θ_2, can be achieved. It is necessary to obtain a set of samples that can well represent the whole process group. Therefore, the process with a normalized dead time equal to the average value of the range is chosen. For example, $\theta_1 = 0.375$ is chosen for case 1, where $0.15 < \theta_1 < 0.6$, and $\theta_1 = 0.08$ is chosen for case 2, where $\theta_1 < 0.15$, as shown in Table 1. An alternative is to obtain training samples from a wide range of processes within the same process group and to leave the whole lot of samples to the neural network for generalization. This approach has not been chosen, because of the large increased number of samples and, thus, the much longer required training time.

As mentioned earlier, the aim of the intelligent tuner is to obtain the fastest possible critically damped response. For each of the processes used for training data set extraction, the optimal gain setting for fastest critical damping is found by exhaustive search throughout the prespecified gain ranges.

For complete PID feedback control, the optimal parameters are denoted as $\tilde{K}_p$, $\tilde{T}_i$, and $\tilde{T}_d$, for proportional gain, integral time, and derivative time, respectively. They are expressed in vector form as:

$$\vec{K}_{opt} = \begin{bmatrix} \tilde{K}_p \\ \tilde{T}_i \\ \tilde{T}_d \end{bmatrix}. \tag{7}$$

To obtain training samples, a large number of combinations of parameters are used, expressed as a vector $\vec{K}$:

$$\vec{K} = \begin{bmatrix} K_p \\ T_i \\ T_d \end{bmatrix}. \tag{8}$$

TABLE I   Samples for Neural Network Training

Open-loop characterization	Recommended feedback control action	Sample plant used to obtain training date set
1. $(0.15 < \theta_1 < 0.6; 2.25 < K_1 < 15)$	PI or PID	$G(s) = e^{-0.375s}/(1 + s)$; $\theta_1 = 0.375$
2. $(\theta_1 < 0.15; K_1 > 15)$	P or PI or PID	$G(s) = e^{-0.08s}/(1 + s)$; $\theta_1 = 0.08$
$(\theta_2 > 0.3; K_2 < 2)$		$G(s) = e^{-s}/s(1 + s)$; $\theta_2 = 1$
3. $(\theta_2 < 0.3; K_2 > 2)$	PD	$G(s) = e^{-0.15s}/s(1 + s)$; $\theta_2 = 0.15$

For the vector $\vec{K}$, the corresponding closed-loop step output is expressed in terms of four variables, R, O, H, E. The normalized values: R_n, O_n, H_n, E_n are then expressed as a vector $\vec{n}$:

$$\vec{n} = \begin{bmatrix} R_n \\ O_n \\ H_n \\ E_n \end{bmatrix}. \tag{9}$$

We can obtain $\vec{K}$ and $\vec{n}$ for each gain combination. Because we also know the optimal gain $\vec{K}_{opt}$, we can constitute a training sample by comparing $\vec{K}$ with $\vec{K}_{opt}$. The input to the neural network is $\vec{n}$, and the output from the neural network is expressed as $\vec{K}_m$, as shown in Eq. (10), where $\vec{K}_m$ is a vector of multiplying factors to the current controller gains in log scale:

$$\vec{K}_m = \begin{bmatrix} \log(mK_p) \\ \log(mT_i) \\ \log(mT_d) \end{bmatrix} = \begin{bmatrix} \log\left(\dfrac{\tilde{K}_p}{K_p}\right) \\ \log\left(\dfrac{\tilde{T}_i}{T_i}\right) \\ \log\left(\dfrac{\tilde{T}_d}{T_d}\right) \end{bmatrix}. \tag{10}$$

The training procedure is to estimate the relationship between $\vec{n}$ and $\vec{K}_m$. After the training is done and a satisfactory performance is achieved, the weights of the neural network are fixed and the network is ready to be used. During the use phase, the gain settings are adjusted iteratively, using the following equation:

$$\vec{K}_{new} = \begin{bmatrix} K_p^{new} \\ T_i^{new} \\ T_d^{new} \end{bmatrix} = \begin{bmatrix} K_p^{old}(mK_p) \\ T_i^{old}(mT_i) \\ T_d^{old}(mT_d) \end{bmatrix}. \tag{11}$$

D. RMS Error

The root mean square (RMS) error is a valuable and common measure of the performance of a neural network during training. The RMS error is given as

$$\text{RMS} = \sqrt{\frac{1}{n_{PE}} \sum_{i=1}^{n_{PE}} (\text{Sample}_i - N_i)^2}, \tag{12}$$

where Sample_i is the actual output value of the training sample of the ith processing element in the output layer, PE_i; N_i is the neural network output value of PE_i; and n_{PE} is the number of PEs in the output layer. This equation is applicable to cases where the epoch size is equal to one. For cases where epoch size is not equal to one, the following equation is used to calculate the RMS:

$$\text{RMS} = \sqrt{\frac{1}{n_{\text{E}} n_{\text{PE}}} \sum_{i=1}^{n_{\text{E}}} \sum_{j=1}^{n_{\text{PE}}} \left(\text{Sample}_{ij} - N_{ij}\right)^2}, \tag{13}$$

where n_{E} is the epoch size; Sample_{ij} and N_{ij} are the actual and neural network output values of PE_j of the ith pattern in the epoch group, respectively.

Because the purpose of the neural network is to capture the appropriate tuning actions, expressed as changes in the controller gains, the ability of the network to suggest right correction directions (increase/decrease) as well as levels is very important. The RMS value is a good indicator of how well the network generalizes the input–output relationship, but it does not take correction directions into account. Therefore, in addition to using the RMS for performance evaluation, percentage of correct directions of both K_p and T_i gains will also be used for the evaluation of neural networks.

E. Results and Discussion

The implementation of neural network tuners for PI control is described in this section. The prime application area of the neural network PI controller tuner is where the normalized dead time is within the range $0.15 < \theta_1 < 0.6$ (case 1 in Table 1). Therefore, this case is used as the first example. To extract the tuning knowledge, a representative set of training data must first be obtained from a representative process. The following first-order process with a pure time delay, where the normalized dead time is equal to the average value of the range ($\theta_1 = 0.375$), has been chosen:

$$G(s) = \frac{e^{-0.375s}}{1 + s}. \tag{14}$$

The fastest critical damping can be achieved roughly at $\tilde{K}_p = 1.1$ and $\tilde{T}_1 = 1.0$. It should be noted that very accurate values for K_p and T_i are not essential because the ultimate goal is to extract knowledge for tuning controllers iteratively for a wide range of processes, but not to obtain exact gain values for a certain process.

The training set should span the entire space of the mapping to enable the neural network to generalize properly. Therefore, a wide range of gain combinations is applied to the process to obtain training samples. The range for both K_p and T_i is set to $[0.1, 50]$, and the multiplying increment is set to 1.1. There are $65^2 = 4225$ gain combinations in total. Among these combinations, 1187 (K_p, T_i) pairs produce oscillatory responses, whereas the other combinations produce either monotonic or unstable responses. During the generation of training samples, the four normalized parameters (R_n, O_n, H_n, E_n) of the oscillatory responses and the corresponding multiplying factors

$(\log(\tilde{K}_p/K_p), \log(\tilde{T}_i/T_i))$ that will bring the current controller gains to their optimal values are recorded, respectively, as input vectors and output vectors to the neural network. In this experiment, the samples are divided into two sets: one learning set and one testing set. The 1187 samples are divided into a learning set with 594 samples and a testing set with 593 samples.

Table 2 specifies the minimum and maximum values for the input and output values presented to the networks during the training procedure. Actual values are scaled to the network ranges for presentation to the network. After the network has produced a network-scaled result, the result is then descaled back to real-world units. The scaling equations for the inputs and outputs are given in Eqs. (5) and (6), respectively.

The neural networks used in this study consist of an input layer that has four nodes (for the four inputs) and an output layer that has two nodes (for the two outputs). The determination of the number of hidden layers and number of nodes in each layer for satisfactory network performance is a trial-and-error process. It was shown in the literature that with a sufficiently large number of nodes in a single middle layer, any arbitrary continuous mapping can be approximated to any required degree of accuracy by a neural network. However, it is a common belief that neural networks with several hidden layers can approximate a function more efficiently (with fewer nodes) for a given accuracy than networks with a single hidden layer. Therefore, in additional to networks with a single hidden layer, a few double hidden layer and triple hidden layer neural networks will also be considered and evaluated.

The sigmoid activation function is used for each node within the network. Initially, all neural networks are untrained and all of the connection weights are randomized in the range $[-0.1, 0.1]$. The cumulative delta rule is used for back-propagation error learning. The parameters used are learning rate $\eta = 0.3$; momentum $\alpha = 0.4$; and epoch size = 20. All networks are trained for a total of 200,000 epochs (iterations). A total of 200,000 epochs has been chosen because the RMS error of the networks usually would have reached a steady state after such a long training period with no apparent improvement. The training results are summarized in Table 3. The average RMS errors and CPU times (486 PC, 33 MHz with 16 MB of RAM) are collected after the first 10,000 epochs and the complete 200,000 epochs. The results in terms of percentage of correct directions are summarized in Table 4.

TABLE 2   Input–Output Scaling

Parameter	Min value	Max value	Normalized scaled ranges
Normalized rise time	2.4	41	[0.0, 1.0]
Normalized overshoot	-0.49	0.96	[0.0, 1.0]
Normalized peak-to-peak height	3e $-$ 6	0.89	[0.0, 1.0]
Normalized final error	-0.44	0.055	[0.0, 1.0]
$\log(mK_p)$	-0.61	1	[0.2, 0.8]
$\log(mT_i)$	-1.7	1	[0.2, 0.8]

■ **TABLE 3** Training Results of Various Neural Network Structures

ANN structure	Average RMS after 10,000 epochs	CPU time of 10,000 epochs (min)	Average RMS after 200,000 epochs	CPU time of 200,000 epochs (h)
4-6-2	0.039	16.8	0.0249	5.6
4-8-2	0.032	19.6	0.0195	6.5
4-10-2	0.036	20.4	0.0144	6.8
4-12-2	0.037	22.0	0.0158	7.3
4-14-2	0.035	23.5	0.0135	7.8
4-16-2	0.044	24.9	0.0174	8.3
4-4-2-2	0.036	20.2	0.0266	6.7
4-6-4-2	0.037	23.4	0.0130	7.8
4-8-6-2	0.030	26.0	0.0125	8.7
4-10-8-2	0.024	30.8	0.0103	10.3
4-12-10-2	0.025	34.3	0.0086	11.4
4-4-4-2-2	0.057	25.0	0.0205	8.3
4-6-6-4-2	0.043	30.3	0.0116	10.1
4-8-8-6-2	0.038	37.0	0.0081	12.3
4-10-10-8-2	0.040	42.4	0.0076	14.13
4-12-12-10-2	0.037	51.3	0.0099	17.1

Among the six single hidden layer networks, the 4-8-2 structure reaches the lowest average RMS error (0.032) after the first 10,000 epochs, followed by the 4-10-2, 4-12-2, and 4-14-2 neural networks. At the two ends, the 4-6-2 and 4-16-2 structures have higher RMS values in the early training stage. After 200,000 epochs, the 4-10-2, 4-12-2, and 4-14-2 neural network reach lower average RMS errors than the other structures; their RMS errors are 0.0144, 0.0158, and 0.0135, respectively. It can be observed that the 4-8-2 network converges a bit faster in the early stage, but because of its limited number of PEs, its performance or generalization capability does not improve as much as the 4-10-2, 4-12-2, or 4-14-2 networks do. The 4-10-2, 4-12-2, and 4-14-2 networks may be considered as "better" single-layer networks because they achieve lower average RMS errors, and further increase in the number of PEs does not lead to any further improvement. The RMS error plot for the first 10,000 epochs of the 4-10-2 network is shown in Fig. 3. Considering the correct direction performance measure expressed as a percentage of outputs with correct directions in both mK_p and mT_i, the 4-10-2 structure excels with 98.32% and 98.31% for the training and testing sets, respectively (Table 4).

Among the five double hidden layer networks, it is observed that lower RMS errors are achieved with more hidden nodes after the same number of epochs. It can be seen that all of the double hidden layer networks, except the 4-4-2-2 structure, can achieve lower RMS errors than all of the single hidden layer networks after 200,000 epochs. By comparison, using the direction performance measure, the 4-10-8-2 structure has the best result of 99.49% and 99.33% for the training and testing sets, respectively. Among the five triple hidden layer networks, the 4-10-10-8-2 network achieves the lowest RMS error (0.0076) after 200,000 epochs. This RMS value is also the lowest of all of the single, double, and triple hidden layer networks. The 4-10-10-8-2

TABLE 4   Number of Incorrect Outputs and Performance of Various Neural Network Structures

ANN structure	Learning Set (594)					Testing Set (593)				
	Number of outputs with incorrect					Number of outputs with incorrect				
	mK_p	mT_i	mK_p and mT_i	Total	Perform-ance (%)	mK_p	mT_i	mK_p and mT_i	Total	Perform-ance (%)
4-6-2	14	8	2	24	95.62	23	11	4	38	93.59
4-8-2	23	6	0	29	95.12	20	3	0	23	96.12
4-10-2	7	3	0	10	98.32	5	5	0	10	98.31
4-12-2	7	4	0	11	98.15	5	7	0	12	97.98
4-14-2	7	4	0	11	98.15	5	7	0	12	97.98
4-16-2	16	10	0	26	95.62	16	10	1	27	95.45
4-4-2-2	8	10	0	18	96.97	8	8	0	16	97.30
4-6-4-2	17	3	0	20	96.63	15	3	0	18	96.96
4-8-6-2	7	1	0	8	98.65	6	7	0	13	97.80
4-10-8-2	1	2	0	3	99.49	2	2	0	4	99.33
4-12-10-2	7	0	0	7	98.82	4	2	0	6	98.99
4-4-4-2-2	18	9	0	27	95.45	15	10	0	25	95.78
4-6-6-4-2	3	4	0	7	98.82	3	6	0	9	98.48
4-8-8-6-2	5	3	0	8	98.65	2	3	0	5	99.16
4-10-10-8-2	3	0	0	3	99.49	2	2	0	4	99.32
4-12-12-10-2	14	0	0	14	97.64	13	0	0	13	97.80

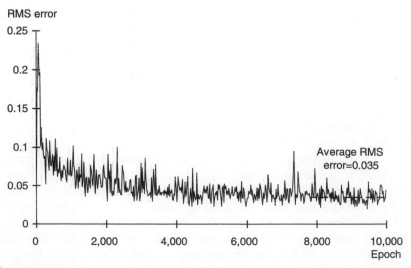

FIGURE 3   Training result of the 4-10-2 neural network, using the following parameters: $\eta = 0.3$, $\alpha = 0.4$, epoch size = 20.

network also achieves the highest direction performance measure with 99.49% and 99.33% for the training and testing sets, respectively. However, these values are just the same as those achieved by the 4-10-8-2 network. The RMS error plot of the 4-10-10-8-2 network for the first 10,000 epochs is shown in Fig. 4. A triple hidden layer network usually requires more epochs before the RMS starts to drop significantly. This phenomenon is seen for all triple hidden layer networks. In general, with comparable hidden nodes in the first hidden layer, such as 4-10-2, 4-10-8-2, and 4-10-10-8-2, the triple-layer network performs better than the double, and the double performs better than the single, after 200,000 epochs, but at the expense of longer training time.

Figure 5 shows the training samples and the trained neural network outputs of the 4-10-10-8-2 network. Similar figures are also obtained for the 4-10-2 and 4-10-8-2 networks. These figures show that the output space basically covers the first three quadrants. The samples in the first quadrant indicate increasing K_p and T_i. The samples in the second quadrant indicate decreasing K_p and increasing T_i, and those in the third quadrant indicate decreasing K_p and T_i. The fourth quadrant is almost empty with only very few samples; this agrees with the fact that increasing K_p and decreasing T_i at the same time would normally increase the oscillation, and this is not the usual action to take for oscillatory responses. The 4-10-2 network can generally suggest the right action, i.e., increasing and/or decreasing K_p and/or T_i, but not to a high accuracy. The performance is particularly worse when $\log(mK_p)$ is very large. However, because the tuning process is iterative, it is acceptable to have a slight improvement after every iteration. The performance of this network is considered satisfactory because it usually suggests the right action, and this guarantees reasonable improvement. The performance of the 4-10-8-2 improves slightly, especially when $\log(mK_p)$ is very small or very large. However, the 4-10-10-8-2 network has the best performance. As shown in Fig. 5, the crosses and dots are very close and

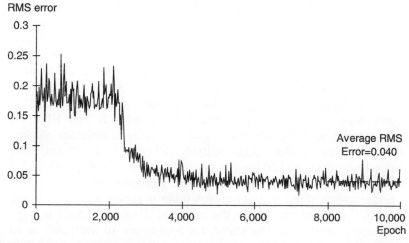

FIGURE 4 Training result of the 4-10-10-8-2 neural network, using the following parameters: $\eta = 0.3$, $\alpha = 0.4$, epoch size = 20.

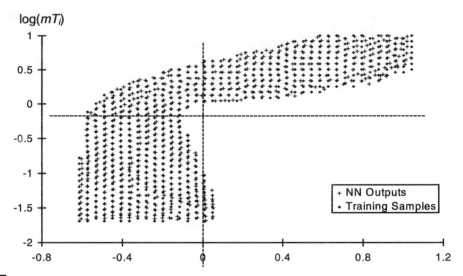

FIGURE 5 Plot of the training samples and the trained neural network outputs (4-10-10-8-2 structure).

sometimes even overlap. This means that the 4-10-10-8-2 network has the best accuracy. Based on the above observations, the 4-10-10-8-2 network is chosen for the PI controller tuner.

Although successful results have been achieved in training the network and reproducing sample outputs (note that the training samples are only obtained from a representative plant), it is necessary to test the network on the family of plants to guarantee its success. In the following, the results from a number of processes are presented. This neural network PI controller tuner is implemented for plants with normalized dead time (θ_1) ranging from 0.15 to 0.6. Two examples representing the two extrema of the range (processes I and II) are tested to ensure that the neural network tuner is applicable to the whole process family. Two additional experiments on non-minimum-phase processes (processes III and IV) are also performed. The characteristics of these four processes are summarized in Table 5.

The neural network tuner has been successful in tuning all four of these processes. However, only graphs for processes II and III are given. The trajectories of the proportional gain and integral time are shown in Figs. 6a and 7a. The results of the initially (Ziegler–Nichols) and finally (neural network) tuned closed-loop systems are shown in Figs. 6b and 7b for processes II and III, respectively. These examples demonstrate the excellent performance of the neural network tuner in tuning a digitally implemented PI controller for plants where the standard Ziegler–Nichols tuning rules yield excessive overshoot. The successful application of the tuner in tuning these four plants also indicates that the tuner is not limited to any specific plant; rather, it can be applied to plants with various process dynamics.

TABLE 5 Process Characterization of the Four Processes Tested

	Process I $\dfrac{Y(s)}{U(s)} = \dfrac{5e^{-0.3s}}{2s+1}$	Process II $\dfrac{Y(s)}{U(s)} = \dfrac{0.5e^{-0.18s}}{0.3s+1}$	Process III $\dfrac{Y(s)}{U(s)} = \dfrac{(1-0.15s)}{(1+s)^3}$	Process IV $\dfrac{Y(s)}{U(s)} = \dfrac{(1-0.6s)}{(1+s)^3}$
Dead time (τ_d)	0.3	0.18	0.95	1.44
Apparant time constant (τ)	2.00	0.3	3.7	3.50
Process gain (K)	5	0.5	1	1
Normalized dead time (θ_1)	0.15	0.6	0.26	0.41
Normalized process gain (K_1)	19.2	2.2	5.7	3.12

IV. INTEGRATED FUZZY-NEURAL NETWORK TUNER

The main advantage of using neural networks is that tuning knowledge can be captured automatically from training samples generated by a representative process for a particular process group. Therefore, unlike the expert system approach, the neural network approach does not require a human expert to provide tuning knowledge, which is generally expressed as linguistic IF–THEN rules.

However, with neural networks, the knowledge is represented numerically as weights, and therefore, it is very difficult, if not impossible, to understand. In real situations, a human expert is always needed to modify and fine-tune the system. But because of the incompatibility between a neural network and linguistic rules, it is impossible for a control expert to modify or incorporate human knowledge into the system.

Fuzzy systems have been an effective approach to the use of linguistic rules, but naturally, they cannot handle numerical data. The motivation in this section is to investigate the possibility of building a common framework that can accommodate both numerical data and linguistic rules through the integration of a neural network and a fuzzy system, so that experimental data and human experience can be combined in developing intelligent systems. The integration scheme will be presented, followed by a study of its application in PI controller tuning.

A. A Fuzzy Rule Generation Scheme Using Neural Networks

A fuzzification scheme and an offline fuzzy rule generation scheme are proposed in this section, employing the following algorithm for developing a neural-fuzzy inference system.

Step 1. Quantize the Input and Output Space Using Fuzzy Sets

Specify the domain intervals in which the values of variables are likely to lie. Then quantize the domain of each variable by using a number of fuzzy

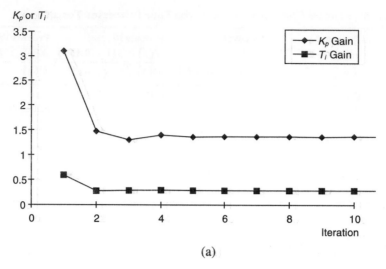

(a)

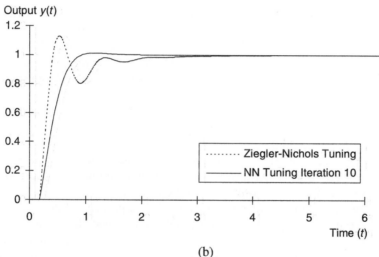

(b)

FIGURE 6 (a) Trajectories of K_p and T_i for process II. (b) Performance of Ziegler–Nichols and neural network tuned closed-loop systems for process II.

subsets. The number of fuzzy sets may be the same or different for different variables. Each fuzzy set is assigned a triangular fuzzy membership function.

Step 2. Fuzzify Input/Output Vectors for Neural Network Training

During the training stage, a fuzzifier is needed to convert both the input and output data from absolute measurements to fuzzy data. These fuzzified data can then be used to train the neural network (Fig. 8a).

The operation of the fuzzifier is simple. Fuzzification is performed by using the central values of fuzzy sets, as proposed in [11]. To explain the fuzzification scheme, consider a variable x that is quantized into $(N + 1)$ fuzzy regions by using $(N + 1)$ fuzzy subsets or membership functions,

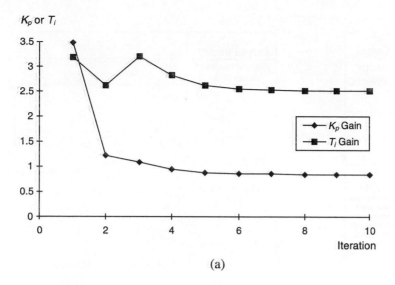

(a)

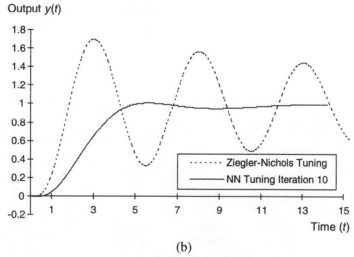

(b)

FIGURE 7 (a) Trajectories of K_p and T_i for process III. (b) Performance of Ziegler–Nichols and neural network tuned closed-loop systems for process III.

$(S_0, S_1, \ldots, S_N)$. These fuzzy subsets overlap significantly, and their intersections are specified as $(L_0, L_1, \ldots, L_{N-1})$. The input to the fuzzifier can be any x value in the continuous x domain. The fuzzifier changes the input to one of the central values of the $(N + 1)$ fuzzy subsets, $(C_0, C_1, \ldots, C_N)$. The fuzzification scheme is as follows:

IF $(x < L_0)$, then output is C_0;
ELSE IF $(L_{n-1} \leq x < L_n, 1 \leq n \leq N - 1)$, THEN output is C_n;
ELSE IF $(x \geq L_{N-1})$, THEN output is C_N.

The same fuzzification scheme is applied to all input and output variables. A large range of network architectures may be tested by using the fuzzified

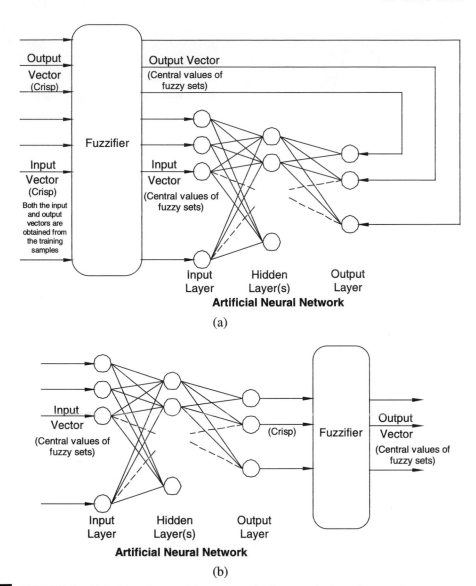

FIGURE 8 (a) Training of a neural-fuzzy system. (b) Fuzzy production rule generation.

input–output data until a satisfactory result is achieved. After sufficient training, the neural network can then be used to generate fuzzy IF–THEN rules.

Step 3. Generate a Fuzzy Rule Base (Offline)

This is an offline procedure, and the rules generated will be stored in a rule base for later use. Ideally, fuzzy rules should be generated for all input combinations so that the entire input space can be covered. We assume that there are m inputs and each input is quantized by n fuzzy subsets. The total number of input combinations is m^n. Therefore, m^n fuzzy production rules should be generated. It is dangerous if not all of the rules are generated, because the fuzzy inference system may not be able to determine an output when an input falls into those missing fuzzy regions or rules. During the rule generation process, the input vectors contain the central values of the corresponding fuzzy subsets, and the output vectors from the neural network contain crisp values. Therefore, the last step is to fuzzify the output vectors by using the same fuzzification scheme (Fig. 8b).

Although it is ideal to generate all of the rules to cover the whole input space, this may not be possible or practical when the computer resource is limited and/or the number of rules is too large. In these situations, it is quite acceptable to generate rules covering only the learning samples. This can be accomplished by generating rules based on the antecedent (premise) parts of all of the samples; however, sufficient training samples should be provided so that most of the accessible input space is covered.

Step 4. Combine Expert Linguistic Rules with Neural Network Generated Fuzzy Rules (Optional)

This step is necessary only if linguistic information is available from a human expert. The human expert can modify and/or delete fuzzy rules generated automatically by the neural network.

Step 5. Assign Weights to All of the Rules (Optional)

This step is optional. Weight assignment is a mechanism for providing final human expert adjustment to the rule base that is either automatically generated from sample data or assigned by a human expert. This step may be performed when *a priori* information about the data set (such as accuracy, reliability, and usefulness) is available. The assigned weight actually represents the belief of the usefulness of a fuzzy production rule. By default, the weight of an automatically generated rule is equal to unity.

Step 6. Fuzzy Inference and Defuzzification

After the rule base is obtained from the given data and the human expert, the fuzzy system is ready to determine a mapping from the input to the output. The correlation product fuzzy inference [12] is used to compute the output fuzzy subsets, to weight them with the user-assigned weights (if any) and to sum them to produce the output fuzzy subsets. The output fuzzy subsets are then defuzzified to obtain a numerical output by using the fuzzy centroid defuzzification scheme.

To illustrate the fuzzy inference and defuzzification procedure, we use the following standard rule representation:

$$\text{For rule } i: (in1_i, in2_i; out_i; w_i),$$

which is equivalent to

$$\text{IF } x_1 = in1_i \text{ AND } x_2 = in2_i \text{ THEN } y = out_i, \text{weight} = w_i.$$

$in1_i$ and $in2_i$ are the antecedent fuzzy subsets; out_i is the consequent fuzzy subset. The input pair (x_1, x_2) activates the antecedent fuzzy subsets $in1_i$ and

$in2_i$ to degrees $m_{in1_i}(x_1)$ and $m_{in2_i}(x_2)$, respectively. The activation value of the consequent part of Rule i is equal to the minimum of the antecedents' conjunction value. In our example, the activation value a_1 is determined as follows:

$$a_i = \min\left(m_{in1_i}(x_1), m_{in2_i}(x_2)\right). \tag{15}$$

The input pair (x_1, x_2) activates the consequent part of each fuzzy rule to a different degree a_i. With correlation product encoding inference, the ith rule yields the weighted output fuzzy subset O_i:

$$m_{o_i}(y) = a_i w_i m_{out_i}(y), \tag{16}$$

or,

$$O_i = a_i w_i out_i. \tag{17}$$

The fuzzy system then sums the output fuzzy subset O_i from each FAM rule to form the combined output fuzzy subset O:

$$m_o(y) = \sum_{i=1}^{N} m_{o_i}(y), \tag{18}$$

or,

$$O = \sum_{i=1}^{N} O_i, \tag{19}$$

where N is the number of fuzzy rules in the FAM bank.

The control output u is equal to the fuzzy centroid of O:

$$u = \frac{\int y m_o(y)\, dy}{\int m_o(y)\, dy}. \tag{20}$$

To reduce computation, we can use the following formula to determine the output u for triangular membership functions:

$$u = \frac{\sum_{i=1}^{N} \overline{out_i} a_i w_i}{\sum_{i=1}^{N} a_i w_i}, \tag{21}$$

where $\overline{out_i}$ denotes the central value of the fuzzy subset out_i.

B. Fuzzy-Neural Network Tuner

To illustrate the application of the integrated neural-fuzzy algorithm to PI controller tuning, we use the same process group (case 1 in Table 1) as in the PI neural network tuners described in Section III.E. The same set of training data has been extracted from the representative process (Eq. 14) and is used during the training of the neural networks. Before the data can be fed into the neural networks for training, it is necessary to define the fuzzy member-

TABLE 6   Fuzzified Input–Output Scaling

Parameter	(Center) Min value	(Center) Max value	Normalized scaled ranges
Normalized rise time	3	30	[0.0, 1.0]
Normalized overshoot	−0.4	0.6	[0.0, 1.0]
Normalized peak-to-peak height	0	0.5	[0.0, 1.0]
Normalized final error	−0.24	0.06	[0.0, 1.0]
$\log(mK_p)$	−0.4	0.6	[0.2, 0.8]
$\log(mT_i)$	−0.4	0.6	[0.2, 0.8]

ship functions for all of the input–output variables, as described in step 1 above, and to fuzzify the data by using the fuzzification algorithm described in step 2 of the scheme.

Triangular fuzzy membership functions can simplify computation and have been chosen for all of the input–output variables. The number of fuzzy subsets is a trade-off between accuracy and computer resources. The larger the number of fuzzy subsets, generally the better the accuracy, but it will also require more computer resources for rule storage. Eleven equal-width triangular fuzzy membership functions have been selected to specify each variable. The variable ranges are shown in Table 6. It should be noted that the ranges specified are smaller than that of the actual training samples (ranges of the actual samples are shown in Table 2). By reducing the range size, narrower fuzzy sets can be obtained, and thus finer control can be achieved around the near-optimal region.

Table 6 specifies the minimum and maximum values for the fuzzified input and output values presented to the networks during the training procedure. Actual input values are scaled and output descaled to the ranges specified in the table. The scaling equations for the inputs and outputs are given in Eqs. 5 and 6, respectively.

C. Results and Discussion

A number of single, double, and triple hidden layer back-propagated neural networks are used in this experiment. As in the case of the previous neural PI tuner, the sigmoid function, the cumulative delta rule, and the same parameters are used. The training results of various neural network structures are shown in Table 7.

As can be seen in the table, all of the neural networks give similar results. From the simplest network, 4-2-2, to the largest network, 4-12-12-10-2, the average RMS errors after 10,000 epochs are in the 0.0738–0.0979 range. The average RMS errors after 200,000 epochs are in the 0.0654–0.0815 range. Increasing the number of hidden layers or nodes does not yield much improvement in accuracy. This shows that the fuzzified input–output relationship is actually a quite simple one. The RMS error plot of the 4-12-10-2 structure for the entire 200,000 epochs is shown in Fig. 9. The error plots for all of these networks are very similar. These plots show that the convergence

TABLE 7   Training Results of Various Neural Network Structures

ANN structure	Average RMS after 10,000 epochs	CPU time of 10,000 epochs (min)	Average RMS after 200,000 epochs	CPU time of 200,000 epochs (h)
4-2-2	0.0823	12.4	0.0815	4.13
4-4-2	0.0776	14.5	0.0696	4.83
4-6-2	0.0769	16.8	0.0706	5.6
4-8-2	0.0738	19.6	0.0722	6.5
4-10-2	0.0850	20.4	0.0698	6.8
4-12-2	0.0862	22.0	0.0715	7.3
4-8-6-2	0.0760	26.0	0.0688	8.7
4-10-8-2	0.0748	30.8	0.0680	10.3
4-12-10-2	0.0752	34.3	0.0676	11.4
4-8-8-6-2	0.0979	37.0	0.0673	12.3
4-10-10-8-2	0.0786	42.4	0.0654	14.13
4-12-12-10-2	0.0791	51.3	0.0655	17.1

rates are very fast, and there are only slight improvements over a long period of training time.

Among the six single hidden layer neural networks, the 4-4-2 network yields the best RMS error (0.0696) after 200,000 epochs. The double and triple hidden layer networks only improve the RMS errors marginally, at the expense of much longer training time. In the pure neural network approach described in the last section, the level of RMS error was about 10 times lower, for example, 0.0076 for the 4-10-10-8-2 structure. Although the RMS errors are much higher in this case, we cannot conclude that the neural-fuzzy

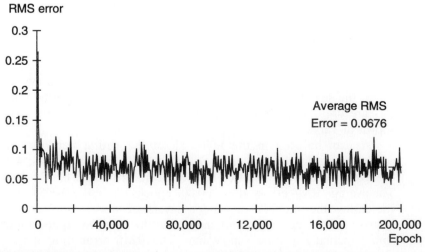

FIGURE 9   Training result of the 4-12-10-2 neural network for the fuzzified input/output data, using the following parameters: $\eta = 0.3$, $\alpha = 0.4$, epoch size = 20.

approach is much worse. The higher level of RMS error is not surprising because conflicting rules may exist in the training sample. However, if the neural-fuzzy system can extract the most accurate rules and eliminate the less accurate rules during training, the resulted tuning system should also be able to perform very well.

To compare using the correct direction performance measure, similar results are again obtained for different network structures (Table 8). For the training set, the best performance (99.66%) is achieved by both the 4-4-2 and 4-6-2 neural networks. For the testing set, the best performance (99.66%) is achieved by the 4-6-2 network, and the second best performance (99.33%) is achieved by the 4-4-2 and 4-12-10-2 networks. Therefore, a larger neural network does not necessarily yield better results.

One more performance measure is added in the neural network evaluation. This measure is the percentage of samples that have the same output fuzzy subsets as that produced by the neural network. If we denote this performance percentage as p, then $q = (1 - p)$ is the percentage of conflicting rules. Therefore, a large p would be an indication of good generalization capability. For the single hidden layer neural networks, the largest p is only 41.75% for the learning set, and 39.80% for the testing set, obtained by the 4-8-2 structure. For the double hidden layer neural networks, p improves by about 10%. The best performance, 50.34% for the learning set and 49.07% for the testing set, is obtained by the 4-12-10-2 network structure. For the triple hidden layer networks, however, the performance is about the same as

TABLE 8 Number of Incorrect Direction Outputs and Performance of Various Neural Network Structures

| ANN structure | Learning Set (594) | | | | | Testing Set (593) | | | | |
| | Number of outputs with incorrect direction | | | | | Number of outputs with incorrect direction | | | | |
	mK_p	mT_i	mK_p and mT_i	Total	Performance (%)	mK_p	mT_i	mK_p and mT_i	Total	Performance (%)
4-2-2	3	21	0	24	95.96	4	23	0	27	95.45
4-4-2	1	1	0	2	99.66	1	3	0	4	99.33
4-6-2	1	1	0	2	99.66	1	1	0	2	99.66
4-8-2	2	21	1	24	95.96	2	28	0	30	94.94
4-10-2	1	22	0	23	96.13	1	24	0	25	95.78
4-12-2	2	23	0	25	95.79	2	27	0	29	95.11
4-8-6-2	2	23	0	25	95.79	2	27	0	29	95.11
4-10-8-2	2	23	0	25	95.79	2	25	0	27	95.45
4-12-10-2	2	1	0	3	99.49	2	2	0	4	99.33
4-8-8-6-2	0	21	0	21	96.46	0	24	0	24	95.95
4-10-10-8-2	2	22	0	24	95.96	2	25	0	27	95.45
4-12-12-10-2	2	1	0	3	99.49	2	3	0	5	99.16

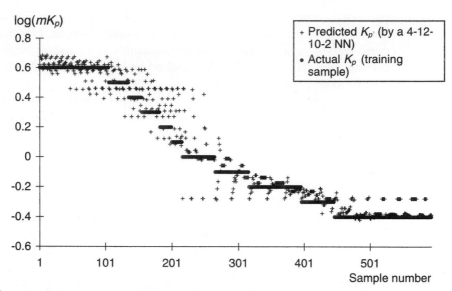

FIGURE 10   Comparison of actual with predicted K_p values by a 4-12-10-2 neural network.

that of the double hidden layer networks, with the highest performance, 50.84% achieved by the 4-12-12-10-2 network.

In addition to the use of the above three performance measures, we can also look at the diagrams comparing the actual and predicted outputs. For the 4-12-10-2 neural network, the actual and neural network predicted K_p and T_i outputs are plotted in Figs. 10 and 11, respectively. It should be noted that the data points shown in these figures are prior to fuzzification. It can be

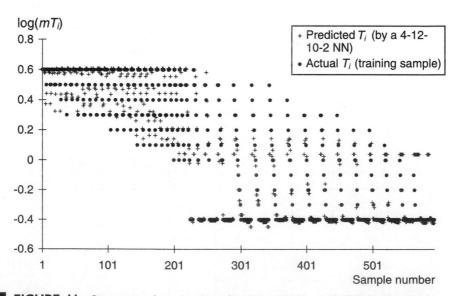

FIGURE 11   Comparison of actual with predicted T_i values by a 4-12-10-2 neural network.

observed that the prediction of T_i is better than that of K_p. Figure 12 shows both the K_p and T_i on the same graph. Most of the predicted outputs concentrate on the bottom left corner and top right corner and are far from the actual outputs. This demonstrates the high percentage of conflicting rules. The fuzzified K_p and T_i outputs are plotted in Figs. 13 and 14, respectively. It can be seen that in these two figures, the neural network can predict T_i better. This may be due to the fact that there are fewer conflicting rules in T_i. In fact, similar graphs are obtained for almost all of the networks tested. Considering all three performance measures, and comparison of the actual and predicted K_p and T_i, the 4-12-10-2 network is chosen for the PI controller tuner.

The last step in evaluating the performance of the neural-fuzzy tuner is to actually test it on different processes. The same processes from the previous examples (Section III.E) are used. The four processes have been specified in Table 5.

The initial gains are set by using the standard Ziegler–Nichols tuning rules. The trajectories of the proportional gain and integral time are shown in Figs. 15a and 16a, respectively. The results of the initially and neural network tuned closed-loop systems are shown in Figs. 15b and 16b for processes II and III, respectively. These examples, representing the extremes of the process group, demonstrate the excellent performance of the neural-fuzzy tuner in tuning PI feedback controllers. Compared with the pure neural network approach, the performance of the neural-fuzzy system is comparable.

Using the integrated neural-fuzzy approach, the input–output relationship is simplified significantly from a large continuous space to a small discrete space. Therefore, a smaller neural network is sufficient to achieve

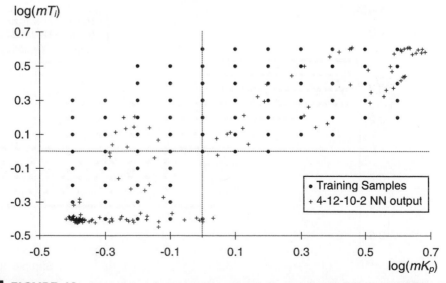

FIGURE 12 Comparison of actual with predicted K_p and T_i values by a 4-12-10-2 neural network.

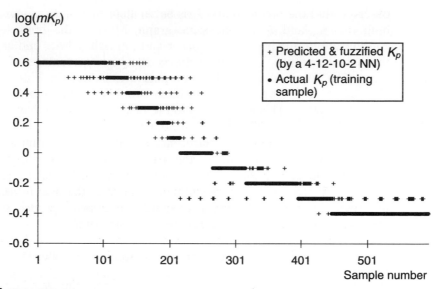

FIGURE 13 Comparison of actual with predicted and fuzzified K_p values by a 4-12-10-2 neural network.

comparable results, and thus shorter development time is required. The integrated approach actually combines all of the advantages of the neural-based and the traditional fuzzy approaches. It represents improvements to the pure neural network approach in the ways that the knowledge is understandable in terms of a list of IF–THEN rules, the knowledge acquisition is

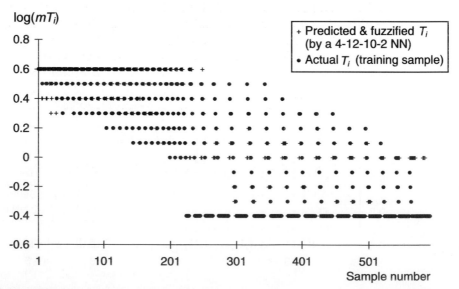

FIGURE 14 Comparison of actual with predicted and fuzzified T_i values by a 4-12-10-2 neural network.

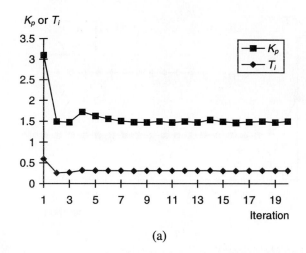

(a)

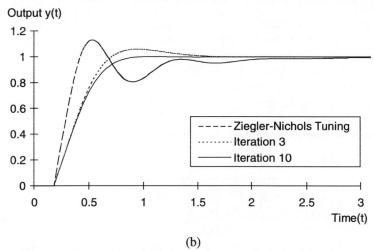

(b)

FIGURE 15 (a) Trajectories of K_p and T_i for process II. (b) Performance of Ziegler–Nichols and neural-fuzzy tuned closed-loop systems for process II.

more flexible (can combine human knowledge with self-organizing knowledge, based on numerical examples), and the knowledge modification is easier (the human expert can actually modify the rules). As shown in the PI tuner example, the performance of a neural-fuzzy system is not much worse than that of a neural network system.

The neural-fuzzy approach has many advantages over the pure neural network approach, but it also inherits a trial-and-error nature in the determination of the most suitable network structure, and the requirement of long training time from the pure neural network approach. Although the training time can be shortened compared to the pure neural network approach, offline training is unavoidable. The scheme involves rule generation using the trained neural network, and the rules generated are stored in the computer for real-time fuzzy inference. However, computer storage may be a problem

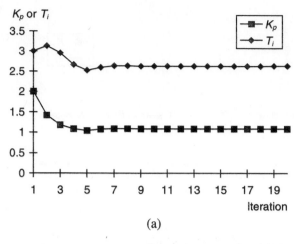

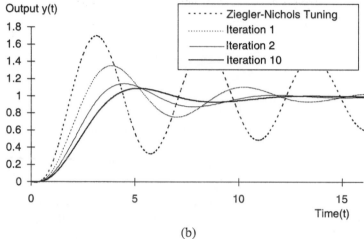

FIGURE 16 (a) Trajectories of K_p and T_i for process III. (b) Performance of Ziegier–Nichols and neural-fuzzy tuned closed-loop systems for process III.

when the number of rules is very large. This neural-fuzzy approach has been applied to PI controller tuning, and very satisfactory results have been achieved.

V. ADAPTIVE VIRTUAL FUZZY TUNER

As discussed in the previous sections, neural networks have been very successful in capturing numerical information, but they are not able to include human knowledge expressed in linguistic rules. On the other hand, fuzzy control has been an effective approach to the use of linguistic rules, but it is not conducive to the use of numerical data. Therefore, the integration of neural networks and fuzzy systems was proposed so that both numerical data and linguistic rules can be used and combined into one framework. Although this approach is feasible, because of the training requirement of a neural network, the knowledge acquisition procedure can be extremely time-consuming.

Because the major bottleneck in the approach lies in the training phase, a logical extension of the development would be to develop a more efficient learning algorithm, replacing the neural network for generating fuzzy rules from numerical data. We have investigated a number of approaches, and the research can be summarized as three main stages:

1. Pure neural network approach: Extract numerical tuning knowledge automatically and tune feedback controllers (Section III).
2. Integrated neural network and fuzzy system approach: Obtain IF–THEN rules through fuzzified numerical input/output data by using neural networks combined with human expert knowledge and fuzzy inference for controller tuning (Section IV).
3. Fuzzy system with a new learning algorithm approach: Extract IF–THEN rules directly from numerical data combined with human expert knowledge and fuzzy inference for controller tuning (this section).

The transition from a pure neural network approach to an integrated neural fuzzy approach is due to the fact that we want to incorporate human knowledge into the system. The transition from an integrated neural fuzzy approach to a fuzzy approach with a new fuzzy learning algorithm is due to the fact that learning is too slow for neural networks, and we want to improve the learning efficiency.

There are two objectives in this work. The first is to develop a new technique to increase the accuracy of fuzzy systems, at a minimum expense of computer resource. The methodology is based on the virtual fuzzy set concept [13]. A virtual fuzzy set, consisting of two consecutive fuzzy sets with different degrees of membership, is an imaginary fuzzy set located at an arbitrary location most appropriate for a rule. By using the virtual fuzzy set representation, the consequent part of a rule is not restricted to those fuzzy sets initially specified by the user, but rather, it can be placed anywhere along the output range. Thus the accuracy of the fuzzy rules can be improved.

The second objective is to overcome the long training requirement of the neural-fuzzy approach and to develop a new methodology for implementing adaptive fuzzy systems based on the virtual fuzzy set concept. The proposed adaptive algorithm is based on the fuzzy rule generation scheme suggested by Wang and Mendel [14]. The approach of Wang and Mendel is a very efficient one-pass build-up procedure. The fuzzy rule base is updated once when a data point is presented during training. Therefore, the time-consuming iterative training, as required in the neural network and the integrated neural-fuzzy approach, is not necessary. The adaptive algorithm generates fuzzy rules automatically from numerical data. In addition, the algorithm also allows users to incorporate human knowledge, in terms of IF–THEN rules, into the fuzzy system. The proposed adaptive algorithm basically follows the steps of the approach of Wang and Mendel, but using the virtual fuzzy sets rather than the traditional fuzzy sets. A feedback controller tuner is then implemented by the adaptive virtual fuzzy algorithm.

A. Concept of a Virtual Fuzzy Set

Consider a single-input single-output mapping $f: x \rightarrow y$, as shown in Fig. 17. Suppose the input space x is quantized using seven identical triangular fuzzy sets $(X0, X1, \ldots, X6)$, and the output space y is quantized into seven similar regions by fuzzy sets $(Y0, Y1, \ldots, Y6)$.

The aim is to create a fuzzy system to estimate the unknown function f: $x \rightarrow y$ according to sample input–output pairs. Numerical samples (x_i, y_i) are required to generate fuzzy production rules. A rule "IF X is X_i, THEN Y is Y_i" is a fuzzy set association (X_i, Y_i). The antecedent term X_i is referred to as the input associate, and the consequent term Y_i as the output associate.

According to the specific example shown in Fig. 17, the training sample (x_i, y_i) activates the fuzzy set association (X_i, Y_i) to a higher degree of membership than the other fuzzy set associations such as $[(X_i, Y_{i-1}), (X_i, Y_{i-2}), \ldots, (X_i, Y_{i+1}), (X_i, Y_{i+2}), \ldots]$. Therefore, the most accurate production rules that can be obtained are $(X_1, Y_1), \ldots, (X_5, Y_5)$. The next step is to use this five-rule fuzzy model to estimate the function $y = f(x)$. According to the five production rules, the best estimate that can be obtained is the straight-line

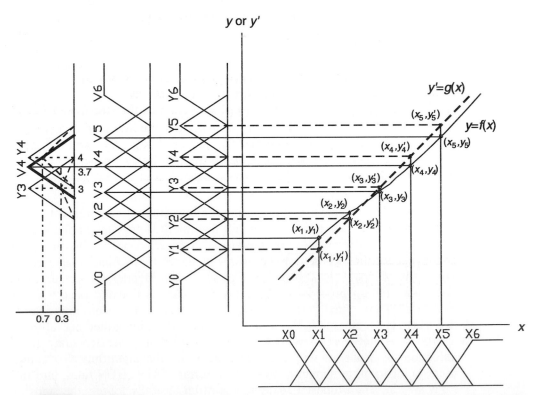

FIGURE 17 Concept of virtual fuzzy sets. Reprinted from *Computers in Industry*, K. C. Chan, Development of a feedback controller tuner using virtual fuzzy sets, 28:219–232, © 1996, with permission from Elsevier Science.

function $y' = g(x)$, but not the true function $y = f(x)$. The inaccuracy is attributed mainly to the locations of the fuzzy membership functions initially defined by the user.

A widely used technique to improve the accuracy of a fuzzy system is to increase the number of fuzzy sets. However, this would increase the complexity of the system and require a significant increase in computation time. In the section below, a new fuzzy rule representation will be proposed to improve the accuracy at the minimum expense of required computer resources.

The fuzzy sets $(X0, X1, \ldots, X6)$ and $(Y0, Y1, \ldots, Y6)$ are user defined. Consider that there are also imaginary fuzzy sets $(V0, V1, \ldots, V6)$, called virtual fuzzy sets. The training sample (x_i, y_i) would activate the fuzzy set association (X_i, V_i) to a degree of 1. Therefore, the rule (X_i, V_i) would be the best representation for the training sample (x_i, y_i).

There are two other issues to be considered. The first is how to represent the virtual fuzzy sets V_is, using the user defined fuzzy sets Y_is. The second is how to obtain fuzzy production rules from numerical training samples based on the concept of virtual fuzzy sets.

B. Virtual Fuzzy Rule Representation

In the new fuzzy rule representation scheme, the antecedent part of a production rule remains the same as that of the traditional fuzzy rule representation; only the consequent part is modified. A consequent term is expressed in terms of two consecutive user-defined fuzzy sets to which degrees are assigned. A virtual fuzzy set is then determined from the centroid of the consecutive fuzzy sets. As shown in Fig. 17, a typical new fuzzy rule representation is:

"IF X is $X4$, THEN Y is ($Y3$ to a degree of 0.3 and $Y4$ to a degree of 0.7),"

or,

"IF X is $X4$, THEN Y is $V4$."

The rule can be alternatively expressed as $(X4; Y3/0.3, Y4/0.7)$. The consequent part of the fuzzy association is a normalized combination of degree values. The center value of the virtual fuzzy set $V4$ is determined by the fuzzy centroid defuzzification method:

$$\bar{\nu}_i = \frac{\sum_{k=i-1}^{i} \bar{y}_k m_{Y_k}(y_i)}{\sum_{k=i-1}^{i} m_{Y_k}(y_i)}, \qquad (22)$$

where $\bar{y}_k$ denotes the center value of fuzzy set Y_k, and $m_{Y_k}(y_i)$ is the degree of membership of y_i assigned to the fuzzy set Y_k. For example, the location of $V4$ is determined as

$$\bar{\nu}_4 = \frac{\bar{y}_3 m_{Y_3}(y_4) + \bar{y}_4 m_{Y_4}(y_4)}{m_{Y_3}(y_4) + m_{Y_4}(y_4)} = \frac{(3)(0.3) + (4)(0.7)}{0.3 + 0.7} = 3.7.$$

C. A Virtual Fuzzy Rule Generation Scheme

The concept of a virtual fuzzy set has been demonstrated to be effective in improving fuzzy system accuracy. Furthermore, the neural-virtual fuzzy set approach has also been shown to be effective in dynamic modeling of a nonlinear process. However, this approach requires a long training time and offline neural network training and thus is not suitable for real-time adaptive systems.

A systematic adaptive learning procedure for the generation of fuzzy rules from numerical training data is proposed in this section. Although an offline nonadaptive system is sufficient for the feedback controller tuning application, this is a general approach and can be applied to many problems, such as identification and time series prediction. It is an extension to the five-step learning algorithm suggested by Wang and Mendel [14], and the concept of a virtual fuzzy set is incorporated into the modified learning algorithm. A two-input one-output system is chosen to explain the algorithm.

Suppose a set of training input–output data is given:

$$\left(x_1^{(1)}, x_2^{(1)}; y^{(1)} \right), \left(x_1^{(2)}, x_2^{(2)}; y^{(2)} \right), \ldots,$$

where x_1 and x_2 are the inputs, and y is the output. The aim is to derive a set of fuzzy rules, from the given input–output data, based on the virtual fuzzy set concept, and to use the resulting rules to determine a mapping f: $(x_1, x_2) \rightarrow y$.

The modified algorithm consists of seven steps.

Step 1. The Input and Output Spaces Are Quantized Using Fuzzy Sets

Specify the domain intervals where the values of variables are likely to lie. Assume that the domain intervals of x_1, x_2, and y are $[x_1^-, x_1^+]$, $[x_2^-, x_2^+]$, and $[y^-, y^+]$, respectively (Fig. 18). Quantize the domain of each variable, using a number of fuzzy sets. The number of fuzzy sets can be different (or the same) for different variables. The fuzzy sets are denoted by symbols (linguistic labels) such as Nn (Negative n), ..., N1 (Negative 1), CE (Center), P1 (Positive 1), ..., Pn (Positive n). Each fuzzy set is assigned a fuzzy membership function. Triangular fuzzy membership functions have been widely used, but other shapes of membership functions can also be used.

Step 2. Generate a Potential Fuzzy Rule from a Data Sample

In the Wang–Mendel approach, all of the samples are processed in one pass. Our approach differs in that only a single data point is processed at one time, thus, minimizing the computer storage requirement. In addition, this allows adaptive learning when new samples become available.

The rule generated in this step is only a potential candidate that may or may not be included in the final rule base, or the FAM bank. If the IF part of the rule is new, the rule is considered to be new and will be included in the FAM bank. If the same IF part of a rule already exists in the FAM bank, the virtual fuzzy set representation for the THEN part of the rule must be

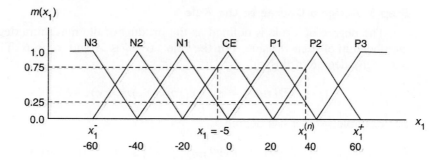

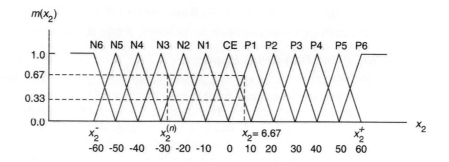

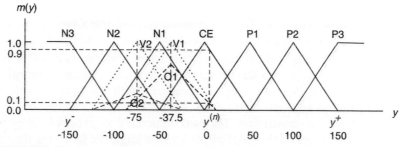

FIGURE 18 Fuzzy membership functions for the input and output spaces.

determined and any conflicts resolved. This procedure will be explained in more detail later in Step 4.

First, the degrees of membership of a given sample $n(x_1^{(n)}, x_2^{(n)}, y^{(n)})$ in different fuzzy sets are determined first. For example, $x_1^{(n)}$ in Fig. 18 has degree 0.25 in P1, degree 0.75 in P2, and zero degrees in all other fuzzy sets. Similarly, $x_2^{(n)}$ has nonzero degree 0.67 in N3 and degree 0.33 in N2; and $y^{(n)}$ has degree 0.9 in CE and degree 0.1 in P1.

To determine a single fuzzy set for each variable, the fuzzy set where the degree of membership is a maximum is chosen. Therefore, the value of $x_1^{(n)}$ is considered to be P2, $x_2^{(n)}$ is N3, and $y^{(n)}$ is CE. Finally, we obtain a fuzzy rule from the sample. In this example, the rule is

Rule n: IF x_1 is P2 and x_2 is N3, THEN y is CE.

Step 3. Assign a Degree to the Rule

The degree of a rule is defined as the product of the maximum degree of membership of each variable. For the rule "IF x_1 is A and x_2 is B, THEN y is C," the degree of this rule is

$$D(rule) = m_A(x_1)m_B(x_2)m_C(y). \qquad (23)$$

For example, *Rule n* has degree

$$D(Rule\ n) = m_{P2}(x_1^{(n)})m_{N3}(x_2^{(n)})m_{CE}(y^{(n)}) = (0.75)(0.67)(0.9) \cong 0.45.$$

Step 4. Assign a Rule to the Rule Base (or to Fill a FAM Cell)

There are three possible cases in assigning a rule to the rule base.

Case 1: The FAM Cell Is Empty.

If the FAM cell is empty (i.e., there is no existing rule with the same antecedent part as the current rule), assign the new rule to the rule base with a normalized rule certainty of 1. It should be noted that the rule degree must be stored in the rule base because it will be required when more samples become available. Table 9 lists a sequence of rule operations and assignments for a set of training samples. As an example for this case, training sample 1 is assigned to an empty FAM cell.

Case 2: The FAM Cell Is Nonempty, and the Consequent Part of the Potential Rule Suggests a New Linguistic Label as Output.

A conflicting rule occurs in this case, and the conflict is resolved by choosing the rule with the highest confidence. In this case, the rule degree of the suggested linguistic label is compared with the maximum degree of the two existing adjacent consequent linguistic labels, and the one with higher degree will be chosen first. The two immediate neighbors (if they exist in the consequent part of the rule) of the chosen linguistic label are then compared, and the one with the higher degree will be chosen as the second fuzzy set. The final step is to normalize the degrees and to express the rule as virtual fuzzy sets. Samples 2–5 in Table 9 are typical examples of this case.

Case 3: The FAM Cell is Nonempty, and the Consequent Part of the Potential Rule Suggests a Linguistic Label That Is the Same as One of the Output Linguistic Labels in the FAM Cell.

In this case, the degree of the suggested linguistic label is compared with that of the same label in the FAM cell, and the maximum degree is kept. Although the consequent terms remain the same in this case, the degrees must be normalized again (samples 6 and 7 in Table 9).

When a new sample is available, steps 2–4 are performed repeatedly, and the FAM bank is modified adaptively.

Step 5. Combine Expert Linguistic Rules with Fuzzy Rules Generated from Numerical Data (Optional)

This step is necessary only if linguistic information is available from a human expert. A linguistic rule can be entered into the FAM bank directly if

███ **TABLE 9** Assignment of Fuzzy Rules in the FAM Bank

Training sample	Maximum degree for x_1	Maximum degree for x_2	Maximum degree for y	Rule degree	Rule operation	Consequent term of the rule	Rule expressed as virtual fuzzy sets
1	(P2/0.75)	(N3/0.67)	(CE/0.90)	0.45	(CE/0.45) (CE/0.45)	(CE; 0.45)	(P2, N3; CE/1)
2	(P2/0.60)	(N3/0.55)	(P1/0.70)	0.23	(P1/0.23) (CE/0.45)	(CE/0.45, P1/0.23)	(P2, N3; CE/0.66, P1/0.34)
3	(P2/0.60)	(N3/0.55)	(P2/0.55)	0.18	(P1/0.23) ~~(P2/0.18)~~ (CE/0.45)	(CE/0.45, P1/0.23)	(P2, N3; CE/0.66, P1/0.34)
4	(P2/0.70)	(N3/0.62)	(P3/0.75)	0.33	(P1/0.23) ~~(P3/0.33)~~ (N1/0.73)	(CE/0.45, P1/0.23)	(P2, N3; CE/0.66, P1/0.34)
5	(P2/0.90)	(N3/0.95)	(N1/0.85)	0.73	(CE/0.45) ~~(P1/0.23)~~ (N1/0.73)	(N1/0.73, CE/0.45)	(P2, N3; N1/0.62, CE/0.38)
6	(P2/0.95)	(N3/0.78)	(N1/0.87)	0.64	(N1/0.64) ~~(CE/0.45)~~ (N1/0.73)	(Nl/0.73, CE/0.45)	(P2, N3; N1/0.62, CE/0.38)
7	(P2/0.85)	(N3/0.77)	(CE/0.90)	0.59	(CE/0.59) ~~(CE/0.45)~~	(N1/0.73, CE/0.59)	(P2, N3; N1/0.55, CE/0.45)

it is a new rule. But if there exists a conflicting rule in the FAM bank, the conflict has to be resolved by the human expert.

Step 6. Assign Weights to all of the Rules (Optional)

This step is optional. Weight assignment is a mechanism to provide final human expert adjustment to the rule base that is either automatically generated from sample data or assigned by a human expert. This step may be performed when *a priori* information about the data set (such as accuracy, reliability, and usefulness) is available. The assigned weight actually represents the belief in the usefulness of a fuzzy production rule.

The weight of an automatically generated rule is equal to unity. When a weight is assigned to a rule, the centroid of the virtual fuzzy set does not change, but the height of the virtual fuzzy set is changed to a value equal to the weight. Consider the last sample listed in Table 9 (P2, N3; N1/0.55, CE/0.45). If a weight of 0.6 is assigned to the rule, the rule representation becomes (P2, N3; N1/0.55, CE/0.45; 0.6), or equivalently, (P2, N3; N1/0.33, CE/0.27).

Step 7. Fuzzy Inference and Defuzzification

After the rule base is obtained from the given data, the fuzzy system is ready to determine a mapping from the inputs (x_1, x_2) to the output y. The correlation product fuzzy inference [12] is used to compute the output fuzzy set. The output fuzzy set is then defuzzified to obtain a numerical output by using the fuzzy centroid defuzzification scheme.

To illustrate the fuzzy inference and defuzzification procedure, we use the following standard rule representation:

$$\text{Rule } i: (in1_i, in2_i; f_i/d_i; s_i/(1 - d_i); w_i), \qquad (24)$$

where $in1_i$ and $in2_i$ are the antecedent fuzzy sets. A degree of membership d_i is assigned to the first consequent fuzzy set f_i, and a degree of membership $(1 - d_i)$ is assigned to the second consequent fuzzy set s_i. The input pair (x_1, x_2) activates the antecedent fuzzy sets $in1_i$ and $in2_i$ to degrees $m_{in1_i}(x_1)$ and $m_{in2_i}(x_2)$, respectively.

To illustrate the fuzzy inference and defuzzification procedure, a numerical example is given. With reference to Fig. 18, we assume that there are two rules in a fuzzy system:

Rule 1: (CE, P1; N1/0.75; CE/0.25; 1)
Rule 2: (N1, CE; N2/0.50; N1/0.50; 1)

The activation value of the consequent part of Rule i is equal to the minimum of the antecedents' conjunction value. In our example, the activation value a_i is determined as follows:

$$a_i = \min(m_{in1_i}(x_1), m_{in2_i}(x_2)), \qquad (25)$$

if the input (x_1, x_2) is equal to $(-5, 6.67)$ and the only rules with nonzero memberships are Rule 1 and Rule 2, for which the activation values are

$$a_1 = \min(m_{CE}(-5), m_{P1}(6.67))$$
$$= \min(0.75, 0.67) = 0.67,$$
$$a_2 = \min(m_{N1}(-5), m_{CE}(6.67))$$
$$= \min(0.25, 0.33) = 0.25.$$

The virtual fuzzy set for rule i is determined by combining the consequent fuzzy sets f_i and s_i as

$$m_{v_i}(y) = d_i m_{f_i}(y) + (1 - d_i)m_{s_i}(y), \qquad (26)$$

or

$$V_i = d_i F_i + (1 - d_i)S_i. \qquad (27)$$

The virtual fuzzy sets for Rules 1 and 2, as shown in Fig. 18, are determined as

$$V_1 = V_1 = 0.75N1 + 0.25CE,$$
$$V_2 = V_2 = 0.50N2 + 0.50N1.$$

The center value of V_i, denoted as $\bar{v}_i$, can be determined as the centroid of the fuzzy sets $d_i F_i$ and $(1 - d_i)S_i$:

$$\bar{v}_i = d_i \bar{f}_i + (1 - d_i)\bar{s}_i, \qquad (28)$$

where $\bar{f}_i$ and $\bar{s}_i$ are the center values of fuzzy sets F_i and S_i, respectively. For Rules 1 and 2, the center values of the virtual fuzzy sets are

$$\bar{\nu}_1 = (0.75 \times - 50) + (0.25 \times 0) = -37.5,$$

$$\bar{\nu}_2 = (0.50 \times - 100) + (0.50 \times - 50) = -75.$$

The input pair (x_1, x_2) activates the consequent part of each FAM rule to a different degree a_i. With correlation product encoding inference, the ith rule yields the weighted output fuzzy set O_i:

$$m_{o_i}(y) = a_i w_i m_{\nu_i}(y), \tag{29}$$

or

$$O_i = a_i w_i V_i. \tag{30}$$

Therefore,

$$O_1 = O1 = 0.67 \times 1 \times V1,$$

$$O_2 = O2 = 0.25 \times 1 \times V2.$$

The fuzzy system then sums the output fuzzy set O_i from each FAM rule to form the combined output fuzzy set O:

$$m_o(y) = \sum_{i=1}^{N} m_{o_i}(y), \tag{31}$$

or

$$O = \sum_{i=1}^{N} O_i, \tag{32}$$

where N is the number of fuzzy rules in the FAM bank. For our example,

$$O = O_1 + O_2 = O1 + O2.$$

The control output u is equal to the fuzzy centroid of O:

$$u = \frac{\int y m_o(y)\, dy}{\int m_o(y)\, dy}. \tag{33}$$

To reduce computation, we can use the following formula to determine the output u for triangular membership functions:

$$u = \frac{\sum_{i=1}^{N} \bar{\nu}_i a_i w_i}{\sum_{i=1}^{N} a_i w_i}. \tag{34}$$

For our example, the output is calculated as

$$u = \frac{(-37.5 \times 0.67 \times 1) + (-75 \times 0.25 \times 1)}{(0.67 \times 1) + (0.25 \times 1)} = -47.69.$$

D. Results and Discussion

This section presents the results of applying the traditional and virtual fuzzy set approach to feedback controller tuning. The modified Wang–Mendel approach has been used to generate banks of fuzzy tuning rules automatically from the training sample. The performance of the tuners is evaluated and compared with the neural network approach described before.

1. Application to PI Controller Tuning

In the modified Wang–Mendel approach, the virtual fuzzy set is applied to tune the same type of processes as in Section 3.E, where the normalized dead time is within the range $0.15 < \theta_1 < 0.6$ (case 1 in Table 1). The same sets of training and testing sample are used.

Each dimension of the input–output space is quantized using 11 fuzzy subsets. Triangular membership functions are used for all fuzzy variables. The membership functions are shown in Fig. 19. The minimum and maximum values for the input and output variables have been specified in Table 2. It should be noted that the ranges of the membership functions are narrower than the complete ranges, so that a more accurate result can be obtained near the center.

There are 1187 data points producing oscillatory responses. These data points are divided randomly into a training set and a learning set. There are 594 data points in the training set and 593 in the testing set. By using the modified Wang–Mendel approach, 157 rules are generated from the 594 sample points. Because our approach is a one-pass build-up procedure and iterative training is not required, the rules are generated instantly from the numerical sample.

The fuzzy system is then used to tune four processes, using both the traditional fuzzy set- and virtual fuzzy set-based rules. The same four processes representing the extremes of the plant group are used (processes I, II, III, and IV in Table 5).

Using the traditional fuzzy set approach, the trajectories of the proportional gain and integral time are shown in Figs. 20a and 21a for processes II and III, respectively. The responses of the tuned systems are shown in Figs. 20b and 21b for processes II and III, respectively.

Using the virtual fuzzy set approach, the trajectories of the proportional gain and integral time are shown in Figs. 22a and 23a for processes II and III, respectively. The responses of the tuned systems are shown in Figs. 22b and 23b for processes II and III, respectively.

From the gain trajectory diagrams, such as Figs. 20a and 22a, the traditional approach appears to be more stable after a number of iterations. The virtual fuzzy set approach has a larger fluctuation in gain values, especially in T_i. However, the performance of the system remains roughly the same, because the K_p and T_i are adjusted in the same direction; therefore, the K_p/T_i ratio remains roughly the same.

It can also be seen that the two gains converge to similar values for both the traditional and virtual fuzzy set approaches, for all four processes. From

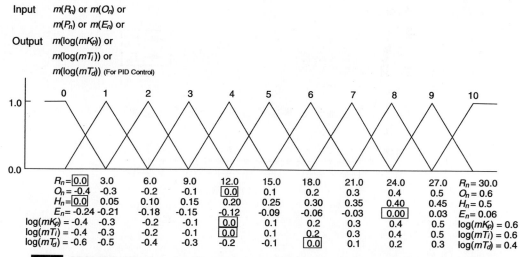

Input $m(R_n)$ or $m(O_n)$ or
$m(P_n)$ or $m(E_n)$ or

Output $m(\log(mK_p))$ or
$m(\log(mT_i))$ or
$m(\log(mT_d))$ (For PID Control)

	0	1	2	3	4	5	6	7	8	9	10
$R_n =$	0.0	3.0	6.0	9.0	12.0	15.0	18.0	21.0	24.0	27.0	$R_n = 30.0$
$O_n =$	-0.4	-0.3	-0.2	-0.1	0.0	0.1	0.2	0.3	0.4	0.5	$O_n = 0.6$
$H_n =$	0.0	0.05	0.10	0.15	0.20	0.25	0.30	0.35	0.40	0.45	$H_n = 0.5$
$E_n =$	-0.24	-0.21	-0.18	-0.15	-0.12	-0.09	-0.06	-0.03	0.00	0.03	$E_n = 0.06$
$\log(mK_p) =$	-0.4	-0.3	-0.2	-0.1	0.0	0.1	0.2	0.3	0.4	0.5	$\log(mK_p) = 0.6$
$\log(mT_i) =$	-0.4	-0.3	-0.2	-0.1	0.0	0.1	0.2	0.3	0.4	0.5	$\log(mT_i) = 0.6$
$\log(mT_d) =$	-0.6	-0.5	-0.4	-0.3	-0.2	-0.1	0.0	0.1	0.2	0.3	$\log(mT_d) = 0.4$

FIGURE 19 Fuzzy membership function definitions for tuner input and output. Fuzzy subsets are named in numerical order; the number in square bracket corresponds to the zeroth fuzzy subset. Reprinted from *Computers in Industry*, K. C. Chan, Development of a feedback controller tuner using virtual fuzzy sets, 28:219–232, © 1996, with permission from Elsevier Science.

the response diagrams, such as Figs. 20b and 22b, the traditional and the virtual fuzzy set approach have similar performances.

Compared with the neural network tuner, the fuzzy tuner has a comparable performance in tuning these four processes. However, the gain adjustments (trajectories) made by the neural network tuner are much smoother and more stable.

2. Application to PID Controller Tuning

In this section, the modified Wang–Mendel virtual fuzzy set is applied to tune a PID feedback controller for the same processes as before. Again, each dimension of the input–output space is quantized by using 11 fuzzy subsets, and triangular membership functions are used.

The minimum and maximum values for the input and output variables are specified in Table 10. There are 1393 data points producing oscillatory responses. These data points are divided into two sets: 680 data points in the training set and 713 data points in the testing set.

Using the modified Wang–Mendel virtual fuzzy set approach, 162 rules are generated from the 680 sample points. Considering the complete input space, 11^4 (14,641) rules are needed to cover the whole input space. The automatically generated 162 rules represent only 1.106% of the input space.

The fuzzy system is then used to tune the four processes, using both the traditional fuzzy set- and virtual fuzzy set-based rules. The same four processes representing the extremes of the plant group are used (processes I, II, III, and IV in Table 5).

Using the traditional fuzzy set approach, the trajectories of the gains (K_p, T_i, T_d) are shown in Figs. 24a and 25a for processes II and III, respec-

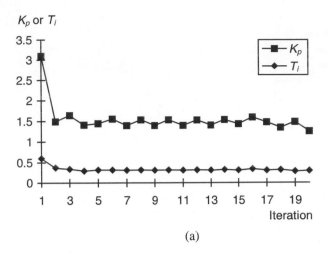

(a)

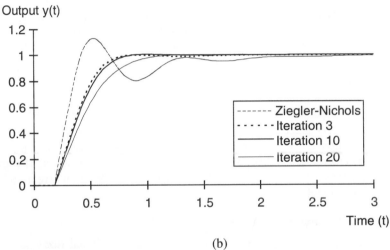

(b)

FIGURE 20 (a) Trajectories of K_p and T_i for process II, obtained with the traditional fuzzy set approach. (b) Performance of Ziegler–Nichols and traditional fuzzy tuning for process II.

tively. The responses of the tuned systems are shown in Figs. 24b and 25b for processes II and III, respectively. In these figures, the responses corresponding to Ziegler–Nichols tuning and the best tuning are given.

Using the virtual fuzzy set approach, the trajectories of the gains (K_p, T_i, T_d) are shown in Figs. 26a and 27a for processes II and III, respectively. The responses of the tuned systems are shown in Figs. 26b and 27b for processes II and III, respectively.

The results show that both the traditional and the virtual fuzzy set approaches can be applied to tune PID feedback controllers successfully. However, the controller gains may converge to different values through the use of the two different approaches.

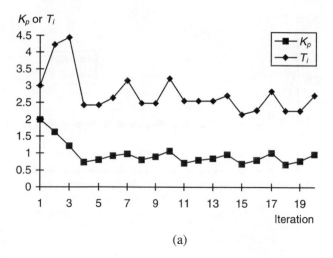

(a)

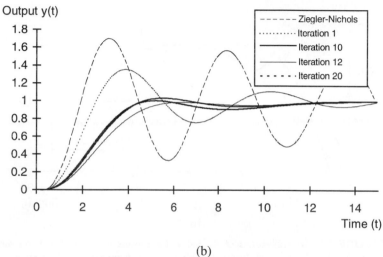

(b)

FIGURE 21 (a) Trajectories of K_p and T_i for process III, obtained with the traditional fuzzy set approach. (b) Performance of Ziegler–Nichols and traditional fuzzy tuning for process III.

In process II, as shown in Figs. 24a and 26a, the tuning performed by the two approaches is quite similar, except that there are "jumps" in the K_p trajectory, because of monotone responses, in the virtual fuzzy set approach.

In process III, as shown in Figs. 25a and 27a, the three gains converge to different levels. For the traditional approach, the gains converge to (K_p = 2.99, T_i = 7.93, T_d = 0.98). For the virtual fuzzy set approach, the gains converge to (K_p = 2.52, T_i = 3.13, T_d = 1.20) with slightly better performance. Although the final gain settings are different, the performance of the responses is similar (Figs. 25b and 27b).

From the response diagrams, the traditional and the virtual fuzzy set approaches have similar and satisfactory performance. The fuzzy system

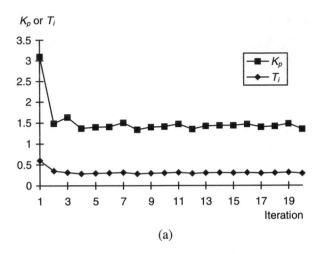

(a)

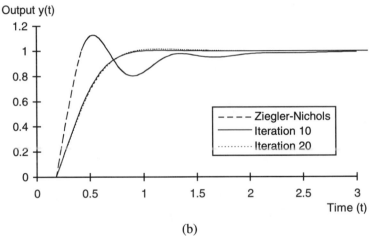

(b)

FIGURE 22 (a) Trajectories of K_p and T_i for process II, obtained with the virtual fuzzy set approach. (b) Performance of Ziegier–Nichols and virtual fuzzy tuning for process II.

tuned responses are less oscillatory than the Ziegler–Nichols tuned responses and have a better performance. The neural network appears to provide smoother tuning actions. However, the neural network approach involves far greater computational effort, is more difficult to modify, and does not provide a structured representation of the system's intelligence in tuning feedback controllers.

Our results show that comparable performance can be achieved using the traditional/virtual fuzzy set approach, as compared to the neural network approach. However, the development of the fuzzy system involves far less computational effort, because of the one-pass built-up learning procedure. A set of structured fuzzy rules is available; we can then adjust the fuzzy tuning system by refining the fuzzy-rule bank with human rules of thumb and with further training data. Explanatory capability is also possible through the use of the IF–THEN rule base.

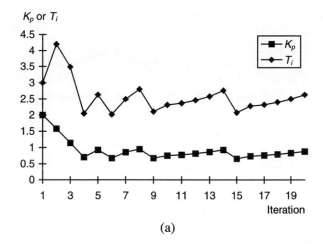

(a)

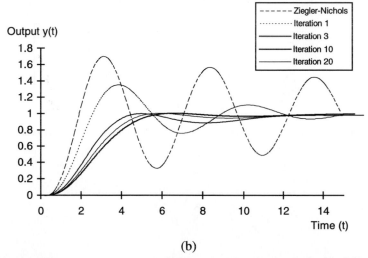

(b)

FIGURE 23 (a) Trajectories of K_p and T_i for the Process III obtained with the virtual fuzzy set approach. (b) Performance of Ziegler–Nichols and virtual fuzzy tuning for process III.

TABLE 10   Input–Output Scaling for PID Tuning

Parameter	Min value	Max value	Normalized scaled ranges
Normalized rise time	1.6	18	[0.0, 1.0]
Normalized overshoot	−0.50	0.97	[0.0, 1.0]
Normalized peak-to-peak height	0.00011	0.87	[0.0, 1.0]
Normalized final error	−0.24	0.099	[0.0, 1.0]
$\log(mK_p)$	−0.35	0.29	[0.2, 0.8]
$\log(mT_i)$	−0.35	0.68	[0.2, 0.8]
$\log(mT_d)$	−0.65	0.70	[0.2, 0.8]

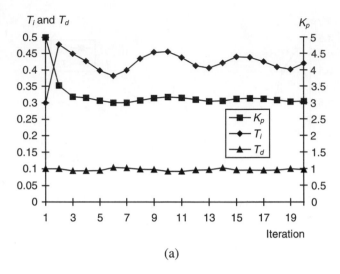

(a)

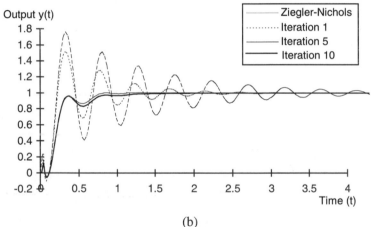

(b)

FIGURE 24   (a) Trajectories of K_p, T_i, and T_d for process II, obtained with the traditional fuzzy set approach. (b) Performance of Ziegler–Nichols and traditional fuzzy tuning for process II. (Note: Iteration 20 is almost identicai to iteration 10.)

VI. SUMMARY

This chapter presents an investigation of artificial intelligent techniques and their applications in tuning feedback controllers. In Section III, back-propagated feed-forward neural networks have been implemented for tuning controllers. Unlike previous expert tuners, which rely totally on human experts to provide tuning rules, the neural network allows automatic knowledge extraction based on the tuning of a representative process of a process group. Therefore, more complete tuning knowledge can be incorporated into the knowledge base. Although a neural network can be trained to approximate the tuning action in terms of a multi-input multi-output mathematical function, there are a few drawbacks of neural-based tuning systems. These

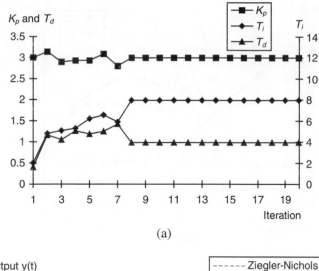

(a)

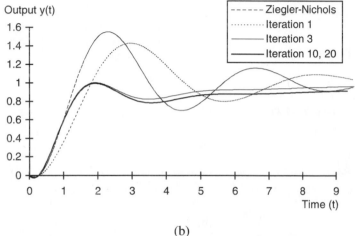

(b)

FIGURE 25 (a) Trajectories of K_p, T_i, and T_d for process III, obtained with the traditional fuzzy set approach. (b) Performance of Ziegler–Nichols and traditional fuzzy tuning for process III.

include:

- a black box approach
- a lack of explanatory capability
- the difficulty of including structured knowledge
- a bias toward quantitative data
- extremely long development (training) time

We are not only interested in obtaining a well-tuned feedback controller; we also want to understand how neural network recommendations are derived. However, because of the inherent disadvantages of neural networks, users do not have any clues to how and why the parameters are adjusted and cannot include or modify the tuning knowledge base. In other words, the neural network tuner is inflexible. Although understanding of the network

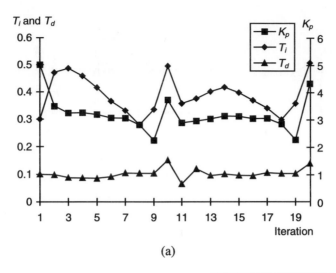

(a)

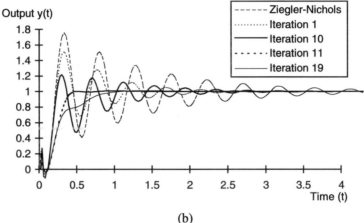

(b)

FIGURE 26 (a) Trajectories of K_p, T_i, and T_d for process II, obtained with the virtual fuzzy set approach. (b) Performance of Ziegler–Nichols and virtual fuzzy tuning for process II.

can be made by analyzing the weights of nodes, the task is difficult if not impossible.

A practical solution to this issue is to integrate fuzzy logic and neural networks in the design of an intelligent feedback tuner. In Section IV, an integration scheme has been proposed. This approach combines all of the advantages of the neural-based and the traditional fuzzy approaches. It represents improvements to the pure neural network approach in the ways that the knowledge is understandable (in rule forms); the knowledge acquisition becomes more flexible (can combine human knowledge with self-organizing knowledge, based on numerical examples); and the knowledge modification is easier (the human expert can actually modify the rules).

Although the neural-fuzzy approach has many advantages over the pure neural network approach, it also inherits the trial-and-error nature of deter-

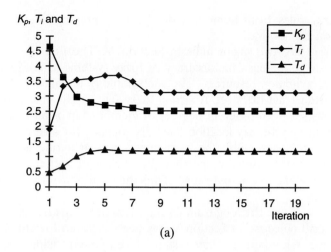

(a)

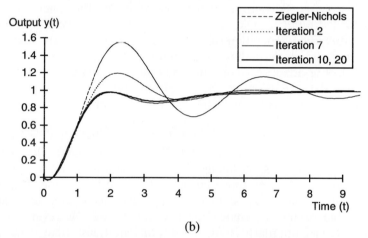

(b)

FIGURE 27 (a) Trajectories of K_p, T_i, and T_d for process III, obtained with the virtual fuzzy set approach. (b) Performance of Ziegler–Nichols and virtual fuzzy tuning for process III.

mining the most suitable network structure and the requirement of a long training time in the neural network approach. Despite the fact that the training time can be shortened compared to the pure neural network approach, offline training is unavoidable.

The proposed offline scheme involves rule generation using the trained neural network, and the rules generated are stored in the computer for real-time fuzzy inference. However, computer storage may be a problem when the number of rules is very large. This neural-fuzzy approach has been applied to PI controller tuning, and very satisfactory results have been achieved.

Because of the training requirement of neural networks, the knowledge acquisition procedure can be extremely time-consuming. Because the major bottleneck in the system lies in learning, a logical extension is to develop a more efficient learning algorithm, replacing the neural network, for generating fuzzy rules from numerical data and combining linguistic rules into a common framework.

There are two major aims in Section V. The first is to develop a new technique to increase the accuracy of fuzzy systems, at a minimum expenditure of computer resources. The methodology is based on a new concept called the virtual fuzzy set. A virtual fuzzy set, which consists of two consecutive fuzzy sets with different degrees of membership, is an imaginary fuzzy set located at an arbitrary location most appropriate for a rule. Using the virtual fuzzy set representation, the consequent part of a rule is not restricted to those fuzzy sets initially specified by the user; rather, it can be placed anywhere along the output range. Thus the accuracy of the fuzzy rules can be improved.

To overcome the long training requirement of the neural-fuzzy approach, the second objective in Section V has been directed toward generalizing the technique for adaptive fuzzy system development, using the concept of a virtual fuzzy set. It is in this area of research that the main contribution to the knowledge in this field has been made. The adaptive algorithm is based on the fuzzy rule generation scheme suggested by Wang and Mendel. It is a one-pass build-up procedure, and time-consuming iterative training is not required. The adaptive algorithm generates fuzzy rules automatically from numerical data. In addition, the algorithm also allows users to incorporate human knowledge, in terms of IF–THEN rules, into the fuzzy system.

The new algorithm was then applied to the development of feedback controllers. As described in Section V.D, the new algorithm has been successfully applied to tune both PI and PID feedback controllers. All of the intelligent algorithms, including the pure neural network approach, the integrated neural-fuzzy approach, and the modified Wang–Mendel (traditional and virtual fuzzy set) approach, have been concluded to achieve comparable results in tuning feedback controllers. However, the modified Wang–Mendel approach is the most efficient (least time consuming) approach to building reliable feedback tuning systems.

REFERENCES

1. Ziegler, J. G. and Nichols, N. B. Optimal settings for automatic controllers. *Trans. ASME* 64:759–768, 1942.
2. Kraus, T. W. and Myron, T. J. Self-tuning PID controller uses pattern recognition approach. *Control Engrg.* 31:106–111, 1984.
3. Litt, J. An expert system for adaptive PID tuning based on pattern recognition techniques. In *Instrumentation in the Chemical and Petroleum Industries, Proceedings of the 1986 Conference*, Secaucus, NJ, 1986, Vol. 18, pp. 87–104.
4. Porter, B., Jones, A. H., and McKeown, C. B. Real-time expert tuners for PI controllers. *IEE Proc. Pt. D* 134:260–263, 1987.
5. Lebow, L. G. Microcomputer based expert control and adaptive PID control. In *Advanced Computing Concepts and Techniques in Control Engineering* (M. J. Denham and A. J. Laub, Eds.), pp. 437–458. Springer-Verlag, Berlin/New York, 1988.
6. Ma, X., and Loh, N. K. Tuning PID controllers by neural networks. In *Proceedings of the Artificial Neural Networks in Engineering (ANNIE '91) Conference*, St. Louis, MO, 1991, pp. 625–630.

7. Willis, M. J. and Montague, G. A. Auto-tuning PI(D) controllers with artificial neural networks. In *Proceedings of the 12th World Congress of the International Federation of Automatic Control*, Sydney, Australia, 1993, Vol. 4, pp. 61–64.

8. Hang, C. C., Åström, K. J., and Ho, W. K. Refinements of the Ziegler–Nichols tuning formula for PID auto-tuners. In *Proceedings of the ISA/88 International Conference and Exhibit*, Houston, Texas, 1988, pp. 1021–1030.

9. Hang, C. C. and Åström, K. J. Refinements of the Ziegler–Nichols tuning formula. *IEE Proc. D* 138:111–118, 1991.

10. Astrom, K. J., Hang, C. C., Persson, P., and Ho, W. K. Towards intelligent PID control. *Automatica*, 28:1–9, 1992.

11. Caudill, M. and Butler, C. *Understanding Neural Networks. Computer Explorations*, Vol. 2, *Advanced Networks*. MIT Press, Cambridge, MA, 1992.

12. Kosko, B. *Neural Networks and Fuzzy Systems*. Prentice Hall, Englewood Cliffs, NJ, 1992.

13. Chan, K. C., Lin, G. C. I., and Leong, S. S. A more accurate adaptive fuzzy inference system. *Comput. Indust. Internat. J.* 26:61–73, 1995.

14. Wang, L. X. and Mendel, J. M. Generating fuzzy rules by learning from examples. In *Proceedings of the 1991 IEEE International Symposium on Intelligent Control*, Arlington, VA, 1991, pp. 263–268.

13

TECHNIQUES AND APPLICATIONS OF FUZZY THEORY VIA H_∞ CONTROL TECHNIQUES FOR TRACKING ALGORITHMS FOR UNCERTAIN NONLINEAR SYSTEMS

BOR-SEN CHEN

YU-MIN CHENG

Department of Electrical Engineering, National Tsing-Hua University, Hsin-Chu, Taiwan, Republic of China

I. INTRODUCTION 345
II. PROBLEM FORMULATION 347
III. A REVIEW OF FUZZY LOGIC SYSTEMS 349
IV. ADAPTIVE FUZZY CONTROL DESIGN 351
    A. H_∞ Tracking Control Design in Indirect Adaptive
       Fuzzy Systems 352
    B. H_∞ Tracking Control Design in Direct Adaptive
       Fuzzy Systems 358
V. ILLUSTRATIVE EXAMPLES 362
VI. CONCLUSION 369
    REFERENCES 372

I. INTRODUCTION

The adaptive control for feedback linearizable nonlinear systems has already been discussed. In [1], the unknown parameters of nonlinear systems are assumed to be linear and the system is free of external disturbances. The nonlinear adaptive problem is transformed into a linear adaptive problem by a feedback linearization method at first, and then the adaptive pole assignments and model matching techniques are used to achieve the asymptotic tracking design. In this work, the H_∞ tracking problem of uncertain nonlinear SISO systems with external disturbance is solved.

Fuzzy control has recently found extensive application for a wide variety of industrial systems and consumer products and has attracted the attention of many control researchers because of its model free approach [2–14]. The fuzzy control algorithms attempt to make use of information from human experts. The expert information is generally represented by fuzzy terms (e.g., small, large, not very large, etc.) for convenience or lack of more precise knowledge, ease of communication, and so forth. However, most of these fuzzy control algorithms are proposed without analytical tools for general design procedures to guarantee basic performance criteria. Generally, these fuzzy control approaches combine expert knowledge with the conventional engineering systems in an *ad hoc* manner. Thereby, simulations are performed to show the validity of the approaches to the specific control problems [2–4].

More recently, an important adaptive fuzzy control system has been developed to incorporate expert information systematically, and stability is guaranteed by theoretical analysis [4, 15, 16]. An adaptive fuzzy system is a fuzzy logic system equipped with a training algorithm, in which the fuzzy logic system is constructed from a collection of fuzzy IF–THEN rules, and the training algorithm adjusts the parameters of the fuzzy logic system according to numerical input–output data. Conceptually, adaptive fuzzy systems combine linguistic information from experts with numerical information from sensors. Linguistic information (in the form of fuzzy IF–THEN rules) can be directly incorporated because fuzzy logic systems are constructed from fuzzy IF–THEN rules; on the other hand, numerical information (in the form of input–output data) is incorporated by training the fuzzy logic system to match the input–output data.

The perfect match via an adaptive fuzzy logic system is generally deemed impossible. In [4, 15, 16], the stability of the adaptive fuzzy control system has been guaranteed. However, the matching error may cause the tracking performance to deteriorate. A prescribed tracking performance cannot be guaranteed in the conventional adaptive fuzzy control systems, owing to the fact that the influence of matching error and external disturbance on tracking error cannot be efficiently eliminated. In most cases, even when the output error is small, the errors of the internal variables are not small in the conventional adaptive fuzzy control designs.

In the last decade, H_∞ optimal control theory has been well developed and extensively applied to efficiently treat robust stabilization and disturbance rejection problems [17–21]. In the conventional H_∞ optimal control, the plant model must be known beforehand. In this study, the H_∞ optimal control design will be extended toward the nonlinear systems with unknown or uncertain models via a fuzzy technique. To our knowledge, however, the H_∞ optimal control design has not yet been used for adaptive control systems or fuzzy control systems.

A new adaptive fuzzy control algorithm is proposed in this study, so that not only the stability of the adaptive fuzzy control system is guaranteed, but the influence of the matching error and external disturbance on the tracking error is attenuated to an arbitrarily designed level via an H_∞ tracking design technique, i.e., the gain (the induced L_2 norm) from both the matching error and external disturbance to tracking error must be less than or equal to a prescribed value. This is the so-called H_∞ tracking problem in adaptive fuzzy control systems. The proposed design method attempts to combine the attenuation technique, fuzzy logic approximation method, and adaptive control algorithm for the robust tracking control design of nonlinear systems with a large uncertainty or unknown variation in plant parameters and structures. Both direct and indirect adaptive fuzzy controls are discussed, and the corresponding design procedures are proposed.

Two simulation examples are finally provided to illustrate the performance of the proposed methods. The adaptive fuzzy approximation technique is used as rough tuning, and the H_∞ disturbance attenuation technique is considered as fine tuning. Therefore, the proposed design method is more suitable for the robust tracking control design of the uncertain nonlinear systems and is an attractive control design concept. Computer simulation results indicate that the effect of both the fuzzy approximation error and external disturbance on the tracking error can be attenuated efficiently by the proposed method. These simulation results illustrate the desired upgrade on both H_∞ and fuzzy control designs.

The paper is organized as follows. The problem is formulated in Section II. Fuzzy logic systems are briefly described in Section III. H_∞ tracking performance designs in both indirect and direct adaptive fuzzy systems are given in Section IV. Two simulation examples are provided to illustrate the performance of the proposed methods in Section V. Concluding remarks are made in Section VI.

II. PROBLEM FORMULATION

Consider the following nth-order SISO nonlinear systems:

$$x^{(n)} = f(x, \dot{x}, \ldots, x^{(n-1)}) + g(x, \dot{x}, \ldots, x^{(n-1)})u + d,$$
$$y = x, \tag{1}$$

where f and g are unknown (uncertain) but bounded continuous functions, and $u \in R$ and $y \in R$ are the input and output of the system, respectively. d denotes the external disturbance, which is unknown but bounded. The external disturbances may be due to system load, external noise, etc. Let $\underline{x} = (x, \dot{x}, \ldots, x^{(n-1)})^T \in R^n$ be the state vector of the system that is assumed to be available. For Eq. (1) to be controllable, we require that $g(\underline{x}) \neq 0$ for $\underline{x}$ in some controllability region $U_c \subset R^n$.

Remark 1. Consider the more general nonlinear system as follows:

$$\dot{z} = f(z) + g(z)u + d',$$
$$y = h(z),$$

where $z \in R^l$, $l \geq n$. Let $L_f h$ and $L_g h$ stand for the Lie derivatives of h with respect to f and g, respectively. If n is the smallest integer such that

$L_g L_f^i h = 0$ for $i = 0, 1, \ldots, n - 2$ and $L_g L_f^{n-1} h \neq 0$, $\forall z \in R^l$. Then we get [22]

$$y^{(n)} = L_f^n h(z) + L_g L_f^{n-1} h(z) u + L_{f+gu+d'}^{n-1} L_{d'} h(z)$$

$$+ \sum_{k=1}^{n-1} L_{f+gu+d'}^{k-1} L_{d'} L_f^{n-k} h(z).$$

Let us denote $y = x$, $\dot{y} = \dot{x}$, $\ddot{y} = \ddot{x}, \ldots$. Then the above equation is of the following form:

$$x^{(n)} = L_f^n x + L_g L_f^{n-1} x u + d$$

$$=: f(\underline{x}) + g(\underline{x}) u + d$$

$$y = x,$$

where

$$f(\underline{x}) := L_f^n x, \quad g(\underline{x}) := L_g L_f^{n-1} x, \quad \text{and}$$

$$d := L_{f+gu+d'}^{n-1} L_{d'} h + \sum_{k=1}^{n-1} L_{f+gu+d'}^{k-1} L_{d'} L_f^{n-k} h.$$

This is of the form in the nonlinear system of Eq. (1).

The control objective is to force y to follow a given bounded desired trajectory y_r. Let us denote the output tracking error e and parameter tracking error $\underline{\tilde{\theta}}$:

$$e = y_r - y,$$

$$\underline{\tilde{\theta}} = \underline{\theta} - \underline{\theta}^*,$$

for some parameter estimate $\underline{\theta}$ of the fuzzy logic system and optimal parameter estimate $\underline{\theta}^*$ of the fuzzy logic system. Let w denote the sum of matching errors owing to fuzzy approximations of $f(\underline{x})$, $g(\underline{x})$, and the external disturbance d. Then our design objective is to impose an adaptive fuzzy control algorithm such that the asymptotically stable tracking

$$e^{(n)} + \alpha_1 e^{(n-1)} + \cdots + \alpha_n e = 0 \tag{2}$$

is achieved while $w = 0$ (i.e., in the case of free of external disturbance and perfect fuzzy approximation. While ~~ appears, the following H_∞ tracking performance is requested ([17–19]):

$$\int_0^T \underline{e}^T Q \underline{e} \, dt \leq \underline{e}^T(0) P \underline{e}(0)$$

$$+ \frac{1}{\gamma} \underline{\tilde{\theta}}^T(0) \underline{\tilde{\theta}}(0) + \rho^2 \int_0^T w^T w \, dt$$

$$\forall T \in [0, \infty), w \in L_2[0, T], \tag{3}$$

for a given weighting matrix $Q = Q^T \geq 0$, an adaption gain $\gamma > 0$, and a prescribed attenuation level ρ, where $\underline{e} = (e, \dot{e}, \ldots, e^{(n-1)})^T$. The matrix P in Eq. (3) is some nonnegative and symmetric matrix that will be assigned later. Both indirect and direct adaptive fuzzy control algorithms are employed in this work to control the nonlinear system in Eq. (1) to guarantee the tracking performance Eq. (3) with a prescribed disturbance attenuation level ρ.

Remark 2. i. The roots of polynomial $h(s) = s^n + \alpha_2 s^{(n-1)} + \cdots + \alpha_{n-1} s + \alpha_n$ in the characteristic equation of Eq. (2) are all in the open left half-plane via an adequate choice of coefficients $\alpha_1, \alpha_2, \ldots, \alpha_n$.

ii. If the system starts with initial conditions $\underline{e}(0) = 0$, $\tilde{\theta}(0) = 0$, then the H_∞ performance in Eq. (3) can be rewritten as

$$\sup_{w \in L_2[0,T]} \frac{\|\underline{e}\|_Q}{\|w\|} \leq \rho,$$

where $\|\underline{e}\|_Q^2 = \int_0^T \underline{e}^T Q \underline{e} \, dt$ and $\|w\|^2 = \int_0^T w^T w \, dt$, i.e., the L_2-gain from w to the tracking error $\underline{e}$ must be equal to or less than ρ.

iii. If $\rho = \infty$, this is the case of minimum error tracking control without disturbance attenuation [17–19].

iv. In the conventional adaptive control systems in the external disturbance case, some adaptive controls with VSS or dead-zone techniques have been employed to guarantee the stability of the adaptive control systems [20, 22]. However, these methods will lead to higher frequency switching control signals and may drive the high-frequency modes of unmodeled dynamics. The control signals are expected here to be smooth.

III. A REVIEW OF FUZZY LOGIC SYSTEMS

The fuzzy logic systems are universal approximations from the viewpoint of human experts and can uniformly approximate nonlinear continuous functions to arbitrary accuracy [2, 3]. In the indirect adaptive fuzzy control case, to achieve the H_∞ control design purpose, nonlinear functions $f(\underline{x})$ and $g(\underline{x})$ will be approximated by tuning the parameters of the corresponding fuzzy logic systems. On the other hand, in the direct adaptive fuzzy control case, the adaptive control algorithm will be approximated by adjusting the parameters of an adequate fuzzy logic system. Therefore, the fuzzy logic systems in Fig. 1 are qualified as building blocks of H_∞ adaptive controllers for nonlinear systems. Furthermore, the fuzzy logic systems are constructed from the fuzzy IF–THEN rules, using some specific inference, fuzzification, and defuzzification strategies. Therefore, linguistic information from human experts can be directly incorporated into controllers [4].

The basic configuration of the fuzzy logic system is shown in Fig. 1. The fuzzy logic system performs a mapping from $U \in R^n$ to R. Let $U = U_1 \times \cdots \times U_n$, where $U_i \subset R$, $i = 1, 2, \ldots, n$. The fuzzy rule base consists of a

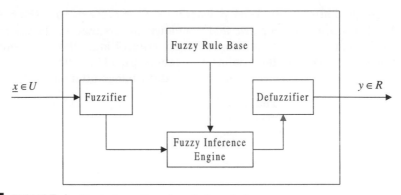

FIGURE 1 The basic configuration of a fuzzy logic system.

collection of fuzzy IF–THEN rules [2–4]:

$$R^{(l)}: \text{IF } x_1 \text{ ix } F_1^l, \text{ and } \cdots \text{ and, } x_n \text{ is } F_n^l,$$

$$\text{THEN } y \text{ is } G^l, \tag{4}$$

where $\underline{x} = (x_1, \ldots, x_n)^T \in U$ and $y \in V \subset R$ are the input and output of the fuzzy logic system, respectively, F_i^l and G^l are labels of fuzzy sets in U_i and R, respectively, and $l = 1, \ldots, M$. The fuzzy inference engine performs a mapping from fuzzy sets in U to fuzzy sets in R, based upon the fuzzy IF–THEN rules in the fuzzy rule base and the compositional rule of inference. The fuzzifier maps a crisp point $\underline{x} = (x_1, \ldots, x_n)^T \in U$ into a fuzzy set A_x in U. The defuzzifier maps a fuzzy set in V to a crisp point in V. More information can be found in [2–4].

The fuzzy logic systems of Fig. 1 comprise a very rich class of static systems mapping from $U \subset R^n$ to $V \subset R$, because many different choices are available within each block, and in addition, many combinations of these choices can result in a useful subclass of fuzzy logic systems. One subclass of fuzzy logic systems is used here as building blocks of our H_∞ adaptive fuzzy controller and is described by the following important result.

LEMMA 1 [4, 15]. *The fuzzy logic systems with center-average defuzzifier, product inference, and singleton fuzzifier are in the following form:*

$$y(\underline{x}) = \frac{\sum_{l=1}^M \bar{y}^l \left(\prod_{i=1}^n \mu_{F_i^l}(x_i) \right)}{\sum_{l=1}^M \left(\prod_{i=1}^n \mu_{F_i^l}(x_i) \right)}, \tag{5}$$

where $\bar{y}^l$ is the point at which μ_{G^l} achieves its maximum value, and we assume that $\mu_{G^l}(\bar{y}^l) = 1$.

Equation (5) can be written as

$$y(\underline{x}) = \underline{\theta}^T \underline{\xi}(\underline{x}), \tag{6}$$

where $\underline{\theta} = (\bar{y}^1, \ldots, \bar{y}^M)^T$ is a parameter vector, and $\underline{\xi}(\underline{x}) = (\xi^1(\underline{x}), \ldots,$

$\xi^M(\underline{x}))^T$ is a regressive vector with the regressor $\xi^l(\underline{x})$ (which is the fuzzy basis function in [16]), defined as

$$\xi^l(\underline{x}) = \frac{\prod_{i=1}^n \mu_{F_i^l}(x_i)}{\sum_{l=1}^M \left(\prod_{i=1}^n \mu_{F_i^l}(x_i)\right)}. \tag{7}$$

There are two main reasons for using the fuzzy logic system Eq. (6) as the basic building block of adaptive fuzzy controllers. First, the fuzzy logic systems in the form of Eq. (6) are proved in [4] and [16] to be universal approximators, i.e., for any given real continuous function f on the compact set U, there exists a fuzzy logic system in the form of Eq. (6) such that it can uniformly approximate f over U to arbitrary accuracy. Therefore, the fuzzy logic systems of Eq. (6) are qualified as building blocks of an adaptive controller for nonlinear systems. Second, the fuzzy logic systems of Eq. (6) are constructed from the fuzzy IF–THEN rules of Eq. (4), using some specific fuzzy inference, fuzzification, and defuzzification strategies. Therefore, linguistic information from a human expert can be directly incorporated into the controllers.

IV. ADAPTIVE FUZZY CONTROL DESIGN

Fuzzy control provides nonlinear control actions for working in situations where a large uncertainty or unknown variation occurs in plant parameters and structures. In general, adaptive control is a useful approach in maintaining the consistent performance of a system in the presence of these uncertainties. Therefore, advanced fuzzy control should be adaptive [4]. An adaptive fuzzy system is a fuzzy logic system equipped with a training (adaptive) algorithm to maintain a consistent performance under plant uncertainties.

The most important advantage of adaptive fuzzy control over conventional adaptive control is that adaptive fuzzy controllers are capable of incorporating linguistic fuzzy information from human operators, whereas conventional adaptive controllers are not. This is especially important for those systems with a high degree of uncertainty. For example, in chemical processes and in the aircraft industry, they are difficult to control from a control theory point of view and are often successfully controlled by human operators [4]. However, human operators successfully control such complex systems without a mathematical model in their mind, but only according to a few control rules in fuzzy terms and some linguistic descriptions regarding the behavior of the system under various conditions that are, of course, also in fuzzy terms. They provide very important information about how to control the system and how the system behaves. Adaptive fuzzy control provides a tool for making use of the fuzzy information in a systematic and efficient manner.

According to the conventional adaptive control [20, 22], adaptive fuzzy control is also divided into two categories [4], i.e., direct and indirect adaptive fuzzy control. An adaptive fuzzy controller that uses fuzzy logic systems as controllers is a direct adaptive fuzzy controller. A direct adaptive fuzzy controller can incorporate fuzzy control rule directly into itself. An adaptive fuzzy controller that uses fuzzy logic systems as a model of the plant is an indirect adaptive fuzzy controller. An indirect adaptive fuzzy controller can directly incorporate fuzzy descriptions about the plant (in terms of fuzzy IF–THEN rules) into itself.

Both direct and indirect adaptive fuzzy control algorithms are developed in this work to facilitate the uncertain nonlinear system in Eq. (1) to achieve the H_∞ tracking performance in Eq. (3). These will be discussed in the following subsections.

A. H_∞ Tracking Control Design in Indirect Adaptive Fuzzy Systems

To begin, let $\underline{\alpha} = (\alpha_n, \ldots, \alpha_1)^T \in R^n$ be the vector of the coefficients of the differential equation in Eq. (2). If the functions f and g in Eq. (1) are known, then the control law

$$u = \frac{1}{g(\underline{x})}\left[-f(\underline{x}) + y_r^{(n)} + \underline{\alpha}^T\underline{e} - u_a\right] \tag{8}$$

can be applied to the nonlinear system in Eq. (1) to achieve the following asymptotically error dynamic system

$$e^{(n)} + \alpha_1 e^{(n-1)} + \cdots + \alpha_n e = u_a - d. \tag{9}$$

Remark 3.   i. The control law u in Eq. (8) is only true over a compact set $U_c \subset R^n$ on which $g(\underline{x}) \neq 0$.

ii. If the system is free of external disturbance $d(t)$ and we let $u_a = 0$, then $e(t) \to 0$ as $t \to \infty$, and its transient behavior of $e(t)$ is determined by specifying the coefficients $\alpha_1, \ldots, \alpha_n$, beforehand.

iii. The control u_a in Eq. (8) is employed to attenuate the external disturbance d and the approximation error due to fuzzy logic systems, whereas fuzzy adaptive control is used in the case without knowledge about $f(\underline{x})$ and $g(\underline{x})$.

The results in Eqs. (8) and (9) are possibly only while $f(\underline{x})$ and $g(\underline{x})$ are well known. However, $f(\underline{x})$ and $g(\underline{x})$ are unknown in our problem. Obtaining a control algorithm similar to Eq. (8) is impossible. In this situation, the approximation by fuzzy logic systems in Section III is employed to treat this tracking control design problem. We replace $f(\underline{x})$ and $g(\underline{x})$ in Eq. (8) with the fuzzy logic systems $\hat{f}(\underline{x}/\underline{\theta}_f)$ and $\hat{g}(\underline{x}/\underline{\theta}_g)$ as Eq. (6), i.e.,

$$\hat{f}(\underline{x}/\underline{\theta}_f) = \underline{\theta}_f^T\underline{\xi}(\underline{x}) = \underline{\xi}^T(\underline{x})\underline{\theta}_f \tag{10}$$

$$\hat{g}(\underline{x}/\underline{\theta}_g) = \underline{\theta}_g^T\underline{\xi}(\underline{x}) = \underline{\xi}^T(\underline{x})\underline{\theta}_g, \tag{11}$$

where $\underline{\xi}(\underline{x})$ is a vector of fuzzy bases, and $\underline{\theta}_f$ and $\underline{\theta}_g$ are the corresponding parameters of fuzzy logic systems. Consequently, the following certainty equivalent controller [20] is obtained, i.e.,

$$u_c = \frac{1}{\hat{g}(\underline{x}/\underline{\theta}_g)} \left[-\hat{f}(\underline{x}/\underline{\theta}_f) + y_r^{(n)} + \underline{\alpha}^T \underline{e} - u_a \right]. \tag{12}$$

Applying Eq. (12) to Eq. (1), after some manipulations, we obtain the tracking error dynamic equation

$$e^{(n)} = -\underline{\alpha}^T \underline{e} + \left[\left(\hat{f}(\underline{x}/\underline{\theta}_f) - f(\underline{x}) \right) \right.$$
$$\left. + \left(\hat{g}(\underline{x}/\underline{\theta}_g) - g(\underline{x}) \right) u_c \right] + u_a - d,$$

or, equivalently,

$$\underline{\dot{e}} = A\underline{e} + Bu_a + b\left[\left(\hat{f}(\underline{x}/\underline{\theta}_f) - f(\underline{x}) \right) \right.$$
$$\left. + \left(\hat{g}(\underline{x}/\underline{\theta}_g) - g(\underline{x}) \right) u_c \right] - Bd, \tag{13}$$

where

$$A = \begin{bmatrix} 0 & 1 & 0 & \cdots & 0 \\ 0 & 0 & 1 & \vdots & 0 \\ \vdots & \vdots & \vdots & \ddots & \vdots \\ -\alpha_n & -\alpha_{n-1} & -\alpha_{n-2} & \cdots & -\alpha_1 \end{bmatrix}, \quad B = \begin{bmatrix} 0 \\ 0 \\ \vdots \\ 1 \end{bmatrix}.$$

Our design objective involves specifying the control u_a and adaptive laws for $\underline{\theta}_f$ and $\underline{\theta}_g$ so that the H_∞ tracking performance in Eq. (3) is achieved.

First, let us define the optimal parameter estimates $\underline{\theta}_f^*$ and $\underline{\theta}_g^*$ as follows:

$$\underline{\theta}_f^* = \arg \min_{\underline{\theta}_f \in \Omega_f} \left[\sup_{\underline{x} \in \Omega_x} \left\| \hat{f}(\underline{x}/\underline{\theta}_f) - f(\underline{x}) \right\| \right]$$

$$\underline{\theta}_g^* = \arg \min_{\underline{\theta}_g \in \Omega_g} \left[\sup_{\underline{x} \in \Omega_x} \left\| \hat{g}(\underline{x}/\underline{\theta}_g) - g(\underline{x}) \right\| \right],$$

where Ω_f, Ω_g, and Ω_x denote the sets of suitable bounds on $\underline{\theta}_f$, $\underline{\theta}_g$, and $\underline{x}$, respectively. We assume that $\underline{\theta}_f$, $\underline{\theta}_g$, and $\underline{x}$ never reach the boundary of Ω_f, Ω_g, and Ω_x. The minimum approximation error is defined as

$$w_e = \left(\hat{f}(\underline{x}/\underline{\theta}_f^*) - f(\underline{x}) \right) + \left(\hat{g}(\underline{x}/\underline{\theta}_g^*) - g(\underline{x}) \right) u_c.$$

Then the tracking error dynamic equation Eq. (13) can be rewritten as

$$\underline{\dot{e}} = A\underline{e} + Bu_a$$
$$+ B\left[\left(\hat{f}(\underline{x}/\underline{\theta}_f) - \hat{f}(\underline{x}/\underline{\theta}_f^*) \right) + \left(\hat{g}(\underline{x}/\underline{\theta}_g) - \hat{g}(\underline{x}/\underline{\theta}_g^*) \right) u_c \right]$$
$$+ B[w_e - d].$$

From Eqs. (10) and (11), the above equation can be rewritten as

$$\dot{e} = Ae + Bu_a + B\left[\underline{\xi}^T(\underline{x})\tilde{\theta}_f + \underline{\xi}^T(\underline{x})\tilde{\theta}_g u_c\right] + Bw, \qquad (14)$$

where $w = w_e - d$, $\tilde{\theta}_f = \underline{\theta}_f - \underline{\theta}_f^*$, and $\tilde{\theta}_g = \underline{\theta}_g - \underline{\theta}_g^*$. Then, the following result can be obtained.

THEOREM 1. *If we select the following adaptive fuzzy control law in the nonlinear system of Eq. (1):*

$$u_c = \frac{1}{\underline{\xi}^T(\underline{x})\underline{\theta}_g}\left[-\underline{\xi}^T(\underline{x})\underline{\theta}_f + y_r^{(n)} + \underline{\alpha}^T\underline{e} - u_a\right], \qquad (15)$$

with

$$u_a = -\frac{1}{r}B^T P\underline{e} \qquad (16)$$

$$\dot{\underline{\theta}}_f = -\gamma_a\underline{\xi}(\underline{x})B^T P\underline{e} \qquad (17)$$

$$\dot{\underline{\theta}}_g = -\gamma_2\underline{\xi}(\underline{x})B^T P\underline{e}u_c, \qquad (18)$$

where r is a positive scalar value, and $P = P^T > 0$ is the solution of the following Riccati-like equation:

$$A^T P + PA + Q - PB\left(\frac{2}{r} - \frac{1}{\rho^2}\right)B^T P = 0, \qquad (19)$$

then the H_∞ tracking performance in Eq. (3) is achieved for a prescribed attenuation level ρ.

Proof. Let us choose a Lyapunov function,

$$V = \frac{1}{2}\underline{e}^T P\underline{e} + \frac{1}{2\gamma_1}\tilde{\theta}_f^T\tilde{\theta}_f + \frac{1}{2\gamma_2}\tilde{\theta}_g^T\tilde{\theta}. \qquad (20)$$

The time derivative of V is

$$\dot{V} = \frac{1}{2}\dot{\underline{e}}^T P\underline{e} + \frac{1}{2}\underline{e}^T P\dot{\underline{e}} + \frac{1}{\gamma_1}\dot{\tilde{\theta}}_f^T\tilde{\theta}_f + \frac{1}{\gamma_2}\dot{\tilde{\theta}}_g^T\tilde{\theta}_g.$$

By the fact that $\dot{\tilde{\theta}}_f = \dot{\underline{\theta}}_f$, $\dot{\tilde{\theta}}_g = \dot{\underline{\theta}}_g$, and Eqs. (14) and Eq. (16), the above equation becomes

$$\dot{V} = \frac{1}{2}\left[\underline{e}^T A^T P\underline{e} - \frac{1}{r}\underline{e}^T PBB^T P\underline{e}\right.$$

$$+ \tilde{\theta}_g^T\underline{\xi}(\underline{x})B^T P\underline{e} + u_c^T\tilde{\theta}_g^T\underline{\xi}(\underline{x})B^T P\underline{e} + w^T B^T P\underline{e}$$

$$+ \underline{e}^T PA\underline{e} - \frac{1}{r}\underline{e}^T PBB^T P\underline{e}$$

$$+\underline{e}^T PB\underline{\xi}^T(\underline{x})\tilde{\underline{\theta}}_f + \underline{e}^T PB\underline{\xi}^T(\underline{x})\tilde{\underline{\theta}}_g u_c + \underline{e}^T PBw \Bigg]$$

$$+\frac{1}{\gamma_1}\dot{\tilde{\underline{\theta}}}_f^T \tilde{\underline{\theta}}_f + \frac{1}{\gamma_2}\dot{\tilde{\underline{\theta}}}_g^T \tilde{\underline{\theta}}_g$$

$$=\frac{1}{2}\underline{e}^T\left(PA + A^T P - \frac{2}{r}PBB^T P\right)\underline{e}$$

$$+\left(\underline{e}^T PB\underline{\xi}^T(\underline{x}) + \frac{1}{\gamma_1}\dot{\underline{\theta}}_f^T\right)\tilde{\underline{\theta}}_f$$

$$+\left(\underline{e}^T PB\underline{\xi}^T(\underline{x})u_c + \frac{1}{\gamma_2}\dot{\underline{\theta}}_g^T\right)\tilde{\underline{\theta}}_g$$

$$+\frac{1}{2}\left(w^T B^T P\underline{e} + \underline{e}^T PBw\right). \tag{21}$$

From the adaptive laws in Eqs. (17)–(18) and the Riccati-like equation (19), we get

$$\dot{V} = -\frac{1}{2}\underline{e}^T Q\underline{e} - \frac{1}{2\rho^2}\underline{e}^T PBB^T P\underline{e}$$

$$+\frac{1}{2}\left(w^T B^T P\underline{e} + \underline{e}^T PBw\right)$$

$$=-\frac{1}{2}\underline{e}^T Q\underline{e} - \frac{1}{2}\left(\frac{1}{\rho}B^T P\underline{e} + \rho w\right)^T\left(\frac{1}{\rho}B^T P\underline{e} - \rho w\right)$$

$$+\frac{1}{2}\rho^2 w^T w$$

$$\leq -\frac{1}{2}\underline{e}^T Q\underline{e} + \frac{1}{2}\rho^2 w^T w.$$

Integrating the above equation from $t = 0$ to $t = T$ yields

$$V(T) - V(0) \leq -\tfrac{1}{2}\int_0^T \underline{e}^T Q\underline{e}\, dt + \tfrac{1}{2}\rho^2\int_0^T w^T w\, dt.$$

Because $V(T) \geq 0$, the above inequality implies the following inequality:

$$\tfrac{1}{2}\int_0^T \underline{e}^T Q\underline{e}\, dt \leq V(0) + \tfrac{1}{2}\rho^2\int_0^T w^T w\, dt.$$

From Eq. (20), the above inequality is equivalent to the following:

$$\frac{1}{2}\int_0^T \underline{e}^T Q\underline{e}\,dt \le \frac{1}{2}\underline{e}^T(0)P\underline{e}(0)$$

$$+\frac{1}{2\gamma_1}\tilde{\underline{\theta}}_f^T(0)\tilde{\underline{\theta}}_f(0) + \frac{1}{2\gamma_2}\tilde{\underline{\theta}}_g^T(0)\tilde{\underline{\theta}}_g(0)$$

$$+\frac{1}{2}\rho^2\int_0^T w^T w\,dt.$$

This is Eq. 3.   ■

Remark 4.   The solvability of H_∞ tracking performance by the adaptive fuzzy control law in Eqs. (15)–(18) is on the existence of positive-semidefinite and symmetric solution P of Eq. (9), which can be rewritten as

$$A^T P + PA + Q - PB\left(\frac{2}{r} - \frac{1}{\rho^2}\right)B^T P = 0.$$

The above Riccati equation has a positive semidefinite solution $P^T = P \ge 0$ if and only if [23]

$$\frac{2}{r} - \frac{1}{\rho^2} \ge 0 \quad \text{or} \quad 2\rho^2 \ge r.$$

Therefore, for a prescribed ρ in H_∞ tracking control, to guarantee the solvability of H_∞ tracking performance by fuzzy adaptive control, the weighting r on the control law u_a of Eq. (16) must satisfy the above inequality.

If the control u_a is also of concern, the H_∞ tracking performance in Eq. (3) should be modified as

$$\int_0^T \left(\underline{e}^T Q\underline{e} + ru_a^2\right) dt \le \underline{e}^T(0)P\underline{e}(0)$$

$$+\frac{1}{\gamma_1}\tilde{\underline{\theta}}_f^T(0)\tilde{\underline{\theta}}_f(0) + \frac{1}{\gamma_2}\tilde{\underline{\theta}}_g^T(0)\tilde{\underline{\theta}}_g(0)$$

$$+\rho^2\int_0^T w^T w\,dt. \tag{22}$$

Then the following result is obtained.

COROLLARY 1.   *For the nonlinear system in Eq. (1), if we choose the fuzzy adaptive law as Eqs. (15)–(18) where $P^T = P \ge 0$ is the solution of the following Riccati-like equation:*

$$A^T P + PA + Q - PB\left(\frac{1}{r} - \frac{1}{\rho^2}\right)B^T P = 0, \tag{23}$$

then the H_∞ tracking performance of Eq. (22) is achieved with a prescribed attenuation level ρ.

Proof. Similar to Theorem 1. ■

Remark 5. i. The H_∞ tracking performance in Eq. (3) without consideration of u_a is a singular H_∞ problem, and the performance in Eq. (22) is a nonsingular H_∞ problem [18].

    ii. The difference in the solutions of the singular H_∞ problem and nonsingular H_∞ problem is the term $-(2/r)PBB^TP$ in Eq. (19) and $-(1/r)PBB^TP$ in Eq. (23).

    iii. The attenuation level ρ in the H_∞ tracking performance can be as small as possible if the positive semidefinite and symmetric solution $P^T = P \geq 0$ in the Riccati-like equation of Eqs. (19) and (23) can be guaranteed.

    iv. From the property of the algebraic Riccati equation [23], Eq. (19) has a solution $P^T = P \geq 0$ if and only if $2\rho^2 \geq r$. However, Eq. (23) has a solution $P^T = P \geq 0$, if and only if $\rho^2 \geq r$. Therefore, for a prescribed disturbance attenuation level ρ, the weighting factor r in u_a must be chosen to satisfy the inequality $2\rho^2 \geq r$ to guarantee the solvability of the H_∞ tracking performance of Eq. (3) in Theorem 1. Similarly, the weighting factor r must be chosen to satisfy $\rho^2 \geq r$ to guarantee the solvability of H_∞ tracking performance of Eq. (22) in Corollary 1.

Remark 6. Notice that Ω_f, Ω_g, and Ω_x need not be known or specified. If the H_∞ attenuation is well designed, $x(t)$ will follow the desired signal $y_r(t)$ well. Because w is bounded by the universal approximation theorem and the assumption of bounded external disturbances, x is bounded. Moreover, additional tools concerning the projection algorithm [4, 20] can be used to analyze the bounded problems of $\underline{\theta}_f$ and $\underline{\theta}_g$. Assume that the constraint sets Ω_f and Ω_g are specified as $\Omega_f := \{\underline{\theta}_f | \|\underline{\theta}_f\| \leq M_f\}$ and $\Omega_g := \{\underline{\theta}_g | \|\underline{\theta}_g\| \leq M_g\}$, respectively, where M_f and M_g are positive constants. Then, the parameter update laws in Eqs. (17) and (18) must be modified as [4, 24]

$$\dot{\underline{\theta}}_f = \begin{cases} -\gamma_1 \underline{\xi}(x) B^T P\underline{e} & \text{if } \|\underline{\theta}_f\| < M_f \\ & \text{or } \left(\|\underline{\theta}_f\| = M_f \text{ and } \underline{e}^T PB\underline{\xi}^T(x)\underline{\theta}_f \geq 0\right) \\ P_f[\cdot] & \text{if } \|\underline{\theta}_f\| = M_f \text{ and } \underline{e}^T PB\underline{\xi}^T(x)\underline{\theta}_f < 0, \end{cases}$$

$$\dot{\underline{\theta}}_g = \begin{cases} -\gamma_2 \underline{\xi}(x) B^T P\underline{e}u_c & \text{if } \|\underline{\theta}_g\| < M_g \\ & \text{or } \left(\|\underline{\theta}_g\| = M_g \text{ and } \underline{e}^T PB\underline{\xi}^T(x)\underline{\theta}_g u_c \geq 0\right) \\ P_g[\cdot] & \text{if } \|\underline{\theta}_g\| = M_g \text{ and } \underline{e}^T PB\underline{\xi}^T(x)\underline{\theta}_g u_c < 0, \end{cases}$$

respectively, where

$$P_f[\cdot] := -\gamma_1 \underline{\xi}(x) B^T P\underline{e} + \gamma_1 \frac{\underline{e}^T PB\underline{\xi}^T(x)\underline{\theta}_f}{\|\underline{\theta}_f\|^2} \underline{\theta}_f$$

and

$$P_g[\cdot] := -\gamma_2 \underline{\xi}(\underline{x}) B^T P \underline{e} u_c + \gamma_2 \frac{\underline{e}^T P B \underline{\xi}^T(\underline{x}) \underline{\theta}_g u_c}{\|\underline{\theta}_g\|^2} \underline{\theta}_g.$$

Because $(\underline{e}^T P B \underline{\xi}^T(\underline{x}) + (1/\gamma_1)\dot{\underline{\theta}}_f^T)\tilde{\underline{\theta}}_f \leq 0$ and $(\underline{e}^T P B \underline{\xi}^T(\underline{x})u_c + (1/\gamma_2)\dot{\underline{\theta}}_g^T)\tilde{\underline{\theta}}_g \leq 0$ in Eq. (21), the H_∞ performance in Eq. (3) can also be guaranteed.

From the above analysis, a design procedure for an indirect H_∞ adaptive fuzzy control algorithm is proposed as follows.

DESIGN ALGORITHM I. Step 1. Select membership functions $\mu_i(\underline{x})$ for $i = 1, 2, \ldots, M$ and compute the fuzzy basis functions $\underline{\xi}(\underline{x})$ as in Eq. (7).

Step 2. Select Q and the desired attenuation level ρ. Give the weighting values γ_1 and γ_2. Select the weighting factor r so that $2\rho^2 \geq r$ in Theorem 1 and $\rho^2 \geq r$ in Corollary 1, to guarantee a positive semidefinite and symmetric solution of the Riccati-like equation.

Step 3. Specify the desired coefficients $\alpha_1, \alpha_2, \ldots, \alpha_n$ as in Eq. (2).

Step 4. Solve the Riccati-like equation in Eq. (19) or Eq. (23).

Step 5. Obtain the fuzzy control law in Eqs. (15)–(18).

B. H_∞ Tracking Control Design in Direct Adaptive Fuzzy Systems

If $g(\underline{x})$ is known in Eq. (1), the direct adaptive fuzzy control in [4, 15], has been used to directly approximate the following control law:

$$\bar{u} = \frac{1}{g(\underline{x})}\left[-f(\underline{x}) + y_r^{(n)} + \underline{\alpha}^T \underline{e}\right].$$

In this paper, an additional H_∞ control u_s is employed to attenuate the effects of fuzzy approximation error and external disturbance. More precisely, when $f(\underline{x})$ and $g(\underline{x})$ in the nonlinear system of Eq. (1) are well known, the following control law would be used:

$$u^* = \bar{u} - \frac{u_s}{g(\underline{x})}, \tag{24}$$

where the H_∞ control u_s will be discussed later. This control law u^* forces the error dynamic to become

$$\dot{\underline{e}} = A\underline{e} + Bu_s - B d.$$

In our case, however, $f(\underline{x})$ is unknown. Therefore, u^* in Eq. (24) cannot be obtained.

Adopt $u = u(\underline{x}/\underline{\theta}) - u_s/g(\underline{x})$, and let the fuzzy control law be in terms of the following fuzzy logic system as Eq. (6):

$$u(\underline{x}/\underline{\theta}) = \underline{\theta}^T \underline{\xi}(\underline{x}) = \underline{\xi}^T(\underline{x})\underline{\theta}.$$

Then we get the control law as

$$u = \underline{\xi}^T(\underline{x})\underline{\theta} - \frac{u_s}{g(\underline{x})}. \tag{25}$$

After some straightforward manipulations, the tracking error dynamic of the nonlinear system of Eq. (1), with the fuzzy control law u in Eq. (25) replacing u^* in Eq. (24), is of the following form:

$$\underline{\dot{e}} = A\underline{e} + Bu_s - Bg(\underline{x})(u(\underline{x}/\underline{\theta}) - \bar{u}) - Bd. \tag{26}$$

Define the optimal parameter vector $\underline{\theta}^*$ for which the fuzzy logic system can approximate the control law $\bar{u}$ optimally:

$$\underline{\theta}^* = \arg \min_{\underline{\theta} \in \Omega_\theta} \left[\sup_{\underline{x} \in \Omega_x} \|u(\underline{x}/\underline{\theta}) - \bar{u}\| \right],$$

where Ω_θ and Ω_x denote the sets of desired bounds on $\underline{\theta}$ and $\underline{x}$, respectively. Let the minimum approximation error be denoted by

$$w_c = -g(\underline{x})(u(\underline{x}/\underline{\theta}^*) - \bar{u}).$$

The error dynamics in Eq. (26) can be rewritten as

$$\underline{\dot{e}} = A\underline{e} + Bu_s - Bg(\underline{x})(u(\underline{x}/\underline{\theta}) - u(\underline{x}/\underline{\theta}^*)) - Bd + Bw_c.$$

From Eq. (25), the above equation becomes

$$\underline{\dot{e}} = A\underline{e} + Bu_s - Bg(\underline{x})\underline{\xi}^T(\underline{x})(\underline{\theta} - \underline{\theta}^*) + B(w_c - d), \tag{27}$$

and then we will specify u_s and an adaptive law for $\underline{\theta}$ to achieve the H_∞ tracking performance. Let us define $\underline{\tilde{\theta}} = \underline{\theta} - \underline{\theta}^*$ and $w = w_c - d$; then Eq. (27) can be rewritten as

$$\underline{\dot{e}} = A\underline{e} + Bu_s - Bg(\underline{x})\underline{\xi}^T(\underline{x})\underline{\tilde{\theta}} + Bw. \tag{28}$$

Assume that the control u_s in Eq. (25) is chosen as

$$u_s = -\frac{1}{r}B^T P\underline{e}, \tag{29}$$

where r is a positive weighting factor, and the matrix $P = P^T \geq 0$ is the solution of the following Riccati-like equation:

$$A^T P + PA + Q - PB\left(\frac{2}{r} - \frac{1}{\rho^2}\right)B^T P = 0. \tag{30}$$

Then we get the following result.

THEOREM 2. *In the nonlinear system in Eq. (1), if the following direct adaptive fuzzy control law is selected*:

$$u = \underline{\xi}^T(\underline{x})\underline{\theta} - u_s/g(\underline{x}) \tag{31}$$

$$u_s = -\frac{1}{r}B^T P\underline{e}$$

$$\dot{\underline{\theta}} = \gamma\underline{\xi}(\underline{x})g(\underline{x})B^T P\underline{e}. \tag{32}$$

then the H_∞ tracking performance in Eq. (3) is achieved for a prescribed attenuation level ρ.

Proof. Let us choose a Lyapunov function,

$$V = \frac{1}{2}\underline{e}^T P\underline{e} + \frac{1}{2\gamma}\tilde{\underline{\theta}}^T\tilde{\underline{\theta}}.$$

Taking the time derivative of the above equation yields

$$\dot{V} = \frac{1}{2}\dot{\underline{e}}^T P\underline{e} + \frac{1}{2}\underline{e}^T P\dot{\underline{e}} + \frac{1}{2\gamma}\dot{\tilde{\underline{\theta}}}^T\tilde{\underline{\theta}} + \frac{1}{2\gamma}\tilde{\underline{\theta}}^T\dot{\tilde{\underline{\theta}}}.$$

By the fact that $\dot{\tilde{\underline{\theta}}} = \dot{\underline{\theta}}$ and Eq. (28), the above equation becomes

$$\dot{V} = \frac{1}{2}\left[\underline{e}^T A^T P\underline{e} + u_s^T B^T P\underline{e}\right.$$

$$- \tilde{\underline{\theta}}^T\underline{\xi}(\underline{x})g(\underline{x})B^T P\underline{e} + w^T B^T P\underline{e} + \frac{1}{\gamma}\dot{\underline{\theta}}^T\tilde{\underline{\theta}}$$

$$+ \underline{e}^T PAe + \underline{e}^T PBu_s - \underline{e}^T PBg(\underline{x})\underline{\xi}^T(\underline{x})\tilde{\underline{\theta}}$$

$$\left. + \underline{e}^T PBw + \frac{1}{\gamma}\tilde{\underline{\theta}}^T\dot{\underline{\theta}}\right].$$

From Eq. (29), we get

$$\dot{V} = \frac{1}{2}\underline{e}^T\left(PA + A^T P - \frac{2}{r}PBB^T P\right)\underline{e}$$

$$- \tilde{\underline{\theta}}^T\left(\underline{\xi}(\underline{x})g(\underline{x})B^T P\underline{e} - \frac{1}{\gamma}\dot{\underline{\theta}}\right)$$

$$+ \frac{1}{2}w^T B^T P\underline{e} + \frac{1}{2}\underline{e}^T PBw.$$

From the adaptive laws in Eq. (32) and the Riccati-like equation (30), we get

$$\dot{V} = -\frac{1}{2}\underline{e}^T Q\underline{e} - \frac{1}{2\rho^2}\underline{e}^T PBB^T P\underline{e}$$

$$+ \frac{1}{2}\left(w^T B^T P\underline{e} + \underline{e}^T PBw\right)$$

$$= -\frac{1}{2}\underline{e}^T Q\underline{e} - \frac{1}{2}\left(\frac{1}{\rho}B^T P\underline{e} - \rho w\right)^T \left(\frac{1}{\rho}B^T P\underline{e} - \rho w\right)$$

$$+ \frac{1}{2}\rho^2 w^T w$$

$$\leq -\frac{1}{2}\underline{e}^T Q\underline{e} + \frac{1}{2}\rho^2 w^T w.$$

By a procedure similar to the proof of Theorem 1, the proof is obtained. ∎

If the control u_s is also of concern, then the H_∞ tracking performance in a direct adaptive fuzzy control system is modified to the following form:

$$\int_0^T \left(\underline{e}^T Q\underline{e} + ru_s^2\right) dt \leq \underline{e}^T(0) P\underline{e}(0) + \frac{1}{\gamma}\tilde{\underline{\theta}}^T(0)\tilde{\underline{\theta}}(0) + \rho^2 \int_0^T w^T w \, dt. \quad (33)$$

COROLLARY 2. *If we adopt the following direct adaptive fuzzy control law for the nonlinear system in Eq. (1):*

$$u = \underline{\xi}^T(\underline{x})\underline{\theta} - u_s/g(\underline{x}) \quad (34)$$

$$u_s = -\frac{1}{r}B^T P\underline{e}$$

$$\dot{\underline{\theta}} = \gamma\underline{\xi}(\underline{x})g(\underline{x})B^T P\underline{e}, \quad (35)$$

where $P^T = P \geq 0$ is the solution of the following Riccati-like equation:

$$A^T P + PA + Q - PB\left(\frac{1}{r} - \frac{1}{\rho^2}\right)B^T P = 0, \quad (36)$$

then the H_∞ tracking performance in Eq. (33) is achieved with a prescribed ρ.

Remark 7. i. The result of H_∞ tracking performance is somewhat different from that in [4] and [15]. In our design objective, the attenuation level ρ can be specified beforehand. In [4] and [15], the stability has been guaranteed. However, the effect due to the approximation error w_c and external disturbance d on the tracking error $\underline{e}(t)$ cannot be attenuated.

ii. Because the adaptive law in Eq. (32) needs to know $g(\underline{x})$ beforehand, the direct adaptive fuzzy control in Theorem 2 is only suitable for those nonlinear systems of which $g(\underline{x})$ is well known. If $g(\underline{x})$ is of the form $g(\underline{x}) = g_0(\underline{x}) + \tilde{g}(\underline{x})$, where $g_0(\underline{x})$ is the well-known part and $\tilde{g}(\underline{x})$ is the uncertain part, then $\tilde{g}(\underline{x})u$ can be considered a part of an external disturbance. In this case, the direct adaptive fuzzy control in Theorem 2 can be applied, and $\tilde{g}(\underline{x})u$ can be attenuated by the proposed H_∞ attenuation design method.

Remark 8. Consider the bounded problem of $\underline{\theta}$. Assume that the constraint set Ω_θ is specified as $\Omega_\theta := \{\underline{\theta} | \|\underline{\theta}\| \le M_\theta\}$, where M_θ is a positive constant. Then the parameter update laws in Eq. (32) must be modified as [4, 24]

$$
\dot{\underline{\theta}} = \begin{cases} \gamma \underline{\xi}(\underline{x}) g(\underline{x}) B^T P \underline{e} & \text{if } \|\underline{\theta}\| < M_\theta \\ & \text{or } \left(\|\underline{\theta}\| = M_\theta \text{ and } \underline{e}^T P B g(\underline{x}) \underline{\xi}^T(\underline{x}) \underline{\theta} \le 0 \right) \\ P_\theta[\cdot] & \text{if } \|\underline{\theta}\| = M_\theta \text{ and } \underline{e}^T P B g(\underline{x}) \underline{\xi}^T(\underline{x}) \underline{\theta} > 0 \end{cases}
$$

where

$$
P_\theta[\cdot] := \gamma \underline{\xi}(\underline{x}) g(\underline{x}) B^T P \underline{e} - \gamma \frac{\underline{e}^T P B g(\underline{x}) \underline{\xi}^T(\underline{x}) \theta}{\|\underline{\theta}\|^2} \underline{\theta}.
$$

Similarly, the H_∞ performance in Eq. (3) can also be guaranteed.

From the above analysis, a design procedure for a direct adaptive fuzzy control algorithm is proposed as follows.

DESIGN ALGORITHM II.   Step 1. Select membership functions $\mu_i(\underline{x})$ for $i = 1, 2, \ldots, M$ and compute the fuzzy basis functions $\underline{\xi}(\underline{x})$ as in Eq. (7).

Step 2. Select Q, γ, r and the desired attenuation level ρ so that $2\rho^2 \ge r$ in Theorem 2 and $\rho^2 \ge r$ in Corollary 2.

Step 3. Specify the desired coefficients α_i, $i = 1, 2, \ldots, n$.

Step 4. Solve the Riccati-like equation in Eq. (30) or Eq. (36).

Step 5. Compute the direct adaptive fuzzy control law in Eqs. (31)–(32) or Eqs. (34)–(35).

V. ILLUSTRATIVE EXAMPLES

Indirect and direct adaptive fuzzy control examples are both given in this section to illustrate the H_∞ tracking performance of the proposed design algorithms.

EXAMPLE 1 (Indirect adaptive fuzzy control approach).   Consider the inverted pendulum system in Fig. 2. Let $x_1 = \theta$ and $x_2 = \dot{\theta}$. The dynamic

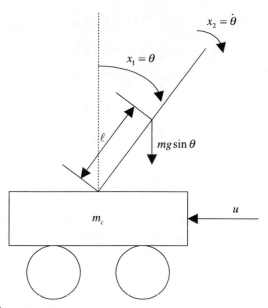

FIGURE 2 The inverted pendulum system.

equations of the inverted pendulum system are [22]

$$\dot{x}_1 = x_2$$

$$\dot{x}_2 = \left[g \sin x_1 - \frac{mlx_2^2 \cos x_1 \sin x_1}{m_c + m} \right] \bigg/ \left[l\left(\frac{4}{3} - \frac{m \cos^2 x_1}{m_c + m} \right) \right]$$

$$+ \left[\frac{\cos x_1}{m_c + m} \right] \bigg/ \left[l\left(\frac{4}{3} - \frac{m \cos^2 x_1}{m_c + m} \right) \right] u + d$$

$$y = x_1,$$

where $g = 9.8 m/s^2$ is the acceleration due to gravity, m_c is the mass of the cart, m is the mass of the pole, l is the half-length of the pole, u is the applied force (control) and d is the external disturbance. For the convenience of simulation, the external disturbance d is assumed to be a square wave with amplitude ± 0.05 and period 2π. The reference signal is assumed here to be $y_r(t) = (\pi/30)\sin t$. We also choose $m_c = 10$, $m = 1$, $l = 3$. How do we use fuzzy adaptive control to achieve the H_∞ tracking performance?

Solution: According to the design procedure 1, the H_∞ tracking design is given by the following steps.

Step 1. The following membership functions are selected:

$$\mu_{F_i^1}(x_i) = \frac{1}{1 + \exp(5 \times (x_i + 0.6))},$$

$$\mu_{F_i^2}(x_i) = \exp\left[-(x_i + 0.4)^2\right],$$

$$\mu_{F_i^3}(x_i) = \exp\left[-(x_i + 0.2)^2\right],$$

$$\mu_{F_i^4}(x_i) = \exp\left[-x_i^2\right],$$

$$\mu_{F_i^5}(x_i) = \exp\left[-(x_i - 0.2)^2\right],$$

$$\mu_{F_i^6}(x_i) = \exp\left[-(x_i - 0.4)^2\right],$$

$$\mu_{F_i^7}(x_i) = \frac{1}{1 + \exp(-5 \times (x_i - 0.6))}.$$

Because we have two state variables, i.e., x_1 and x_2, 14 fuzzy rules of the following form are included in the fuzzy rule bases:

$$R^{(l)}: \text{IF } x_1 \text{ is } F_1^j, \text{THEN } y \text{ is } G_1^j$$

$$\text{for } j = 1, \ldots, 7, l = 1, \ldots, 7,$$

$$R^{(l)}: \text{IF } x_2 \text{ is } F_2^j, \text{THEN } y \text{ is } G_2^j$$

$$\text{for } j = 1, \ldots, 7, l = 8, \ldots, 14.$$

Denote

$$D = \sum_{j=1}^{7} \prod_{i=1}^{2} \mu_{F_i^j}(x_i)$$

$$\underline{\xi}(\underline{x}) = \left[(\mu_{F_1^1} \mu_{F_2^1})/D \cdots (\mu_{F_1^7} \mu_{F_2^7})/D\right]^T$$

$$\underline{\theta}_f = \left[\theta_{f1} \cdots \theta_{f7}\right]^T$$

$$\underline{\theta}_g = \left[\theta_{g1} \cdots \theta_{g7}\right]^T.$$

Then we have

$$\hat{f}(\underline{x}/\underline{\theta}_f) = \underline{\theta}_f^T \underline{\xi}(\underline{x}) = \underline{\xi}^T(\underline{x})\underline{\theta}_f \quad \text{and}$$

$$\hat{g}(\underline{x}/\underline{\theta}_g) = \underline{\theta}_g^T \underline{\xi}(\underline{x}) = \underline{\xi}^T(\underline{x})\underline{\theta}_g.$$

The initial conditions are selected to be $x_1(0) = x_2(0) = 0.2$, $\underline{\theta}_f(0) = \underline{0}$, and $\underline{\theta}_g(0) = 0.2 I_{7 \times 1}$.

Step 2. Select

$$Q = \begin{bmatrix} 10 & 0 \\ 0 & 10 \end{bmatrix}.$$

For high penalty on initial parameter errors $\tilde{\underline{\theta}}_f(0)$ and $\tilde{\underline{\theta}}_g(0)$, select $\gamma_1 = 0.1$ and $\gamma_2 = 0.01$. Consider three cases of prescribed attenuation levels $\rho = 0.05, 0.1, 0.2$ and set $r = 0.005, 0.02, 0.08$, respectively.

Step 3. Select the coefficients $\alpha_1 = 2$ and $\alpha_2 = 1$.
Step 4. Solve the Riccati equation (19)

$$A^T P + PA + Q - PB\left(\frac{2}{r} - \frac{1}{\rho^2}\right)B^T P = 0.$$

Then,

$$P = \begin{bmatrix} 15 & 5 \\ 5 & 5 \end{bmatrix}$$

for all cases.

Step 5. Obtain the indirect adaptive law as

$$u_c = \frac{1}{\xi^T(\underline{x})\underline{\theta}_g}\left[-\underline{\xi}^T(\underline{x})\underline{\theta}_f - \frac{\pi}{30}\sin t + e + 2\dot{e} + \frac{5}{r}(e + \dot{e})\right]$$

$$\underline{\dot{\theta}}_f = -0.5(e + \dot{e})\underline{\xi}(\underline{x})$$

$$\underline{\dot{\theta}} = -0.05(e + \dot{e})u_c\underline{\xi}(\underline{x}).$$

The MATLAB command "ode23" is used to simulate the overall control system with step size 0.001. From the computer simulation, the H_∞ tracking performance is illustrated in Figs. 3–7. The curves of $x_1(t)$ under different attenuation levels are given in Fig. 3; meanwhile, the curves of $x_2(t)$ under different attenuation levels are given in Fig. 4. The tracking control using a conventional adaptive fuzzy logic system does not attack the problem of attenuation of the effect of both the fuzzy approximate error and eternal disturbance on the tracking error. Furthermore, under low attenuation (ρ is large, e.g., $\rho = 0.2$), the H_∞ tracking performance is often poor too. The simulation result in Fig. 5 indicates that the integrals of error signals under different suitably prescribed attenuation levels are decreased obviously. Notice the important feature that even in the computer simulation, the matching errors are unknown. However, the effect of both the fuzzy approximation error and external disturbance on the tracking error has been attenuated efficiently by indirect H_∞ adaptive fuzzy control. The control signals are given in Figs. 6 and 7. As expected, Fig. 6 indicates that the control effort at a higher attenuation is observed to be larger than that at low ones. This is due to the high gain nature under which good performance of the tracking control is desired.

EXAMPLE 2 (Direct adaptive fuzzy approach). Consider the Duffing forced oscillation system [4],

$$\dot{x}_1 = x_2$$

$$\dot{x}_2 = -0.1x_2 - x_1^3 + 12\cos t + u + d$$

$$y = x_1.$$

The system is chaotic if $u(t) = 0$ and $d(t) = 0$. For the convenience of simulation, the reference signal selected here is $y_r(t) = \sin t$, and the exter-

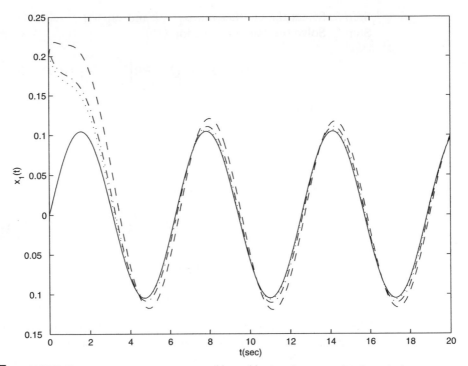

FIGURE 3 Families of response curves $y(t) = x_1(t)$ of tracking control in Example 1.

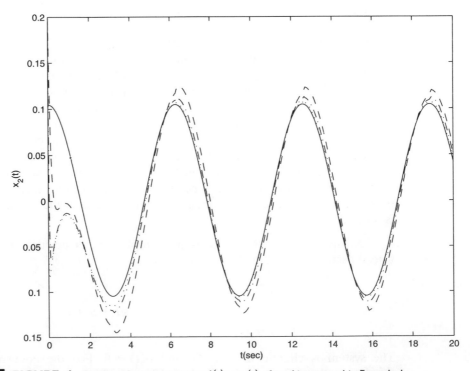

FIGURE 4 Families of response curves $\dot{y}(t) = x_2(t)$ of tracking control in Example 1.

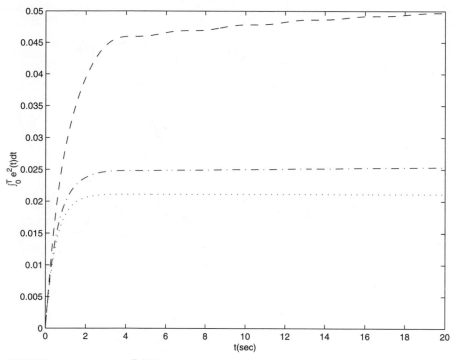

FIGURE 5 Families of $\int_0^T e^2(t)\,dt$ in Example 1.

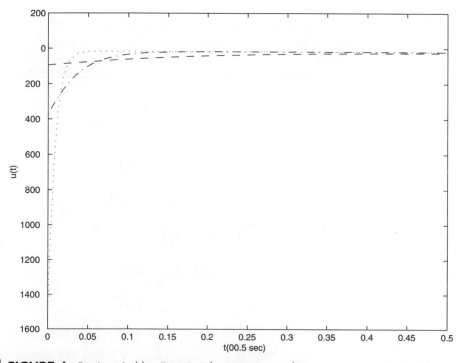

FIGURE 6 Families of $u(t)$ in Example 1 (transient response).

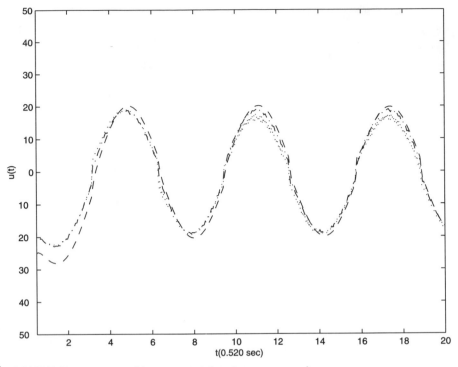

FIGURE 7   Families of $u(t)$ in Example 1 (steady-state response).

nal disturbance d is a square wave with amplitude ± 1 and period 2π. Next, our direct adaptive fuzzy controller is used to achieve the H_∞ tracking performance.

Solution: Step 1. The fuzzy membership functions are chosen as follows:

$$\mu_{F_i^1}(x_i) = \frac{1}{1 + \exp(5 \times (x_i + 2.5))},$$

$$\mu_{F_i^2}(x_i) = \exp\left[-(x_i + 1.5)^2\right],$$

$$\mu_{F_i^3}(x_i) = \exp\left[-(x_i + 0.5)^2\right],$$

$$\mu_{F_i^4}(x_i) = \exp\left[-x_i^2\right],$$

$$\mu_{F_i^5}(x_i) = \exp\left[-(x_i - 0.5)^2\right],$$

$$\mu_{F_i^6}(x_i) = \exp\left[-(x_i - 1.5)^2\right],$$

$$\mu_{F_i^7}(x_i) = \frac{1}{1 + \exp(-5 \times (x_i - 2.5))}.$$

The fuzzy basis conditions and fuzzy approximations are determined as in Example 1. The initial conditions are chosen to be $x_1(0) = x_2(0) = 2$ and $\underline{\theta}(0) = \underline{0}$.

Step 2. Assume that Q is the same as in Example 1 and choose $\gamma = 1$. Consider also three cases of prescribed attenuation levels $\rho = 0.05, 0.1, 0.2$ and select $r = 0.005, 0.02, 0.08$, respectively.

Step 3. Assume that the coefficients α_1, α_2 are the same as in Example 1.

Step 4. The solution of the Riccati-like equation (30) is the same as before for all cases, i.e.,

$$P = \begin{bmatrix} 15 & 5 \\ 5 & 5 \end{bmatrix}.$$

Step 5. Obtain the direct adaptive law as

$$u = \underline{\xi}^T(\underline{x})\underline{\theta} + \frac{1}{r}(e + \dot{e})$$

$$\underline{\dot{\theta}} = 5(e + \dot{e})\underline{\xi}(\underline{x}).$$

From the computer simulation, the H_∞ tracking performance is illustrated in Figs. 8–12. The effects of both the fuzzy logic approximation error and external disturbance on the tracking error are also attenuated efficiently by the proposed method. In Fig. 8, the curves of $x_1(t)$ under different attenuation levels are given. Meanwhile, the curves of $x_2(t)$ under different attenuation levels are given in Fig. 9. If the H_∞ design method is not exploited, even $x_1(t)$ is near the desired sin curve, and the deviation of $x_2(t)$ from the desired sin curve is large. The simulation result in Fig. 10 indicates that the integrals of error signals under different prescribed attenuation levels are decreased sequentially. The control signals under different attenuation levels are given in Figs. 11 and 12. Figure 11 indicates that the initial control effort at a high attenuation is much larger than that at low ones. These phenomena are due to the high gain effect, which is similar to that in Fig. 7.

VI. CONCLUSION

The H_∞ control technique has been combined with an adaptive control algorithm and a fuzzy approximation method in this study to achieve the desired attenuation of disturbance due to the approximation error and external noise in a class of nonlinear system under a large uncertainty or unknown variation in plant parameter and structure. Both indirect and direct adaptive fuzzy control algorithms have been employed to treat this H_∞ tracking design problem. The solvability conditions and the corresponding adaptive fuzzy control laws are both derived. Our results indicate that an arbitrarily small attenuation level can be achieved if the magnitude of

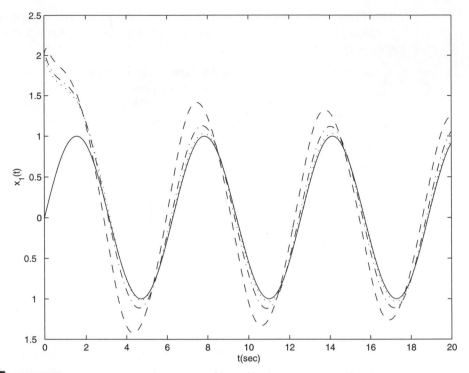

FIGURE 8   Families of response curves $y(t) = x_1(t)$ of tracking control in Example 2.

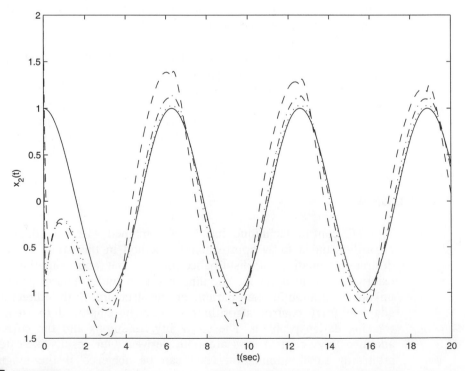

FIGURE 9   Families of response curves $\dot{y}(t) = x_2(t)$ of tracking control in Example 2.

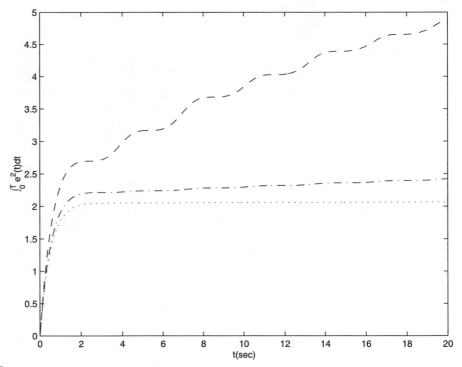

FIGURE 10 Families of $\int_0^T e^2(t)\,dt$ in Example 2.

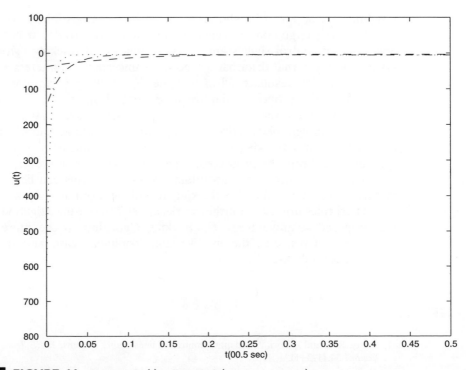

FIGURE 11 Families of $u(t)$ in Example 2 (transient response).

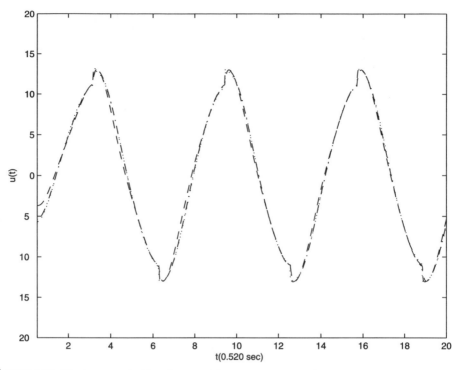

FIGURE 12 Families of $u(t)$ in Example 2 (steady-state response).

weighting factor r can be chosen to be as small as possible. However, this may lead to a large control signal. This situation is a trade-off between the amplitude of control signals and the performance of tracking error for the systems with external disturbance and large uncertain parameters and structures. Simulation results confirm that the H_∞ tracking performance can be achieved via the proposed adaptive fuzzy control algorithms. These results make an important upgrade on both H_∞ and fuzzy control designs. From the H_∞ control design perspective, the H_∞ control method can be extended toward the robust tracking design of the uncertain linear systems via fuzzy logic method. From the fuzzy control design perspective, however, the robust tracking performance of the uncertain nonlinear systems can be guaranteed in fuzzy control via the H_∞ tracking technique even if only a few fuzzy IF–THEN rules are used. Further works are still under investigation to apply the proposed adaptive fuzzy H_∞ tracking algorithm to the more general nonlinear systems, e.g., the multivariable nonlinear case and the output feedback control case.

REFERENCES

1. Sastry, S. S. and Isidori, A. Adaptive control of linearizable systems. *IEEE Trans. Automat. Control* 34:1123–1131, 1989.

2. Kosko, B. *Neural Network and Fuzzy Systems*. Prentice Hall, Englewood Cliffs, NJ, 1992.
3. Jamshidi, M., Vadiee, N., and Ress, T. J. *Fuzzy Logic and Control*. Prentice Hall, Englewood Cliffs, NJ, 1993.
4. Wang, L. X. Adaptive Fuzzy Systems and Control: Design and Stability Analysis. Prentice Hall, Englewood Cliffs, NJ, 1994.
5. Buckley, J. J. Theory of fuzzy controller: An introduction. *Fuzzy Sets and Systems* 51:249–258, 1992.
6. Hwang, G. C. and Lin, S. C. A stability approach to fuzzy control design for nonlinear systems. *Fuzzy Sets and Systems* 48:279–287, 1992.
7. Gramand, B. P. and Newell, R. B. Fuzzy adaptive control of a first order process. *Fuzzy Sets and Systems* 31:47–65, 1989.
8. Lakov, D. Adaptive robot under fuzzy control. *Fuzzy Sets and Systems* 17:1–8, 1985.
9. Lee, C. C. Fuzzy logic in control systems: Fuzzy logic controller. Part I and II. *IEEE Trans. Systems Man Cybernet.* 20:404–435, 1990.
10. Kong, S. G. and Kosko, B. Adaptive fuzzy systems for backing up a truck and trailer. *IEEE Trans. Neural Networks* 3:211–23, 1992.
11. Maeda, M. and Murakami, S. A self-tuning fuzzy controller. *Fuzzy Sets and Systems* 51:29–40, 1992.
12. Raju, G. V. S. and Zhou, J. Fuzzy logic adaptive algorithm to improve robustness in a stream generator level controller. *Control Theory Adv. Tech.* 8:479–493, 1992.
13. Ying, H., Siler, W., and Buckley, J. J. Fuzzy control theory: A nonlinear case. *Automatica* 26:513–520, 1990.
14. Zadeh, L. A. Fuzzy sets. *Information Control* 8:338–353, 1965.
15. Wang, L. X. Stable adaptive fuzzy control of nonlinear systems. *IEEE Trans. Fuzzy Systems* 1:146–155, 1993.
16. Wang, L. X. and Mendel, J. M. Fuzzy basis function, universal approximation, and orthogonal least square learning. *IEEE Trans. Neural Networks* 3:807–814, 1992.
17. Francis, B. A. *A Course in H_∞ Control Theory. Lecture Notes in Control and Information Sciences*, Vol. 8. Springer-Verlag, Berlin, 1987.
18. Stoorvogel, A. *The H_∞ Control Problem: A State Space Approach*. Prentice Hall, Englewood Cliffs, NJ, 1992.
19. Doyle, J., Glover, K., Khargonekar, P. P., and Francis, B. A. State space solution to standard *H*_∞ control problems. *IEEE Trans. Automat. Control* 34:831–847, 1989.
20. Narendra, K. S. and Annaswamy, A. M. *Stable Adaptive Systems*. Prentice Hall, Englewood Cliffs, NJ, 1989.
21. Chen, B. S., Lee, T. S., and Feng, J. H. A nonlinear *H*_∞ control design in robotic systems under parameter perturbation and external disturbance. *Internat. J. Control* 59:439–461, 1994.
22. Slotine, J. E. and Li, W. *Applied Nonlinear Control*. Prentice Hall, Englewood Cliffs, NJ, 1991.
23. Anderson, B. D. O. and Moore, J. B. *Optimal Control: Linear Quadratic Methods*. Prentice Hall, Englewood Cliffs, NJ, 1990.
24. Goodwin, G. C. and Mayne, D. Q. A parameter estimation perspective of continuous time model reference adaptive control. *Automatica* 23:57–70, 1987.
25. Yesildirek, A. and Lewis, F. L. A neural network controller for feedback linearization. In *Proceedings of the 33rd Conference on Decision and Control*. Lake Buena Vista, Florida, 1994, pp. 2494–2498.
26. Chen, B. S. and Lee, C. H. *H*_∞ adaptive fuzzy tracking control design in a class of uncertain nonlinear systems. In *IFAC Automation Conference*. Beijing, YAC, 1995, pp. 407–512.

14

TECHNIQUES AND APPLICATIONS OF FUZZY SMOOTHING ALGORITHMS FOR CONTROL SYSTEMS

YEAN-REN HWANG

Department of Mechanical Engineering, National Central University, Chung-Li, Taiwan, Republic of China

I. INTRODUCTION   375
II. FUZZY SETS AND FUZZY RULE-BASED CONTROL   377
    A. Fuzzy Sets   378
    B. Fuzzy Set Operations   380
    C. Defuzzification   380
    D. Fuzzy Rule-Based Control   382
III. SIGNED DISTANCE OF VARIABLE STRUCTURE
    SYSTEMS   382
IV. FUZZY TUNING LOW-PASS FILTERING CONTROL   385
V. FUZZY REGIONS AND FUZZY BOUNDARIES   394
VI. CONCLUSION   402
    REFERENCES   402

I. INTRODUCTION

Real-life systems are inherently nonlinear, with their dynamics changing at different operation points. When the system's operation range is small, some real systems can be approximated by linear dynamic systems. Based on these linear dynamic systems, linear controllers can be designed to utilize a variety of linear control algorithms. Linear controllers in general have good performance for real systems with a small operation range. However, when a system's operation range is large or its dynamics are not linearizable, linear controllers may perform poorly and sometimes even become unstable. For such systems, nonlinear controllers may be a better choice for controlling the systems. Currently, the gain scheduling and sliding mode controllers are two commonly used nonlinear controllers in industry. Both controllers divide the space of the operation points into several regions and have different control rules in different regions. The control rules are switched when the operation point changes from one region to another. At the boundaries of two regions, the switching of control rules sometimes create nonsmoothed or discontinuous control input, which may excite high-order unmodelled dynamics and induce high-frequency chattering of the systems. Although there are many mathematic smoothing algorithms for gain scheduling and sliding mode controllers, fuzzy smoothing algorithms provide simple but very effective ways to solve the unsmoothed control input problems. In this paper, we will discuss two fuzzy smoothing algorithms and show their applications to automotive suspensions and robot arms.

Design and analysis of linear controllers are mature subjects with many powerful theories and algorithms. However, these linear control theories and algorithms rely on the assumption of small-range operation for linear models to be valid. When the required operation range is large, a real system behaves like a nonlinear system. Traditional linear controllers, which are designed based on linear dynamic systems, are likely to fail or perform poorly for nonlinear systems. This suggests that nonlinear controllers may be better choices for the systems with a large operation range. Although there is no general algorithm for designing nonlinear controllers, the gain scheduling and sliding mode controllers are widely used by researchers and engineers because of their simplicity.

The gain scheduling control [1] attempts to apply linear control methodologies to nonlinear systems. Hence the nonlinear systems have to be linearizable to apply the gain scheduling control. The idea of gain scheduling is to divide the space of operation points into many regions. The nonlinear system is approximated by a linear model in each region. Based on this linear model, a linear controller is designed for that particular region. Although the controller is linear in each region, the overall controller is obviously nonlinear. When the operation changes from one region to another, the gain scheduling controller switches to the new controller accordingly. Because the operation point either belongs or does not belong to one region in gain scheduling control, the control input may suddenly vary at the boundary of two regions. This may have a large impact on the mechanical systems and damage the machines.

When a system is not linearizable at some operation points, the gain scheduling control cannot be applied. Another approach is the sliding mode control [2, 3]. A sliding mode controller defines a sliding surface, which divides the state space into two regions. The dynamics of the sliding surface are exponentially convergent to the origin. The control rules in each region are designed to move all of the states in that region such that they approach the sliding surface. Ideally, the state trajectory will first move from the starting surface to the sliding surface and then approach the origin along the sliding surface. However, the actual state trajectory usually moves forward and backward through the sliding surface in application. Because of this, high-frequency chattering usually happens when the state is close to the sliding surface. The chattering should be eliminated because it will damage the machines and hurt the position accuracy. Several smoothing algorithms had been proposed for sliding mode controllers, such as the smoothing function [4], the boundary layer method [2], the fuzzy rule-based method [5–7], etc.

By applying the fuzzy set idea, two smoothing algorithms were proposed to improve the gain scheduling [8, 9] and sliding mode controllers [9, 10] in the past few years. The improvement can be made in two ways:

1. A fuzzy tuning low-pass filter is added to the original gain scheduling controllers. The bandwidth of the low-pass filter is tuned according to the distance between the operation point and the boundaries of regions.
2. The regions are defined as fuzzy sets instead of crisp sets. Because one state may belong to many fuzzy regions, the control input is calculated by fuzzy operations of the control actions generated from several regions.

It will be shown that the abrupt change in the control input is eliminated by utilizing these two smoothing algorithms.

The remainder of this paper is organized as follows. The basic concepts of fuzzy sets and fuzzy logic control are discussed in Section II; the mathematical description of the signed distance of variable structure systems is discussed in Section III; the fuzzy tuning low-pass filter and its application to automotive suspensions are discussed in Section IV; the fuzzy region controller and its application to a robot arm are discussed in Section V; and the conclusion is given in Section VI.

II. FUZZY SETS AND FUZZY RULE-BASED CONTROL

The fuzzy set theory was proposed by Zadeh [11] in 1967 and has attracted the attention of researchers and engineers since then. One major application of fuzzy set theory is fuzzy logic control, which has undergone a tremendous increase in industrial applications in the past two decades. The advantages of applying fuzzy set theory to control systems can be roughly classified into three categories as follows:

1. It implements experts' experience and knowledge for better automatic control. The experts' experience and knowledge can be expressed as linguistic IF–THEN rules. Based on the fuzzy set operations, these rules can be used to generate control actions.
2. It reduces development and maintenance time. The designers can develop new controllers based on the previous systems by slight changes of the fuzzy sets and inference rules. The maintenance procedures can be created by operator adjustment of the parameters of fuzzy rules.
3. It helps marketing, especially for home appliance products. Fuzzy rules are expressed in linguistic terms, which in general is more acceptable by users. Because of the better user interface, the products with fuzzy logic control usually attract more buyers.

A. Fuzzy Sets

A fuzzy set can be considered as a generalization of a crisp set. An element can either belong or not belong to a crisp set. For example, if a crisp set is defined as $D = \{0\}$, then an element $x_1 = 0$ belongs to D, but $x_2 = 0.001$ does not belong to D. In contrast, a fuzzy set considers elements having a certain degree of membership as belonging to a particular fuzzy set. For example, consider a fuzzy set $\tilde{A} = \{x | x \in R,$ and x is close to $0\}$, and an element $x_2 = 0.001$ can be considered to belong to $\tilde{A}$ to some degree. This degree of belonging to a fuzzy set $\tilde{A}$ ranges from 0 to 1. This means that the fuzzy set $\tilde{A}$ can be characterized by a membership function that is a mapping from the universe of discourse U to the interval $[0, 1]$, namely,

$$\mu_{\tilde{A}}(x) : U \rightarrow [0, 1].$$

Several researchers, such as Procyk and Mamdani [12], Mizumoto [13], and Pedrycz [14], have studied the advantages and disadvantages of different membership functions. Although the membership functions can take any suitable form as long as the range interval is $[0, 1]$, the triangular/trapezoidal and the bell-shaped functions are most commonly used because of their simplicity. Details about fuzzy membership functions are discussed in [15–17]. A triangle membership function, as shown in Fig. 1, can be represented by a Λ-function with three parameters as follows:

$$\Lambda(x; \alpha, \beta, \gamma) = \begin{cases} 0 & x < \alpha \\ (x - \alpha)/(\beta - \alpha) & \alpha \leq x \leq \beta \\ (\gamma - x)/(\beta - \alpha) & \beta \leq x \leq \gamma \\ 0 & x > \gamma, \end{cases}$$

where α, β, and γ, as shown in Fig. 1, are parameters defining the shape of this triangle function. A trapezoidal function, as shown in Fig. 1, can be represented by a Π-function with four parameters as follows:

$$\Pi(x; \beta, \gamma, \sigma, \zeta) = \begin{cases} 0 & x < \beta \\ (x - \beta)/(\gamma - \beta) & \beta \leq x \leq \gamma \\ 1 & \gamma \leq x \leq \sigma \\ (\zeta - x)/(\zeta - \sigma) & \sigma \leq x \leq \zeta \\ 0 & x > \zeta. \end{cases} \tag{1}$$

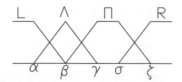

FIGURE 1   Triangular and trapezoidal membership function.

Two other commonly used functions, as shown in Fig. 1, are the L-function and the R-function:

$$L(x; \alpha, \beta) = \begin{cases} 1 & x < \alpha \\ (\beta - x)/(\beta - \alpha) & \alpha \le x \le \beta \\ 0 & x > \beta \end{cases}$$

$$R(x; \sigma, \zeta) = \begin{cases} 0 & x < \sigma \\ (x - \sigma)/(\zeta - \sigma) & \sigma \le x \le \zeta \\ 1 & x > \zeta. \end{cases}$$

Similar to the L- and R-functions, S_L- and S_R-functions can be defined as follows:

$$S_L(x; \alpha, \beta, \gamma) = \begin{cases} 0 & x < \alpha \\ 1 - 2\left(\dfrac{x - \alpha}{\gamma - \alpha}\right)^2 & \alpha \le x \le \beta \\ 2\left(\dfrac{x - \gamma}{\gamma - \alpha}\right)^2 & \beta \le x \le \gamma \\ 1 & x > \beta \end{cases}$$

$$S_R(x; \zeta, \sigma, \tau) = \begin{cases} 0 & x < \zeta \\ 2\left(\dfrac{x - \zeta}{\tau - \zeta}\right)^2 & \zeta \le x \le \sigma \\ 1 - 2\left(\dfrac{x - \tau}{\tau - \zeta}\right)^2 & \sigma \le x \le \tau \\ 1 & x > \sigma, \end{cases}$$

where α, β, γ, δ, ζ, σ, and τ are defined in Fig. 2. Zadeh's bell-shaped function is defined based on the S_R-function:

$$\Phi(x; \epsilon, \delta) = \begin{cases} S_R(x; \delta - \epsilon, \delta - \epsilon/2, \delta) & x \le \delta \\ 1 - S_R(x; \delta, \delta + \epsilon/2, \delta + \epsilon) & x > \delta, \end{cases}$$

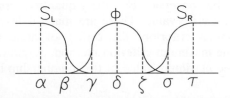

FIGURE 2 Bell-shaped membership functions.

where ϵ is equal to $\delta - \beta$ in Fig. 2. The bell-shaped functions are smoother than the triangle functions. However, they require more computation for set operations than triangle functions.

B. Fuzzy Set Operations

Fuzzy set operations have been defined by researchers in different forms (see [14, 16] for examples). Because fuzzy sets are characterized by their membership functions, the set operations on fuzzy sets are also defined in terms of membership functions. Zadeh proposed the AND and OR fuzzy set operations as follows:

$$\mu_{\tilde{A} \cap \tilde{B}}(x) = \min(\mu_{\tilde{A}}(x), \mu_{\tilde{B}}(x)),$$

$$\mu_{\tilde{A} \cup \tilde{B}}(x) = \max(\mu_{\tilde{A}}(x), \mu_{\tilde{B}}(x)),$$

where $\mu_{\tilde{A}}(x)$ and $\mu_{\tilde{B}}(x)$ are the degree of membership that x belongs to $\tilde{A}$ and $\tilde{B}$, respectively, and $\mu_{\tilde{A} \cap \tilde{B}}(x)$ and $\mu_{\tilde{A} \cup \tilde{B}}(x)$ are the degree of membership that x belongs to "$\tilde{A}$ AND $\tilde{B}$" and "$\tilde{A}$ OR $\tilde{B}$," respectively. Other widely used definitions for AND and OR operations on fuzzy sets are the algebraic product and algebraic sum.

$$\mu_{\tilde{A} \cap \tilde{B}}(x) = \mu_{\tilde{A}}(x) \cdot \mu_{\tilde{B}}(x)$$

$$\mu_{\tilde{A} \cup \tilde{B}}(x) = \mu_{\tilde{A}}(x) + \mu_{\tilde{B}}(x) - \mu_{\tilde{A}}(x) \cdot \mu_{\tilde{B}}(x).$$

Details about fuzzy set operations can be found in [14–16].

The fuzzy set operations are the mathematical tool required to process the inference rules in a fuzzy rule-based controller. These inference rules are presented as linguistic statements such as

$$\text{IF } \tilde{p} \text{ is } \tilde{A}_i, \quad \text{AND} \quad \text{IF } \tilde{q} \text{ is } \tilde{B}_i, \quad \text{THEN } \tilde{u} \text{ is } \tilde{U}_i,$$

where $i = 1, 2, \ldots$ is the rule index, $\tilde{p}$ and $\tilde{q}$ are linguistic variables representing the process state variables or outputs, $\tilde{u}$ is a linguistic variable representing the control variable, and $\tilde{A}_i$, $\tilde{B}_i$, and $\tilde{U}_i$ are fuzzy sets over $\tilde{p}$, $\tilde{q}$, and $\tilde{u}$, respectively. The truth value of a rule is the degree to which the antecedent is true. This value can be calculated by the fuzzy set operations.

C. Defuzzification

The defuzzification process is the conversion of a fuzzy quantity, represented by a membership function, to a crisp value. There are many defuzzification strategies such as the maximum criterion, the center of area, and the weighted average methods. The maximum criterion method, as shown in Fig. 3, defines the crisp value as the maximum value of the membership function,

$$u = \max_{u \in U}(\mu_U(u)),$$

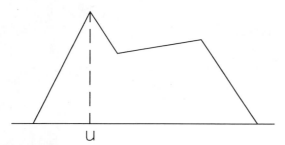

FIGURE 3 Maximum criterion method.

where $\mu_U(u)$ is the fuzzy membership function representing the fuzzy quantity.

The center-of-area method, as shown in Fig. 4, defines the crisp value as the gravity center of the membership function:

$$u = \frac{\int \mu_U(u)\dot{u}\,du}{\int \mu_U(u)}$$

The weighted average method defines the crisp value as the weighted average of membership functions, as shown in Fig. 5.

This method is valid only for the case where the output membership funciton is an union result of several fuzzy quantities:

$$u = \frac{\sum_{i=1}^{m} \mu_i u_i}{\sum_{i=1}^{m} \mu_i}, \tag{2}$$

where μ_i is the maximum value of the ith fuzzy quantity and u_i is chosen as the quantity at which $\mu_{\tilde{U}_i}$ equals its maximum value.

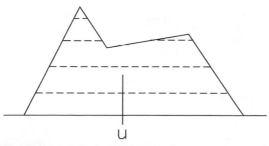

FIGURE 4 Center-of-area method.

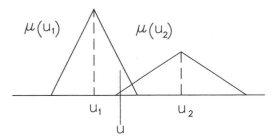

FIGURE 5 Weighted average method.

D. Fuzzy Rule-Based Control

A typical fuzzy rule-based controller is shown in Fig. 6. It contains four major blocks, i.e., rule data base, the fuzzification, inference engine, and defuzzification blocks. The rule data base stores the control rules for the inference engine. The fuzzification block converts the sensors' crisp data to fuzzy quantities represented by their membership functions. These fuzzy quantities are processed in the inference engine by applying the fuzzy set operations described in Section II.B and result in some fuzzy outputs. These fuzzy outputs are then changed back into crisp values through the defuzzification process described in Section II.C.

III. SIGNED DISTANCE OF VARIABLE STRUCTURE SYSTEMS

The gain schedule and sliding controllers have different control rules in different regions of the space. This means that these two controllers can be classified as variable structure systems. The dynamics of the system depends on the location of the operation point and its relation to the regions' boundaries of the variable structure systems. This relation can be defined by the "signed distance" introduced in this section.

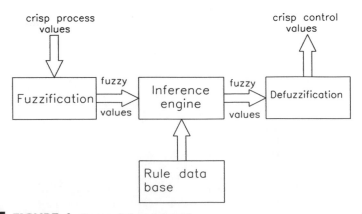

FIGURE 6 Fuzzy rule-based control.

A continuous system with a gain schedule controller or a sliding controller can be expressed by the following equations:

$$\dot{\mathbf{x}} = f(\mathbf{x}, u, t)$$

and

$$\begin{cases} u = g_1(\mathbf{x}) & \text{if } \mathbf{x} \in \Sigma_1 \\ u = g_2(\mathbf{x}) & \text{if } \mathbf{x} \in \Sigma_2 \\ \quad \vdots \\ u = g_m(\mathbf{x}) & \text{if } \mathbf{x} \in \Sigma_m, \end{cases} \tag{3}$$

where $\mathbf{x} \in R^n$, $u \in R$, and Σ_i's are mutually exclusive regions of the state space, and $g_i(\mathbf{x})$ is the control rule in Σ_i. Note that the control rule $g_i(\mathbf{x})$ is designed for the corresponding region Σ_i based on some particular criteria. It could be any linear or nonlinear control algorithms. Because $g_i(\mathbf{x})$ is different in different regions, the overall dynamic system is a variable structure system.

The union of all regions is equal to the entire state space R^n, i.e.,

$$\Sigma_1 \cup \Sigma_2 \cup \cdots \cup \Sigma_m = R^n.$$

These regions are crisp sets because the state can only either be in or not in one particular region. Furthermore, each state belongs to one and only one region. These regions are divided by their boundary manifolds, which represent the checking conditions of the variable structure system. Let S_{ij} denote the boundary manifold between the regions Σ_i and Σ_j. In general, the degree of freedom of each region is equal to n, and the degree of freedom of each manifold is equal to $n - 1$.

That a state belongs to a particular region can be checked by the "sign distance" (denoted by ζ) between the state and the boundary manifolds of that region. There are many ways to define ζ; two of them are given below:

1. The l_p form signed distance $d_p(\mathbf{x}, S_{ij})$, defined as

$$\zeta = d_p(\mathbf{x}, S_{ij}) = sign_{ij}(\mathbf{x}) \cdot \min_{\mathbf{y} \in S_{ij}} \{\|\mathbf{x} - \mathbf{y}\|_p\}$$

$$= sign_{ij}(\mathbf{x}) \cdot \min_{\mathbf{y} \in S_{ij}} \{|(x_1 - y_1)^p + (x_2 - y_2)^p$$

$$+ \cdots + (x_n - y_n)^p|^{1/p}\},$$

where $sign_{ij}(\mathbf{x})$ is positive if $\mathbf{x}$ is on one particular side of S_{ij} and negative if $\mathbf{x}$ is on the other side of S_{ij}. The most common use is the $d_2(\mathbf{x}, S_{ij})$ function, because it represents the distance in a Cartesian sense:

$$\zeta = d_2(\mathbf{x}, S_{ij})$$

$$= sign_{ij}(\mathbf{x}) \cdot \min_{\mathbf{y} \in S_{ij}} \left\{ \sqrt{(x_1 - y_1)^2 + (x_2 - y_2)^2} \right\}.$$

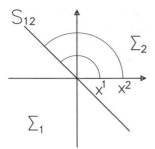

FIGURE 7 Example of the signed distance.

If S_{ij} is a linear manifold, then $d_2(\mathbf{x}, S_{ij})$ is equal to $\hat{s}^T\mathbf{x}$, where $\hat{s}$ is the unit normal vector of S_{ij}. Figure 7 shows an example of a two-dimensional sliding mode control problem. The sliding surface, which in general is a line passing through the origin, divides the entire state space (i.e., R^2) into two regions. If $d_2(\mathbf{x}, S_{12})$ is chosen as the "signed distance" to S_{12},

$$d_2(\mathbf{x}, S_{12}) = sign_{12}(\mathbf{x})\,\frac{\mathbf{x}^T\mathbf{n}}{|\mathbf{n}|},$$

where $\mathbf{n}$ is the normal vector to S_{12}. The sign function $sign_{12}(\mathbf{x})$ is equal to 1 when $\mathbf{x}$ is in Σ_1, and $sign_{12}(\mathbf{x})$ is equal to -1 when $\mathbf{x}$ is in Σ_2. Because of its physical meaning and simple calculation, $d_2(\mathbf{x}, S_{ij})$ is frequently used, (e.g., Kawaji and Matsunaga [6].)

2. The angle between $\mathbf{x}$ and the normal vector of S_{ij}. For all linear systems, the time required for a state to travel to any linear manifold containing the origin is independent of the modulus of the state. For example, if a variable structure system is defined as

$$\dot{\mathbf{x}} = \begin{pmatrix} -1 & 0 \\ 2 & -1 \end{pmatrix}\mathbf{x} + \begin{pmatrix} 1 \\ -1 \end{pmatrix}u, \tag{4}$$

and the control law is

$$u = (1, -1)^T\mathbf{x} \qquad \text{if } \mathbf{x} \in \Sigma_1$$

$$u = (-1, 1)^T\mathbf{x} \qquad \text{if } \mathbf{x} \in \Sigma_2,$$

where Σ_1 and Σ_2, shown in Fig. 7, are two regions divided by the boundary line,

$$(1, -1)^T\mathbf{x} = 0.$$

Consider two initial states,

$$\mathbf{x}^1 = (1, 0)^T$$

$$\mathbf{x}^2 = (2, 0)^T.$$

Although the modulus of x^2 is twice that of x^1, the times required for the initial states x^1 and x^2 to reach the boundary line will be the same. This fact is important for a discrete time control system because the sampling time is usually constant. The previous definition of $d_p(x, S_{ij})$ is not suitable because, for the same d_p values, x with a larger modulus requires less time to hit the manifold than $\bar{x}$ with a smaller modulus. In this case, the angle θ between x and the normal vector of the boundary manifold, n, and its cosine value are suitable measurements for the "signed distance":

$$\cos\theta = \frac{x^T n}{|x|\,|n|}$$

or

$$\theta = \cos^{-1}\left(\frac{x^T n}{|x|\,|n|}\right).$$

When the state x is close to the boundary manifold, θ is close to $90°$ and $\cos\theta$ is close to zero.

Note that the sign of the "signed distance" ζ shows the side of the manifold that the state belongs to. Because a variable structure system has different dynamics at different sides of the manifold, the sign of ζ is meaningful and should not be neglected.

For variable structure systems, the state can only be in one particular region. When the state passes through a boundary manifold, the state enters a different region instantly and the control input also switches instantly from one law to another. These switchings generally are not smooth and may cause undesired dynamics of the variable structure system. A typical example is the chattering phenomenon of a sliding mode control system that has discontinuous control inputs on different sides of the sliding surface. To smooth the control input, we propose two methods for designing a fuzzy smoothing controller. First, a low-pass filter with fuzzy tuning bandwidth is used. Second, a fuzzy logic controller with fuzzy regions is proposed to replace the original crisp regions of a variable structure system.

IV. FUZZY TUNING LOW-PASS FILTERING CONTROL

The first type of fuzzy smoothing algorithm is one in which a low-pass filter is added to smooth the control input. This type of algorithm has been successfully applied to semiactive automobile suspensions.

A typical passive quarter car suspension system mainly consists of four major elements: a sprung mass, an unsprung mass, an absorber, and a spring, as shown in Fig. 8. The suspension's characteristics are determined by these elements. Once these elements are selected, the suspension's characteristics cannot be modified during the operation.

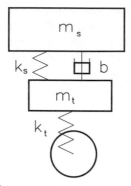

FIGURE 8 Passive quarter car suspension model.

The values of m_s, m_u, k_s, k_u, b, $u_{\min}$, and $u_{\max}$ are chosen according to the Nissan Infiniti model [18]:

$$m_s = 240 \text{ kg}$$

$$m_u = 36 \text{ kg}$$

$$k_s = 16{,}000 \text{ NT/m}$$

$$k_t = 160{,}000 \text{ NT/m}$$

$$b = 1000 \text{ NT s/m}.$$

A fully active suspension system has an actuator (in general, a hydraulic actuator) between the sprung and unsprung masses. The characteristics of the fully active suspension may be arbitrarily adjusted by choosing different control rules of the actuator input. Although the active suspension system is known to improve performance and is available in several luxury models, it is very expensive and may cause the suspension system to become unstable.

A much cheaper and safer alternative is to install a damper with a variable damping coefficient as shown in Fig. 9, which is referred to as a semiactive suspension system. Note that the variable damper is effective for generating a pushing force for $\dot{z}_s - \dot{z}_u < 0$ and a pulling force for $\dot{z}_s - \dot{z}_u > 0$. The semiactive suspension performs much better than a passive suspension and worse than a fully active suspension. However, a semiactive suspension is much cheaper than a fully active suspension. Therefore, more and more car manufacturers are interested in developing semiactive suspension systems for their new models, such as the Nissan Infiniti.

Figure 9 shows the dynamic model of a quarter car semiactive suspension system, which has the variable damping coefficient u as its control input. The state equation of a quarter car semiactive suspension system is described as

$$\dot{\mathbf{x}} = A\mathbf{x} + (N\mathbf{x})u + Lw, \tag{5}$$

where $A \in R^{4 \times 4}$, $N \in R^{4 \times 4}$ are constant matrices, $L \in R^4$ is a constant vector, $w \in R$ is the road disturbance, and $u \in R$ is the control input and

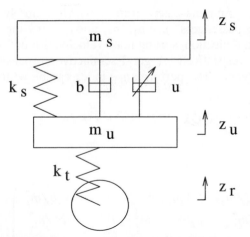

FIGURE 9 Semiactive quarter car suspension model. m_s = sprung mass = 240 kg; m_u = unsprung mass = 36 kg; k_s = suspension spring = 16,000 Nt/m; k_t = tire spring = 160,000 Nt/m; b = constant damping = 1000 Nt s/m; u = variable damping; z_s = vertical position of sprung mass; z_u = vertical position of unsprung mass; z_r = vertical position of road; $w = \dot{z}_r$ = vertical velocity of road; $x_1 = z_s - z_u$, $x_2 = \dot{z}_s$; $x_3 = z_u - z_s$, $x_4 = \dot{z}_u$. Reprinted with permission from Y. R. Hwang and S. C. Wang, *IEEE Trans. Fuzzy Systems* 2:277–284 (© 1994 IEEE).

corresponds to the adjustable damping coefficient taking a value in $[u_{\min}, u_{\max}]$:

$$
A = \begin{pmatrix}
0 & 1 & 0 & -1 \\
-k_s/m_s & -b/m_s & 0 & b/m_s \\
0 & 0 & 0 & 1 \\
k_s/m_u & b/m_u & -k_t/m_u & -b/m_u
\end{pmatrix};
$$

$$
N = \begin{pmatrix}
0 & 0 & 0 & 0 \\
0 & -1/m_s & 0 & 1/m_s \\
0 & 0 & 0 & 0 \\
0 & 1/m_u & 0 & -1/m_u
\end{pmatrix};
$$

$$
L = \begin{pmatrix} 0 & 0 & -1 & 0 \end{pmatrix}^T.
$$

The $u_{\min}$ and $u_{\max}$ are chosen to be 0 and 3000 NT s/m, respectively, in this paper.

The state $\mathbf{x} = (x_1, x_2, x_3, x_4)^T$, and x_1, x_2, x_3, and x_4 represent the suspension deflection, sprung mass velocity, tire deflection, and unsprung mass velocity, respectively. These physical values, i.e., x_1, x_2, x_3, and x_4, determine the performance of the semiactive suspension system. Hence the performance index of the semiactive suspension system can be defined as

$$
J = \lim_{t_f \to \infty} \tfrac{1}{2} \int_0^{t_f} \{\dot{x}_2^2 + \rho_1 x_1^2 + \rho_2 x_2^2 + \rho_3 x_3^2 + \rho_4 x_4^2\}\, dt,
$$

where $\dot{x}_2$ represents the sprung mass acceleration, which is closely related to the force acting on the passengers. ρ_1, ρ_2, ρ_3, and ρ_4 are the weighting factors on the suspension deflection, sprung mass velocity, tire deflection, and unsprung mass velocity, respectively. A good semiactive suspension should have a small J and vice versa. The performance index can be written as

$$J = \lim_{t_f \to \infty} \tfrac{1}{2} \int_0^{t_f} \{\mathbf{x}^T Q \mathbf{x} + 2\mathbf{x}^T S \mathbf{x} u + \mathbf{x}^T R \mathbf{x} u^2\}\, dt$$

and

$$Q = \begin{pmatrix} \rho_1 + k_s^2/m_s^2 & k_s b/m_s^2 & 0 & -k_s b/m_s^2 \\ k_s b/m_s^2 & \rho_2 + b^2/m_s^2 & 0 & -b^2/m_s^2 \\ 0 & 0 & \rho_3 & 0 \\ -k_s b/m_s^2 & -b^2/m_s^2 & 0 & \rho_4 + b^2/m_s^2 \end{pmatrix};$$

$$S = \begin{pmatrix} 0 & 0 & 0 & 0 \\ k_s/m_s^2 & b/m_s^2 & 0 & -b/m_s^2 \\ 0 & 0 & 0 & 0 \\ -k_s/m_s^2 & -b/m_s^2 & 0 & b/m_s^2 \end{pmatrix};$$

$$R = \begin{pmatrix} 0 & 0 & 0 & 0 \\ 0 & 1/m_s^2 & 0 & -1/m_s^2 \\ 0 & 0 & 0 & 0 \\ 0 & -1/m_s^2 & 0 & 1/m_s^2 \end{pmatrix}.$$

The suboptimal control law, derived by applying dynamic programming techniques [9, 18], is described as

$$u = u_{\min} \quad \text{if} \quad \begin{cases} \alpha^T x > 0 \text{ and } \gamma_1^T x > 0 \\ \alpha^T x < 0 \text{ and } \gamma_1^T x < 0 \end{cases}$$

$$u = u^o \quad \text{if} \quad \begin{cases} \alpha^T x > 0 \text{ and } \gamma_1^T x \le 0 \le \gamma_2^T x \\ \alpha^T x < 0 \text{ and } \gamma_1^T x \ge 0 \ge \gamma_2^T x \end{cases}$$

$$u = u_{\max} \quad \text{if} \quad \begin{cases} \alpha^T x > 0 \text{ and } \gamma_2^T x < 0 \\ \alpha^T x < 0 \text{ and } \gamma_2^T x > 0, \end{cases}$$

where

$$u^o = -\frac{\beta^T \mathbf{x}}{\alpha^T \mathbf{x}} - b$$

is the optimal control for nonconstrained input (i.e., $u^o \in [-\infty, \infty]$), b is the damping coefficient of the passive damper (see Fig. 9), and

$$\alpha = \begin{pmatrix} 0 & 1/m_s & 0 & -1/m_s \end{pmatrix}^T;$$

$$\gamma_1 = (u_{\min} + b)\alpha + \beta;$$

$$\gamma_2 = (u_{\max} + b)\alpha + \beta;$$

$$\beta = H\eta + \zeta;$$

$$\eta = \begin{pmatrix} 0 & -1/m_s & 0 & 1/m_u \end{pmatrix}^T;$$

$$\zeta = \begin{pmatrix} k_s/m_s^2 & 0 & 0 & 0 \end{pmatrix}^T;$$

and H is a positive symmetric matrix satisfying the following algebra Riccati equation:

$$H\bar{A} + \bar{A}^T H + \bar{Q} - H\eta \bar{R}^{-1} \eta^T H = 0,$$

where

$$\bar{A} = A - m_s^2 \eta (b\alpha + \zeta)^T$$

$$\bar{Q} = m_s^2 (b\alpha + \zeta)(b\alpha + \zeta)^T$$

and $\bar{R} = 1/m_s^2$. α, γ_1, and γ_2 are the normal vectors of the following three linear boundary manifolds:

$$\alpha^T x = 0; \qquad \text{with normal vector} = \alpha$$

$$\gamma_1^T x = 0; \qquad \text{with normal vector} = \gamma_1$$

$$\gamma_2^T x = 0; \qquad \text{with normal vector} = \gamma_2.$$

Note that $\gamma_1 = (v_{\min} + b)\alpha + \beta$, $\gamma_2 = (u_{\max} + b)\alpha + \beta$, and α are all in the same second-order subspace defined by α and β. Therefore, these three boundary manifolds intersect at the same R^2 subspace and divide the state space into six regions.

The conceptual picture of the three boundary manifolds, their normal vectors, and the six regions divided by them are shown in Fig. 10. The control input u is assigned to be $u_{\min}$ in Region I (where $\alpha^T x > 0$ and $\gamma_1^T x > 0$) and Region IV (where $\alpha^T x < 0$ and $\gamma_1^T x < 0$). This is because u^o is smaller than $u_{\min}$ in Regions I and IV, and u has to be within the range bounded by $u_{\min}$ and $u_{\max}$. Similarly, u^o is larger than $u_{\max}$ in Region III (where $\alpha^T x > 0$ and $\gamma_2^T x < 0$) and Region VI (where $\alpha^T x < 0$ and $\gamma_2^T x > 0$). Therefore, u is assigned to be $u_{\max}$ in Regions III and VI for this suboptimal control law. In Region II (where $\alpha^T x > 0$ and $\gamma_1^T x \leq 0 \leq \gamma_2^T x$) and Region V (where $\alpha^T x < 0$ and $\gamma_1^T x \geq 0 \geq \gamma_2^T x$), u^o is within the range $[u_{\min}, u_{\max}]$, and u is assigned

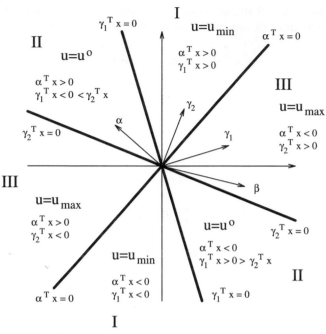

FIGURE 10 Geometric interpretation of suboptimal control law. Reprinted with permission from Y. R. Hwang and S. C. Wang, *IEEE Trans. Fuzzy Systems* 2:277–284 (© 1994 IEEE).

to be u^o. Note that the system's dynamics are linear within each region and all boundary manifolds are not attractive sets:

$$\dot{\mathbf{x}} = \left(A + u_{\min} m_s^2 \eta \alpha^T \right) \mathbf{x} \qquad \text{if } \mathbf{x} \in \text{Regions I and IV}$$

$$\dot{\mathbf{x}} = \left(A - m_s^2 (\beta + b\alpha)^T \right) \mathbf{x} \qquad \text{if } \mathbf{x} \in \text{Regions II and V}$$

$$\dot{\mathbf{x}} = \left(A + u_{\max} m_s^2 \eta \alpha^T \right) \mathbf{x} \qquad \text{if } \mathbf{x} \in \text{Regions III and VI.}$$

Apparently, the entire system is a variable structure system. As the state $\mathbf{x}$ crosses these boundary manifolds, the controller switches from one law to another and the dynamic system switches from one linear system to another. It has been shown by simulation that high sprung mass jerk happens when $\mathbf{x}$ crosses the manifold $\alpha^T \mathbf{x} = 0$. This means that a large "force impact" is acting on the passengers at this moment. To eliminate large force impact, the jerk must be reduced and the control input must be smoothed.

A common approach to smoothing the control input is

$$\dot{u} = -\lambda(u - u_i) \qquad \text{if } \mathbf{x} \in \Sigma_i, \tag{6}$$

where λ is the bandwidth of the filter, u_i is the control command obtained from the suboptimal control law in different regions, and u is the filtered input of this system. When λ is small, abrupt changes in u can be prevented. However, if λ is too small, the difference between u and u_i will become large and the performance achieved by the original suboptimal control law may be lost. Intuitively, when the state is extremely close to one of the boundary manifolds, λ should be small because the change in u is expected to be abrupt. Otherwise, λ can be made large so that the performance achieved by the original suboptimal controlled semiactive suspension system will not be lost. Therefore, the following fuzzy rules are used to select λ:

$$\text{IF} \quad \zeta_1 \text{ is } \tilde{M}_i \quad \text{THEN } \lambda \text{ is } \tilde{N}_i, \tag{7}$$

where ζ_1 is the "signed distance" presented in the previous section, and $\tilde{M}_i$'s and $\tilde{N}_i$'s are fuzzy sets over ζ_1 and λ, respectively. According to Eq. (2), the crisp value for λ can be calculated by the following equation:

$$\lambda = \frac{\sum_{i=1}^{m} \mu_i \lambda_i}{\sum_{i=1}^{m} \mu_i}, \tag{8}$$

where λ_i is the value such that $\mu_{\tilde{N}_i}(\lambda_i) = \max\{\mu_{\tilde{N}_i}\}$.

The discrete time version of this low-pass filter is

$$\Delta u(k) = -\lambda \Delta t (u(k-1) - u_i(k-1)), \tag{9}$$

where Δt represents the sampling time and $\Delta u(k) = u(k) - u(k-1)$. Therefore, the fuzzy inference laws for selecting λ can be generalized to the fuzzy inference laws for controlling $\Delta u(k)$. Because $\Delta u(k)$ is equal to the multiplication of $\lambda \Delta t$ and $(u(k-1) - u_i(k-1))$, the inference laws for controlling $\Delta u(k)$ may be further generalized to depend both on $\zeta_1(k-1)$ and $(u(k-1) - u_i(k-1))$ as follows:

$$\text{IF } \zeta_1(k-1) \text{ is } \tilde{M}_i \quad \text{AND}$$

$$\text{IF } \left(u(k-1) - u_i(k-1)\right) \text{ is } \tilde{P}_i, \quad \text{THEN } \Delta u(k) \text{ is } \tilde{Q}_i,$$

where $\tilde{M}_i$, $\tilde{P}_i$, and $\tilde{Q}_i$ are suitable fuzzy sets. Note that this type of inference rule is often used when a fuzzy logic controller is designed.

Because the overall system (Eqs. (3) and (6)) is one order higher than the original system, the low-pass filter introduces additional delay into the system. When applied to a sliding mode control system whose boundary manifold is an attractor [19] and control input switches at very high frequency, this type of smoothing controller may make the system less robust and even cause instability. However, when the boundary manifold is not an attractor and the state always passes through the boundary manifold, this type of low-pass filter can be effective. The boundary manifold $\alpha^T \mathbf{x} = 0$ represents the condition $x_2 - x_2 = 0$. The state vector $\mathbf{x}$ can only pass through this manifold. Therefore, the additional filter will not hurt the system's robustness.

In constructing the fuzzy tuning low-pass (FTLP) filter for the suboptimal controller of the semiactive suspension system, the signed distance, ζ_1, is defined as the cosine value of the angle between $\mathbf{x}$ and α. The membership functions for $\tilde{M}_i$'s and $\tilde{N}_i$'s are defined as those shown in Figs. 11 and 12, respectively, the inference rules are given in Table 1, and λ is calculated by Eq. (8).

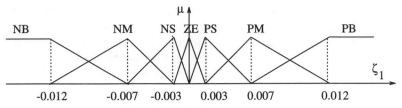

FIGURE 11   Membership function for $\bar{M}_i$.

During the first simulation, the initial condition is chosen as $\mathbf{x}_0 = (0, 0, 0.02, 0)^T$, and the disturbance w is set to be zero. The control inputs of two different controllers for semiactive suspension are shown in Fig. 13. The dashed line shows the control input of the unsmoothed suboptimal controller, and the solid line shows the control input with a FTLP filter. Apparently, the control input is smoothed by the FTLP filter. Figure 14 shows that the suspension deflection x_1 changes only slightly for the suboptimal controller with FTLP filter. Similarly, Fig. 15 shows that the sprung mass acceleration does not change much between these two controllers. However, Fig. 16 shows the reduction of the maximum sprung mass jerk from 800 m/s^3 to 300 m/s^3 (or 62% reduction) when the fuzzy controller is used. The sprung mass jerk, which is equal to the derivative of the sprung mass acceleration, is closely related to the force impact on passengers. Large sprung mass jerk represents a large force impact on passengers and less riding comfort. This proves that large improvement of the riding comfort for the semiactive suspension has been achieved by the FTLP filter.

In the last simulation, only the initial condition is considered and the road disturbance w is set to be zero throughout the simulation. Here we design the second simulation, which mimics the situation of the vehicle traveling over a hole. The hole is modeled as a cosine function:

$$w(t) = \begin{cases} -0.01 \sin(5\pi(t - 0.1)) & \text{when } t \in [0.1, 0.3] \\ 0 & \text{otherwise.} \end{cases} \qquad (10)$$

The initial condition of the suspension is set to zero. Figure 17 shows the control input of the suboptimal controller with and without a FTLP filter

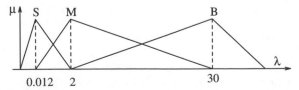

FIGURE 12   Membership function for $\bar{N}_i$. B, big; M, medium; S, small; P, positive; N, negative; ZE, zero.

TABLE 1   Fuzzy Inference Laws for Selecting λ

	ζ						
	NB	NM	NS	ZE	PS	PM	PB
λ	B	M	S	S	S	M	B

P, Positive; *N*, negative; *S*, small; *M*, medium; *B*, big.

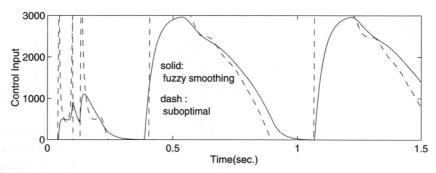

FIGURE 13   Control input for initial condition $x = (0, 0, 0.02, 0)$.

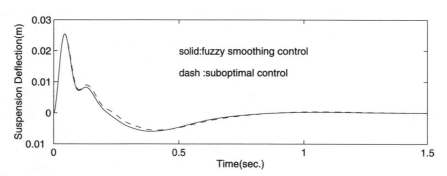

FIGURE 14   Time response of suspension deflection for initial condition $x = (0, 0, 0.02, 0)$.

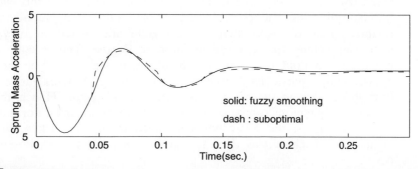

FIGURE 15   Time response of sprung mass acceleration for initial condition $x = (0, 0, 0.02, 0)$.

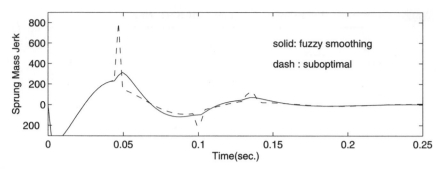

FIGURE 16   Time response of sprung mass jerk for initial condition $x = (0, 0, 0.02, 0)$.

when the vehicle travels over this hole. The dashed line shows the control input of the controller without a FTLP filter, and the solid line shows the control input of the controller with a FTLP filter. Figures 18, 19, and 20 show the time response of the suspension deflection, sprung mass acceleration, and sprung mass jerk, respectively. Similar to the previous simulation results, the control input is smoothed by the FTLP filter, and the maximum absolute sprung mass jerk is reduced from 3.2 m/s^3 to 2.2 m/s^3 (about a 30% reduction) at the price of increasing the maximum absolute sprung mass acceleration from 0.0622 m/s^2 to 0.0628 m/s^2 (about a 1% increment).

V. FUZZY REGIONS AND FUZZY BOUNDARIES

Another approach to smoothing the control input of a variable structure system is to define each region as a fuzzy set instead of a crisp set.

The regions of a variable structure system are crisp sets because the state can only be in or not in one particular region. Furthermore, each state belongs to one and only one region. Therefore, the control rule is switched when the state crosses one boundary manifold. Because of the switching of the control rules, the control input in general is not smooth when the boundary manifold is crossed. If the regions are treated as fuzzy sets, the state can belong to each region to some degree. This means that each boundary manifold does not strictly divide two regions. For those states approaching a particular boundary, the belonging of states varies smoothly (not abruptly) from the first region to the second region. Hence, the boundaries of fuzzy regions can be considered as fuzzy boundaries. For the example shown in Fig. 7, the fuzzy boundary becomes a band when the "signed distance" ζ is chosen as $d_2(x, S_1 2)$ and a cone when ζ is chosen as θ or $\cos \theta$.

The control rules of the variable structure system in Eq. (3) are defined for crisp regions. When these regions are treated as fuzzy regions, these

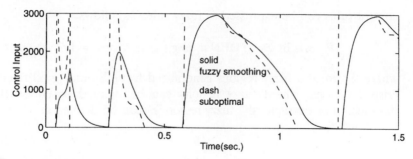

FIGURE 17 Control input for traveling over a hole.

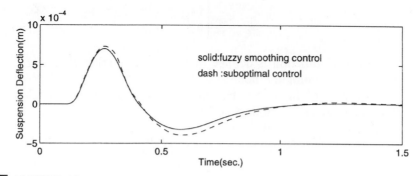

FIGURE 18 Time response of suspension deflection for traveling over a hole.

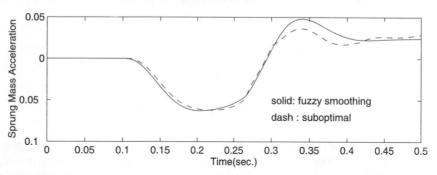

FIGURE 19 Time response of sprung mass acceleration for traveling over a hole.

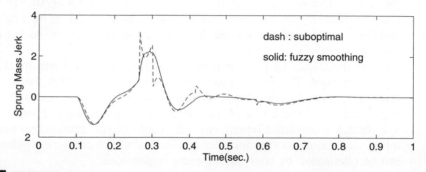

FIGURE 20 Time response of sprung mass jerk for traveling over a hole.

control rules can be expressed as fuzzy rules as follows:

$$\text{IF} \quad \text{x is in } \tilde{\Sigma}_i \quad \text{THEN } u_i \text{ is } g_i(\mathbf{x}) \qquad \text{for } i = 1, 2, \ldots, m,$$

where $\tilde{\Sigma}_i$ are the fuzzy regions generalized from Σ_i, and $g_i(\mathbf{x})$ remain to be crisp functions. These fuzzy regions can be defined based on the "sign distance." For example, the fuzzy region $\tilde{\Sigma}_i$ can be defined as

$$\mu_{\tilde{\Sigma}_i} = \Pi(\zeta; \zeta_{1i}, \zeta_{2i}, \zeta_{3i}, \zeta_{4i}) = \begin{cases} 0 & \zeta < \zeta_{1i}, \\ (\zeta - \zeta_{1i})/(\zeta_{2i} - \zeta_{1i}) & \zeta_{1i} \leq \zeta \leq \zeta_{2i} \\ 1 & \zeta_{2i} \leq \zeta \leq \zeta_{3i} \\ (\zeta_{4i} - \zeta)/(\zeta_{4i} - \zeta_{3i}) & \zeta_{3i} \leq \zeta \leq \zeta_{4i} \\ 0 & \zeta > \zeta_{4i}, \end{cases}$$

where the Π function is defined by Eq. (1) in Section II, and $\zeta_{1i} \ldots \zeta_{4i}$ are parameters defining the boundary of $\tilde{\Sigma}_i$.

The control input can be obtained by applying the weighted average method, as described in Section II:

$$u = \frac{\sum_{i=1}^{m} \mu_i u_i}{\sum_{i=1}^{m} \mu_i}, \tag{11}$$

where $u_i = g_i(\mathbf{x})$ is the control input in region Σ_i.

The fuzzy regions and fuzzy boundaries method is tested for a single-axis robot arm whose model is

$$\dot{\mathbf{x}}(t) = A\mathbf{x}(t) + B(u(t) + d(t))$$

$$= \begin{bmatrix} 0 & 1 \\ 0 & -\dfrac{b}{J} \end{bmatrix} \begin{pmatrix} x_1(t) \\ x_2(t) \end{pmatrix} + \begin{bmatrix} 0 \\ \dfrac{K}{J} \end{bmatrix} (u(t) + d(t)), \tag{12}$$

where $x_1(t)$ and $x_2(t)$ represent, respectively, the angular position and velocity of the arm; $u(t)$ is the control input; $d(t)$ is the disturbance, which is bounded by $[-d_{\max}, d_{\max}]$; K is the torque constant; b is the damping coefficient; J_n is the nominal inertia of the arm; and J is the inertia of the robot arm. The values of b and K are assumed to be 1.4 N/(m · s) and 39.0 Nm/V, respectively. The nominal inertia J_n is assumed to be 1.89 kg · m^2, and the variation of the inertia $\Delta J = J - J_n$ is assumed to be bounded by 1.06 kg · m^2.

The digital sliding mode controller was developed by Langari [21] for the position control. This sliding mode controller guaranteed that the state would reach a cone around the sliding surface. However, the state may jump out this cone because of the effect of the disturbance. When this happens, the state will chatter around the boundary of the cone. In this section, we will review this digital sliding mode controller and show that the chattering phenomenon can be eliminated by applying the fuzzy region idea.

Following [20], the sliding surface is defined as $s = 0$, where

$$s = C^T \mathbf{x}$$

and

$$C = [15, 1]^T.$$

When the input u is preceded by a zero-order hold, the following discrete time model can be obtained:

$$x(k + 1) = A_d x(k) + B_d u(k) + \int_0^{\Delta t} e^{A\tau} d(\tau) B \, d\tau,$$

where

$$A_d = e^{A \Delta t}$$

and

$$B_d = \int_0^{\Delta t} e^{A\tau} B \, d\tau.$$

The sampling time Δt is chosen as 4 ms.

A cone Ψ_C and a ball Ψ_B are defined in [21] as shown in Fig. 21:

$$\Psi_C = \{\mathbf{x} | \nu M(\mathbf{x}) \geq |C^T \mathbf{x}|\} = \{\mathbf{x} | \nu \|C\| \|\mathbf{x}\| \cos \theta_c \geq |C^T \mathbf{x}|\},$$

$$\Psi_B = \{\mathbf{x} | \phi \geq |\mathbf{x}|\},$$

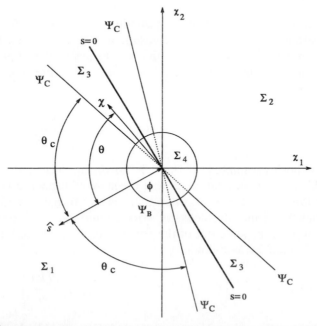

FIGURE 21 Ψ_C and Ψ_B and four regions for a direct-drive robot arm. Reprinted with permission from Y. R. Hwang and S. C. Wang, *IEEE Trans. Fuzzy Systems* 2:277–284 (© 1994 IEEE).

where

$$M(\mathbf{x}) = \|C\| \, \|\mathbf{x}\| [e^{\gamma \Delta t} - 1]$$

$$\nu = \frac{(C^T B_d)_{\min} + (C^T B_d)_{\max}}{2(C^T B_d)_{\min}}$$

$$\theta_c = \cos^{-1}(e^{\gamma \Delta t} - 1)$$

$$\phi = 2(C^T B_d)_{\max} d_{\max},$$

and γ is the maximum singular value of A. The state space is then partitioned into four regions as follows (see Fig. 21):

$$\Sigma_1 = \{\mathbf{x} | \, |\mathbf{x}| > \phi \text{ and } \cos \theta > \cos \theta_c\}$$

$$\Sigma_2 = \{\mathbf{x} | \, |\mathbf{x}| > \phi \text{ and } \cos \theta < -\cos \theta_c\}$$

$$\Sigma_3 = \{\mathbf{x} | \, |\mathbf{x}| > \phi \text{ and } -\cos \theta_c \leq \cos \theta \leq \cos \theta_c\}$$

$$\Sigma_4 = \{\mathbf{x} | \, |\mathbf{x}| \leq \phi\},$$

where θ is the angle between the state vector and the normal vector of the sliding surface. The control laws in Regions Σ_1 and Σ_2 are defined according to [20] as follows:

$$u(k) = -M(\mathbf{x}(k)) - d_{\max} \qquad \text{if } s > 0 \text{ (i.e., } \mathbf{x} \in \Sigma_1)$$

$$u(k) = M(\mathbf{x}(k)) + d_{\max} \qquad \text{if } s < 0 \text{ (i.e., } \mathbf{x} \in \Sigma_2).$$

Note that this control law satisfies the conditions, developed by Jabbari in [20], which guarantee that $|s(\cdot)|$ will decrease if $\mathbf{x}$ is in Σ_1 and Σ_2. However, these two control laws do not guarantee that $\mathbf{x}$ approaches the sliding surface if the state is in Regions Σ_3 and Σ_4. Therefore, we define the following control law for Region Σ_3:

$$u(k) = -\frac{1}{C^T B_d} \left(C^T A_d \mathbf{x}(k) - s(k) + K_1 \Delta t \, \frac{\cos \theta(k)}{\nu \cos \theta_c} \right),$$

where K_1 is a positive number. If there is no disturbance and parameter uncertainty, this control law guarantees that the state will stay inside the cone and approach the origin. However, when there are disturbance and parameter uncertainties, this controller does not work around the origin because the disturbance will push the state out of the cone. Therefore, we define another control law in Region Σ_4 as follows:

$$u(k) = -\frac{1}{C^T B_d} \left(C^T A_d \mathbf{x}(k) - s(k) + K_2 \Delta t \, \frac{s(k)}{\phi} \right),$$

where K_2 is a positive number. Note that K_1 and K_2 should be large enough to cover the disturbance and parameter uncertainties.

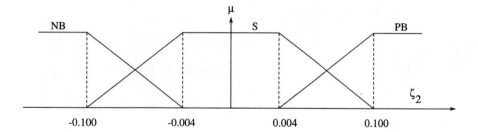

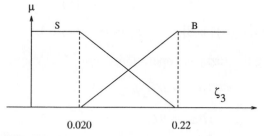

FIGURE 22 Membership functions for characterizing ζ_2 and ζ_3. B, big; S, small; P, positive; N, negative.

To design the fuzzy smoothing controller, we define the following fuzzy regions based on Σ_1 to Σ_4:

$$\tilde{\Sigma}_1 = \{\mathbf{x}| \zeta_2(\mathbf{x}) \text{ is PB AND } \zeta_3(\mathbf{x}) \text{ is B}\}$$

$$\tilde{\Sigma}_2 = \{\mathbf{x}| \zeta_2(\mathbf{x}) \text{ is NB AND } \zeta_3(\mathbf{x}) \text{ is B}\}$$

$$\tilde{\Sigma}_3 = \{\mathbf{x}|\, | \zeta_2(\mathbf{x})|\, \text{ is S AND } \zeta_3(\mathbf{x}) \text{ is B}\}$$

$$\tilde{\Sigma}_4 = \{\mathbf{x}| \zeta_3(\mathbf{x}) \text{ is S}\},$$

where

$$\zeta_2(\mathbf{x}) = \frac{C^T \mathbf{x}}{\|C\| \, \|\mathbf{x}\|}$$

$$\zeta_3(\mathbf{x}) = \|\mathbf{x}\|,$$

and B and S represent linguistic terms "big" and "small," respectively, and P and N represent "positive" and "negative." Note that ζ_2 is the cosine value of the angle between $\mathbf{x}$ and the normal vector of the sliding surface. The membership functions for characterizing ζ_2 and ζ_3 are chosen as shown in Fig. 22. The fuzzy sets $\tilde{\Sigma}_1$, $\tilde{\Sigma}_2$, and $\tilde{\Sigma}_3$ can be represented as the intersection of two fuzzy sets, i.e.,

$$\tilde{\Sigma}_1 = \tilde{E}_{PB} \cap \tilde{F}_B,$$

$$\tilde{\Sigma}_2 = \tilde{E}_{NB} \cap \tilde{F}_B,$$

$$\tilde{\Sigma}_3 = \tilde{E}_S \cap \tilde{F}_B,$$

where

$$\tilde{E}_{PB} = \{\mathbf{x} | \zeta_2(\mathbf{x}) \text{ is PB}\}, \qquad \tilde{E}_{NB} = \{\mathbf{x} | \zeta_2(\mathbf{x}) \text{ is NB}\},$$

$$\tilde{E}_S = \{\mathbf{x} | | \zeta_2(\mathbf{x}) | \text{ is S}\}, \qquad \tilde{F}_B = \{\mathbf{x} | \zeta_3(\mathbf{x}) \text{ is B}\}.$$

Therefore, by applying the fuzzy set operation,

$$\mu_{\tilde{\Sigma}_1} = \mu_{\tilde{E}_{PB}} \mu_{\tilde{F}_B}$$

$$\mu_{\tilde{\Sigma}_2} = \mu_{\tilde{E}_{NB}} \mu_{\tilde{F}_B}$$

$$\mu_{\tilde{\Sigma}_3} = \mu_{\tilde{E}_S} \mu_{\tilde{F}_B}.$$

The final control input can be obtained with Eq. (11).

The initial condition $\mathbf{x}(0)$ is chosen as $[-2, -2]^T$, the disturbance $d(t)$ is a random signal with its maximum magnitude as 0.1, ΔJ is assumed to be $-1.06 \text{ kg} \cdot \text{m}^2$, and K_1 and K_2 are chosen to be 2 and 1, respectively, for the simulation. As shown in Fig. 23, the reaching time is almost the same for both systems. Figures 24 and 25 show that the control input of the variable

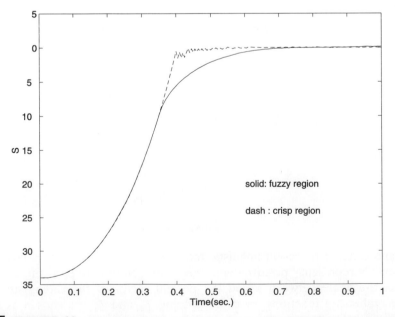

FIGURE 23 Time response of S. Solid line, fuzzy region; dashed line, crisp region.

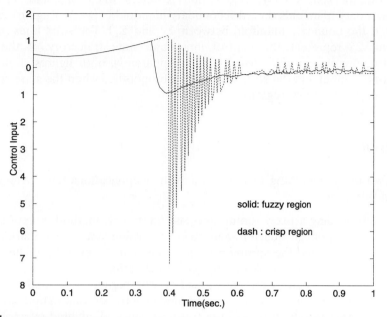

FIGURE 24 Control input of a direct-drive robot arm. Solid line, fuzzy region; dashed line, crisp region.

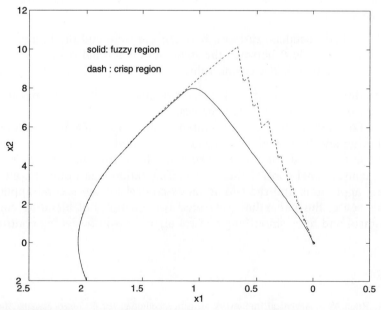

FIGURE 25 Phase plot of a direct-drive robot arm. Solid line, fuzzy region; dashed line, crisp region.

structure system with crisp regions chatters when the state crosses two boundary manifolds (i.e., the boundary manifold between Regions Σ_1 and Σ_3 and the boundary manifold between Σ_3 and Σ_4). The solid lines in Figs. 24 and 25 represent the control input and state trajectory of the variable structure system with fuzzy regions. Apparently, both figures show that the control input and the state trajectory are smoothed when the state crosses the boundaries of two regions.

VI. CONCLUSION

Two fuzzy smoothing techniques and their applications have been discussed in this paper:

1. Adding a fuzzy tuning low-pass filter. This method is applied to a suboptimal quarter semiactive suspension system. Simulation results show that the sprung mass jerk was reduced by 62% at the price of slightly increased sprung mass acceleration.
2. Defining fuzzy regions instead of crisp regions. This method is applied to a robot arm with a sliding mode controller. Simulation results show that the chattering of the control input is eliminated by applying this method.

Both of these controllers are proved to smooth the control input during the switching of the control laws.

In developing these two fuzzy smoothing algorithms, the "signed distance" measurement is proposed. Two commonly used definitions for the "signed distance" are

1. The Euclidean distance between the state and the boundary
2. The angle θ between the state and the normal vector of the boundary manifold or the cosine value of θ

For discrete time control of linear dynamic systems with constant sampling time, θ and $\cos \theta$ are better choices.

For both applications presented in this paper, the fuzzy tuning low-pass filtering and fuzzy region controllers were used to enhance the performance of the traditional controllers. The controllers' structures were derived based on conventional control theoretic consideration (the optimal control for the first application and the sliding mode control for the second application). In this sense, this paper has presented two examples of blending conventional control and fuzzy smoothing control algorithms in designing control systems.

REFERENCES

1. Rugh, W. J. Analytical framework for gain scheduling, *IEEE Control Systems Magazine* 11(1), 1991.
2. Slotine, J. E. and Li, W. *Applied Nonlinear Control*. Prentice Hall, Englewood Cliffs, NJ, 1991.

3. Utkin, V. I. Variable structure systems with sliding modes. *IEEE Trans. Automat. Control* AC-22:212–222, April 1977.

4. Burton, T. and Zinober, A. Continuous approximation of variable structure control. *Internat. J. Systems Sci.* 17:876–885, 1986.

5. Glower, J. S. and Munighan, J. Designing fuzzy controllers from a variable structures standpoint. *IEEE Trans. Fuzzy Systems* 5(1):138–144, 1997.

6. Kawaji, S. and Matsunaga, M. Fuzzy control of vss type and its robustness. In *Proceedings of IFSA World Congress*, Engineering, 1991, pp. 81–84.

7. Zhao, Z. Y., Tomizuka, M., and Isaka, S. Fuzzy gain scheduling of PID controllers. *IEEE Trans. Systems Man Cybernet.* 23(5):1392–1398, 1993.

8. Hwang, Y. R. and Tomizuka, M. Fuzzy smoothing algorithms for semi-active suspension systems. In *Proceedings of the IEEE Workshop on Neural Network and Fuzzy Logic Control*, December 1992.

9. Hwang, Y. R. and Tomizuka, M. Fuzzy smoothing algorithms for variable structure systems. *IEEE Trans. Fuzzy Systems* 2:277–284, 1994.

10. Hwang, Y. R. and Wang, S. C. Fuzzy smoothing controller design of half car semi-active suspension system. In *Proceedings of the 13th National Conference on Mechanical Engineering: The Chinese Society of Mechanical Engineers*, 1996.

11. Zadeh, L. A. Fuzzy sets. *Inform. and Control* 8:338–353, 1965.

12. Procyk, T. J. and Mamdani, E. H. A linguistic self-organizing process controller. *Automatica* 15:15–30, 1979.

13 Mizumoto, M. Fuzzy reasoning and fuzzy control. *Computol* 28:32–45, 1989.

14. Pedryc, W. *Fuzzy Control and Fuzzy Systems*. Wiley, New York, 1989.

15. Kosko, B. *Neutral Networks and Fuzzy Systems*. Prentice Hall, Englewood Cliffs, NJ, 1992.

16. Lee, C. C. Fuzzy logic in control system: Fuzzy logic controller. Parts i and ii. *IEEE Trans. Systems Man Cybernet.* 20(2):404–435, 1990.

17. Mamdani, E. H. Applications of fuzzy algorithms for simple dynamic plant. *Proc. IEE* 121(12):1565–1588, 1974.

18. Butsuen, T. and Hedrick, J. K. Optimal semi-active suspensions for automotive vehicles: The 1/4 car model. *Adv. Automotive Technol. ASME, WAM*, DSC-13:305–319, 1989.

19. Guckenheimer, J. and Holmes, P. *Nonlinear Oscillations, Dynamical Systems, and Bifurcations of Vector Fields*. Springer-Verlag, New York, 1985.

20. Jabbari, A. and Tomizuka, M. Robust discrete-time control of continuous time plant. In *Japan–USA Symposium on Flexible Automation*, July 1990.

21. Langari, G. and Tomizuka, M. Analysis and synthesis of fuzzy linguistic control systems. In *ASME Winter Annual Meeting*, 1990.

15

TECHNIQUES AND APPLICATIONS OF FUZZY THEORY IN THE VALIDITY OF COMPLEXITY REDUCTION BY MEANS OF DECOMPOSITION OF MULTIVARIABLE FUZZY SYSTEMS

P. G. LEE

Department of Computer Control Enginerring, Uiduk Univeristy, Kyungju, Kyungpook 780-910

G. J. JEON

School of Electronics and Electrical Engineering, Kyungpook National University, Taegu 702-701 and Engineering Research Center for Advanced Control and Instrumentation, Seoul National University, Seoul, Korea

K. K. LEE

School of Electronics and Electrical Engineering, Kyungpook National University, Taegu 702-701

I. INTRODUCTION   406
II. FUZZY RELATION MATRIX   406
   A. Fuzzy Sets   407
   B. Fuzzy Relation   407
III. COMPLEXITY REDUCTION   408
IV. INDEX OF APPLICABILITY   411
V. APPLICATIONS   415
   A. Multivariable Open-Loop Fuzzy Systems   415
   B. Multivariable Closed-Loop Fuzzy Systems   416
   C. Parallel Distributed Multivariable Fuzzy Systems   418
   D. Numerical Examples   421
VI. SUMMARY   428
   REFERENCES   429

Portions are reprinted, with permission, from *Fuzzy Sets and Systems* Vol. 71, G. J. Jeon and P. G. Lee, Structure of multivariable fuzzy control systems with a coordinator, pp. 85–94, © 1995 Elsevier Science, and from P. G. Lee, K. K. Lee, and G. J. Jeon, *IEEE Trans. Fuzzy Systems* 3:364–369 (© 1995 IEEE).

I. INTRODUCTION

Since the fuzzy set theory was introduced by Zadeh [1], fuzzy logic has been used successfully in a number of control systems with single input and single output [2, 3]. However, many complex industrial processes are multivariable systems that have several inputs and outputs. It is difficult to infer the proper control input for these systems, because the dimension of the relation matrix composed of the control rules is very large. The high dimensionality of these relation matrices accompanies computational difficulties. Therefore, recent studies on the application of fuzzy set theory to the design of control systems are directed at the description of the multivariable structure of these systems. Shakouri *et al.* [4] suggested a fuzzy control algorithm for a multivariable system that was represented by a conventional state space model. Cheng *et al.* [5, 6] introduced intersection coefficients and a formula for a multivariable fuzzy controller in terms of these coefficients. Czogala and Zimmermann [7] proposed random intersection coefficients and a structure for a multidimensional probabilistic controller. Trojan and Kiszka [8] introduced a method for obtaining solutions of multivariable fuzzy equations. Gupta *et al.* [9] proposed a fuzzy control algorithm where the multivariable fuzzy system was decomposed into a set of one-dimensional systems.

The decomposition of multivariable control rules is preferable because it alleviates the complexity of the problem. Decomposition of complex multivariable fuzzy systems has the following advantages. First, the computer implementation algorithm is fast and flexible. Second, it reduces computation time and storage requirement. Third, with the decomposition method, the design procedure of multivariable systems is simple. The main disadvantage that arises in the use of such a decomposition method for the multivariable fuzzy system, however, is that the inference error is large in some cases because of its approximate nature. We have observed that a large inference error is generated when Gupta's decomposition method is applied to the ENOR gate model [10]. To measure the validity of complexity reduction by means of decomposition of multivariable fuzzy systems, we have defined an index of applicability which judges whether the decomposition method can be applied to multivariable fuzzy systems [11].

This chapter is organized as follows. In Section II, some definitions and notations of fuzzy sets and fuzzy relations are summarized, and in Section III a complexity reduction procedure using the decomposition of multivariable fuzzy rules is described. We define an index of applicability for the decomposition of multivariable fuzzy rules as a measure of validity of complexity reduction in Section IV. Section V presents some applications to multivariable fuzzy systems and corresponding numerical examples, and Section VI contains the summary.

II. FUZZY RELATION MATRIX

For the convenience of the reader, we shall briefly summarize some definitions and notations of fuzzy sets and fuzzy relations that will be utilized in the following sections.

A. Fuzzy Sets

Let U be a collection of objects denoted generally by $\{y\}$, which could be discrete or continuous. U is called with the universe of discourse and y represents the generic element of U. A fuzzy set A in a universe of discourse U is characterized by a membership function μ_A in the interval $[0,1]$ which represents the grade of membership of y in A. Thus a fuzzy set A in U may be represented as a set of ordered pairs of a generic element y and its value of the membership function $A = \{(y, \mu_A(y))/y \in U\}$. When U is not finite, a fuzzy set A can be written as $A = \int_U \mu_A(y_i)/y_i$. If the fuzzy set A has a finite support $\{y_1, y_2, \ldots, y_n\}$, then A may be represented in the form

$$A = \sum_{i=1}^{n} \mu_A(y_i)/y_i. \tag{1}$$

As an illustration, if U is the collection $\{7, 8, \ldots, 13\}$ of integers, then the fuzzy set A in U labeled *integer approximately equal to* 10 may be represented as

$$A = 0.2/7 + 0.5/8 + 0.8/9 + 1.0/10 + 0.8/11 + 0.5/12 + 0.2/13. \tag{2}$$

B. Fuzzy Relation

DEFINITION 1. Cartesian product: If $A_1, A_2, \ldots, A_n$ are fuzzy sets in $U_1, U_2, \ldots, U_n$, the Cartesian product of $A_1, A_2, \ldots, A_n$ is a fuzzy set in the product space $U_1 \times U_2 \times \cdots \times U_n$ with the membership function

$$\mu_{A_1 \times, A_2 \times \cdots \times A_n}(y_1, y_2, \ldots, y_n) = \min(\mu_{A_1}, \ldots, \mu_{A_n}) \tag{3}$$

or

$$\mu_{A_1 \times, A_2 \times \cdots \times A_n}(y_1, y_2, \ldots, y_n) = \mu_{A_1}(y_1) \cdot, \ldots, \cdot \mu_{A_n}(y_n). \tag{4}$$

A fuzzy relation R from a set X to a set Y is a fuzzy set of the Cartesian product $X \times Y$, which is the collection of the ordered pairs (x, y), $x \in X$, $y \in Y$. R is characterized by a membership function $\mu_{R(x,y)}$ and is expressed as

$$R_{X \times Y}(x, y) = \{\mu_R(x, y)/(x, y)\}. \tag{5}$$

More generally, for an n-ary fuzzy relation R that is a fuzzy set of $X_1 \times X_2 \times \cdots \times X_n$, we have

$$R_{X_1 \times X_2 \times \cdots \times X_n}(x_1, x_2, \ldots, x_n) = \{\mu_R(x_1, x_2, \ldots, x_n)/(x_1, x_2, \ldots, x_n)$$

$$\in X_1 \times X_2 \times \cdots \times X_n\}. \tag{6}$$

For a finite universe of discourse, a matrix notation is useful. In this case, fuzzy relation R is treated as a matrix $R(x_i, y_j)$, with the entries taking values between 0 and 1. As an example, if $X = \{x_i\}$ and $Y = \{y_j\}$, $i = j = 2$, then a fuzzy relation of similarity between members of X and Y might be expressed as

$$similarity = 0.2/(x_1, y_1) + 0.7/(x_1, y_2) + 0.5/(x_2, y_1) + 0.3/(x_2, y_2). \tag{7}$$

In this case, a fuzzy relation may be represented as a relation matrix

$$R_{X \times Y} = \begin{bmatrix} 0.2 & 0.7 \\ 0.5 & 0.3 \end{bmatrix}, \tag{8}$$

in which the (i, j)th element is the value of $\mu_R(x_i, y_j)$.

III. COMPLEXITY REDUCTION

Analysis and design procedures for multivariable fuzzy systems are very difficult, mainly because of the multidimensionality of the relation matrices. Therefore, for the design of multivariable fuzzy control systems, decomposition of control rules is preferable, because it alleviates the complexity of the problem. Consider a multivariable system with N input variables and M output variables, as shown in Fig. 1. The linguistic control rules of this system are given by

$$
\begin{aligned}
&IF\ X_1\ is\ X_{1(1)}\ AND\ \cdots\ X_N\ is\ X_{N(1)} \\
&\quad THEN\ Y_1\ is\ Y_{1(1)}\ AND\ \cdots\ Y_M\ is\ Y_{M(1)} \\
&ALSO \\
&\qquad\qquad\qquad \vdots \\
&IF\ X_1\ is\ X_{1(k)}\ AND\ \cdots\ X_N\ is\ X_{N(k)} \\
&\quad THEN\ Y_1\ is\ Y_{1(k)}\ AND\ \cdots\ Y_M\ is\ Y_{M(k)} \\
&ALSO \\
&\qquad\qquad\qquad \vdots \\
&IF\ X_1\ is\ X_{1(K)}\ AND\ \cdots\ X_N\ is\ X_{N(K)} \\
&\quad THEN\ Y_1\ is\ Y_{1(K)}\ AND\ \cdots\ Y_M\ is\ Y_{M(K)},
\end{aligned} \tag{9}
$$

where the input and the output variables are normalized fuzzy sets, and

FIGURE I   Structure of the multivariable system

$X_{N(k)}$, $Y_{M(k)}$ are the membership functions of the Nth input variable and the Mth output variable of the kth rule, respectively. Let the dimension of the discrete fuzzy inputs and outputs be $\dim[X_n] = q_n$, $\dim[Y_m] = p_m$, $n = 1, 2, \ldots, N$, $m = 1, 2, \ldots, M$. Then the following conditions are satisfied because the elements $(i_1, i_2, \ldots, i_N)$ and $(j_1, j_2, \ldots, j_M)$ obtained from quantizing physical values of the input and the output, respectively, are some elements in the universe of discourse:

$$1 \le i_n \le q_n, \quad 1 \le j_m \le p_m. \tag{10}$$

The fuzzy relation matrix of the mth output is defined by

$$R_m = \bigvee_{k=1}^{K} \{X_{1(k)} \wedge X_{2(k)} \wedge \cdots \wedge X_{N(k)} \wedge Y_{m(k)}\}, \quad 1 \le m \le M, \tag{11}$$

where $\vee$ is the max-operator and $\wedge$ is the min-operator, and the dimension of R_m is

$$\dim[R_m] = q_1 \times q_2 \times \cdots \times q_N \times p_m. \tag{12}$$

Given the current inputs $(X_1, X_2, \ldots, X_N)$, the present mth output Y_m is obtained by the following compositional rule:

$$Y_m = X_1 \circ X_2 \circ X_3 \circ \cdots \circ X_N \circ R_m, \quad 1 \le m \le M, \tag{13}$$

where $\circ$ is the max-min composition operator of fuzzy relations. When the number of inputs is large, a direct inference, which is expressed by Eq. (13), accompanies computational difficulties such as computing time and storage space because of the high dimensionality of the relation matrix in Eq. (11).

To overcome this problem, Gupta et al. [9] proposed a decomposition method that uses two-dimensional relation matrices only. The two-dimensional relation matrix R_{nm} represents the cause-effect relationship between the nth input and the mth output, defined as

$$R_{nm} = \bigvee_{k=1}^{K} \{X_{n(k)} \wedge Y_{m(k)}\}, \quad 1 \le n \le N, \quad 1 \le n \le N. \tag{14}$$

By Gupta's method, the outputs are inferred from the following equations:

$$Y_1 = X_1 \circ R_{11} \wedge X_2 \circ R_{21} \wedge \cdots \wedge X_N \circ R_{N1}$$

$$\vdots \tag{15}$$

$$Y_M = X_1 \circ R_{1M} \wedge X_2 \circ R_{2M} \wedge \cdots \wedge X_N \circ R_{NM}.$$

These equations can be written in a compact matrix form,

$$
\begin{bmatrix} Y_1 \\ Y_2 \\ \vdots \\ Y_M \end{bmatrix} = \begin{bmatrix} X_1 & X_2 & \cdots & X_N \end{bmatrix} * \begin{bmatrix} R_{11} & \cdots & R_{1M} \\ R_{21} & \cdots & R_{2M} \\ \vdots & \ddots & \vdots \\ R_{N1} & \cdots & R_{NM} \end{bmatrix}, \qquad (16)
$$

where $*$ is the $(\circ, \wedge)$ operator. The structure of the multivariable fuzzy system of Gupta's method is shown in Fig. 2.

To analyze the loss of accuracy in the inference result, we shall compare the mth output Y_m inferred by original relation matrix with the output by Gupta's method. The mth output Y_m of Eq. (13) is obtained in the following way. Replacing the composition operator $\circ$ with max($\vee$) and min($\wedge$) operators in Eq. (13), we find that

$$
Y_m = \bigvee_{X^1} X_1 \wedge \bigvee_{X^2} X_2 \wedge \cdots \wedge \bigvee_{X^N} X_m \wedge \bigvee_{k=1}^{K} \{ X_{1(k)} \wedge X_{2(k)} \wedge \cdots \wedge X_{m(k)} \}
$$

$$
= \bigvee_{k=1}^{K} \left\{ \bigvee_{X^1} X_1 \wedge X_{1(k)} \wedge Y_{m(k)} \wedge \bigvee_{X^2} X_2 \wedge X_{2(k)} \wedge Y_{m(k)} \right.
$$

$$
\left. \wedge \cdots \bigvee_{X^N} X_N \wedge X_{N(k)} \wedge Y_{m(k)} \right\}, \qquad (17)
$$

where X^k is the universe of discourse of the kth input variable. Then relation of equality between the Y_m in Eq. (17) and the mth output by

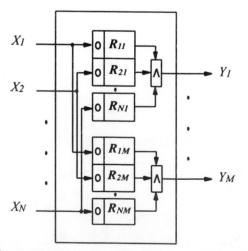

FIGURE 2 Structure of a multivariable fuzzy system using two-dimensional relation matrices.

Gupta's method becomes an inclusion as

$$
Y_m \subset \bigvee_{X^1} X_1 \wedge \bigvee_{k=1}^{K} \{X_{1(k)} \wedge Y_{m(k)}\} \wedge \bigvee_{X^2} X_2 \wedge \bigvee_{k=2}^{K} \{X_{2(k)} \wedge Y_{m(k)}\}
$$

$$
\wedge \cdots \wedge \bigvee_{X^N} X_N \wedge \bigvee_{k=N}^{K} \{X_{N(k)} \wedge Y_{m(k)}\}, \tag{18}
$$

where the right-hand side of Eq. (18) is the output by Gupta's method. The inclusion sign in Eq. (18) means that the output Y_m by the original relation matrix is contained in the output fuzzy set obtained by neglecting the terms in the expansion in Eq. (17). The decomposition of multivariable control rules is preferable because it alleviates the complexity of the problem. However, the main disadvantage in the use of the decomposition by two-dimensional relation matrices is that the inference error is large in some cases.

IV. INDEX OF APPLICABILITY

In any case, there exists some inference error in Gupta's method because of its approximate nature. Therefore, we will analyze the error generated by the decomposition of control rules and introduce an index of applicability of the decomposition method in this section.

Consider the following linguistic control rules of a system with two inputs and one output:

$$IF\ X_1\ is\ X_{1(1)}\ AND\ X_2\ is\ X_{2(1)}\ THEN\ Y_1\ is\ Y_{(1)}$$
$$ALSO$$

$$\vdots$$

$$IF\ X_1\ is\ X_{1(k)}\ AND\ X_2\ is\ X_{2(k)}\ THEN\ Y_1\ is\ Y_{(k)} \tag{19}$$
$$ALSO$$

$$\vdots$$

$$IF\ X_1\ is\ X_{1(K)}\ AND\ X_2\ is\ X_{2(K)}\ THEN\ Y_1\ is\ Y_{(K)}.$$

Then the relation matrix is composed as follows:

$$
R = \bigvee_{k=1}^{K} \{X_{1(k)} \wedge X_{2(k)} \wedge Y_{(k)}\}. \tag{20}
$$

On the other hand, two dimensional relation matrices, R_{11} and R_{21}, which relate one input and one output each, are expressed as

$$
R_{11} = \bigvee_{k=1}^{K} \{X_{1(k)} \wedge Y_{(k)}\} \tag{21}
$$

$$
R_{21} = \bigvee_{k=1}^{K} \{X_{2(k)} \wedge Y_{(k)}\}. \tag{22}
$$

Given the current inputs X_1 and X_2, the output computed by Gupta's decomposition method is described as follows:

$$Y = X_1 \circ R_{11} \wedge X_2 \circ R_{21}$$

$$= X_1 \circ A_2 \circ \left\{ \bigvee_{k=1}^{K} X_{1(k)} \wedge A_{2(k)} \wedge Y_{(k)} \right\}$$

$$\wedge A_1 \circ X_2 \circ \left\{ \bigvee_{k=1}^{K} A_{1(k)} \wedge X_{2(k)} \wedge Y_{(k)} \right\}$$

$$\geq X_1 \circ X_2 \circ \left\{ \bigvee_{k=1}^{K} X_{1(k)} \wedge X_{2(k)} \wedge Y_{(k)} \right\}, \tag{23}$$

where A_1 and A_2 are fuzzy sets whose membership functions are identical to ones over the universe of discourse, and the last term of Eq. (23) is the inference output from the original relation matrix. This equation shows the possible inference error generated by the decomposition method.

To illustrate this point let us assume that the inputs X_1 and X_2 are fuzzy singletons at the i_1th and i_2th elements in the universe of discourse, respectively. Then the value of the membership function of the inference output at the pth element is obtained as follows:

$$Y(p) = X_1 \circ A_2 \circ \left\{ \bigvee_{k=1}^{K} X_{1(k)} \wedge A_{2(k)} \wedge Y_{(k)} \right\}(p)$$

$$\wedge A_1 \circ X_2 \circ \left\{ \bigvee_{k=1}^{K} A_{1(k)} \wedge X_{2(k)} \wedge Y_{(k)} \right\}(p)$$

$$= \left\{ [0, \dots, 1, \dots, 0] \circ [1, \dots, 1, \dots, 1] \circ R \right\}$$
$$\qquad\quad (i_1\text{th element})$$

$$\wedge \left\{ [1, \dots, 1, \dots, 1] \circ [0, \dots, 1, \dots, 0] \circ R \right\}$$
$$\qquad\qquad\qquad\qquad (i_2\text{th element})$$

$$= \max_{i_1(p)} \wedge \max_{i_2(p)}, \tag{24}$$

where $\max_{i_1(p)}$ represents the projection along the variable X_2, which is the maximum value of the relation matrix whereas the input variable X_1 is fixed to i_1 and the output variable is fixed to p. The $\max_{i_2(p)}$ is similarly obtained. Figure 3 depicts the steps needed to obtain the pth element of output, and the directions of the arrows indicate the projections. Let the dimension of the universe of discourse Y be $\dim[Y] = P$. Then pth element of the inference output, $Y(p)$ holds following inequalities:

$$Y(p) = \max_{i_1(p)} \wedge \max_{i_2(p)} \geq R(i_1, i_2, p), \qquad 1 \leq p \leq P. \tag{25}$$

Thus the inference output by the decomposition method may be bigger than that from the original relation matrix.

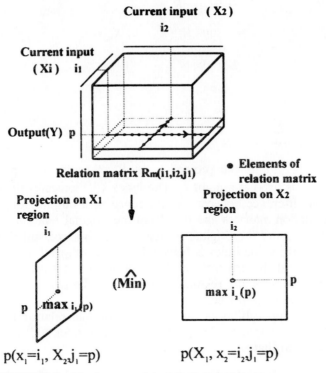

FIGURE 3 Inferring steps of the pth element of output.

To quantify the inference error generated by the decomposition method, we define indices of applicability $C_m^{(2)}$ and $C_m^{(3)}$ in that order. $C_m^{(2)}$ represents the index of two-dimensional relation matrices resulting from the decomposition of control rules into one input and the mth output, and $C_m^{(3)}$ is the index of three-dimensional relation matrices of two inputs and the mth output. The index $C_m^{(2)}$ is defined by the orthogonal projection $P_{(X_n=i_n, Y_m=j_m)}$ as follows [10].

$$C_m^{(2)} = \frac{\Sigma_{i_1,\ldots,i_N,j_m}\left\{\bigwedge_{n=1}^{N} P_{(X_n=i_n, Y_m=j_m)} - R_m(i_1, i_2, \ldots, i_N, j_m)\right\}}{q_1 \times q_2 \times \cdots \times q_N \times p_m}, \quad (26)$$

where

$$0 \leq C_m^{(2)} < 1, \qquad m = 1, \ldots, M$$

$$\bigwedge_{n=1}^{N} P_{(X_n=i_n, Y_m=j_m)} = P_{(X_1=i_1, Y_m=j_m)} \wedge P_{(X_2=i_2, Y_m=j_m)}$$

$$\wedge \cdots \wedge P_{(X_N=i_N, Y_m=j_m)}$$

$$P_{(X_n=i_n,Y_m=j_m)} = proj_{X_n=i_n}R_m$$

$$= \sup_{\substack{x_1 \in X_1 \\ x_2 \in X_2 \\ \vdots \\ x_N \in X_N}} R_m(x_1,\ldots,X_n=i_n,\ldots,x_N,Y_m=j_m),$$

$$\sum_{i_1,\ldots,i_N,j_m} = \sum_{j_m=1}^{p_m}\sum_{i_N=1}^{q_N}\cdots\sum_{i_2=1}^{q_2}\sum_{i_1=1}^{q_1},$$

where $P_{(X_n=i_n,Y_m=j_m)}$ is the projection of R_m through all input variables, except that x_n is fixed at a point i_n. The index $C_m^{(2)}$ indicates the normalized average area difference between the inference mth output from the two-dimensional relation matrices and that from the original relation matrix. Thus the decomposition of the original relation matrix is successful if the index is close to zero. But if the index is bigger than a preassigned upper bound, we may conclude that the error of the decomposition is not acceptable. In that case, the dimension of the relation matrix should be increased by 1 successively to avoid large inference error.

To test the applicability of the three-dimensional relation matrix that is obtained from two inputs and the mth output, the index $C_m^{(3)}$ is defined in a manner similar to that of $C_m^{(2)}$, as follows [11]:

$$C_m^{(3)} = \frac{\sum_{i_1,\ldots,i_N,j_m}\left\{\bigwedge_{l=1}^{N-1}\bigwedge_{n=l+1}^{N}P_{(X_l=i_l,X_n=i_n,Y_m=j_m)} - R_m(i_1,\ldots,i_N,j_m)\right\}}{q_1 \times q_2 \times \cdots \times q_N \times p_m},$$

$$(27)$$

where

$$0 \le C_m^{(3)} < 1, \qquad m = 1,\ldots,M$$

$$\bigwedge_{l=1}^{N-1}\bigwedge_{n=l+1}^{N} P_{(X_l=i_l,X_n=i_n,Y_m=j_m)}$$

$$= P_{(X_1=i_1,X_2=i_2,Y_m=j_m)} \wedge \cdots \wedge P_{(X_1=i_1,X_N=i_N,Y_m=j_m)}$$

$$\wedge P_{(X_2=i_2,X_3=i_3,Y_m=j_m)} \wedge \cdots \wedge P_{(X_2=i_2,X_N=i_N,Y_m=j_m)}$$

$$\cdots$$

$$\wedge P_{(X_{N-2}=i_{N-2},X_{N-1}=i_{N-1},Y_m=j_m)} \wedge P_{(X_{N-2}=i_{N-2},X_N=i_N,Y_m=j_m)}$$

$$\wedge P_{(X_{N-1}=i_{N-1},X_N=i_N,Y_m=j_m)}$$

$$P_{(X_n=i_n,Y_m=j_m)} = proj_{X_l=i_l,X_n=i_n}R_m$$

$$= \sup_{\substack{x_1 \in X_1 \\ x_2 \in X_2 \\ \vdots \\ x_N \in X_N}} R_m(x_1,\ldots,X_l=i_l,\ldots,X_n=i_n,\ldots,x_N,Y_m=j_m).$$

The general form of the index can be defined in a similar way, but it is omitted because of its complicated notation.

V. APPLICATIONS

A. Multivariable Open-Loop Fuzzy Systems

Open-loop fuzzy systems are those systems in which the outputs have no effect upon the input signals. Taking advantage of the index of applicability to these systems, we propose a new decomposition method of the relation matrix to reduce the inference error possibly resulting from the oversimplification of the linguistic rules. In the method we adopt a coordinator that determines the dimension of the relation matrix in the upper level.

Now the procedure for inferring output in the multivariable open-loop fuzzy systems with a coordinator can be summarized as follows:

Step 1. Assign an upper bound (ζ) of the indices of applicability and set the number of dimension $r = 2$.

Step 2. Compute the indices $C_i^{(r)}$, $i = 1, \ldots, M$, corresponding to each output from the original relation matrix of the fuzzy system.

Step 3. Check whether the indices are within the preassigned upper bound (ζ). If they are not within the bound, increase the dimension of the corresponding relation matrices, r, by 1 and go to step 2. Otherwise, set the dimension of the relation matrices to r, and decompose the rules according to the dimension r for each output.

As an explanatory example, we consider a system with three inputs and two outputs. Suppose that the indices $C_1^{(2)}$ and $C_2^{(3)}$ are less than the preassigned upper bound and $C_2^{(2)}$ is greater than that. In this example the output Y_1 can be inferred with tolerable inference error by Gupta's decomposition method, whereas the output Y_2 should be inferred from the three-dimensional relation matrices. The block diagram of the proposed decomposition method is shown in Fig. 4. With the operators $\circ$ and $\wedge$, the proposed method is expressed by the following equations:

$$Y_1 = X_1 \circ R_{11} \wedge X_2 \circ R_{21} \wedge X_3 \circ R_{31} \tag{28}$$

$$Y_2 = X_1 \circ X_2 \circ R_{122} \wedge X_1 \circ X_2 \circ R_{122} \wedge X_1 \circ X_2 \circ R_{122}, \tag{29}$$

where

$$R_{uv} = \bigvee_{k=1}^{K} \{X_{u(k)} \wedge Y_{v(k)}\}, \qquad u = 1,2,3, \quad v = 1,2$$

$$R_{huv} = \bigvee_{k=1}^{K} \{X_{h(k)} \wedge X_{u(k)} \wedge Y_{v(k)}\}, \qquad h = 1,2,3.$$

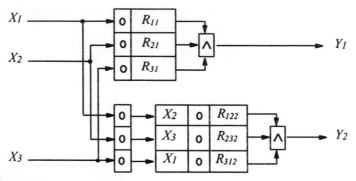

FIGURE 4 Block diagram of the multivariable fuzzy system by the proposed decomposition method.

B. Multivariable Closed-Loop Fuzzy Systems

In implementing a practical system, the large inference error of the fuzzy controller could be a main source of degradation of the performance. To reduce the inference error, we propose a new structure of fuzzy controller with a coordinator in the upper level, as shown in Fig. 5. In this figure, S_i denotes a desired value of the ith output of the system and E_i^* and ΔE_i^* represent the error and error change of the ith output, respectively.

The design procedure for a fuzzy logic controller with a coordinator is similar to that of the multivariable open-loop fuzzy systems described in the previous subsection. As an example, we consider a fuzzy controller with four inputs $(E_1, \Delta E_1, E_2, \Delta E_2)$ and two outputs (U_1, U_2). Suppose that the index $C_1^{(2)}$ is greater than the upper bound and $C_1^{(3)}$ and $C_2^{(2)}$ are less than that.

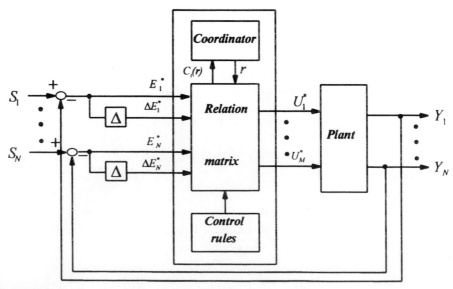

FIGURE 5 Structure of the fuzzy logic controller with a coordinator.

Then Fig. 6 shows the block diagram of a fuzzy controller by the proposed method.

Using the operators $\circ$ and $\wedge$, inference equations that calculate outputs of the fuzzy controller by the proposed method have the following form:

$$U_1 = E_1 \circ \Delta E_1 \circ R_{111} \wedge E_1 \circ E_2 \circ R^*_{121} \wedge \Delta E_1 \circ E_2 \circ R_{121}$$

$$\wedge\, \Delta E_1 \circ \Delta E_2 \circ R^{**}_{121} \wedge E_2 \circ \Delta E_2 \circ R_{221} \wedge \Delta E_2 \circ E_1 \circ R_{211} \quad (30)$$

$$U_2 = E_1 \circ R_{12} \wedge \Delta E_1 \circ R^*_{12} \wedge E_2 \circ R_{22} \wedge \Delta E_2 \circ R^*_{22}, \quad (31)$$

where

$$R_{111} = \bigvee_{k=1}^{K} \{E_{1(k)} \wedge \Delta E_{1(k)} \wedge U_{1(k)}\}$$

$$R_{121} = \bigvee_{k=1}^{K} \{\Delta E_{1(k)} \wedge E_{2(k)} \wedge U_{1(k)}\}$$

$$R^*_{121} = \bigvee_{k=1}^{K} \{E_{1(k)} \wedge E_{2(k)} \wedge U_{1(k)}\}$$

$$R^{**}_{121} = \bigvee_{k=1}^{K} \{\Delta E_{1(k)} \wedge \Delta E_{2(k)} \wedge U_{1(k)}\}$$

$$R_{221} = \bigvee_{k=1}^{K} \{E_{2(k)} \wedge \Delta E_{2(k)} \wedge U_{1(k)}\}$$

$$R_{211} = \bigvee_{k=1}^{K} \{\Delta E_{2(k)} \wedge E_{1(k)} \wedge U_{1(k)}\}$$

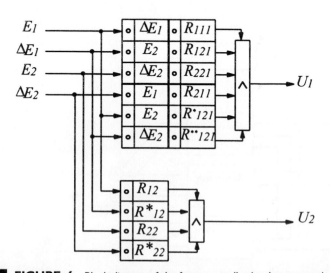

FIGURE 6 Block diagram of the fuzzy controller by the proposed method.

$$R_{12} = \bigvee_{k=1}^{K} \{E_{1(k)} \wedge U_{2(k)}\}$$

$$R_{12}^* = \bigvee_{k=1}^{K} \{\Delta E_{1(k)} \wedge U_{2(k)}\}$$

$$R_{22} = \bigvee_{k=1}^{K} \{E_{2(k)} \wedge U_{2(k)}\}$$

$$R_{22}^* = \bigvee_{k=1}^{K} \{\Delta E_{2(k)} \wedge U_{2(k)}\}$$

C. Parallel Distributed Multivariable Fuzzy Systems

In an industrial process a complex system can be divided into several subsystems based on the physical structure of the process. To avoid the computational difficulty due to the complexity of the global system and reduce the inference error and yet retain the coupling effects, a hierarchical distributed structure can be used. Consider the hierarchical structure of a parallel distributed fuzzy system that is composed of three subsystems with three inputs and two outputs, as shown in Fig. 7. Assume that the three subsystems are expressed by the following linguistic rules.

Fuzzy subsystem A:

$$\begin{aligned} &IF\ X_1\ is\ X_{1(k)}\ AND\ X_2\ is\ X_{2(k)}\ AND\ X_3\ is\ X_{3(k)} \\ &THEN\ Z_1\ is\ Z_{1(k)}\ AND\ Z_2\ is\ Z_{2(k)}, \qquad k = 1, 2, \ldots, K. \end{aligned} \tag{32}$$

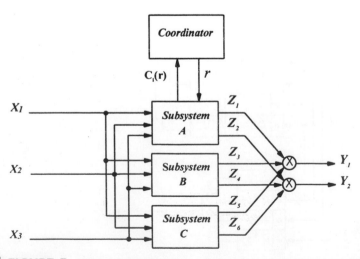

FIGURE 7 Hierarchical structure of a parallel distributed multivariable fuzzy system.

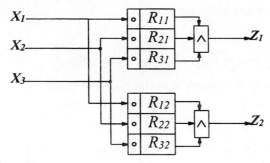

FIGURE 8   Structure of subsystems A and C.

Fuzzy subsystem B:

$$IF\ X_1\ is\ X_{1(k)}\ AND\ X_2\ is\ X_{2(k)}\ AND\ X_3\ is\ X_{3(k)}$$
$$THEN\ Z_3\ is\ Z_{3(k)}\ AND\ Z_4\ is\ Z_{4(k)}, \qquad k = 1, 2, \ldots, K. \tag{33}$$

Fuzzy subsystem C:

$$IF\ X_1\ is\ X_{1(k)}\ AND\ X_2\ is\ X_{2(k)}\ AND\ X_3\ is\ X_{3(k)}$$
$$THEN\ Z_5\ is\ Z_{5(k)}\ AND\ Z_6\ is\ Z_{6(k)}, \qquad k = 1, 2, \ldots, K, \tag{34}$$

where $X_{i(k)}$ and $Z_{j(k)}$ are the membership functions of the ith input and the jth output of each subsystem in the kth rule, respectively.

The explanation of the procedure of inference of the distributed fuzzy system is similar to that already given. Assume that the indices $C_3^{(2)}$ and $C_4^{(2)}$ are greater than the upper bound and $C_1^{(2)}$, $C_2^{(2)}$, $C_5^{(2)}$, $C_6^{(2)}$, $C_3^{(3)}$, and $C_4^{(3)}$ are less than that. Then the structures of the subsystems by the proposed method are shown in Figs. 8 and 9. The inference equations of the subsystems with

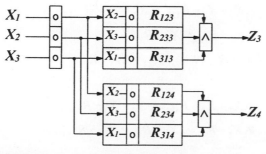

FIGURE 9   Structure of subsystem B.

reduced inference error are given by

Subsystem A:

$$Z_1 = X_1 \circ R_{11} \wedge X_2 \circ R_{21} \wedge X_3 \circ R_{31} \tag{35}$$

$$Z_2 = X_1 \circ R_{12} \wedge X_2 \circ R_{22} \wedge X_3 \circ R_{32}, \tag{36}$$

where

$$R_{uv} = \bigvee_{k=1}^{K} \{X_{u(k)} \wedge Y_{v(k)}\} \quad u = 1,2,3, \quad v = 1,2.$$

Subsystem B:

$$Z_3 = X_1 \circ X_2 \circ R_{123} \wedge X_2 \circ X_3 \circ R_{233} \wedge X_3 \circ X_1 \circ R_{313} \tag{37}$$

$$Z_4 = X_1 \circ X_2 \circ R_{124} \wedge X_2 \circ X_3 \circ R_{234} \wedge X_3 \circ X_1 \circ R_{314}, \tag{38}$$

where

$$R_{huv} = \bigvee_{k=1}^{K} \{X_{h(k)} \wedge X_{u(k)} \wedge Y_{v(k)}\}, \quad h = u = 1,2,3, \quad v = 3,4.$$

Subsystem C:

$$Z_5 = X_1 \circ R_{15} \wedge X_2 \circ R_{25} \wedge X_3 \circ R_{35} \tag{39}$$

$$Z_6 = X_1 \circ R_{16} \wedge X_2 \circ R_{26} \wedge X_3 \circ R_{36}. \tag{40}$$

Then the outputs of global system are calculated as

$$Y_1 = Z_1 \otimes Z_3 \otimes Z_5 \tag{41}$$

$$Y_2 = Z_2 \otimes Z_4 \otimes Z_6, \tag{42}$$

where $\otimes$ is an appropriate mathematical operation on fuzzy sets. To obtain the outputs of the global system in this example, the extension principle is applied to Eqs. (41) and (42) as follows:

$$Y_1 = \sup_{\substack{z_1, z_3, z_5 \\ y_1 = f(z_1, z_3, z_5)}} \{X_1 \circ R_{11} \wedge X_2 \circ R_{21} \wedge X_3 \circ R_{31}$$

$$\wedge X_1 \circ X_2 \circ R_{123} \wedge X_2 \circ X_3 \circ R_{233} \wedge X_3 \circ X_1 \circ R_{313}$$

$$\wedge X_1 \circ R_{15} \wedge X_2 \circ R_{25} \wedge X_3 \circ R_{35}\} \tag{43}$$

$$Y_2 = \sup_{\substack{z_2, z_4, z_6 \\ y_2 = f(z_2, z_4, z_6)}} \{X_1 \circ R_{12} \wedge X_2 \circ R_{22} \wedge X_3 \circ R_{32}$$

$$\wedge X_1 \circ X_2 \circ R_{124} \wedge X_2 \circ X_3 \circ R_{234} \wedge X_3 \circ X_1 \circ R_{314}$$

$$\wedge X_1 \circ R_{16} \wedge X_2 \circ R_{26} \wedge X_3 \circ R_{36}\}, \tag{44}$$

where f is a mapping function from a Cartesian product of universes

$Z_1 \times Z_2 \times Z_3$ and $Z_2 \times Z_4 \times Z_6$ to the universes of global system Y_1 and Y_2, respectively.

D. Numerical Examples

To illustrate the approach used in the decomposition of the linguistic rules of the multivariable fuzzy systems, it is instructive to consider some numerical examples. They give a better insight into the method discussed.

Consider now the case where the decomposition method can be applied. Suppose that four fuzzy rules have been formulated for a system as follows:

> IF X_1 is Z AND X_2 is ZS THEN Y_1 is ZS
> ALSO
> IF X_1 is ZS AND X_2 is BMB THEN Y_1 is S
> ALSO $\qquad$ (45)
> IF X_1 is M AND X_2 is Z THEN Y_1 is M
> ALSO
> IF X_1 is B AND X_2 is BMB THEN Y_1 is BMB.

The membership functions of the input and output variables are given in Table 1. Assume current inputs $X_1 = ZS$ and $X_2 = BMB$. Then the output by the decomposition method is calculated as follows:

$$Y = X_1 \circ R_{11} \wedge X_2 \circ R_{21} = \begin{bmatrix} 0 & 0.5 & 1 & 0.5 & 0.5 & 0 \end{bmatrix}, \qquad (46)$$

where

$$R_{11} = \bigvee_{k=1}^{4} \{X_{1(k)} \wedge Y_{1(k)}\}$$

$$= \begin{bmatrix} 0.5 & 1 & 0.5 & 0.5 & 0 & 0 \\ 0.5 & 0.5 & 1 & 0.5 & 0.5 & 0 \\ 0 & 0.5 & 1 & 1 & 0.5 & 0 \\ 0 & 0.5 & 1 & 1 & 0.5 & 0 \\ 0 & 0.5 & 0.5 & 0.5 & 0.5 & 0.5 \\ 0 & 0 & 0 & 0.5 & 1 & 0.5 \end{bmatrix}$$

TABLE I   Membership Functions of the Input and Output Variables

Linguistic sets	Quantized levels					
	0	1	2	3	4	5
Z	1	0.5	0	0	0	0
ZS	0.5	1	0.5	0	0	0
S	0	0.5	1	0.5	0	0
M	0	0.5	1	1	0.5	0
BMB	0	0	0	0.5	1	0.5
B	0	0	0	0	0.5	1

$$R_{21} = \bigvee_{k=1}^{4} \{X_{2(k)} \wedge Y_{1(k)}\}$$

$$= \begin{bmatrix} 0.5 & 0.5 & 1 & 1 & 0.5 & 0 \\ 0.5 & 1 & 0.5 & 0.5 & 0.5 & 0 \\ 0.5 & 0.5 & 0.5 & 0 & 0 & 0 \\ 0 & 0.5 & 0.5 & 0.5 & 0.5 & 0.5 \\ 0 & 0.5 & 1 & 0.5 & 1 & 0.5 \\ 0 & 0.5 & 0.5 & 0.5 & 0.5 & 0.5 \end{bmatrix}.$$

For comparison, the output inferred by the original relation matrix is given by

$$Y_1 = X_1 \circ X_2 \circ R_1 = \begin{bmatrix} 0 & 0.5 & 1.0 & 0.5 & 0 & 0 \end{bmatrix}, \qquad (47)$$

where

$$R_1 = \bigvee_{k=1}^{4} \{X_{1(k)} \wedge X_{2(k)} \wedge Y_{1(k)}\}.$$

These results are graphically expressed in Fig. 10. The index is calculated by Eq. (26) as $C_1^{(2)} = 0.053$. The value of this index means that the difference of the normalized average area is 5.3%. In this case, we know that Gupta's decomposition method may be applied because the difference in the normalized average area is as small as 5.3%.

To better understand the fact that the decomposition method may result in a considerable inference error, we apply this method to an exclusive nor (ENOR) gate model shown in Fig. 11. Fig. 12 shows the membership function of the input and output variables of the ENOR gate. The linguistic descriptions of control rules for the ENOR gate model are given by

$$
\begin{array}{ll}
IF\ X_1\ is\ L\ AND\ X_2\ is\ L & THEN\ Y_1\ is\ H \\
ALSO & \\
IF\ X_1\ is\ L\ AND\ X_2\ is\ H & THEN\ Y_1\ is\ L \\
ALSO & \\
IF\ X_1\ is\ H\ AND\ X_2\ is\ L & THEN\ Y_1\ is\ L \qquad (48)\\
ALSO & \\
IF\ X_1\ is\ H\ AND\ X_2\ is\ H & THEN\ Y_1\ is\ H.
\end{array}
$$

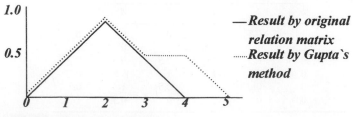

FIGURE 10   Result of inference.

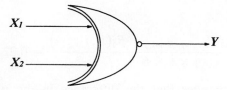

FIGURE 11 Logic symbol of an ENOR gate.

It can be shown that the inference outputs from Gupta's decomposition method for any combinations of inputs X_1 and X_2 are identical. This implies that the decomposition method cannot distinguish one input from the other and generates very large inference error in this case. Suppose that the current inputs X_1 and X_2 are both the high states H. Then the output inferred from the decomposition method is computed as follows:

$$Y = X_1 \circ R_{11} \wedge X_2 \circ R_{21} = [1.0 \quad 1.0 \quad 0.8 \quad 0.8 \quad 1.0 \quad 1.0], \quad (49)$$

where

$$R_{11} = \bigvee_{k=1}^{4} X_{1(k)} \wedge Y_{1(k)}$$

$$= \begin{bmatrix} 1.0 & 1.0 & 0.8 & 0.8 & 1.0 & 1.0 \\ 1.0 & 1.0 & 0.8 & 0.8 & 1.0 & 1.0 \\ 0.8 & 0.8 & 0.8 & 0.8 & 0.8 & 0.8 \\ 0.8 & 0.8 & 0.8 & 0.8 & 0.8 & 0.8 \\ 1.0 & 1.0 & 0.8 & 0.8 & 1.0 & 1.0 \\ 1.0 & 1.0 & 0.8 & 0.8 & 1.0 & 1.0 \end{bmatrix}.$$

$$R_{21} = \bigvee_{k=1}^{4} \{X_{2(k)} \wedge Y_{1(k)}\}$$

$$= \begin{bmatrix} 1.0 & 1.0 & 0.8 & 0.8 & 1.0 & 1.0 \\ 1.0 & 1.0 & 0.8 & 0.8 & 1.0 & 1.0 \\ 0.8 & 0.8 & 0.8 & 0.8 & 0.8 & 0.8 \\ 0.8 & 0.8 & 0.8 & 0.8 & 0.8 & 0.8 \\ 1.0 & 1.0 & 0.8 & 0.8 & 1.0 & 1.0 \\ 1.0 & 1.0 & 0.8 & 0.8 & 1.0 & 1.0 \end{bmatrix}.$$

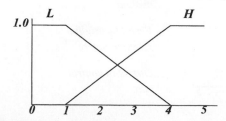

FIGURE 12 Membership function of input and output variables.

The index is obtained as

$$C_1^{(2)} = \frac{\sum_{i_1,i_2,j_1}\left\{\wedge_{n=1}^2 P_{(X_n=i_1,Y_1=j_1)} - R_1(i_1,i_2,j_1)\right\}}{6 \times 6 \times 6} = 0.32. \qquad (50)$$

The index in this example implies that the difference in the normalized average area is as large as 32%. Fig. 13 shows the inference result of the decomposition method. From this result, we know that the discrepancies between the inference results from the original relation matrix and those from the decomposition method are very large, as expected.

For another example of the fuzzy logic controller, let us consider a system described as follows:

$$Y_n = 1.4Y_{n-1} - 0.4Y_{n-2} + 0.165U_{n-1}. \qquad (51)$$

The linguistic control rules of the fuzzy logic controller are given by

$$
\begin{aligned}
&IF\ E_1\ is\ LP\ AND\ CE_1\ is\ ANY\quad THEN\ U_1\ is\ LP\\
&ALSO\\
&IF\ E_1\ is\ SP\ AND\ CE_1\ is\ SP\quad THEN\ U_1\ is\ SP\\
&ALSO\\
&IF\ E_1\ is\ ZE\ AND\ CE_1\ is\ ANY\quad THEN\ U_1\ is\ ZE\\
&ALSO\\
&IF\ E_1\ is\ ZE\ AND\ CE_1\ is\ SP\quad THEN\ U_1\ is\ SP \qquad (52)\\
&ALSO\\
&IF\ E_1\ is\ SN\ AND\ CE_1\ is\ ANY\quad THEN\ U_1\ is\ SN\\
&ALSO\\
&IF\ E_1\ is\ LN\ AND\ CE_1\ is\ ANY\quad THEN\ U_1\ is\ LN\\
&ALSO\\
&IF\ E_1\ is\ ZE\ AND\ CE_1\ is\ LN\quad THEN\ U_1\ is\ SN.
\end{aligned}
$$

The membership functions of fuzzy sets for the error E and the error change CE are shown in Table 2.

From the definition, Eq. (26), the index of this system is calculated as $C_1^{(2)} = 0.087$. And the output responses of the system by the original three-dimensional relation matrix and by Gupta's decomposition method are shown

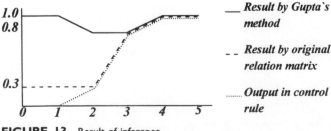

FIGURE 13   Result of inference.

TABLE 2 Membership Function of the Inputs and Outputs for the Controller

Linguistic sets	Quantized levels								
	−4	−3	−2	−1	0	1	2	3	4
LN	1	0.7	0.3	0	0	0	0	0	0
SN	0.3	0.7	1	0.5	0	0	0	0	0
ZE	0	0	0	0.7	1	0.7	0.3	0	0
SP	0	0	0	0	0	0.5	1	0.7	0.3
LP	0	0	0	0	0	0	0.3	0.7	1
ANY	1	0.6	1	0.6	1	0.6	1	0.6	1

in Figs. 14 and 15, respectively. Comparing these two output responses, we might conclude that Gupta's decomposition method was applicable to the control system without much degradation of control performance. This result was predicted from the fact that the index was as small as 0.087.

As an another example, consider a parallel distributed fuzzy system that is composed of three subsystems with the following linguistic rules.

Subsystem A:

$$
\begin{aligned}
&IF\ X_1\ is\ B\ AND\ X_2\ is\ Z \quad THEN\ Z_1\ is\ Z\ AND\ Z_2\ is\ M \\
&ALSO \\
&IF\ X_1\ is\ AZ\ AND\ X_2\ is\ Z \quad THEN\ Z_1\ is\ B\ AND\ Z_2\ is\ AZ \\
&ALSO \\
&IF\ X_1\ is\ B\ AND\ X_2\ is\ BMB \quad THEN\ Z_1\ is\ B\ AND\ Z_2\ is\ BMB \\
&ALSO \\
&IF\ X_1\ is\ AZ\ AND\ X_2\ is\ BMB \quad THEN\ Z_1\ is\ AZ\ AND\ Z_2\ is\ S.
\end{aligned}
\tag{53}
$$

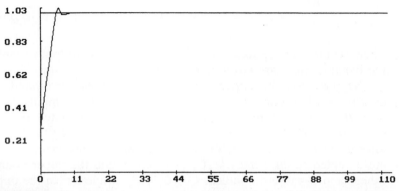

FIGURE 14 Unit step response by the original three-dimensional relation matrix.

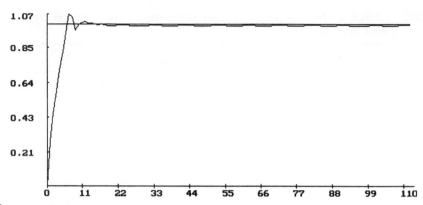

FIGURE 15 Unit step response by Gupta's method.

Subsystem B:

> *IF X_1 is AZ AND X_2 is Z   THEN Z_3 is AZ AND Z_4 is S*
> *ALSO*
> *IF X_1 is B AND X_2 is Z   THEN Z_3 is B AND Z_4 is AZ*
> *ALSO*   (54)
> *IF X_1 is B AND X_2 is BMB   THEN Z_3 is Z AND Z_4 is BMB*
> *ALSO*
> *IF X_1 is AZ AND X_2 is BMB   THEN Z_3 is B AND Z_4 is M.*

Subsystem C:

> *IF X_1 is B AND X_2 is Z   THEN Z_5 is Z AND Z_6 is AS*
> *ALSO*
> *IF X_1 is AZ AND X_2 is Z   THEN Z_5 is BMB AND Z_6 is S*
> *ALSO*   (55)
> *IF X_1 is B AND X_2 is BMB   THEN Z_5 is BMB AND Z_6 is B*
> *ALSO*
> *IF X_1 is AZ AND X_2 is BMB   THEN Z_5 is Z AND Z_6 is BMB.*

Then the membership functions of the input and the output variables used in the linguistic rules are given in Table 3.

Suppose that the upper bounds of the indices corresponding to each output of subsystems are set to 0.1 and the current inputs are given by $X_1 = B$, $X_2 = BMB$, respectively. The indices are computed as $C_1^{(2)} = 0.297$, $C_2^{(2)} = 0.046$, $C_3^{(2)} = 0.287$, $C_4^{(2)} = 0.045$, $C_5^{(2)} = 0.312$, $C_6^{(2)} = 0.052$. Now to infer the outputs Z_2, Z_4, Z_6 of subsystems, two-dimensional relation matrices are accepted because the $C_2^{(2)}, C_4^{(2)}, C_6^{(2)}$ are within the upper bound. On the other hand, because the indices $C_1^{(3)}, C_3^{(3)}, C_5^{(3)}$ are less than the upper bound, it is concluded that three-dimensional relation matrices are acceptable to

▆▆▆ **TABLE 3   Membership Functions of the Input and
Output Variables**

Linguistic	Quantized levels					
sets	0	1	2	3	4	5
Z	1	0.7	0.3	0	0	0
ZS	1	1	0.7	0.3	0	0
S	0.7	1	0.7	0.3	0	0
M	0.3	0.7	1	0.7	0.3	0
BMB	0	0.3	0.7	1	1	0.7
B	0	0	0.3	0.7	1	1

infer Z_1, Z_3, Z_5. Then the outputs of the global system are calculated as

$$Y_1 = Z_1 \otimes Z_3 \otimes Z_5$$

$$= [0.3 \quad 0.3 \quad 0.3 \quad 0.3 \quad 0.3 \quad 0.7 \quad 0.7 \quad 1 \quad 1 \quad 1$$

$$0.7 \quad 0.7 \quad 0.3 \quad 0.3 \quad 0.3 \quad 0.3] \quad (56)$$

$$Y_2 = Z_2 \otimes Z_4 \otimes Z_6$$

$$= [0.3 \quad 0.3 \quad 0.3 \quad 0.3 \quad 0.7 \quad 0.7 \quad 0.7 \quad 0.7 \quad 0.7 \quad 0.7$$

$$1 \quad 1 \quad 1 \quad 1 \quad 0.7 \quad 0.7]. \quad (57)$$

For the comparison, the outputs Y_1^0, Y_2^0 by the original relation matrix and the outputs Y_1^G, Y_2^G inferred from the two-dimensional relation matrices only are computed and given as follows:

$$Y_1^0 = [0.3 \quad 0.3 \quad 0.3 \quad 0.3 \quad 0.3 \quad 0.7 \quad 0.7 \quad 1 \quad 1 \quad 1$$

$$0.7 \quad 0.7 \quad 0.3 \quad 0.3 \quad 0.3 \quad 0.3] \quad (58)$$

$$Y_2^0 = [0.3 \quad 0.3 \quad 0.3 \quad 0.3 \quad 0.3 \quad 0.3 \quad 0.3 \quad 0.7 \quad 0.7 \quad 0.7$$

$$1 \quad 1 \quad 1 \quad 1 \quad 0.7 \quad 0.7] \quad (59)$$

$$Y_1^G = [1 \quad 1 \quad 0.7 \quad 1 \quad 1 \quad 1 \quad 0.7 \quad 1 \quad 1 \quad 1 \quad 1 \quad 1 \quad 1 \quad 1 \quad 1 \quad 0.7]$$

$$(60)$$

$$Y_2^G = [0.3 \quad 0.3 \quad 0.3 \quad 0.3 \quad 0.7 \quad 0.7 \quad 0.7 \quad 0.7 \quad 0.7 \quad 0.7$$

$$1 \quad 1 \quad 1 \quad 1 \quad 0.7 \quad 0.7]. \quad (61)$$

Figures 16 and 17 illustrate the comparison of outputs inferred by the different methods, respectively. Notice that the result of the proposed method in Fig. 16 is very close to that of the original relation matrix. Furthermore, comparing the inference outputs in Fig. 17, we conclude that Gupta's method is applicable to the system because $C_2^{(2)}, C_4^{(2)}, C_6^{(2)}$ are within the upper bound.

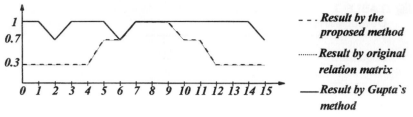

FIGURE 16   Results of inference output Y_1.

From these results we know that the performances of the proposed method with a coordinator in the upper level are better than that of Gupta's method.

VI. SUMMARY

In this chapter we have shown that a large inference error is generated when Gupta's decomposition method is applied to the ENOR gate model which is used as a counterexample. And we define the indices of applicability that can be used to determine whether Gupta's method can be applied to multivariable fuzzy systems. These indices represent the difference in the normalized average area between the inference output membership functions from the decomposition method and those from the original relation matrix. We have also proposed a decomposition method with a coordinator of multivariable fuzzy systems. The coordinator in the upper level calculates the inference error of the decomposition by the indices of applicability and decides the dimension of the relation matrices, which are employed to infer outputs. Simulation results of examples show that the index is applicable to the decomposition of the relation matrix in conjunction with the inference error and control performance and that the coordinator prevents a large amount of inference error induced by Gupta's decomposition method.

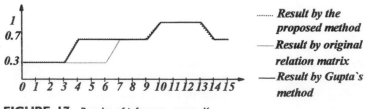

FIGURE 17   Results of inference output Y_2.

REFERENCES

1. Zadeh, L. A. Fuzzy sets. *Inform. Control* 8:338–353, 1965.
2. King, P. J. and Mamdani, E. H. The application of fuzzy control system to industrial processes. *Automatica* 13:235–242, 1977.
3. Li, Y. F. and Lau, C. C. Development of fuzzy algorithms for servo systems. *IEEE Control Syst. Mag.* 65–71, 1989.
4. Shakouri, A., van den Bosch, P., van Nauta Lemke, H., and Dijkman, J. Fuzzy control for multivariable systems. In *Proceedings of the Second IFAC Symposium on Computer-Aided Design of Multivariable Technological Systems*, 1982.
5. Cheng, W., Ren, S., Wu, C., and Tsuei, T. Fuzzy information and decision processes. In *An Expression for Fuzzy Controller* (M. M. Gupta and E. Sanchez, Eds.), North-Holland, Amsterdam, 1982.
6. Cheng, W., Ren, S., Wu, C., and Tsuei, T. The intersection of fuzzy subsets and the robustness of fuzzy control. In *Proceedings of the 1980 International Conference of the Cybernetics Society*, 1980.
7. Czogala, E. and Zimmermann, H. J. Some aspects of synthesis of probabilistic fuzzy controllers. *Fuzzy Sets and Systems* 13:169–177, 1984.
8. Trojan, G. M. and Kiszka, J. B. Solution of multivariable fuzzy equations. *Fuzzy Sets and Systems* 23:271–279, 1987.
9. Gupta, M. M., Kiszka, J. B., and Trojan, G. M. Multivariable structure of fuzzy control systems. *IEEE Trans. Syst. Man, Cybern.* 16:638–655, 1986.
10. Lee, P. G., Lee, K. K., and Jeon, G. J. An index of applicability for the decomposition method of multivariable fuzzy systems. *IEEE Trans. Fuzzy Systems* 3:364–369, 1995.
11. Jeon, G. J. and Lee, P. G. Structure of multivariable fuzzy control systems with a coordinator. *Fuzzy Sets and Systems* 71:85–94, 1995.

16

TECHNIQUES AND APPLICATIONS OF CONTROL SYSTEMS BASED ON KNOWLEDGE-BASED INTERPOLATION

FRANK KLAWONN

Department of Electrical Engineering and Computer Science, FH Ostfriesland,
University of Applied Sciences, Constantiaplatz 4, D-26723 Emden, Germany

RUDOLF KRUSE

Department of Computer Science, Otto-von-Guericke University, Universitätsplatz 2,
D-39106 Magdeburg, Germany

I. INTRODUCTION 431
II. FUZZY SETS, VAGUE ENVIRONMENTS,
   AND INDISTINGUISHABILITY 433
III. APPLICATION TO FUZZY CONTROL 444
IV. A REGRESSION TECHNIQUE FOR FUZZY CONTROL 450
V. CONCLUSIONS 457
   REFERENCES 459

I. INTRODUCTION

When L. A. Zadeh introduced the notion of a fuzzy set in his seminal paper [1] in 1965, his intention was the development of a concept for the representation of the vagueness inherent in common linguistic statements like "The temperature should be kept low, high pressure should be avoided," etc. Of course, at that time probabilistic, statistic, and stochastic models were already well known and applied in many fields. About the same time that Zadeh introduced fuzzy sets, R. E. Moore published his book on interval analysis [2]. Handling imperfect knowledge and information is the common aim of probabilistic models, interval analysis, and fuzzy sets.

However, these paradigms aim at modeling different aspects of imperfect knowledge. Probability is usually understood as the uncertainty about whether a certain (well-defined) event occurs. An event can, for instance, be a real number, the outcome of a measuring experiment. Interval analysis is concerned with imprecision, i.e., instead of crisp values, intervals are considered without making further assumptions about the specific number in a considered interval. The imprecision itself is considered to be precise, because the interval boundaries are assumed to be exact.

Vague values and vague intervals are used to build the underlying interpretation of fuzzy sets in many applications. Linguistic concepts like *slow*, *young*, or *small* incorporate vagueness in the sense of a valuated imprecision, meaning that these concepts do not represent crisp values, but ranges or intervals with boundaries that cannot be specified exactly. This viewpoint clarifies the fact that when dealing with vagueness we have to explain how we want to model values or intervals with nonsharp boundaries.

In later years of fuzzy set theory, L. A. Zadeh introduced the notion of computing with words, based on the perception that human thinking is not based on handling numbers, but on operating with linguistic concepts. Nevertheless, for computers processing numbers is essential, starting with input data in the form of real numbers. Thus fuzzy set theory tries to build a bridge between human-like thinking and processing on computers.

The elementary concept in fuzzy set theory is the notion of membership degree that allows for graded memberships of elements in sets. This idea can be used to model vague concepts because it enables us to assign a (vague) property like *fast* to an object like a *car* to a certain degree. Membership degrees as a generalization of the crisp membership degrees 0 and 1 are often considered as elementary concepts in fuzzy set theory. However, without providing an explanation of what the concrete meaning of a membership value is, there are no unique canonical operations for handling fuzzy sets, and a naive approach can lead to contradictions.

For fuzzy systems the meaning of the membership degree is a crucial point. In [3] three different semantics for fuzzy sets are mentioned:

- Preference: The unit interval is considered as nothing more than a linear ordering. The membership degrees are used to formulate flexible constraints in the sense that besides the values 0 (forbidden) and 1 (completely allowed), intermediate values are admitted [4].
- Uncertainty in the sense of possibility: Fuzzy sets are considered as derived concepts from possibility or necessity measures [5–8].
- Similarity, indistinguishability, indiscernibility etc.: These properties are understood in a gradual way, and a fuzzy set is induced, for instance, by one element as the (fuzzy) set of those elements that are indistinguishable from the considered element [9, 10].

Maybe a fourth interpretation should also be mentioned:

- Partial contradiction or partial inconsistency: This interpretation was proposed in [11] and further pursued in [12] on the basis of the Ulam game [13], which deals with an unreliable information source.

In this paper we concentrate on the interpretation of fuzzy sets from the viewpoint of similarity, indistinguishability, or indiscernibility.

The paper is organized as follows. Section II introduces the notion of vague environments and explains how fuzzy sets can be viewed as points or magnitudes in vague environments. Section III explains fuzzy control from the viewpoint of vague environments and the consequences for fuzzy controllers. Finally, we use the concept of vague environments to describe a regression technique for learning a fuzzy controller from data in Section IV.

II. FUZZY SETS, VAGUE ENVIRONMENTS, AND INDISTINGUISHABILITY

Many authors distinguish between a fuzzy set M and its associated membership function μ_M. However, they only use M as a name for μ_M and operate always with μ_M. Therefore, we do not distinguish between M and μ_M here. A fuzzy set is a mapping $\mu: X \to [0, 1]$ from a universe of discourse X to the unit interval. $\mu(x) \in [0, 1]$ is considered as the degree of membership of the object or element $x \in X$ in the fuzzy set μ. In many applications the universe of discourse X is a real interval, the real numbers, and sometimes a Cartesian product of real intervals. When looking at applications, especially in fuzzy control, one can see that nonarbitrary fuzzy sets or membership functions $\mu: \mathbb{R} \to [0, 1]$ are used. For instance, no one would really want to consider a fuzzy set on the unit interval like

$$\mu(x) = \begin{cases} x & \text{if } x \text{ is a rational number} \\ 0 & \text{otherwise.} \end{cases}$$

Fuzzy sets on real intervals as real-valued functions are almost always "well behaved" in the sense that they are continuous and unimodal (having only one local maximum or a range with maximum membership degree). This is of course reasonable, but the pure mathematical definition of a fuzzy set does not require any such property.

As we have already mentioned, for a consistent handling of fuzzy sets a concrete interpretation of the membership degrees is essential. So in the following we explain how fuzzy sets can be seen as induced concepts, when we start with the elementary and canonical notion of distance.

In engineering applications we have in general to deal with real-valued measurements and control actions in quantified form. We should be aware of the fact that the involved real numbers can never be exact. Of course, in many applications the inexactness is small enough that we do not have to worry about it.

In the following we will provide a model that is able to represent this inexactness to handle problems connected to this phenomenon. Two different forms of inexactness can be distinguished:

- Enforced inexactness of measurement and control values, which is caused by the limited precision of measuring or other instruments or by properties of the physical environment that make an exact measurement impossible.
- Intended imprecision, where we are not interested in arbitrary exactness or where it even does not make sense. As an example, consider the room temperature. A difference of 0.0000001°C in the temperature is not of interest to a human being, nor should it influence the air conditioning system. Because human beings do not usually think in

terms of exact numbers but more in the sense of magnitudes, intended imprecision is very much in the spirit of computing with words.

A very simple model of the above-mentioned phenomena of inexactness identifies values whose distance is less than an error- or tolerance-bound $\varepsilon > 0$. This identification can cause problems because it does not satisfy the law of transitivity; i.e., although x_1 and x_2 as well as x_2 and x_3 are identified according to $|x_1 - x_2| \leq \varepsilon$ and $|x_2 - x_3| \leq \varepsilon$, it is possible that x_1 and x_3 would not be identified because $|x_1 - x_3| > \varepsilon$. The following situation illustrates this nontransitivity property. The decision to buy a certain luxury car does not depend in general on an increase in the price of \$1. But it is, of course, not allowed to iterate this argument; otherwise we would accept any price.

Because of this nontransitivity, we cannot define adequate equivalence classes of indistinguishable or identifiable numbers. Enforcing artificial equivalence classes by partitioning the real numbers into disjoint intervals of a certain length and identifying values that fall in the same interval leads automatically to an incoherent treatment of values that are near the boundary of an interval.

Another question is whether we can specify an appropriate tolerance bound ε so that we consider exactly those values as indistinguishable whose difference is less than ε. Choosing ε to be too small leads to a very tedious and inefficient model that is difficult to handle. If ε is too large, the model becomes too rough. Therefore we do not restrict ourselves to just one value of ε, but to an interval of possible values for ε. We assume that the interval of possible values for ε is the unit interval. Although this might look like an arbitrary choice, we will see later on when we introduce the concept of scaling that it is sufficient to concentrate on the unit interval.

A very simple approach derived from the above consideration is the following. We replace the exact real numbers (that do actually not come from exact measurements or specifications) with the (fuzzy) set of indistinguishable values. We interpret indistinguishability as the dual concept to distance. Therefore, we define the degree of indistinguishability or similarity between two real number as 1 minus the absolute value of their difference. To avoid negative degrees of indistinguishability, we define a fuzzy equivalence relation $E(x, y) = 1 - \min\{|x - y|, 1\}$. In this way, a real number is indistinguishable from itself to the degree 1 (completely indistinguishable) and indistinguishable to the degree 0 (completely distinguishable) from all real numbers that differ by more than 1 from the considered number.

A real value $x_0 \in \mathbb{R}$ induces the fuzzy set of all real numbers that are indistinguishable from x_0 by

$$\mu_{x_0}(x) = E(x_0, x) = 1 - \min\{|x_0 - x|, 1\}.$$

It is quite remarkable that this fuzzy set μ_{x_0} representing the magnitude or approximate value x_0 has a symmetric triangular membership function with its maximum at x_0, reaching the membership degree 0 at $x_0 - 1$ and

$x_0 + 1$. So we have an interpretation and justification for the use of such triangular membership functions, not just because of their simplicity.

However, this very simple approach is too restrictive to be really useful for applications, because it neglects certain important aspects by just defining indistinguishability on the basis of the canonical metric on the real numbers.

The answer to the question of whether two values should be considered more or less the same magnitude is, of course, problem dependent. But it also depends on the measurement unit. For keeping an airplane on a certain flight route, a distance of less than 3 feet is not of importance, so that we can identify positions (coordinates) whose distance is less than $\varepsilon = 3$ (feet). But we must not stick to this tolerance bound when we measure in miles instead of feet.

The same situations appears when we consider temperatures in Celsius or in Fahrenheit. For this reason we are allowed to introduce a scaling factor $c \geq 0$ [14] and consider two values as ε-distinguishable if their distance times c is greater than ε. So when thinking in terms of Celsius instead of Fahrenheit, we have to take a scaling factor $c = 1.8$ into account, according to the formula of transformation between Fahrenheit and Celsius, $F = 1.8C + 32$.

Let us assume that we have to deal with measurements in the interval $X = [a, b]$. Introducing a scaling factor corresponds to a linear transformation of the interval $[a, b]$ to another interval of length $c(b - a)$, say, to the interval $[0, c \cdot (b - a)]$. The transformation is determined by

$$t_c : [a, b] \to [0, \infty), \quad x \mapsto c \cdot (x - a). \tag{1}$$

The distinguishability of two values $x_1, x_2 \in [a, b]$ is now determined on the basis of the distance of the transformed values, i.e., $|t_c(x_1) - t_c(x_2)|$.

The use of a single scaling factor c enables us to overcome the problem of different scalings, as in the case of Fahrenheit and Celsius. However, we are not able to model the fact that a measurement instrument might provide quite precise values in a certain range, whereas out of this range the measured values are less reliable. This phenomenon corresponds to the enforced inexactness mentioned above. Also in the case of intended imprecision, we might wish to distinguish between values in a certain range very carefully, but for other ranges we are not interested in precise values. To solve this problem of differing precision for different ranges, we introduce varying scaling factors for the ranges. A scaling factor $c > 1$ implies a weak indistinguishability (strong distinguishability) for values in the corresponding range, whereas a scaling factor $c < 1$ leads to a strong indistinguishability. A typical example for varying indistinguishability for different ranges can be experienced in many control applications. When the system to be controlled is very far from the desired set value, then strong control actions are necessary. In this case the concrete value of the error—the difference between the actual value and the desired set value—is not very important. It is sufficient just to know the direction in which the system is out of order. Therefore, we do not intend to distinguish between measurement values in this range. The situation is completely different when the system has almost

TABLE I   Scaling Factors for the Room Temperature

Error e	Scaling factor	Interpretation
$-10 > e$	0	Don't care about the concrete value of the error, just counteract strongly
$-10 \geq e > -5$	0.5	Still a large error, but a little bit of care has to be taken
$-5 \geq e \leq 5$	1.5	Very sensitive control is necessary to avoid strong overshoots
$5 < e \leq 10$	0.5	Same as in the range $-10 \geq e > -5$
$10 < e$	0	Same as in the range $e < -10$

reached the desired set value. Then careful and very sensitive control actions have to be taken. It is very important to know whether the error is positive or negative. This means that we distinguish very carefully in the range where the error is near zero.

A possible choice of scaling factors for this problem is shown in Table 1.

The great scaling factor 1.5 for the range around zero, i.e., for the error values between -5 and $+5$, indicates that these values should be distinguished very carefully to give a sensitive control action. For the ranges of a medium-sized absolute value of the error between 5 and 10, it is important to distinguish between these error values, but because we are still far from the set point, we do not have to be very sensitive. For absolute error values greater than 10, we have to react with the highest possible (positive or negative) value for the control action, regardless of whether the absolute value of the error is 12 or 19.

Let us assume that $X = [-20, 20]$ is the set of possible error values. The function $c: X \rightarrow [0, \infty)$, assigning to each error value the corresponding scaling factor, is shown in Fig. 1. The corresponding transformation induced by these scaling factors is illustrated in Fig. 2.

It is easy to check that the piecewise linear transformation in Fig. 2 from $X = [a, b] = [-20, 20]$ to $[0, 20]$ can be computed by

$$t_c: [a, b] \rightarrow [0, \infty), \quad x \mapsto \int_a^x c(s) \, ds, \tag{2}$$

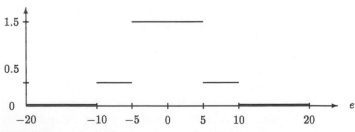

FIGURE I   The scaling factor function for the error value.

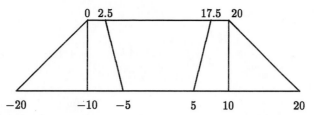

FIGURE 2 The transformation induced by the scaling function in Fig. 1.

where the function c is given by

$$c: X \to [0, \infty), \quad s \mapsto \begin{cases} 0 & \text{if } -20 \leq s < -10 \\ 0.5 & \text{if } -10 \leq s < -5 \\ 1.5 & \text{if } -5 \leq s < 5 \\ 0.5 & \text{if } 5 \leq s < 10 \\ 0 & \text{if } 10 \leq s < 20. \end{cases} \tag{3}$$

If we choose a tolerance bound $\varepsilon = 0.5$ then the error values 0.8 and 1.2 are distinguishable (w.r.t. this tolerance bound), whereas 9 and 11 are indistinguishable. By Eq. (2) the transformed values for 0.8, 1.2, 9, and 11 are 11.2, 11.8, 19.5, and 20, respectively.

Note that Eq. (2) coincides with Eq. (1) when we choose a constant scaling function c.

The scaling function in Fig. 1 reflects the idea that we distinguish values near the optimal set value very carefully, whereas the distinguishability decreases the farther we are from this value. The piecewise linear function was only chosen to elucidate the principle of different scaling factors and to have a simple transformation function. In the most general case we associate with each value x of our set $X = [a, b]$ a scaling factor $c(x) \geq 0$. The function c does not have to be piecewise linear. All we have to assume is that c is integrable. For the transformation induced by such a general scaling function, Eq. (2) is still valid. To understand this, we assume that there is an integrable scaling function $c: [a, b] \to [0, \infty)$. In the neighborhood of the point x_0, the transformed (directed) distance $\delta_c^{\text{dir}}(x_0, x)$ between x_0 and a point x very near x_0 should be approximately $c(x_0) \cdot (x - x_0)$, i.e., $\delta_c^{\text{dir}}(x_0, x) \approx c(x_0) \cdot (x - x_0)$, or

$$\frac{\delta_c^{\text{dir}}(x_0, x)}{x - x_0} \approx c(x_0).$$

For $x \to x_0$ we assume

$$\lim_{x \to x_0} \frac{\delta_c^{\text{dir}}(x_0, x)}{x - x_0} = c(x_0).$$

Thus we have

$$\frac{\partial \delta_c^{\mathrm{dir}}(x_0, x)}{\partial x} = c(x_0),$$

so that we obtain

$$\delta_c^{\mathrm{dir}}(x_0, x) = \int_{x_0}^{x} c(s)\, ds + \mathrm{constant}.$$

because $\delta_c^{\mathrm{dir}}(x_0, x_0) = 0$, we conclude that constant $= 0$ and obtain

$$\delta_c(x_1, x_2) = \left| \delta_c^{\mathrm{dir}}(x_1, x_2) \right| = \left| \int_{x_1}^{x_2} c(s)\, ds \right|. \qquad (4)$$

x_1 and x_2 are considered to be ε-distinguishable with respect to the scaling function c if their "transformed distance" $\delta_c(x_1, x_2)$ is greater than ε.

We now turn to the problem of representing a *vague* environment that is characterized by a distance function δ_c of the above-mentioned type. We do not consider only one fixed value ε, but a whole set of values for ε, each of them leading to a different ε-distinguishability. We consider all numbers from the unit interval as possible values for ε. If one would prefer to have a smaller or larger interval as possible values for ε, this can be amended by an appropriate choice of the scaling function c. If, for example, the scaling function c is replaced with the scaling function $\hat{c} = \lambda \cdot c$, then ε-distinguishability with respect to c corresponds to (ε / λ)-distinguishability with respect to $\hat{c}$. In this sense, allowing all values from the unit interval for ε already covers the most general case.

For each $\varepsilon \in [0, 1]$ we associate with the value $x_0 \in X$ all values $x \in X$ that are not ε-distinguishable from x_0 (with respect to the scaling function c), i.e., the set

$$S_{x_0, \varepsilon} = \{ x \in X \mid \delta_c(x, x_0) \leq \varepsilon \}. \qquad (5)$$

A more convenient representation of this family of sets is described by the mapping

$$\mu_{x_0} \colon X \to [0, 1], \quad x \mapsto 1 - \min\{ \delta_c(x, x_0), 1 \}, \qquad (6)$$

so that we have

$$S_{x_0, \varepsilon} = \{ x \in X \mid \mu_{x_0}(x) \geq 1 - \varepsilon \}.$$

$\mu_{x_0}(x)$ can be interpreted intuitively as the degree to which x can be identified with x_0. Therefore we can understand μ_{x_0} as the fuzzy set of values that are indistinguishable from x_0. Note that in the most simple case, where we have the same scaling factor $c > 0$ for all $x \in X$, i.e., a constant scaling function, we obtain a triangular membership function with slope c taking its maximum at x_0 as the fuzzy set μ_{x_0}, which represents the value x_0 in the vague environment X.

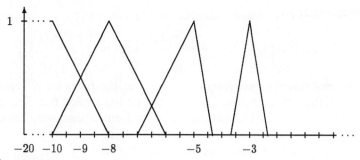

FIGURE 3 The fuzzy sets μ_{x_0} associated with the values $x_0 = -10, -8, -5, -3$ in the vague environment induced by the scaling function in Eq. (3).

Note that the α-cut $\{x \in X \,|\, \mu_{x_0} \geq \alpha\}$ of the fuzzy set μ_{x_0} is equal to the set $S_{x_0, 1-\alpha}$ of elements that are $(1 - \alpha)$-indistinguishable from x_0.

Let us return to the vague environment characterized by the scaling function introduced in Eq. (3). Figure 3 illustrates the fuzzy sets μ_{x_0} that are associated with the values $x_0 = -10, -8, -5, -3$ in this vague environment, i.e., the fuzzy sets of real numbers that are indistinguishable from these values.

The fuzzy sets in Fig. 3 are all of triangular or trapezoidal type. This is not necessarily the case as Fig. 4 illustrates, where the fuzzy sets μ_{x_0} associated with the values $x_0 = -9$ and $x_0 = -4.5$ are shown. (Note the different scaling on the x-axis in comparison to Fig. 3.)

Because the scaling function in Eq. (3) is piecewise constant, the fuzzy set μ_{x_0} associated with a value x_0 will always be piecewise linear.

To obtain other shapes for the fuzzy set associated with the value x_0, an appropriate scaling function has to be defined. As an example, let us consider a bell-shaped fuzzy set of the form

$$\mu: \mathbb{R} \to [0,1], \quad x \mapsto \exp\left(-\frac{1}{2}\left(\frac{x - x_0}{\sigma}\right)^2\right). \tag{7}$$

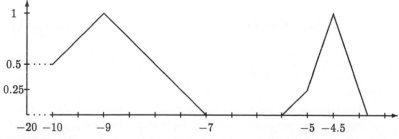

FIGURE 4 The fuzzy sets μ_{x_0} associated with the values $x_0 = -9$ and $x_0 = -4.5$ in the vague environment induced by the scaling function in Eq. (3).

Choosing

$$c: \mathbb{R} \to [0, \infty), \quad x \mapsto \frac{|x - x_0|}{\sigma^2} \cdot \exp\left(-\frac{1}{2}\left(\frac{x - x_0}{\sigma}\right)^2\right) \tag{8}$$

as the scaling function, we obtain $\mu = \mu_{x_0}$, i.e., the fuzzy set μ represents the value x_0 in the vague environment induced by the scaling function c. This is a direct consequence of Eq. (4) for the transformed distance, because (8) is simply the absolute value of the first derivative of (7).

More generally, we have the following theorem.

THEOREM 1.   *Let $\mu: \mathbb{R} \to [0, 1]$ be a fuzzy set such that there exists $x_0 \in \mathbb{R}$* with

   (i) $\mu(x_0) = 1$.
   (ii) μ *is a nondecreasing function on* $(-\infty, x_0]$.
   (iii) μ *is a nonincreasing function on* $[x_0, \infty)$.
   (iv) μ *is continuous.*
   (v) μ *is almost everywhere differentiable.*

Then there exists a scaling function $c: \mathbb{R} \to [0, \infty)$ such that μ coincides with the fuzzy set μ_{x_0}, which is associated with the value x_0 in the vague environment induced by c.

   Proof.   Choose $c(x) = |d\mu(x)/dx|$ as the scaling function.   ∎

It is obvious that the reverse of Theorem 1 also holds, which means that, given a scaling function $c: \mathbb{R} \to [0, \infty)$ and a value x_0, the fuzzy μ_{x_0} associated with x_0 in the vague environment induced by c satisfies conditions (i)–(v) of Theorem 1.

Conditions (i)–(iii) guarantee that the fuzzy set is fuzzy convex (i.e., all of its α-cuts are convex), so that it can be considered as the representation of a single value in a vague environment. Fuzzy sets that are not fuzzy convex cannot appear in vague environments when fuzzy sets stand only for single values. It is possible, of course, to associate a fuzzy set in a vague environment not just with a single value, but to associate with any set of values a corresponding fuzzy set by generalizing Eqs. (5) and (6) for a set $M \subseteq X$ by

$$S_{M, \varepsilon} = \left\{ x \in X | \exists x_0 \in M: \delta_c(x, x_0) \le \varepsilon \right\}$$

and

$$\mu_M: X \to [0, 1], \quad x \mapsto 1 - \min\left\{ \inf_{x_0 \in M}\{\delta_c(x, x_0)\}, 1 \right\},$$

respectively. Figure 5 illustrates an example of such a fuzzy set associated with the set $M = \{2, 4\}$ in the vague environment induced by the constant scaling function $c = 0.5$. This fuzzy set is not fuzzy convex.

Up to now we have only considered fuzzy sets as representations of crisp values in vague environments that were described by scaling functions. In this way a fuzzy partition (set of fuzzy sets), as in Fig. 3, is induced by a set of crisp values together with a scaling function. We now turn to the question of

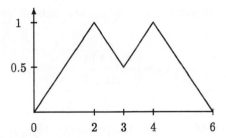

FIGURE 5 The fuzzy set $\mu_{\{2,4\}}$ in the vague environment induced by the constant scaling function $c = 0.5$.

whether we can provide a scaling function for a given set of fuzzy sets such that the corresponding fuzzy sets can be interpreted as representations of crisp values in the vague environment characterized by the scaling function.

THEOREM 2. *Let* $(\mu_i)_{i \in I}$ *be an at most countable family of fuzzy sets on* $\mathbb{R}$, *and let* $(x_0^{(i)})_{i \in I}$ *be a family of real numbers such that* $\mu_i(x_0^{(i)}) = 1$ *holds and conditions* (i)–(v) *of Theorem 1 are satisfied for all* $i \in I$. *There exists a scaling function* $c\colon \mathbb{R} \to [0, \infty)$ *such that* μ_i *coincides with the fuzzy set* $\mu_{x_0^{(i)}}$ (*for each* $i \in I$) *induced by the value* $x_0^{(i)}$ *in the vague environment induced by the scaling function* c, *if and only if*

$$\min\{\mu_i(x), \mu_j(x)\} > 0 \quad \Rightarrow \quad \left|\frac{d\mu_i(x)}{dx}\right| = \left|\frac{d\mu_j(x)}{dx}\right| \tag{9}$$

holds almost everywhere for all $i, j \in I$.

Proof. Assume that (9) is satisfied. Define the scaling function

$$c\colon \mathbb{R} \to [0, \infty), \quad x \mapsto \begin{cases} \left|\dfrac{d\mu_i(x)}{dx}\right| & \text{if } \mu_i(x) > 0 \\ 0 & \text{otherwise.} \end{cases}$$

Equation (2) guarantees that c is well defined almost everywhere. Theorem 1 yields that $\mu_i = \mu_{x_0^{(i)}}$ holds for all $i \in I$. Note that it is sufficient for the proof of Theorem 1 that the scaling function coincides with the derivate of the fuzzy set only on the support of the fuzzy set.

To prove the reverse implication, we assume now that there is a scaling function $c\colon \mathbb{R} \to [0, \infty)$ such that $\mu_i = \mu_{x_0^{(i)}}$ holds for all $i \in I$. Let $i, j \in I$ and let $x \in \mathbb{R}$ with $\min\{\mu_i(x), \mu_j(x)\} > 0$. By definition we have

$$\mu_k(x) = 1 - \left|\int_{x_0^{(k)}}^{x} c(s)\, ds\right|$$

for $k \in \{i, j\}$, which implies

$$\left|\frac{d\mu_k(x)}{dx}\right| = c(x)$$

if μ_k is differentiable at x. Because μ_i and μ_j are almost everywhere differentiable, we obtain

$$\left| \frac{d\mu_i(x)}{dx} \right| = c(x) = \left| \frac{d\mu_j(x)}{dx} \right|$$

almost everywhere.   ■

Theorem 2 simply states that we can find a corresponding scaling function for a given fuzzy "partition" if for each real number $x \in \mathbb{R}$ the absolute value of the slope at x is the same for all fuzzy sets in the fuzzy partition whenever x belongs to the support of the fuzzy set.

A very common type of fuzzy partition is obtained by choosing crisp values $x_1 < x_2 < \cdots < x_n$ and defining the fuzzy set μ_i (for $1 < i < n$) by a triangular membership function that takes its maximum at x_i and reaches the value zero at x_{i-1} and x_{i+1}, respectively. Such a fuzzy partition is illustrated in Fig. 6.

For such fuzzy partitions the corresponding scaling function can be defined as the piecewise constant function

$$c(x) = \frac{1}{x_{i+1} - x_i} \qquad \text{if } x_i < x < x_{i+1},$$

so that the fuzzy sets μ_i represent the fuzzy sets induced by the values x_i in the vague environment induced by the scaling function c.

When we interpret fuzzy sets on the basis of underlying scaling functions, narrow fuzzy sets indicate that in the range of their support we have to be quite careful about the exact value, whereas wider fuzzy sets should be used in regions where the exact value is so important.

The use of scaling functions does not admit arbitrary fuzzy partitions. The fuzzy sets have to be coherent in the sense of Theorem 2, i.e., the absolute values of the first derivative of any two fuzzy sets have to be equal on the intersection of the supports of these fuzzy sets (almost everywhere). This corresponds to the above mentioned observation that the first derivative of a fuzzy set characterizes the indistinguishability in a region. Therefore, fuzzy sets with different values for the first derivative in the same position would mean that we have low and high indistinguishability in this place at the same time, which would make no sense.

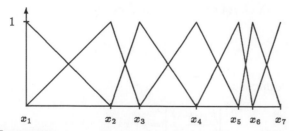

FIGURE 6 A typical fuzzy partition for which a scaling function can be defined easily.

So far we have restricted our considerations to a single real-valued domain. In general we have to deal with multidimensional spaces in control applications, with the dimension depending on the number of input and output variables. Indistinguishability is of course inherent in multidimensional spaces as well. A possible starting point for the $\mathbb{R}^p$ could be the metric induced by a norm, i.e., $\delta(x, y) = \|x - y\|$ for $x, y \in \mathbb{R}^p$. $\|.\|$ could, for instance, be the Euclidean norm. However, the concept of scaling cannot be transferred to multidimensional spaces in a simple way, because there are infinitely many possible directions in any point and infinitely many possible candidates for the shortest (scaled) ways between two points. We would immediately run into problems of Riemann geometry. Another problem is that humans usually do not think in terms of multidimensional spaces. Operating on single variables and combining them later on is a more convenient approach to handling multidimensional spaces. Thus we carry out the specification of the vague environment in the single spaces by scaling functions and aggregate them in the product space. There are various possible ways to carry out this aggregation. Here we mention only two. For a more general treatment we refer to [15].

When we want to define the degree of indistinguishability between the two vectors $(x_1, \ldots, x_p) \in \mathbb{R}^p$ and $(y_1, \ldots, y_p) \in \mathbb{R}^p$, a simple approach is to take the minimum of the indistinguishability degrees in each dimension. So if $\delta_{c_i}(x_i, y_i)$ is the scaled distance in the ith dimension, i.e., the indistinguishability degree between x_i and y_i is $E_i(x_i, y_i) = 1 - \min\{\delta_{c_i}(x_i, y_i), 1\}$, then the indistinguishability degree between the vectors is

$$E\big((x_1, \ldots, x_p), (y_1, \ldots, y_p)\big) = \min\{E_i(x_i, y_i)\}. \tag{10}$$

This indistinguishability can be considered as induced by the metric

$$\delta\big((x_1, \ldots, x_p), (y_1, \ldots, y_p)\big) = \max\{\delta_{c_i}(x_i, y_i)\},$$

i.e., $E = 1 - \min\{\delta, 1\}$. Using this definition, the indistinguishability of two vectors depends only on the lowest indistinguishability of their components.

An alternative to the minimum is based on the Łukasiewicz-t-norm $\alpha * \beta = \max\{\alpha + \beta - 1, 0\}$, defining the indistinguishability by

$$E\big((x_1, \ldots, x_p), (y_1, \ldots, y_p)\big) = \sum_{i=1}^{p} E_i(x_i, y_i) - (p - 1).$$

This indistinguishability can be considered as induced by the metric

$$\delta\big((x_1, \ldots, x_p), (y_1, \ldots, y_p)\big) = \sum_{i=1}^{p} \delta_{c_i}(x_i, y_i).$$

This approach takes all of the indistinguishabilities in the single dimensions for the aggregation into account.

Nevertheless, in any of these approaches the underlying assumption is that the indistinguishabilities in the single domains can be specified independently. In the following section we will see that this assumption applies to fuzzy control as well, although it is usually not satisfied in control applications. Consider, for instance, a controller that uses the error e and the change in the error Δe as inputs. We consider the indistinguishability for the change of the error. If the error is very large, then a strong counteracting control action has to be carried more or less independent of the value of Δe. Therefore we could conclude that we can be content with a high indistinguishability for Δe, because (in the case of a large error) the exact value of Δe is not so important. However, when the error is almost zero, then it really is necessary to be very careful about the value of Δe. For the control action it is crucial to distinguish very well at least in the neighborhood of zero, because we have to know the sign of Δe, indicating the direction in which the system is changing. These considerations show that the indistinguishability in the domain of Δe depends on the value of the variable e, a phenomenon that we cannot model when we characterize the indistinguishabilities on the single domains by scaling functions. Although this might be an argument against our model and even against fuzzy control, this is not really true, as the successful applications show. In engineering applications we usually have to find a compromise between an almost perfect but very complicated model with sophisticated parameters and a simplifying approach that is very easy to handle.

III. APPLICATION TO FUZZY CONTROL

We now provide a framework for fuzzy control based on the concept of vague environments and indistinguishability and establish the connection to Mamdani's fuzzy control model.

Let us first describe the (simplified) formalization of the control problem that we want to consider here. We are given n input variables $\xi_1, \ldots, \xi_n$ taking values in the sets $X_1, \ldots, X_n$, respectively. For reasons of simplicity we assume that we have one output or control variable η with values in the set Y. The problem we have to solve is to find an adequate control function φ: $X_1 \times \cdots \times X_n \to Y$ that assigns to each input tuple $(x_1, \ldots, x_n) \in X_1 \times \cdots \times X_n$ an appropriate output value $y = \varphi(x_1, \ldots, x_n) \in Y$.

Before we introduce knowledge-based control making use of the notion of vague environments, we briefly recall how Mamdani's fuzzy control model [16] is defined, because we will see later on that we also obtain Mamdani's model as a result of our approach. The control function φ is specified by k linguistic control rules R_r in the form

$$R_r: \text{if } \xi_1 \text{ is } A^{(1)}_{i_1, r} \text{ and } \cdots \text{ and } \xi_n \text{ is } A^{(n)}_{i_n, r}, \text{ then } \eta \text{ is } B_{i_r} \qquad (r = 1, \ldots, k),$$

where each linguistic term $A^{(1)}_{i_1, r}, \ldots, A^{(n)}_{i_n, r}, B_{i_r}$ is associated with a fuzzy set $\mu^{(1)}_{i_1, r}, \ldots, \mu^{(n)}_{i_n, r}, \mu_{i_r}$ on $X_1, \ldots, X_n, Y$, respectively. If we are given the input tuple $(x_1, \ldots, x_n) \in X_1 \times \cdots \times X_n$, the output of Mamdani's fuzzy controller is the fuzzy set

$$\mu^{\text{output}}_{x_1, \ldots, x_n} : Y \to [0, 1], \quad y \mapsto \max_{r \in \{1, \ldots, k\}} \left\{ \min\left\{ \mu^{(1)}_{i_1, r}(x_1), \ldots, \mu^{(n)}_{i_n, r}(x_n), \mu_{i_r}(y) \right\} \right\}$$

on Y. To obtain a crisp output value, the fuzzy set $\mu^{\text{output}}_{x_1, \ldots, x_n}$ has to be defuzzified. A very common defuzzification strategy is the center-of-area method, but the mean of maximum and the max criterion method are also applied (see, for example, [17]). For our purposes, it is sufficient to know that these defuzzification strategies compute a crisp value from a fuzzy set; the exact algorithm for each method is not of importance in this section.

We now come to the presentation of a concept of knowledge-based control based on vague environments, which looks at first glance completely different from Mamdani's model, although there are parallels to the rationale behind Zadeh's compositional rule of inference [18, 19]. But it turns out that the same computations are carried out. Thus we are able to provide a reasonable semantics for Mamdani's model.

The first step in the design of a controller based on vague environments is the specification of appropriate scaling functions $c_1, \ldots, c_n, c$ on the sets $X_1, \ldots, X_n, Y$, respectively. These scaling functions are intended to model the indistinguishability or similarity of values as described in the previous section and induce the fuzzy equivalence relations

$$E_i(x_i, x_i') = 1 - \min\left\{ \left| \int_{x_i}^{x_i'} c_i(s)\, ds \right|, 1 \right\}$$

and

$$F(y, y') = 1 - \min\left\{ \left| \int_{y}^{y'} c(s)\, ds \right|, 1 \right\}$$

on the spaces X_i and Y, respectively.

Remember that there are two different concepts of indistinguishability. In fuzzy control we mainly have to deal with intended imprecision, which is not enforced by difficulties in measuring exact values, but which is intended to model the fact that arbitrary precision is not needed. Later on we can make use of this fact, because it will be sufficient to specify controller outputs only for certain "typical" values. Taking the (intended) indistinguishability into account, we can extend this partially defined control function to a fully defined one.

Scaling functions are very appealing, because they have a reasonable interpretation. Small scaling factors imply a low distinguishability, meaning that in this area the control action changes only slowly with varying input values. A greater scaling factor indicates a high distinguishability, i.e., even small variations in the inputs might lead to greater alterations in the control action.

In the next step a control expert has to provide a set of input-output tuples, i.e., tuples $((x_1^{(r)}, \ldots, x_n^{(r)}), y^{(r)}) \in (X_1 \times \cdots \times X_n) \times Y$ $(r = 1, \ldots, k)$. The tuple $((x_1^{(r)}, \ldots, x_n^{(r)}), y^{(r)})$ simply means that $y^{(r)}$ is the appropriate output value for input $(x_1^{(r)}, \ldots, x_n^{(r)})$. These k input-output tuples correspond to a partial specification of the control function, because they can be understood as the function

$$\varphi_0: \{(x_1^{(r)}, \ldots, x_n^{(r)})|r \in \{1, \ldots, k\}\} \to Y, \qquad (x_1^{(r)}, \ldots, x_n^{(r)}) \mapsto y^{(r)},$$

which is a partial mapping from $X_1 \times \cdots \times X_n$ to Y.

Our task is now to determine an appropriate output value $\varphi(x_1, \ldots, x_n) = y \in Y$ for an arbitrary input $(x_1, \ldots, x_n) \in X_1 \times \cdots \times X_n$ from the knowledge given by the indistinguishabilities induced by the scaling functions and the partial control function. We extend the indistinguishabilities given by the scaling functions on the domains $X_1, \ldots, X_n, Y$ to an indistinguishability on the product space $X_1 \times \cdots \times X_n \times Y$, say in the form of Eq. (10), i.e.,

$$\begin{aligned}
E((x_1, &\ldots, x_n, y), (x_1', \ldots, x_n', y')) \\
&= \min\{E_1(x_1, x_1'), \ldots, E_n(x_n, x_n'), F(y, y')\}.
\end{aligned}$$

Now we consider the fuzzy set of points in the product space $X_1 \times \cdots \times X_n \times Y$ that are indistinguishable from one of the points $((x_1^{(r)}, \ldots, x_n^{(r)}), y^{(r)})$:

$$\mu_{\varphi_0}(x_1, \ldots, x_n, y) = \max_{r \in \{1, \ldots, k\}} \{E((x_1^{(r)}, \ldots, x_n^{(r)}, y^{(r)}), (x_1, \ldots, x_n, y))\}$$

on $X_1 \times \cdots \times X_n \times Y$. To obtain an "output" for a given input tuple $(x_1, \ldots, x_n)$, we compute the projection of this fuzzy set at $(x_1, \ldots, x_n)$, which leads to the fuzzy set

$$\mu_{\varphi_0}^{(x_1, \ldots, x_n)}(y) = \max_{r \in \{1, \ldots, k\}} \{\min\{E_1(x_1^{(r)}, x_1), \ldots, E_n(x_n^{(r)}, x_n), F(y^{(r)}, y)\}\} \tag{11}$$

on Y. Remembering that the fuzzy set $E(., x_0)$ stands for the points that are indistinguishable from the point x_0 with respect to the vague environment induced by the corresponding scaling function, we can rewrite the fuzzy set (11) in the form

$$\mu_{\varphi_0}^{(x_1, \ldots, x_n)}(y) = \max_{r \in \{1, \ldots, k\}} \{\min\{\mu_{x_1^{(r)}}(x_1), \ldots, \mu_{x_n^{(r)}}(x_n), \mu_{y^{(r)}}(y)\}\}, \tag{12}$$

because $E_1(x_1^{(r)}, .), \ldots, E_n(x_n^{(r)}, .), F(y^{(r)}, .)$ corresponds to the fuzzy set $\mu_{x_1^{(r)}}, \ldots, \mu_{x_n^{(r)}}, \mu_{y^{(r)}}$, respectively.

Now we are able to see the connection to Mamdani's model. For our approach we started with the specification of scaling functions on the sets $X_1, \ldots, X_n, Y$ and a partial control mapping $\varphi_0: X_1 \times \cdots \times X_n \to Y$ in the form of the set

$$\{((x_1^{(r)}, \ldots, x_n^{(r)}), y^{(r)})|r \in \{1, \ldots, k\}\}. \tag{13}$$

Taking into account that because of the indistinguishability characterized by the scaling functions we are working in vague environments, this partial mapping can be interpreted as k control rules of the form

$$R_r: \text{if } \xi_1 \text{ is } approximately\, x_1^{(r)} \text{ and } \dots \text{ and } \xi_n \text{ is } approximately\, x_n^{(r)}, \\ \text{then } \eta \text{ is } approximately\, y^{(r)} \quad (r = 1, \dots, k), \quad (14)$$

where $approximately\, x_1^{(r)}, \dots, approximately\, x_n^{(r)}, approximately\, y^{(r)}$ is represented by the fuzzy set $\mu_{x_1^{(r)}}, \dots, \mu_{x_n^{(r)}}, \mu_{y^{(r)}}$, respectively. Taking the above control rules together with these fuzzy sets, we can define a fuzzy controller in the sense of Mamdani and obtain for the input $(x_1, \dots, x_n)$ and the fuzzy set $\mu_{x_1, \dots, x_n}^{\text{output}}$ as output. This fuzzy set coincides with the fuzzy set $\mu_{\varphi_0}^{(x_1, \dots, x_n)}$, which is the output derived in our knowledge-based control model based on vague environments. In this way we can translate our control approach to Mamdani's model and obtain in both models the same output (before defuzzification).

The obvious question that turns up is whether a fuzzy controller in the sense of Mamdani can be translated to a controller based on vague environments. The answer is yes if the fuzzy partitions used in Mamdani's model satisfy the conditions mentioned in Theorem 2.

Viewing Mamdani's model in the light of fuzzy equivalence relations also provides explanations for the choice of typical fuzzy partitions. The scaling functions should be chosen depending on how sensitively the process reacts when the corresponding value changes. But how should the interpolation points for the partial control function be chosen? Of course, it might be reasonable to specify as many interpolation points as possible. However, we stick here to the philosophy that the expert tries to define the fewest interpolation points necessary for a satisfactory description of the function. This method frees the expert from specifying redundant knowledge and leads to a very information-compressed representation of the function to be interpolated. Let us assume that the output $y_0^{(i)}$ for the imprecisely known input $x_0^{(i)}$ is given. The fuzzy equivalence relation E induced by the scaling function c on X enables us to get information about the output corresponding to the value x, as long as $E(x, x_0^{(i)}) > 0$ holds. Thus the next imprecisely known interpolation points $x_0^{(i-1)}$ and $x_0^{(i+1)}$ should be chosen such that $E(x_0^{(i-1)}, x_0^{(i)}) = 0 = E(x_0^{(i+1)}, x_0^{(i)})$ and $E(x, x_0^{(i)}) > 0$ for all $x_0^{(i-1)} < x < x_0^{(i+1)}$. If we follow this minimality philosophy, we obtain a fuzzy partition from the imprecisely known values $x_0^{(i)}$ that satisfies the condition $\mu_i(x) + \mu_{i+1}(x) = 1$ for all $x_0^{(i)} < x < x_0^{(i+1)}$. Thus we can provide an interpretation for such typical fuzzy partitions in terms of a "lazy" expert who specifies the fewest interpolation points necessary.

To elucidate our new theoretical and semantic approach to fuzzy control, we briefly review an application to engine idle speed control [20] in the light of scaling functions. We do not go into the technical details of engine idle speed control, but restrict ourselves to the formal specification of the fuzzy controller and how it can be reformulated in terms of scaling functions.

Two input variables are used for the controller, namely the deviation dREV of the number of revolutions to the target rotation speed of the engine and the gradient gREV of the number of revolutions (to be understood as the difference in numbers of revolutions w.r.t. two measurement points). The change in current dAARCUR for the auxiliary air regulator serves as the output variable used to influence the rotation speed.

The fuzzy sets defined on the domains of these variables are shown in Fig. 7. The underlying domains are $X^{(\mathrm{dREV})} = [-70, 70]$ (rotations per minute), $X^{(\mathrm{gREV})} = [-40, 40]$ (rotations per minute), and $Y^{(\mathrm{dAARCUR})} = [-25, 25]$, where the last one is to be interpreted as a linear transformation of the real value of the current change dAARCUR.

The rule base of the fuzzy controller is given in Table 2.

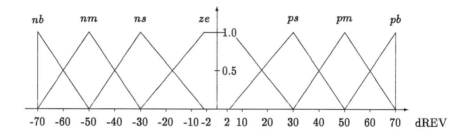

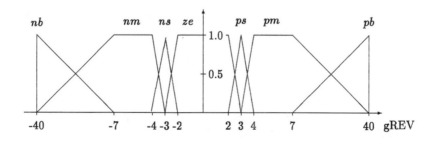

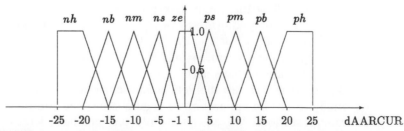

FIGURE 7   The fuzzy partitions for the variables dREV, gREV, and dAARCUR.

TABLE 2   The Rules for Controlling the Engine Idle Speed

		gREV						
		nb	nm	ns	ze	ps	pm	pb
dREV	nb	ph	pb	pb	pm	pm	ps	ps
	nm	ph	pb	pm	pm	ps	ps	ze
	ns	pb	pm	ps	ps	ze	ze	ze
	ze	ps	ps	ze	ze	ze	nm	ns
	ps	ze	ze	ze	ns	ns	nm	nb
	pm	ze	ns	ns	nm	nb	nb	nh
	pb	ns	ns	nm	nb	nb	nb	nb

For the fuzzy partitions Theorem 2 is applicable, so that we obtain the scaling functions

$$
c_{\text{dREv}} : X^{(\text{dREV})} \to [0, \infty), \quad x \mapsto
\begin{cases}
\frac{1}{20}, & \text{if } -70 \leq x < -30 \\
\frac{1}{28}, & \text{if } -30 \leq x < -2 \\
0, & \text{if } -2 \leq x < 2 \\
\frac{1}{28}, & \text{if } 2 \leq x < 30 \\
\frac{1}{20}, & \text{if } 30 \leq x < 70
\end{cases}
$$

$$
c_{\text{gREV}} : X^{(\text{gREV})} \to [0, \infty), \quad x \mapsto
\begin{cases}
\frac{1}{33}, & \text{if } -40 \leq x < -7 \\
0, & \text{if } -7 \leq x < -4 \\
1, & \text{if } -4 \leq x < -2 \\
0, & \text{if } -2 \leq x < 2 \\
1, & \text{if } 2 \leq x < 4 \\
0, & \text{if } 4 \leq x < 7 \\
\frac{1}{33}, & \text{if } 7 \leq x < 40
\end{cases}
$$

$$
c_{\text{dAARCUR}} : X^{(\text{dAARCUR})} \to [0, \infty), \quad x \mapsto
\begin{cases}
0, & \text{if } -25 \leq x < -20 \\
\frac{1}{5}, & \text{if } -20 \leq x < -5 \\
\frac{1}{4}, & \text{if } -5 \leq x < -1 \\
0, & \text{if } -1 \leq x < 1 \\
\frac{1}{4}, & \text{if } 1 \leq x < 5 \\
\frac{1}{5}, & \text{if } 5 \leq x < 20 \\
0, & \text{if } 20 \leq x < 25.
\end{cases}
$$

In this connection it should be emphasized that the imprecision of the measured dREV values suggests the choice of minor distinguishability in an environment of 0 to avoid control actions that refer to stochastic error processes rather than to important state changes.

Table 3 specifies for each fuzzy set a corresponding value, so that the fuzzy set represents the set of all values indistinguishable from the considered

TABLE 3   The Fuzzy Sets Corresponding to the Points in the Vague Environments

	nh	nb	nm	ns	ze	ps	pm	pb	ph
$\chi^{(dREV)}$	—	−70	−50	−30	0	30	50	70	—
$\chi^{(gREV)}$	—	−40	−5	−3	0	3	5	40	—
$\chi^{(dAARCUR)}$	−20	−15	−10	−5	0	5	10	15	20

value in the vague environment induced by the corresponding scaling function.

In this way, we can transform the rule base provided in Table 2 into the partial control mapping shown in Table 4.

The idea of fuzzy control as interpolation in vague environments was further developed by S. Kovács and L. T. Kóczy. First of all, they propose to use an approximate scaling function in case the prerequisites of Theorem 2 are not satisfied [21]. In addition, the scaling functions are used directly to interpolate the output value. The idea is to choose an output value whose (scaled) distance to a reference output point coincides with the (scaled) distance of the actual input to a reference input point.

In [22] a method is discussed for reducing the number of rules of a fuzzy controller by using an approximate scaling function. Kovács *et al.* [23] describes a successful application of fuzzy control on the basis of vague environments to the control of an automated guided vehicle.

IV. A REGRESSION TECHNIQUE FOR FUZZY CONTROL

In the previous section fuzzy control was explained on the basis of scaling functions characterizing vague environments. Instead of the specification of suitable fuzzy sets and control rules in the context of vague environments, a partial control mapping and scaling functions have to be determined.

There are three principal approaches to the design of a fuzzy controller:

- A control expert is able to formulate appropriate control rules and fuzzy sets based on his experience and his knowledge about the

TABLE 4   The Partial Mapping φ_0 for Idle Speed Control

		gREV						
	φ_0	−40	−6	−3	0	3	6	40
	−70	20	15	15	10	10	5	5
	−50	20	15	10	10	10	5	0
	−30	15	10	5	5	5	0	0
dREV	0	5	5	0	0	0	−10	−5
	30	0	0	0	−5	−5	−10	−15
	50	0	−5	−5	−10	−15	−15	−20
	70	−5	−5	−10	−15	−15	−15	−15

process. This is usually a very tedious task, and at least a fine-tuning of the fuzzy sets (by hand) cannot be avoided.

- An experimental environment (the real process or a simulation of it) is available. Good control actions can be distinguished from bad ones. Then a neuro-fuzzy model like the one described in [24] or techniques based on evolutionary algorithms as in [25, 26] can be used to automatically generate a fuzzy controller.
- Data from observations of a control expert who is able to handle the process are available. In this case neuro-fuzzy methods (for an overview see [27]) as well as evolutionary algorithms (see for instance [28]) are applicable.

In this section we develop a regression technique that is tailored for fuzzy control in the spirit of vague environments for the above-mentioned third case, when observation data are available.

We consider a simple Sugeno-type fuzzy controller here that can be interpreted as a slightly modified version of Mamdani's model. The rules are of the form

$$R: \text{if } \xi_1 \text{ is } A_1^{(R)} \text{ and } \dots \text{ and } \xi_n \text{ is } A_n^{(R)} \text{ then } \eta \text{ is } b_R,$$

where each linguistic term $A_1^{(R)}, \dots, A_n^{(R)}$ is associated with a fuzzy set $\mu_1^{(R)}, \dots, \mu_n^{(R)}$ on $X_1, \dots, X_n$, respectively. b_R is a (crisp) output value assigned to the rule R.

For the aggregation of the premises of the rules, we only assume that it is carried out by the t-norm $\odot$ (a commutative, associative, monotone increasing binary operation on the unit interval with 1 as the unit; see, for instance, [29]), for instance, the minimum or the product. The output value for the input tuple $(x_1, \dots, x_n)$ is defined by the formula

$$f(x_1, \dots, x_n) = \frac{\sum_R \left(\odot_{\nu=1}^n \mu_n^{(R)}(x_\nu) \right) \cdot b_R}{\sum_R \odot_{\nu=1}^n \mu_\nu^{(R)}(x_\nu)}, \qquad (15)$$

where the sum in the nominator and denominator is defined for a finite set $\mathscr{R}$ of rules $R \in \mathscr{R}$.

Let us assume we have a set

$$D = \left\{ \left(x_1^{(1)}, \dots, x_n^{(1)}, y^{(1)} \right), \dots, \left(x_1^{(s)}, \dots, x_n^{(s)}, y^{(s)} \right) \right\}$$

of sample data where the output $y^{(i)}$ is assigned to the input $(x_1^{(i)}, \dots, x_n^{(i)})$. Let us for the moment consider the fuzzy sets μ_ν^R as given. For any set of parameters b_R $(R \in \mathscr{R})$ we can compute the quadratic error that is caused by the fuzzy controller with respect to the data set:

$$E = \sum_{l=1}^s \left(f\left(x_1^{(l)}, \dots, x_n^{(l)} \right) - y^{(l)} \right)^2. \qquad (16)$$

To minimize E, we have to choose the parameters b_R appropriately. To determine the b_R we take the partial derivatives of E with respect to the b_{R_0}

$(R_0 \in \mathcal{R})$ and require them to be zero:

$$\frac{\partial E}{\partial b_{R_0}} = 0 \qquad (R_0 \in \mathcal{R}). \tag{17}$$

We obtain

$$\frac{\partial E}{\partial b_{R_0}} = \sum_{l=1}^{s} 2 \cdot \left(f\left(x_1^{(l)}, \ldots, x_n^{(l)}\right) - y^{(l)} \right) \cdot \frac{\partial f\left(x_1^{(l)}, \ldots, x_n^{(l)}\right)}{\partial b_{R_0}}.$$

Taking the partial derivative of (15) and replacing f with (15), we get

$$\begin{aligned}
\frac{\partial E}{\partial b_{R_0}} &= \sum_{l=1}^{s} 2 \cdot \left(f\left(x_1^{(l)}, \ldots, x_n^{(l)}\right) - y^{(l)} \right) \cdot \frac{\bigodot_{\nu=1}^{n} \mu_\nu^{(R_0)}\left(x_\nu^{(l)}\right)}{\sum_R \bigodot_{\nu=1}^{n} \mu_\nu^{(R)}\left(x_\nu^{(l)}\right)} \\
&= 2 \cdot \left(\sum_{l=1}^{s} \frac{\bigodot_{\nu=1}^{n} \mu_\nu^{(R_0)}\left(x_\nu^{(l)}\right)}{\left(\sum_R \bigodot_{\nu=1}^{n} \mu_\nu^{(R)}\left(x_\nu^{(l)}\right)\right)^2} \cdot \sum_R b_R \left(\bigodot_{\nu=1}^{n} \mu_\nu^{(R)}\left(x_\nu^{(l)}\right) \right) \right. \\
&\qquad\qquad \left. - \sum_{l=1}^{s} \frac{\bigodot_{\nu=1}^{n} \mu_\nu^{(R_0)}\left(x_\nu^{(l)}\right)}{\sum_R \bigodot_{\nu=1}^{n} \mu_\nu^{(R)}\left(x_\nu^{(l)}\right)} \cdot y^{(l)} \right) \\
&= 2 \cdot \left(\sum_R b_R \sum_{l=1}^{s} \frac{\bigodot_{\nu=1}^{n} \mu_\nu^{(R_0)}\left(x_\nu^{(l)}\right)}{\left(\sum_{R'} \bigodot_{\nu=1}^{n} \mu_\nu^{(R')}\left(x_\nu^{(l)}\right)\right)^2} \cdot \bigodot_{\nu=1}^{n} \mu_\nu^{(R)}\left(x_\nu^{(l)}\right) \right. \\
&\qquad\qquad \left. - \sum_{l=1}^{s} \frac{\bigodot_{\nu=1}^{n} \mu_\nu^{(R_0)}\left(x_\nu^{(l)}\right)}{\sum_R \bigodot_{\nu=1}^{n} \mu_\nu^{(R)}\left(x_\nu^{(l)}\right)} \cdot y^{(l)} \right) \\
&= 0. \tag{18}
\end{aligned}$$

Thus (16) provides the following system of linear equations, from which we can compute the b_{R_0} ($R_0 \in \mathcal{R}$):

$$\sum_R b_R \sum_{l=1}^{s} \frac{\bigodot_{\nu=1}^{n} \mu_\nu^{(R_0)}\left(x_\nu^{(l)}\right)}{\left(\sum_{R'} \bigodot_{\nu=1}^{n} \mu_\nu^{(R')}\left(x_\nu^{(l)}\right)\right)^2} \cdot \bigodot_{\nu=1}^{n} \mu_\nu^{(R)}\left(x_\nu^{(l)}\right) = \sum_{l=1}^{s} \frac{\bigodot_{\nu=1}^{n} \mu_\nu^{(R_0)}\left(x_\nu^{(l)}\right)}{\sum_R \bigodot_{\nu=1}^{n} \mu_\nu^{(R)}\left(x_\nu^{(l)}\right)} \cdot y^{(l)}. \tag{19}$$

A similar least-squares method was proposed in [30] for the identification of nonlinear dynamic systems. We want to go a step further and tune the fuzzy sets in addition to the parameters b_R. Of course, this will not be possible by linear regression, because we would immediately run into a nonlinear optimization problem. Nevertheless, remembering the concept of fuzzy control based on vague environments, we can provide good heuristics for tuning the fuzzy sets. We assume that the fuzzy sets on the domains $X_1, \ldots, X_n$ represent vague values, i.e., each of them corresponds to a fuzzy set of values that are indistinguishable from a considered value. The vague environments are characterized by scaling functions that are constant between two neighboring values that are used to build the fuzzy sets so that we obtain fuzzy partitions as already illustrated in Fig. 6.

The idea is the following. We start with a constant scaling factor leading to a homogeneous fuzzy partition. Then we compute the corresponding b_{R_0} from (19). In an area where we have a large error, it is necessary to specify the function f in more detail, i.e., we need more and narrower fuzzy sets. Therefore, we have to choose a larger scaling factor for such an area. Of course, we have to accept smaller scaling factors for areas with a small error. Otherwise we would increase the number of fuzzy sets that we do not want to consider here. This concept reflects the philosophy of large scaling factors for areas where the process (control function) is very sensitive to small changes and small scaling factors where the process (control function) does not change drastically with a variation of the parameters. Figure 8 illustrates the idea of contracting and moving fuzzy sets in an area with a large error.

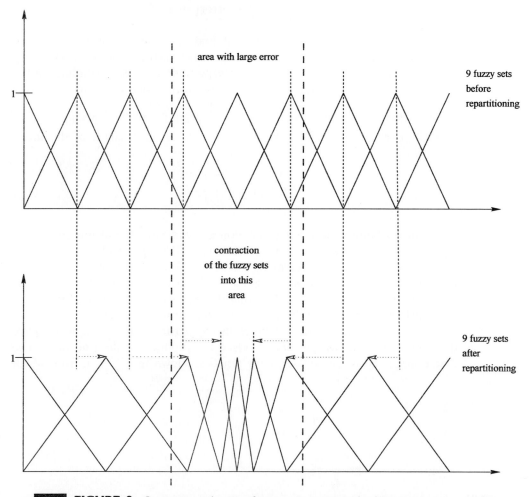

FIGURE 8 Contracting and moving fuzzy sets in an area with a large error.

To determine the change in the scaling factors we compute for each domain X_ν and each area the error of the regression function. By an area we mean the interval between two neighboring points that induce the corresponding fuzzy sets appearing in the rules.

$$\text{error}(\text{area}_i) = \sum_{(x_1^{(l)}, \ldots, x_n^{(l)}): \, x_\nu^{(l)} \in \text{area}_i} \left(y^{(l)} - f\left(x_1^{(l)}, \ldots, x_n^{(l)} \right) \right)^2. \quad (20)$$

area_i is the interval between the $(i - 1)$th and ith points in the domain X_ν that induce the fuzzy sets appearing in the rules (compare Fig. 9).

Now the area_i should be contracted, resulting in a larger scaling factor, when the error is relatively high, whereas it can be stretched when the error is small. If L_i^{old} is the length of area_i, then we define the new (relative) length of the ith area by

$$L_i^{\text{rel}} = \frac{L_i^{\text{old}}}{\text{const.} + \text{error}(\text{area}_i)},$$

where const. is a positive constant that first of all avoids devision by zero when the error is zero for a certain area. const. also determines how strong the contraction or stretching of the corresponding area is, depending on the error. If const. is small in comparison to the error, this will result in drastic changes, whereas a large constant allows only very small variations.

Finally, it is necessary to normalize the relative length of each area so that the overall length is the same as before, i.e., the length of the interval X_ν:

$$L_i = L_i^{\text{rel}} \cdot \frac{\sum_j L_j^{\text{old}}}{\sum_j L_j^{\text{rel}}}. \quad (21)$$

Figure 10 illustrates a fuzzy partition and indicates the magnitude of the error of the regression function for each area. The result of computing the updated length of each interval by formula (21) and building the new fuzzy partition by taking the induced fuzzy sets of the points at the boundaries of the areas is shown in Fig. 11. Note that the corresponding scaling function is piecewise constant, taking the value $1/L_i$ on the interval area_i.

Let us illustrate by two examples the approximation technique we have developed. In both cases we have chosen the product for the t-norm $\odot$. The first example is the piecewise linear function shown in Fig. 12a, from which

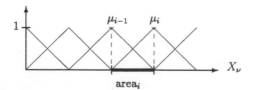

FIGURE 9 An area with a constant scaling factor.

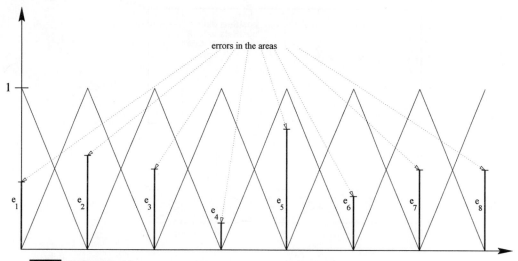

FIGURE 10   A fuzzy partition and errors for the corresponding areas.

we take 17 equidistant sample points. We start with the homogeneous fuzzy partition at the bottom of Fig. 12a. Then the values b_R in the conclusions in the rules are determined as the solution of the system of linear equations (19). After that we compute the error of the regression function for each subinterval (area) and obtain a new fuzzy partition by stretching or contracting the subintervals, depending on the magnitude of the error. Then we determine new values b_R and obtain new errors for the (new) areas. This procedure is iterated until the error is sufficiently small or no significant

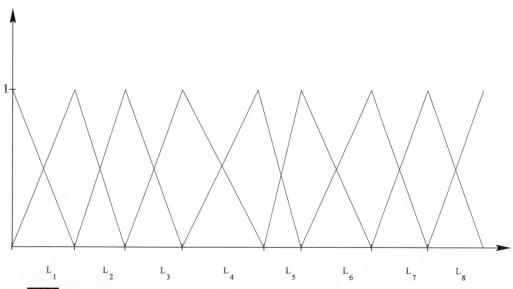

FIGURE 11   The new fuzzy partition.

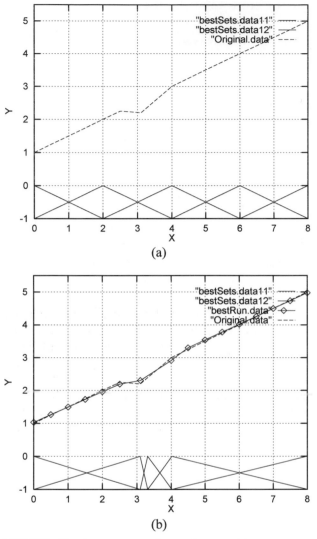

FIGURE 12   Approximation of a one-dimensional function.

improvements of the overall error were achieved during the last iteration steps. Finally, we end up with the fuzzy partition shown at the bottom of Fig. 12b and the (nearly) perfect approximation (see the corresponding graph in the figure). Note that in the ranges where the function remains linear over a long interval a few wide fuzzy sets (small scaling factors) are chosen, whereas in ranges where the function varies, more and narrower fuzzy sets (greater scaling factors) appear.

The second example is the two-dimensional function $\sin(x) \cdot \cos(y)$ (Fig. 13) in the range $[0, 2\pi]^2$ from which we take 1849 equidistant sample data. We start with a homogeneous fuzzy partition with eight fuzzy sets on each of the two domains, resulting in a rule base of $8 \cdot 8 = 64$ rules. Already after two iterations we obtain the modified fuzzy partitions illustrated in Fig. 14 with

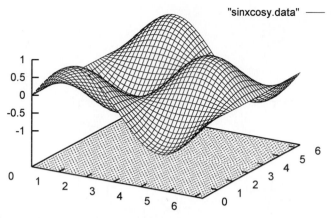

"sinxcosy.data" ——

FIGURE 13 The function $\sin(x) \cdot \cos(y)$.

the resulting regression function in Fig. 15, which reduced the error to 10% of the error of the first regression function with homogeneous fuzzy partitions.

It should be emphasized that we do not intend to develop a new approximation technique that is superior to standard methods. The main point is that we are interested in rules that describe the function approximately and are understandable and interpretable for staff without strong mathematical training or background. Thus the aim is knowledge extraction instead of best approximation.

V. CONCLUSIONS

The semantics for fuzzy sets we have provided in this paper leads to a restriction of the possible parameter choices (e.g., arbitrary fuzzy partitions are not allowed) and a better understanding of the underlying assumptions (e.g., the indistinguishabilities in the different domains are considered to be independent). This can definitely simplify the design process of a fuzzy controller. We have also introduced a regression technique based on our considerations that enables us to learn a fuzzy controller from data. Other techniques that are also related to this idea are based on fuzzy clustering (see, for instance, [31]; for an overview see [32]). Fuzzy clusters are also fuzzy sets induced by a point, the so-called prototype or cluster center. The membership degree is a function decreasing with increasing distance from the cluster center.

Even if the semantics of vague environments may not always seem to be appropriate for a certain application, the concept of indistinguishability is inherent in fuzzy sets and cannot be avoided when operating with fuzzy sets [33].

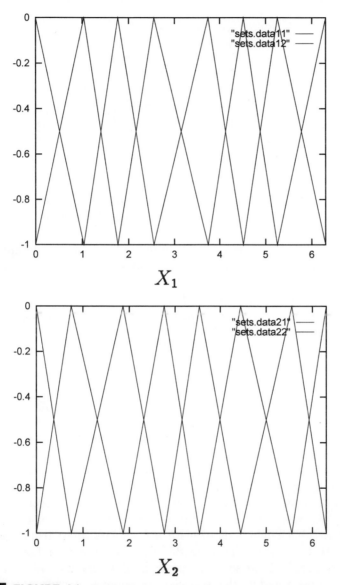

FIGURE 14 Modified fuzzy partitions after two iteration steps.

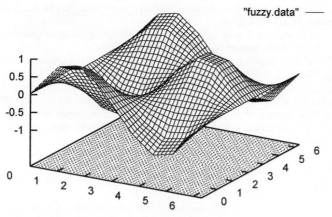

FIGURE 15 The resulting regression function.

Another advantage of vague environments is that the transformations induced by the scaling functions can be exploited in order to get more efficient computation schemes for fuzzy controllers [34].

REFERENCES

1. Zadeh, L. A. Fuzzy sets. *Inform. and Control* 8:338–353, 1965.
2. Moore, R. E. *Interval Analysis*. Prentice Hall, Englewood Cliffs, NJ, 1966.
3. Dubois, D. and Prade, H. Similarity-based approximate reasoning. In *Computational Intelligence Imitating Life* (J. M. Zurada, R. J. Marks, II, and C. J. Robinson, Eds.), pp. 69–80. IEEE Press, New York, 1994.
4. Bellman, R. E. and Zadeh, L. A. Decision making in a fuzzy environment. *Management Sci.* 17:141–164, 1970.
5. Dubois, D. and Prade, H. Fuzzy sets in approximate reasoning, Part 1. Inference with possibility distributions. *Fuzzy Sets and Systems* 40:143–202, 1991.
6. Gebhardt, J. and Kruse, R. The context model—an integrating view of vagueness and uncertainty. *Internat. J. Approx. Reason.* 9:283–314, and 1993.
7. Kruse, R., Schwecke, E., and Heinsohn, J. *Uncertainty and Vagueness in Knowledge Based Systems: Numerical Methods*. Springer, Berlin, 1991.
8. Zadeh, L. A. Fuzzy sets as a basis for a theory of possibility. *Fuzzy Sets and Systems* 1:3–28, 1978.
9. Bellman, R. E., Kalaba, R., and Zadeh, L. A. Abstraction and pattern classification. *J. Math. Anal. Appl.* 13:1–7, 1966.
10. Zadeh, L. A. Similarity relations and fuzzy orderings. *Inform. Sci.* 3:177–200, 1971.
11. Mundici, D. Ulam games, Łukasiewicz logic, and AF C^*-Algebras. *Fund. Inform.* 18:151–161, 1993.
12. Klawonn, F. and Kruse, R. A Łukasiewicz logic based prolog. *Mathware Soft Comput.* 1:5–29, 1994.
13. Ulam, S. M. *Adventures of a Mathematician*. Scribner's, New York, 1976.
14. Klawonn, F. Fuzzy sets and vague environments. *Fuzzy Sets and Systems* 66:207–221, 1994.
15. Klawonn, F. and Novák, V. The relation between inference and interpolation in the framework of fuzzy systems. *Fuzzy Sets and Systems* 81:331–354, 1996.
16. Mamdani, E. H. Application of fuzzy logic to approximate reasoning using linguistic systems. *IEEE Trans. Comput.* 26:1182–1191, 1977.

17. Lee, C. C. Fuzzy logic in control systems: Fuzzy logic controller. Part II. *IEEE Trans. Systems Man Cybernet.* 20:419–435, 1990.
18. Zadeh, L. A. Outline of a new approach to the analysis of complex systems and decision processes. *IEEE Trans. Systems Man Cybernet.* 3:28–44, 1973.
19. Zadeh, L. A. The concept of a linguistic variable and its application to approximate reasoning. Parts 1, 2, 3. *Inform. Sci.* 8:199–249, 301–357, 9:43–80, 1975.
20. Klawonn, F., Gebhardt, J., and Kruse, R. Fuzzy control on the basis of equality relations—with an example from idle speed control. *IEEE Trans. Fuzzy Systems* 3:336–350, 1995.
21. Kovács, S. New aspects of interpolative reasoning. In *Proceedings of the 6th International Conference on Information Processing and Management of Uncertainty in Knowledge-Based Systems* (IPMU '96), Granada, 1996, pp. 477–482.
22. Kovács, S. and Kóczy, L. T. Approximate fuzzy reasoning based on interpolation in the vague environment of the fuzzy rulebase as a practical alternative of the classical CRI. In *Proceedings of the 7th International Fuzzy Systems Association World Congress* (IFSA '97), Prague, 1997, Vol. II, pp. 144–149.
23. Kovács, S. and Kóczy, L. T. Application of the approximate fuzzy reasoning based on interpolation in the vague environment of the fuzzy rulebase in the fuzzy logic controlled path tracking strategy of differential steered AGVs. In *Computational Intelligence: Theory and Applications* (B. Reusch, Ed.), pp. 456–467. Springer, Berlin, 1997.
24. Nauck, D. Building neural fuzzy controllers with NEFCON-I. In *Fuzzy Systems in Computer Science* (R. Kruse, J. Gebhardt, and R. Palm, Eds.), pp. 141–151. Vieweg, Braunschweig, Germany, 1994.
25. Hopf, J. and Klawonn, F. Learning the rule base of a fuzzy controller by a genetic algorithm. In *Fuzzy Systems in Computer Science* (R. Kruse, J. Gebhardt, and R. Palm, Eds.), pp. 63–74. Vieweg, Braunschweig, 1994.
26. Kinzel, J., Klawonn, F., and Kruse, R. Modifications of genetic algorithms for designing and optimizing fuzzy controllers. In *Proceedings of the IEEE Conference on Evolutionary Computation*, IEEE, Orlando, 1994, pp. 28–33.
27. Nauck, D., Klawonn, F., and Kruse, R. *Foundations of Neuro-Fuzzy Systems*. Wiley, Chichester, 1997.
28. Surmann, H., Kanstein, A., and Goser, K. Self-organizing and genetic algorithms for an automatic design of fuzzy control and decision systems. In *Proceedings of the First European Congress on Intelligent Techniques and Soft Computing* (EUFIT '93), Aachen, Germany, 1993, pp. 1097–1104.
29. Kruse, R., Gebhardt, J., and Klawonn, F. *Foundations of Fuzzy Systems*. Wiley, Chichester, 1994.
30. Kecman, V. and Pfeiffer, B.-M. Exploiting the structural equivalence of learning fuzzy systems and radial basis function neural networks. In *Proceedings of the Second European Congress on Intelligent Techniques and Soft Computing* (EUFIT '94), Aachen, Germany, 1994, pp. 58–66.
31. Klawonn, F. and Kruse, R. Constructing a fuzzy controller from data. *Fuzzy Sets and Systems* 85:177–193, 1997.
32. Höppner, F., Klawonn, F., Kruse, R., and Runkler, T., *Fuzzy Cluster Analysis*. Wiley, Chichester, 1999.
33. Klawonn, F. and Castro, J. L. Similarity in fuzzy reasoning. *Mathware Soft Comput.* 2:197–228, 1995.
34. Klawonn, F. and Kruse, R. Fuzzy partitions and transformations. In *Proceedings of the 3rd IEEE Conference on Fuzzy Systems*, IEEE, Orlando, 1994, pp. 1269–1273.

17

TECHNIQUES AND APPLICATIONS OF FUZZY THEORY TO AN ELEVATOR GROUP CONTROL SYSTEM

HYUNG LEE-KWANG
CHANGBUM KIM

Department of Computer Science, KAIST, Taejon, 305-701, Korea

I. ELEVATOR GROUP CONTROL SYSTEM   461
II. AREA WEIGHT GENERATION   465
   A. The Area Weight   466
   B. A Fuzzy Model to Determine the Area Weight   467
III. FUZZY CLASSIFICATION OF PASSENGER TRAFFIC   471
IV. HALL CALL ASSIGNMENT   475
   A. Calculation of Input Variables   476
   B. Fuzzy Inference   478
V. CONCLUSIONS   480
   REFERENCES   480

I. ELEVATOR GROUP CONTROL SYSTEM

The elevator group control system (EGCS) is a control system that manages multiple elevators in a building to efficiently transport the passengers. The performance of EGCS is measured by several criteria, such as the average waiting time of passengers, the percentage of passengers waiting more than 60s, and power consumption [1–5]. The EGCS manages elevators to minimize the evaluation criteria; however, it is difficult to satisfy all criteria at the same time. There are trade-offs between evaluation criteria. Nowadays, system managers define the importance of evaluation criteria of EGCS by the patterns of passenger traffic. For example, the average waiting time is most important during rush hour, and power consumption is important at other times.

An EGCS consists of *hall call* buttons, *car call* buttons, elevators, and a group controller. If a passenger wants to go to another floor, he or she presses a direction (hall call) button and waits for an elevator to arrive, then enters the elevator and presses a floor (car call) button in the elevator. At this time, the group controller selects a suitable elevator when a passenger presses the hall call button. In this case, the group controller considers the current situation of the building to select the most appropriate elevator in the group.

In the EGCS, it is not easy to select a suitable elevator for the following reasons. First, the EGCS is very complex. If a group controller manages n elevators and assigns p hall calls to the elevators, the controller considers n^p cases. Second, the controller must consider the hall calls that will be generated in the near feature. Third, it must consider many uncertain factors, such as the number of passengers at the floors where hall calls and car calls are generated. Fourth, it must be possible for a system manager to change the control strategy. Some managers want to operate the system to minimize the passenger waiting time, and others want to reduce power consumption.

To understand the EGCS, consider an example of the elevator group control process. There are a couple of hall call (up, down) buttons on a floor and multiple elevators, as shown in Fig. 1. When a person arrives at a hall, he or she will press a hall call button to call an elevator. The EGCS (elevator group control system) selects an elevator for the passenger by the event of hall call. The selected elevator moves to the floor where the hall call occurred to serve the passenger.

1. A passenger who is going to the 15th floor from the 2nd floor presses the up hall call button
2. The hall call signal is transmitted to the EGCS.
3. The EGCS selects an elevator to service the passenger.
4. The EGCS sends a message to the selected elevator.
5. The selected elevator moves to the 2nd floor and the passenger gets on it.
6. The passenger presses the car call button of 15.

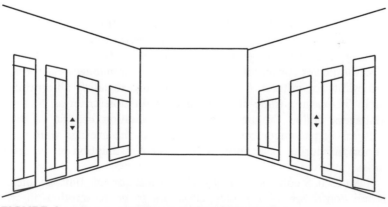

FIGURE 1 A floor in a building where the EGCS is installed.

7. The elevator sends a message to the EGCS and moves to the 15th floor.

8. The elevator arrives at the 15th floor and the passenger gets out of the elevator.

The EGCS repeats the process of selecting service elevators for hall calls. We call the selection the *hall call assignment*. In the EGCS, the hall call assignment is important, and the performance depends on the hall call assignment method.

Many evaluation criteria are used to estimate the performance of the EGCS [6, 7]. Average waiting time, the percentage of passengers waiting more than 60 s, and power consumption are the most important criteria of EGCS.

1. AWT (average waiting time): Waiting time is the time spent waiting on a floor until the service elevator arrives at the floor after a passenger presses a hall call button. AWT is the average of all passenger waiting times in a unit of time.

2. LWP (long waiting percentage): LWP is the percentage of the passengers who wait more than 60 s in a unit of time.

3. RNC (RuN count): In the elevator system, most energy is consumed to start or stop elevators. RNC is the number of the elevator moves in a unit of time, and it is used to estimate the power consumption of a system.

Figure 2 shows the general structure of the EGCS. In Fig. 2, the EGCS manages eight elevators in a building. Each elevator has its own controller, represented by the CC (car controller), and communicates with the elevator group controller. The EGCS consists of three main parts and several modules. The three main parts are the *traffic data management*, the *control strategy generation*, and the *hall call assignment*. The traffic data management part collects various statistics on traffic data such as the number of hall calls, car calls, and passengers getting in or out. This part learns the traffic data and predicts the future traffic [8, 9]. The control strategy generation part classifies the traffic into one of several modes and determines the hall call assignment method that is suitable for the classified traffic mode. Finally, the hall call assignment part selects service elevators for hall calls by the method determined in the control strategy generation part. In addition, there are several small modules that support the main parts, such as task management, data management, and communication management.

The EGCS generally tests the degree of suitability of each elevator for a hall call and selects an elevator with the best suitability. If we consider the above three evaluation criteria, then the suitability can be represented by their combination. Let ϕ_i be the suitability function for the ith elevator and $T_{\mathrm{AWT}}(i)$, $T_{\mathrm{LWP}}(i)$, and $T_{\mathrm{RNC}}(i)$ be the evaluation values of the ith elevator for AWT, LWP, and RNC, respectively. $T_{\mathrm{AWT}}(i)$ is the waiting time of the passenger, $T_{\mathrm{LWP}}(i)$ is the probability of a long wait, and $T_{\mathrm{RNC}}(i)$ is the number of moves in a time interval when the ith elevator is assigned to

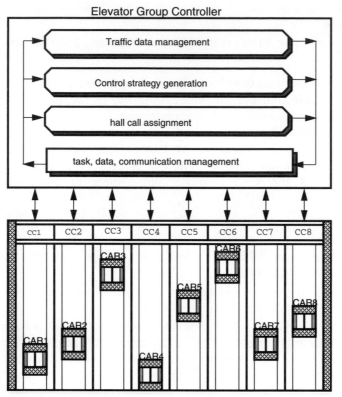

FIGURE 2   General structure of the EGCS.

the hall call. The suitability of ith elevator for a hall call can be represented by the following equation:

$$\phi(i) = v_1 \cdot T_{\text{AWT}}(i) + v_2 \cdot T_{\text{LWP}}(i) + v_3 \cdot T_{\text{RNC}}(i).$$

In this equation, v_1, v_2, and v_3 are weighting factors of each evaluation criterion. The weighting factors can be fixed by managers or determined dynamically by the control strategy generation part. The elevator with the minimum value of the function is selected to serve passengers.

The most important parts of EGCS are the control strategy generation and hall call assignment. In the control strategy generation, EGCS classifies the passenger traffic of building and decides how to assign the hall calls for each traffic pattern. The importance of each evaluation criterion and weighting factors for the hall call assignment are generated in this part. In the hall call assignment part, hall calls are assigned to elevators by assignment algorithms by the state of elevators and values generated in the control strategy generation part.

Recently a lot of significant progress has been made in the EGCS by applying AI (artificial intelligence) techniques such as fuzzy logic and neural networks. There are three approaches in the intelligent EGCS. The first approach is to generate a control strategy by employing AI techniques to improve the performance of EGCS. Kim studied the weighting factors that are used in conventional ECGS and proposed a fuzzy model to increase the performance of EGCS [1]. Hitachi, Leppala, and Kim proposed the fuzzy classication method of passenger traffic [10–12]. The second approach is to use AI techniques to improve the hall call assignment. Hitachi and Kim proposed hall call assignment methods that use fuzzy logic [12, 13]. The last approach is to combine the control strategy generation and hall call assignment. Mitsubishi and Markon developed a neural network-based EGCS in which hall calls are assigned by the output of neural network from state values of each elevator [14, 15]. Kim proposed an intelligent EGCS in which fuzzy logic is used to generate a control strategy and to assign hall calls [16].

I will show three applications of fuzzy theory to elevator group control systems in the following three sections. In beginning studies on FEGCSs (fuzzy elevator group control systems), the weighting factors of conventional methods are dynamically generated by employing fuzzy theory. First, I will show a fuzzy approach to EGCS to determine the area weight, which is a weighting factor. Area weight represents the range (floors) that can be served easily by each elevator. Second, I will show the fuzzy classification method of passenger traffic. Finally, a hall call assignment method employing fuzzy theory will be presented.

II. AREA WEIGHT GENERATION

The most important task of an elevator group control system is selecting a suitable elevator for each passenger's hall call (up, down). The selection is made to minimize the average waiting time of passengers, the long wait probability (the probability that a passenger will wait for a long time), and power consumption [7]. The selection method is called the *hall call assignment method*. In this method, an *evaluation function* is used to achieve the above multiple objectives.

The function is evaluated for each elevator, and the elevator with the smallest value is selected. Let $\phi(i)$ be the evaluating function for the kth elevator; then this function is represented by the following formula;

$$\phi(i) = T(i) - \alpha \cdot T_\alpha(i).$$

When a new hall call is given on floor n, the function is evaluated for the ith elevator, where $i = 1, \ldots, n$. In the above formula, $T(i)$ is the estimated arrival time of the ith elevator, which is the waiting time of the passenger when we assign the ith elevator to the new hall call. We call $T_\alpha(i)$ the *area value* and α the *area weight*, where $T_\alpha(i)$ is determined by the positions of assigned calls for each elevator. The *area value* represents the minimum distance from the floor of new hall call to floors where the elevator is assigned to stop by calls (hall call, car call). *Area weight* is a weighting factor for area value, and it affects the performance of EGCS by the patterns of passenger traffic.

A. The Area Weight

Figure 3 shows a situation in a four-elevator system. In this figure the arrow represents the direction of each elevator. A black circle indicates the car call that was generated by passengers in the elevator. A black triangle represents the hall call that will be served by the elevator. Finally, new hall calls generated on a floor are marked by white triangles. In Fig. 3, the first elevator is going down, has a car call on floor $N - 2$, and is assigned a hall call on floor N for passengers who want to go down.

Let's consider a new hall call for up from floor N. The area values of elevators 1 and 4 are 0 because of directions of the elevators and hall calls are different. The area value of elevator 3 is larger than elevator 2 because elevator 2 has a car call on floor $N + 3$ and elevator 3 has a car call on floor $N + 1$. In general the area is defined in the form of a triangle or a trapezoid, as shown in Fig. 4a. In Fig. 4b, the trapezoidal area of elevator i on floor n is given. We can see that the area value $T_\alpha(i) = 1$ for the floor n; $n + 1, n - 1$, $T_\alpha(i) = 0.5$ for floor $n + 2, n - 2$; and $T_\alpha(i) = 0$ for the others. The area value $T_\alpha(i)$ is defined for the floors where there is a call (hall and car).

In Fig. 3, elevators 1 and 4 are going down and $T(1)$ and $T(4)$ must be large, so I will show elevators 2 and 3, being assigned a new hall call. The following formula shows evaluating function values of elevators 2 and 3 for the selection of a service elevator. In this case the area value $T_\alpha(i)$ for elevator 2 is zero, i.e., $T_\alpha(2) = 0$:

$$\phi(2) = T(2) - \alpha \cdot T_\alpha(2)$$

$$= T(2)$$

$$\phi(3) = T(3) - \alpha \cdot T_\alpha(3).$$

Here, $T(2)$ and $T(3)$ can be estimated and $T(2) < T(3)$. If $\alpha \cdot T_\alpha(3) < T(3) - T(2)$, then elevator 2 is assigned, otherwise elevator 3 is assigned.

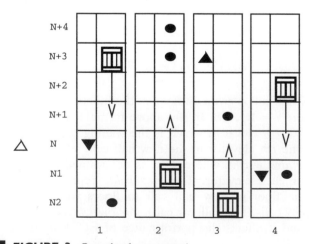

FIGURE 3   Example of area control.

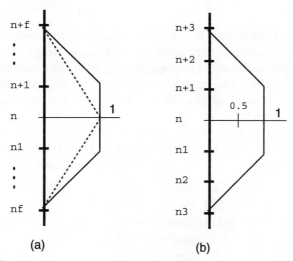

FIGURE 4 Example of the area value.

In this example, we can see that $T_\alpha(3)$ is the same for passenger traffic, and $T(2)$ and $T(3)$ are estimated values. However, value α is calculated whenever a call is generated. Therefore the determination of value α is important in the hall call assignment method.

If α is large, the possibility that the elevator close to the relevant floor will be selected is increased. Consequently, the transportation load may be assigned to a specific elevator that is in the area. Furthermore, the average waiting time increases and the total running frequency of elevators decreases. Therefore the power consumption is reduced. If α is small, the average waiting time decreases and the running frequency increases.

In [17], the predefined area weight is used, and it is defined according to the traffic modes classified by the passenger traffic pattern. In this case, it is difficult to reflect variation in passenger traffic in the same traffic mode, and this method cannot consider some important characteristics of different buildings. In [11], a simulation method is used to determine the area weight. In this method, the future situation of the system is simulated with some predefined area weights and then the area weight giving the best performance is selected. As this method simulates with a fixed number of area weights, we cannot expect a precise control. Furthermore, because the hall call data used in the simulation are forecasted, the simulation may differ from the real situation, and the uncertainty of the data limits the reliability of the model. In this study we use the fuzzy approach to model the uncertain situation of the system and to determine the area weight.

B. A Fuzzy Model to Determine the Area Weight

To handle the uncertainty of an elevator group control system, we use the fuzzy approach [18–21]. In this approach we design a fuzzy model to represent the fuzzy knowledge and use the fuzzy inference method to determine the area weight. Fuzzy knowledge is generally formulated in the form of rules.

I. Fuzzy Rules

We classify the facts related to the determination of area weight into two groups. The first group includes the up-going and down-going traffic amount, and the second group includes the average waiting time, long wait probability, and power consumption. The up-going and down-going traffic are major factors in determining area weight, but we must consider the other factors related to the evaluation criteria.

Let UP, DN, and α' be the up-going traffic, down-going traffic, and area weight, respectively. We can represent the traffic (UP, DN) knowledge related to the area weight in the form of fuzzy rules as follows:

If UP is VL and DN is SM, then α' is VS.
If UP is MD and DN is SM, then α' is MD.
If UP is MD and DN is SM, then α' is VL.
If UP is SM and DN is VL, then α' is SM.
If UP is SM and DN is SM, then α' is VS.
⋮
If UP is MD and DN is MD, then α' is SM.

In the fuzzy rules words such as SM (small), MD (medium), LR (large), and VL (very large) are fuzzy sets defined on the variables UP and DN. The terms VS (very small), SM (small), MD (medium), LR (large), and VL (very large) are defined on α' as shown in Fig. 5.

When the traffic is fixed, the average waiting time and long wait probability decrease as the area weight becomes smaller, whereas the power consumption decreases if the area weight increases. After the first step of

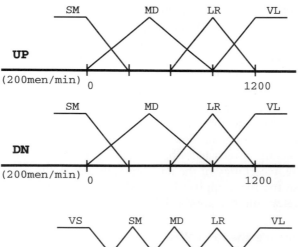

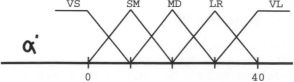

FIGURE 5 The membership functions (UP, DN, α').

determination by the first group factors, the area weight is adjusted according to the second group factors.

Let *AWT* be the average waiting time, *PC* be the power consumption, and *LWP* be the long wait probability. We introduce an adjustment value k that represents the influence of the second group factors. This adjustment value will be added to the area weight α' and give a new area weight α (that is, $\alpha = \alpha' + k$). Then we can represent fuzzy rules as follows.

- average waiting time
  If *AWT* is VL, then k is NL.
  If *AWT* is SM, then k is PL.
  $\vdots$
  If *AWT* is MD, then k is ZE.
- power consumption
  If *PC* is VL, then k is PL.
  If *PC* is SM, then k is NL.
  $\vdots$
  If *PC* is MD, then k is ZE.
- long wait probability
  If *LWP* is VL, then k is NL.
  If *LWP* is SM, then k is PL.
  $\vdots$
  If *LWP* is SM, then k is ZE.

In the fuzzy rules words such as VS (very small), SM (small), MD (medium), LR (large), and VL (very large) are fuzzy sets defined on the variables *AWT*, *PC*, and *LWP*. The terms NL (negative large), NM (negative medium), ZE (zero), PM (positive medium), and PL (positive large) are defined on k as shown in Fig. 6.

2. A Fuzzy Model

The fuzzy model for determining the area weight consists of two steps. The area weight is predetermined (α') in the first step, and the area weight is finally determined by the adjustment value (k). Such a two-step inference mechanism improves the system's stability during external accidents and reduces the complexity of the system. Figure 7 shows the two-step fuzzy inference mechanism.

In step 1 of the fuzzy inference engine, the predetermined area weight α' is calculated by using the up-going (UP) and down-going (DN) traffic. In this step the Mamdani max-min inference method [22] is used with the rules presented previously. The result of the inference is obtained through defuzzification using the center of gravity method [6].

In step 2 the adjustment value k is determined through the fuzzy inference mechanism, using the average waiting time, the long wait probability, and the power consumption values. After the defuzzification of the result of the second inference, this value is added to the predetermined area weight.

Figure 8 shows an inference example of the proposed fuzzy model when the up-going traffic is 200, the down-going traffic is 800, the average waiting

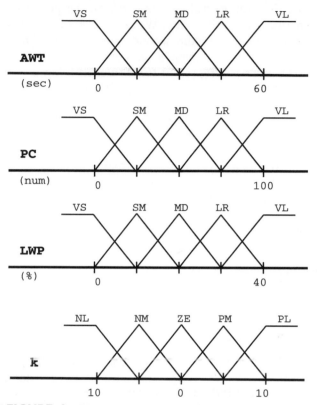

FIGURE 6 The membership functions (*AWT, PC, LWP, k*).

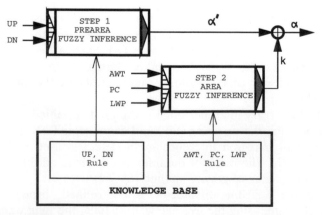

FIGURE 7 The fuzzy model used to determine the area weight. Horizontal stripes, fuzzification; gray, defuzzification.

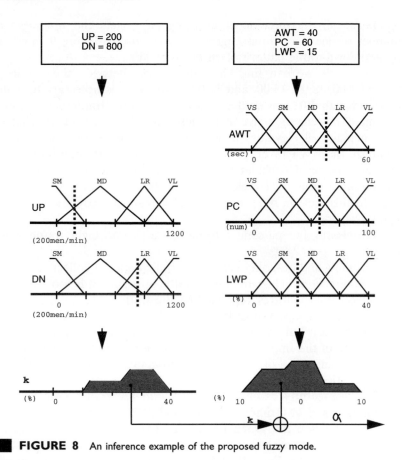

FIGURE 8   An inference example of the proposed fuzzy mode.

time is 40, the long wait probability is 15, and the power consumption is 60. In this example the computed values are $\alpha' = 26$, $k = -4$, and $\alpha = 22$.

III. FUZZY CLASSIFICATION OF PASSENGER TRAFFIC

The traffic pattern of passengers shows distinct characteristics according to the period of time in which it is observed, and it is efficient to manage elevators by traffic patterns [20]. For example, it is more useful to divide the elevators into two groups in the up peak time [2] and to distribute the elevators during the business time. Therefore, the EGCS changes the strategies of hall call assignment to increase the performance in terms of passenger traffic. The classification of passenger traffic is important for determining the strategy of hall call assignment at a proper time, and it affects the performance of the EGCS.

In this section, I will show a fuzzy classification method of passenger traffic. The traffic is classified into eight modes according to their characteristics.

Let's call the number of up-going passengers the *up traffic* and the number of down-going passengers the *down traffic*. Figure 9 shows a typical up and down traffic pattern of an office building.

In Fig. 9, we can find clear differences between the traffic patterns around 9:00, 12:00, 13:00, and 18:00 hours. So it is necessary to manage the elevators with different strategies according to the traffic pattern. For example, many passengers arrive at the lobby around 9 o'clock. At that time, we must manage all elevators to go back to the lobby as soon as possible to minimize the AWT (average waiting time) and LWP (long wait percentage). However, we must consider the RNC (RuN Count) to minimize the power consumption around 11 o'clock, because the total traffic is medium at that time. In this section, we propose a traffic pattern recognition method based upon fuzzy logic.

We classify the passenger traffic patterns into eight traffic modes (Table 1).

Incoming and outgoing passenger traffic data are collected at every floor for both directions. This amount of traffic information is too large to handle, but we can find features representing the traffic. The number of up/down going passengers is the most importance feature, and the degrees of centralization and distribution of passengers on floors give much information. The time is one of the important features because some traffic patterns occur at nearly the same time of day. Table 2 shows five important features of traffic characteristics.

The five features can be categorized into TA (traffic amount), TP (traffic percentage), and TM (time). The *UPT* and *DNT* belong to TA, the *CITP* and *DOTP* to TP, and the *TIME* belongs to TM. Figure 10 shows the linguistic terms and their membership functions used in the variables representing the traffic features. In Fig. 10, TA_S, TA_M, and TA_L represent membership functions (small, medium, large) for the *TA*-type features and TP_S, TP_M, and TP_L represent the *TP* type. TM_{UP}, TM_{BT}, TM_{LT}, and TM_{DN} are membership functions for *TM*.

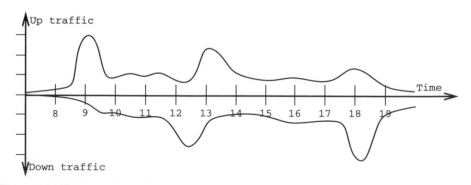

FIGURE 9 Up and down traffic of an office building.

TABLE 1   Passenger Traffic Patterns

BT	Total traffic is medium
(business time)	Before and after noon
UP	A lot of passengers come into the building
(up peak)	Office-going hour
DP	A lot of passengers leave the building
(down peak)	Closing hour
LT-A	Most passengers go the restaurant
(lunch time A)	Beginning of lunch time
IT	Total traffic is small
(inactive time)	Night
LT-B	Most passengers go back to their office
(lunch time B)	End of lunch time
BTH	Total traffic is large
(business time and heavy traffic)	Any time
HT	Many passengers gather on a floor
(heavy traffic)	and scatter from a floor
	Any time

The five features are used as input variables in the fuzzy inferencing model, which gives the traffic mode as its output. Some fuzzy rules are given in the following, and the rules are grouped by mode.

- BT
  - R_{11}) If *UPT* is TA_M and *DNT* is TA_M and
        CITP is TP_S and *DOTP* is TP_M and
        TIME is TM_{BT}
            then *belongs to BT*
  - R_{1j}) ... ...
  - R_{1n_1}) If *UPT* is TA_M and *DNT* is TA_M and
        CITP is TP_S and *DOTP* is TP_L and
        TIME is TM_{BT}
            then *belongs to BT*
- *UP*

TABLE 2   Features of Passenger Traffic

UPT	Number of up-going passengers
(up traffic)	
DNT	Number of down-going passengers
(down traffic)	
CITP	The ratio of incoming passengers
(centralized in traffic percentage)	on the most crowded floor to those of all floors
DOTP	The ratio of outgoing passengers on
(distributed out traffic percentage)	all floors except the most crowded floor to those of all floors
TIME	Current time

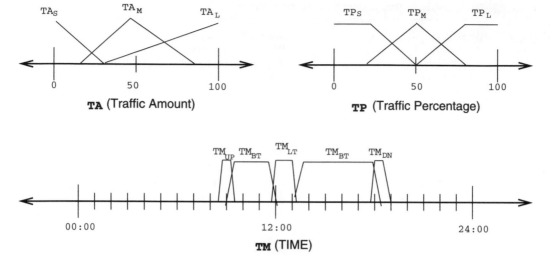

FIGURE 10 Membership functions for the traffic feature.

- R_{21}) If *UPT* is TA_L and *DNT* is TA_S and
  CITP is TP_L and *DOTP* is TP_M and
  TIME is TM_{UP}
    then *belongs to UP*
- R_{2j}) ... ...
- R_{2n_2}) If *UPT* is TA_L and *DNT* is TA_S and
  CITP is TP_L and *DOTP* is TP_L and
  TIME is TM_{UP}
    then *belongs to UP*

Figure 11 shows the structure of the fuzzy inference. Five features of current traffic are given for the fuzzy inferencing to classify the traffic pattern. Next, the fuzzy inference engine infers the possibility of each mode becoming the current traffic mode. After the inference, the traffic mode with the maximum possibility (α_{max}) is selected. If the possibility (α_{max}) of the selected traffic mode is more than α_{mode}, the control strategy is changed for this mode. If not, the previous control strategy is preserved. α_{mode} is a threshold that is a predefined constant and protects the oscillation of the traffic modes.

Let R_i be the rules for the ith traffic mode (*BT, UP, DP, LT-A, IT, LT-B, BTH, HT*). The maximum possibility (α_{max}) is fuzzy inferred and *max_mode* is determined. Here m is the number of features, and n_i is the number of rules of R_i. α_i is the result of fuzzy inference for the ith traffic mode, and α_{ij} is the intermediate variable that represents the matching degree of the jth rule in the ith traffic mode. $\mu_{A_{ijk}}(x_k)$ is the matching degree of the ith mode, jth rule, and kth feature. $\mu_{A_{ijk}}(x_k)$ is an abstract presentation of membership functions in TA, TP and TM. x_k is the kth feature

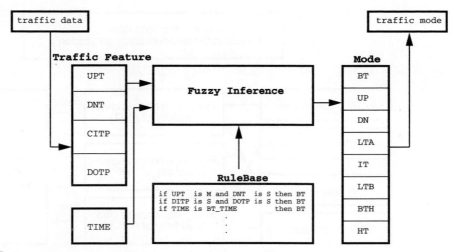

FIGURE II Fuzzy classification of the passenger traffic.

$(UPT, DNT, CITP, DOTP, TIME)$.

$$\alpha_{ij} = \prod_{k=1}^{m} \mu_{A_{ijk}}(x_k)$$

$$\alpha_i = \max_{j=1}^{n_i} \alpha_{ij}$$

$$= \max_{j=1}^{n_i} \left\{ \prod_{k=1}^{m} \mu_{A_{ijk}}(x_k) \right\}$$

$$\alpha_{\max} = \max(\alpha_1, \alpha_2, \ldots, \alpha_{n_i})$$

$$max\_mode = k \; (\alpha_k = \alpha_{\max}).$$

The *max_mode* is determined by the comparison with the maximum possibility ($\alpha_{\max}$). Let *mode* be the traffic mode determining the control strategy. The *mode* is set by the following equation. The threshold α_{mode} is used to protect the oscillation of traffic mode when the traffic mode is changed from one mode to another:

$$mode = max\_mode, \quad \text{if } \alpha_{\max} > \alpha_{\mathrm{mode}}.$$

IV. HALL CALL ASSIGNMENT

The hall call assignment part assigns hall calls to suitable elevators whenever a new hall call is received. Three fuzzy inferences are implemented to test the suitability of each evaluation criterion, and an OWA (ordered weighted average [23]) operation is employed to obtain the overall suitability of each elevator. Finally, the elevator with the greatest overall suitability is selected to service the new hall call.

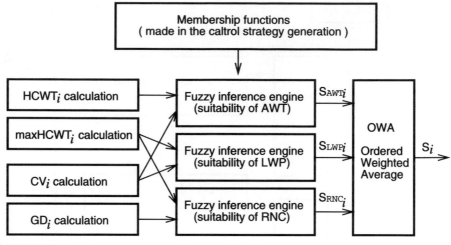

FIGURE 12   Inference structure of the overall suitability.

Figure 12 shows the inference structure of the overall suitability of the ith elevator. In Fig. 12, the suitability of each evaluation criterion (S_{AWT_i}, S_{LWP_i}, S_{RNC_i}) is inferred, and then the overall suitability (S_i) is calculated.

A. Calculation of Input Variables

The hall call assignment part considers data such as the direction and floor of hall calls, condition of elevators, and future hall calls. These data reflect the suitability of each elevator according to the evaluation criteria. Because the evaluation criteria can be computed after the hall calls are assigned, the *AWT*, *LWP*, and *RNC* are estimated using the *HCWT*, *maxHCWT*, *CV*, and *GD*.

The input variables of the fuzzy inference are *HCWT* (hall call waiting time), *maxHCWT* (maximum hall call waiting time), *CV* (coverability), and *GD* (gathering degree). Brief definitions of input variables are shown in Table 3. The input variables are very closely related to the evaluation criteria and can be estimated.

The $HCWT_i$ is related to the *AWT*, the $maxHCWT_i$ is related to the *LWP* and *RNC*, the *CV* is related to the *AWT* and *LWP*, and GD_i is related to the *RNC*. Two membership functions (small, large) for each input variable are used; these are shown in Fig. 13.

Each input variable is calculated for each elevator when a hall call is received. In the calculation, we assume that the hall call is assigned to the ith elevator. Let f be the number of floors and n be the number of elevators. The following are the calculation methods.

- $HCWT_i$. Waiting time until arrival the ith elevator at the floor for a hall call from the current position.

■ **TABLE 3   Input Variables**

HCWT$_i$ (hall call waiting time)	Waiting time until arrival of the *i*th elevator at the floor for a hall call from the current position
maxHCWT$_i$ (maximum hall call waiting time)	Maximum of *HCWT$_i$* for assigned hall calls to the *i*th elevator
CV$_i$ (coverability)	The capability of the elevator group control system for future hall calls to the *i*th elevator
GD$_i$ (gathering degree)	Minimum distance from a new hall call to the hall and car calls assigned to the *i*th elevator

- $HCWT_i$ = moving_time + stop_time
- moving_time = normal_speed_floors · normal_time_per_floor
  + speed_up_floors · speed_up_time
  + speed_down_floors · speed_down_time
- stop_time = Σ(door_opening_time + in_out_time + door_closing_time)
- *maxHCWT$_i$*. Maximum waiting time of hall calls for the *i*th elevator.
  - maxHCWT$_i$ = max(assigned_hall_call_waiting_times, new_hall_call_waiting_time)
    ◇ assigned_hall_call_waiting_times
      = HCWT for the assigned hall calls
    ◇ new_hall_call_waiting_time = $HCWT_i$
- *CV$_i$*. The capability of the elevator group control system for future hall calls to the *i*th elevator.
  - CV$_i$ = 1 − *err_traffic/in_traffic*
    - in_traffic = number of passengers who try to get on elevator in an unit time
      = $\sum_{j=1}^{f}$ incoming passengers at *j*th floor

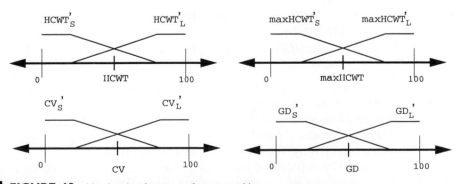

■ **FIGURE 13**   Membership functions of input variables.

- err_traffic = number of passengers that cannot be serviced in an unit time

$$= \textstyle\sum_{j=1}^{f}\sum_{k=1}^{n}(\text{in\_traffic}_j - \text{capacity}_{jk})$$

  capacity_{jk} = the number of persons could get on kth elevator on jth floor

- GD_i. Minimum distance between the position of the new hall call and hall calls and car calls assigned to the ith elevator.
  - $GD_i = 1/\text{min\_distance}$
    - $\text{min\_distance} = min_{j=1,\ldots,number\_of\_assigned\_calls}$
      Distance(new_hall_call, assigned_call)

B. Fuzzy Inference

Three fuzzy rule sets are derived to infer the suitability of each evaluation criterion. The suitability of AWT_i is inferred using the input values $HCWT_i$ and CV_i, the LWP_i is inferred using $maxHCWT_i$ and CV_i, and the RNC is inferred using $maxHCWT_i$ and GD_i.

The fuzzy rule sets used to infer the suitabilities are as follows:

- Suitability of the AWT_i
  - If $HCWT_i$ is $HCWT_S$ and CV_i is CV_L, then S_{AWT_i} is LARGE.
  - If $HCWT_i$ is $HCWT_S$ and CV_i is CV_S, then S_{AWT_i} is MEDIUM.
  - If $HCWT_i$ is $HCWT_L$ and CV_i is CV_L, then S_{AWT_i} is MEDIUM.
  - If $HCWT_i$ is $HCWT_L$ and CV_i is CV_S, then S_{AWT_i} is SMALL.
- Suitability of the LWP_i
  - If $maxHCWT_i$ is $maxHCWT_S$ and CV_i is CV_L, then S_{LWP_i} is LARGE.
  - If $maxHCWT_i$ is $maxHCWT_S$ and CV_i is CV_S, then S_{LWP_i} is MEDIUM.
  - If $maxHCWT_i$ is $maxHCWT_L$ and CV_i is CV_L, then S_{LWP_i} is MEDIUM.
  - If $maxHCWT_i$ is $maxHCWT_L$ and CV_i is CV_S, then S_{LWP_i} is SMALL.
- Suitability of the RNC_i.
  - If $maxHCWT_i$ is $maxHCWT_S$ and GD_i is GD_L, then S_{RNC_i} is LARGE.
  - If $maxHCWT_i$ is $maxHCWT_S$ and GD_i is GD_S, then S_{RNC_i} is MEDIUM.
  - If $maxHCWT_i$ is $maxHCWT_L$ and GD_i is GD_L, then S_{RNC_i} is MEDIUM.
  - If $maxHCWT_i$ is $maxHCWT_L$ and GD_i is GD_S, then S_{RNC_i} is SMALL.

In the rule sets, SMALL, MEDIUM, and LARGE are linguistic terms defined in the suitability domain and are commonly used in three inferences. Figure 14 shows the linguistic terms SMALL, MEDIUM, and LARGE. In the fuzzy inferences, Mamdani's method and the center-of-gravity method are used, respectively, for inferencing and defuzzification.

The OWA (ordered weighted average [23]) operator is used to aggregate the suitabilities of evaluation criteria. The OWA operator is an aggregation

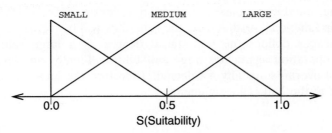

FIGURE 14 Linguistic terms defined in the suitability domain.

operator that forms an overall decision function in a multicriterion problem. The OWA is defined as follows.

DEFINITION. A mapping F from

$$I^n \rightarrow I \qquad (\text{where } I = [0, 1])$$

is called an OWA operator of dimension n if associated with F is a weighting vector W,

$$W = \begin{bmatrix} W_1 \\ W_2 \\ . \\ W_n \end{bmatrix},$$

such that $W_i \in (0, 1)$, $\Sigma W_i = 1$, and where

$$F(a_1, a_2, \ldots, a_n) = W_1 b_1 + W_2 b_2 + \cdots + W_n b_n,$$

where b_i is the ith largest element in the collection $a_1, a_2, \ldots, a_n$.

In this paper, $W^T = [0.5, 0.3, 0.2]$ is used to emphasize the criteria with the largest suitability. S_{AWT_i}, S_{LWP_i}, and S_{RNC_i} are the suitability of each evaluation criterion, and S_i is the overall suitability of the ith elevator. Therefore the S_i is aggregated by the following:

$$S_i = F(S_{AWT_i}, S_{LWP_i}, S_{RNC_i}).$$

By the above methods, the overall suitabilities of elevators $S_1, S_2, \ldots, S_n$ are inferred, where the n is the number of elevators in the elevator group control system.

Finally, the hall call assignment part selects an elevator with the largest overall suitability. Let e be a selected elevator to service the new hall call. The e is defined as

$$e \in 1, 2, \ldots, n \qquad \text{such that } S_e = \max(S_1, S_2, \ldots, S_n).$$

In this selection mechanism, the importance degrees defined by the system manager are affected in the following way. First, the membership functions defined by the input variables are made by the importance degrees.

Second, the suitability of each evaluation criterion is inferred using the membership functions. At that time, the suitability will be increased when the system manager defines the importance degree as a large value. Third, the OWA operation aggregates three suitabilities. Finally, an elevator with the greatest overall suitability is selected. Therefore the importance degrees are applied to the hall call assignment.

V. CONCLUSIONS

In this chapter, the EGCS (elevator group control system) is introduced as a way to schedule elevators by passenger hall calls. Techniques and applications of fuzzy theory to the elevator group control system are introduced and three applications to EGCS are reviewed. First, an approach to control of the weighting factor by employing fuzzy theory is introduced. In this approach, area weight, which effects to the performance of EGCS, is introduced, and an example of control of area weight is shown. Second, fuzzy classification of passenger traffic by the use of fuzzy logic is presented. The traffic patterns of a typical building and features of passenger traffic are introduced, and a fuzzy classification method is shown. Finally, a study of hall call assignment is presented, in which we study the structure of hall call assignment and the effects of input variables ($HCWT, maxHCWT, CV, GD$). Hall calls are assigned according to the results of fuzzy inferencing using input variables.

REFERENCES

1. Kim, C. B., Seong, K. A., and Lee-Kwang, H. Fuzzy approach to elevator group control system. In *Proceedings of the 5th Congress*, 1993, Vol. 2, pp. 1218–1221.
2. Kim, C. B., Seong, K. A., and Lee-Kwang, H. A fuzzy approach to elevator group control system. *IEEE Trans. Systems Man Cybernet.* 25(5), 1995.
3. Peters, R. D. The theory and practice of general analysis lift calculations. In *Elevcon Proceedings*, 1992, pp. 197–206.
4. Barney, G. C. and dos Santos, S. M. *Elevator Traffic Analysis, Design, and Control*. Peter Peregrinus, 1985.
5. Strakosch, G. R. *Vertical Transportation: Elevators and Escalators*. Wiley-Interscience, New York, 1983.
6. Song, Y. S. A comparative study of fuzzy implication operators: an application to elevator systems. M.S. Thesis, CSD KAIST, 1992.
7. Lee, J. B. A study on the development of optimal model for determining configurations of the elevator systems. M.S. Thesis, Hanyang University, 1988.
8. Lutfi, R. and Barney, G. C. The Inverse S-P Method Deriving Lift Traffic Patterns from Monitored Data. Control Systems Centre Report no. 745, 1991.
9. Lutfi, R. and Barney, G. C. The use of Moving Average in Filtering Lift Traffic Patterns. Control Systems Centre Report no. 747, 1991.
10. Siikonen, M. L. and Leppala, J. Elevator traffic pattern recognition. In *Proceedings of the 4th IFSA Congress*, 1991, pp. 195–198.
11. Hitachi, Japan, Ltd. Elevator Group Control System. Patent document 208, japan 90-6397, 1990.
12. Kim, C. B., Seong, K. A., Lee-Kwang, H., and Kim, J. O. Design and implementation of FEGCS: fuzzy elevator group control system. In *Proceedings of NAPFIS'96*, 1996, pp. 1218–1221.
13. Hitachi, Japan, Ltd. The Fuzzy Elevator Group Supervisory System. Japan Patent, 2-52875, 1992.

14. Mitsubishi, Japan, Ltd. Group Supervisory System of Elevator Cars Based on Neural Network. Japan Patent, 4-32472, 1992.
15. Markon, S., Kita, H., and Nishikawa, Y. Adaptive optimal elevator group control by use of neural network. *Trans. Inst. Systems Control Inform. Engineers* 7(12):487–497, 1994.
16. Kim, C. B., Seong, K. A., Lee-Kwang, H., and Kim, J. O. Design and implementation of FEGCS. *IEEE Trans. Systems Man Cybernet.* 28(part A, 3), 1998.
17. Hitachi, Japan, Ltd. Elevator Group Control System. Patent document 1273, japan 87-555, 1987.
18. Sugeno, M. and Kang, G. T. Structure identification of fuzzy model. *Fuzzy Sets and Systems* 28:15–33, 1988.
19. Araki, S., Nomura, H., Hayashi, I., and Wakami, N. A self-generating method of fuzzy inference rules. In *IFES91 Proceedings* 1991, pp. 1047–1058.
20. Nomura, H., Hayashi, I., and Wakami, N. A learning method of fuzzy inference rules by descent method. In *1992 IEEE* 1992, pp. 203–210.
21. Takagi, T. and Sugeno, M. Fuzzy identification of systems and its applications to modeling and control. *IEEE Trans. Systems Man Cybernet.* SMC-15:116–132, 1985.
22. Lee, C. C. Fuzzy logic in control systems: fuzzy logic controller. Part II. *IEEE Trans. Systems Man Cybernet.* 20:419–435, 1990.
23. Yager, R. R. On ordered weighted averaging aggregation operators in multicriteria decision-making. *IEEE Trans. Systems Man Cybernet.* 18(1):183–190, 1988.